Oil and Gas Processing Handbook

Editor: Andy Margo

www.callistoreference.com

Callisto Reference,
118-35 Queens Blvd., Suite 400,
Forest Hills, NY 11375, USA

Visit us on the World Wide Web at:
www.callistoreference.com

© Callisto Reference, 2017

This book contains information obtained from authentic and highly regarded sources. Copyright for all individual chapters remain with the respective authors as indicated. All chapters are published with permission under the Creative Commons Attribution License or equivalent. A wide variety of references are listed. Permission and sources are indicated; for detailed attributions, please refer to the permissions page and list of contributors. Reasonable efforts have been made to publish reliable data and information, but the authors, editors and publisher cannot assume any responsibility for the validity of all materials or the consequences of their use.

ISBN: 978-1-63239-889-5 (Hardback)

The publisher's policy is to use permanent paper from mills that operate a sustainable forestry policy. Furthermore, the publisher ensures that the text paper and cover boards used have met acceptable environmental accreditation standards.

Trademark Notice: Registered trademark of products or corporate names are used only for explanation and identification without intent to infringe.

Printed in the United States of America.

Cataloging-in-Publication Data

Oil and gas processing handbook / edited by Andy Margo.
p. cm.
Includes bibliographical references and index.
ISBN 978-1-63239-889-5
1. Petroleum. 2. Natural gas. 3. Petroleum engineering. 4. Gas engineering. I. Margo, Andy.
TN870 .O35 2017
665.5--dc23

Table of Contents

Preface

This book on oil and gas processing discusses the various processes that are involved in petroleum refining such as distillation, dimerization and fluid cracking. Oil refineries can have different configurations depending on the end product. Such plants also require adequate storage facilities, electricity and cooling water for proper functioning. Those in search of information to further their knowledge will be greatly assisted by this book. The various sub-fields of oil and gas processing along with technological progress that have future implications are glanced at. This book is a vital tool for all researching or studying oil and gas processing as it gives incredible insights into emerging trends and concepts.

This book is the end result of constructive efforts and intensive research done by experts in this field. The aim of this book is to enlighten the readers with recent information in this area of research. The information provided in this profound book would serve as a valuable reference to students and researchers in this field.

At the end, I would like to thank all the authors for devoting their precious time and providing their valuable contributions to this book. I would also like to express my gratitude to my fellow colleagues who encouraged me throughout the process.

Editor

1

Sequence stratigraphy of the petroliferous Dariyan Formation (Aptian) in Qeshm Island and offshore (southern Iran)

P. Mansouri-Daneshvar[1] · R. Moussavi-Harami[1] · A. Mahboubi[1] · M. H. M. Gharaie[1] · A. Feizie[2]

Abstract After sea level rises during the Early Cretaceous, upper parts of the Khami Group sediments (Fahliyan, Gadvan, and Dariyan Formations) deposited over Jurassic sediments. The Lower Cretaceous (Aptian) Dariyan Formation (equivalent to the Shu'aiba Formation and Hawar Member of the Arabian Plate) carbonates, which have hydrocarbon reservoir potential, form the uppermost portion of the Khami Group that unconformably overlays the Gadvan Formation and was unconformably covered by the Kazhdumi Formation and Burgan sandstones. Detailed paleontological, sedimentological, and well log analysis were performed on seven wells from Qeshm Island and offshore in order to analyze the sequence stratigraphy of this interval and correlate with other studies of the Dariyan Formation in this region. According to this study, the Dariyan Formation contains 14 carbonate lithofacies, which deposited on a ramp system that deepened in both directions (NE—wells 5, 6 and SW—wells 1, 2). Sequence stratigraphy led to recognition of 5 Aptian third-order sequences toward the Bab Basin (SW—well 1) and 4 Aptian third-order sequences toward Qeshm Island (NE—wells 5 and 6) so these areas show higher gamma on the gamma ray logs and probably have higher source rock potential. Other wells (wells 2–4 and 7) mainly deposited in shallower ramp systems and contain 3 Aptian third-order sequences. On the other hand, rudstone and boundstone lithofacies of studied wells have higher reservoir potential and were deposited during Apt 3 and Apt 4 sequences of the Arabian Plate. The Dariyan Formation in Qeshm Island (well 6) and adjacent well (well 5) was deposited in an intrashelf basin that should be classified as a new intrashelf basin in future Aptian paleogeographic maps. We interpret that salt-related differential subsidence, crustal warping, and reactivation of basement faults of the Arabian Plate boundary were responsible for the creation of the intrashelf basin in the Qeshm area.

Keywords Arabian Plate · Sequence stratigraphy · Qeshm Island · Aptian and Dariyan formation

1 Introduction

The Persian Gulf contains 55–68 % of the world's recoverable oil and more than 40 % of gas reserves (Konyuhov and Maleki 2006). Cretaceous carbonate platforms encompass approximately 16 % of world's hydrocarbon reservoirs developed in the Tethyan region including the Persian Gulf (Scott et al. 1993). The Lower Cretaceous (Aptian) Dariyan Formation comprises the uppermost portion of the Upper Jurassic to Lower Cretaceous Khami Group (James and Wynd 1965). This formation is predominantly composed of limestones deposited on a passive margin (Sharland et al. 2001) and its upper part is equivalent to the Shu'aiba Formation in the Arabian Plate region. In the Arabian Plate, the Hawar Member and Shu'aiba Formation (in this study both equivalent to the Dariyan Formation) unconformably overlay the Kharaib Formation (in this study equivalent to the Gadvan Formation) and were unconformably covered by the Nahr Umr

✉ P. Mansouri-Daneshvar
p.mansouri.daneshvar@gmail.com

[1] Department of Geology, Faculty of Sciences, Ferdowsi University of Mashhad, Mashhad, Iran

[2] Exploration Directorate, National Iranian Oil Company, Tehran, Iran

Edited by Jie Hao

Formation (equivalent to the Kazhdumi Formation) and Burgan sandstones after relative sea-level fall resulting in exposure of the platform with a strong diachronous boundary (Van Buchem et al. 2010b; Motiei 1993) (Fig. 1). The Dariyan Formation has potential for hydrocarbon reservoirs and produced oil from the Alpha and Reshadat (Raksh) fields, and oil and gas in the Resalat (Rostam) Field (Alsharhan and Nairn 1997). Because of the

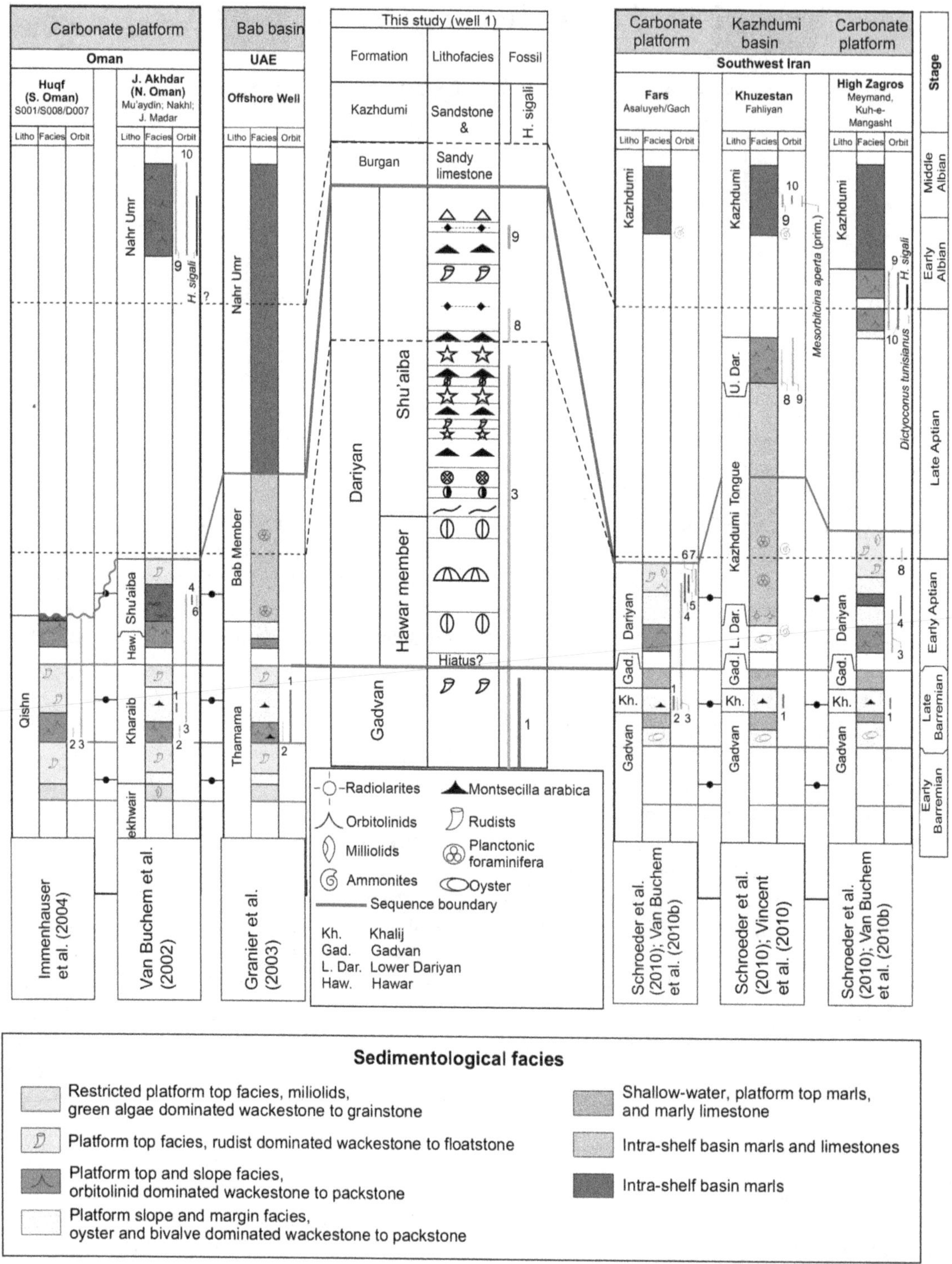

Fig. 1 Chronostratigraphy of the Barremian–Aptian with the orbitolinid distribution in the southern Arabian Plate (Oman, UAE, southwest Iran) (after Schroeder et al. 2010) that correlated with well 1 of this study, *1 Montseciella arabica*, *2 Eopalorbitolina transiens*, *3 Palorbitolina lenticularis*, *4 Palorbitolina ultima*, *5 Palorbitolinoides cf. orbiculata*, *6 Palorbitolina cormy*, *7 Palorbitolina wienandsi*, *8 Mesorbitolina parva*, *9 Mesorbitolina texana*, *10 Mesorbitolina subconcava*, for lithofacies symbols of this study see Fig. 5

importance of such reservoirs, sequence stratigraphic studies can play a significant role in reservoir studies and access its connectivity around the study area (Ainsworth 2006; Van Wagoner et al. 1988). Van Buchem et al. (2010b) and Vincent et al. (2010) proposed a new sequence stratigraphic framework for the Gadvan, Dariyan, and Kazhdumi Formations in SW Iran but there is no detailed work in the Iranian part of the Persian Gulf and no correlations performed with adjacent Arabian parts according to log data. The objectives of this study are to interpret depositional sequences by collecting petrographical studies of cores, cuttings, and log data and correlation of sequences between the Iranian part and the Arabian Plate in a vast area of Qeshm Island and offshore in southern Iran (eastern Persian Gulf) (Fig. 2) that can give a better clue in paleogeography during the Aptian time.

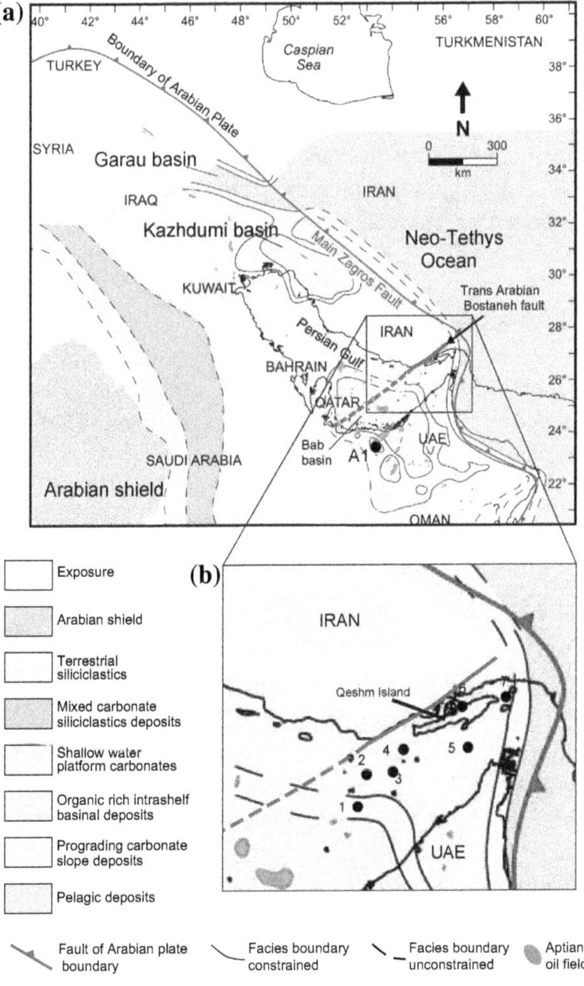

Fig. 2 **a** Early Aptian paleogeographic map of the eastern Arabian Plate (from Van Buchem et al. 2010a) with locations of the Persian Gulf, Trans-Arabian Bostaneh basement fault (Bahroudi and Koyi 2004) and studied area, A1: location of one of Abu Dhabi's fields containing wells which were used in Fig. 10; **b** location of studied wells (1–7)

2 Geological setting

The Persian Gulf Basin (Fig. 2) developed in the offshore part of Zagros Belt in the northeast of the Arabian Plate (Edgell 1996; Alavi 2004). This basin is situated in the Arabian Peninsula (Konyuhov and Maleki 2006) and its eastern part is called the Shilaif Basin or the northern part of Rub al-Khali and Ras al-Khaimah subbasins (Alsharhan and Nairn 1997). The Persian Gulf and Zagros Mountains have a similar sedimentary record encompassing strata ranging in age from the latest Precambrian to the Recent (James and Wynd 1965; Berberian and King 1981; Motiei 1993; Alavi 2004). The total thickness of these successions from the deformed Fars region in the north to the nearly undeformed strata of Persian Gulf region reaches about 7 km (Alavi 2007). Following Neotethys opening in Middle to Late Permian between the Cimmerian continental blocks and the eastern margin of the Arabian Plate (e.g., Bechennec et al. 1990; Ruban et al. 2007), a stable passive carbonate platform (lasting for over 160 Myrs) was established during the Triassic to late Cretaceous (Cenomanian) that simultaneously subsided due to post-rift thermal subsidence (Glennie 2000; Sharland et al. 2001; Piryaei et al. 2011; Ali et al. 2013).

After rising sea level during Early Cretaceous time, the upper part of the Khami Group sediments (Fahliyan, Gadvan, and Dariyan Formations) were deposited on a gently eastward dipping carbonate ramp over Jurassic evaporites (Hith Formation) (Alsharhan and Nairn 1997) and exhibit significant variation in thicknesses. During the onset of Neotethys subduction in the Aptian (Glennie 2000), the Dariyan Formation has been deposited in a carbonate platform to intrashelf basin in the northeastern passive margin of the Arabian Plate (Schroeder et al. 2010; Van Buchem et al. 2010b; Vincent et al. 2010) that was followed by deposition of Burgan sandstone equivalent sediments of latest Aptian to Early Albian age. This formation is equivalent to the Shu'aiba Formation and Hawar Member (Aptian carbonate sediments) of the Arabian Plate and deepened toward the Kazhdumi intrashelf basin in southwest Iran and the Bab intrashelf basin in UAE containing high levels of organic matter (Schroeder et al. 2010; Van Buchem et al. 2010b) (Figs. 1, 2). The Burgan sandstone deposited during a long break in the carbonate sedimentation in the Fars area (adjacent to the studied area) and other parts of the Arabian Plate (Van Buchem et al. 2010a, b).

At the other side, halokinetic tectonism produced by salt movements of Late Pre Cambrian-Early Cambrian age Hormuz and equivalent series has further effects on this region (Ala 1974; Sadooni 1993; Motiei 2001; Jahani et al. 2007). Salt movements have considerable effects on facies changes and shallowing or deepening of Cretaceous sedimentary basins in the studied region (e.g., Piryaei et al. 2011).

3 Study methods

Detailed paleontological, sedimentological, and well log analysis were performed on 7 wells throughout Qeshm Island and offshore area (Fig. 2) and correlated with other studies of the Dariyan Formation in this region. These wells are from a large area across the eastern part of the Persian Gulf and the thickness of the Dariyan Formation ranges from 70 to 122 m. Three hundred core and cutting thin sections were studied by petrographic microscopy to determine the lithofacies and fossil components. Carbonate rocks were classified according to the Dunham (1962), and Embry and Klovan (1971) schemes. Lateral and vertical lithofacies changes were determined using well logs, cores, and cutting data, and depositional environments were interpreted using methods such as those presented by Flugel (2010). Along with biostratigraphic and lithofacies studies, gamma ray and acoustic well logs have an important role for stratigraphic correlation in the Dariyan. Similar to regions to the south such as Oman and UAE (e.g., Vahrenkamp 2010; Droste 2010), the gamma log profile has a similar shape over most of studied area. This profile starts with a high gamma spike at the base of the Hawar Member followed by a sharply decreasing trend, and a more or less constant slightly increasing trend upward. With regard to a lower amount of argillaceous material in the lower parts of the Dariyan Formation, gamma ray logs over the Lower Shu'aiba can reflect variations of uranium content of organic materials under reducing conditions (e.g., Wignall and Myers 1988). On the other hand, the high gamma ray content in the Hawar and top of the Dariyan Formation probably is due to propagation of exposure surfaces and may have been related to concentration by groundwater movement (Ehrenberg and Svånå 2001; Droste 2010). In all wells, the boundary between the Burgan and Dariyan Formations was characterized by a high gamma spike and high acoustic response due to clastic influx during deposition of these formations. Applying essential concepts to determine sequence boundaries and comparing gamma ray and sonic logs as well as paleolog data with Arabian Plate sequences, we interpret depositional sequences and sea-level changes during deposition of the Dariyan Formation.

4 Biostratigraphy

Schroeder et al. (2010) produced a chronostratigraphic correlation of Oman, UAE, and southwest Iran according to Orbitolinids (Fig. 1) that helps us in correlating stratigraphic units of the Persian Gulf. According to this correlation, the Early to Late Aptian Dariyan Formation is equivalent to both the Hawar Member and Shu'aiba Formation, and the Barremian Gadvan Formation correlates with the Kharaib Formation. In this study, we used all paleolog data (Jalali 1969; Nayebi 2000, 2001; Azimi 2010a, b; Hadavandkhani 2010; Fonooni 2010) and revised orbitolinid biostratigraphic zonation for the Barremian–Aptian of the eastern Arabian Plate (Schroeder et al. 2010) to identify biozones in studied wells. In paleologs, the Dariyan and Gadvan Formations with *Choffatella decipiens* and *Palorbitolina lenticularis* (Late Barremian to the Early/Late Aptian age) show the *Palorbitolina lenticularis* biozone and intervals containing *Dictyoconus arabicus* (*Montseciella arabica*) can be differentiated as a *Montseciella arabica* subzone. This subzone in eastern Arabian and northeastern African plates appeared approximately from the middle to end of the Late Barremian time. In this study, similar to southwest Iran (Van Buchem et al. 2010b), this subzone exists in the Barremian age Gadvan Formation (Kharaib equivalent). In paleolog of well 1, Early Late Aptian age can be related to the time range from appearance of *Mesorbitolina parva* to *Mesorbitolina texana*. Occurrence of *Mesorbitolina texana* coincides with a late Late Aptian age for the upper part of the Dariyan Formation in this well. The presence of *Hemicyclammina sigali* of Albian age in Kazhdumi and Burgan sandstones in studied wells is a key to separating the Aptian–Albian boundary as mentioned by Schroeder et al. (2010).

5 Lithofacies analysis

14 carbonate lithofacies (Figs. 3, 4; Table 1), which were grouped into three lithofacies associations, were identified in the Dariyan Formation similar to the Shu'aiba and Hawar Formations in Arabian parts (e.g., Raven et al. 2010; Van Buchem et al. 2010a, b). These were deposited in a carbonate ramp system to intrashelf basin (Fig. 5) as follows:

5.1 Shallow inner ramp lithofacies association

This lithofacies association consists of sandy mudstone, milliolidae wackestone/pellet grainstone, benthic foraminifera wackestone/pellet grainstone, and conical Orbitolina wackestone/packstone (Fig. 3; Table 1). This lithofacies mainly deposited during deposition of the lower part of the Dariyan Formation on a shallow carbonate ramp. The abundance of mud-dominated textures, existing milliolidae benthic foraminifers, the low abundance, and diversity of open marine fauna/flora are consistent with deposition within a restricted, shallow-water, low-energy lagoonal setting (Flugel 2010; Pittet et al. 2002). The milliolids are intolerant to oxygen-deficient conditions and increase in abundance according to food availability, when oxygen is not a limiting factor (Halfar and Ingle 2003). In grainstone lithofacies, shallow marine higher energy

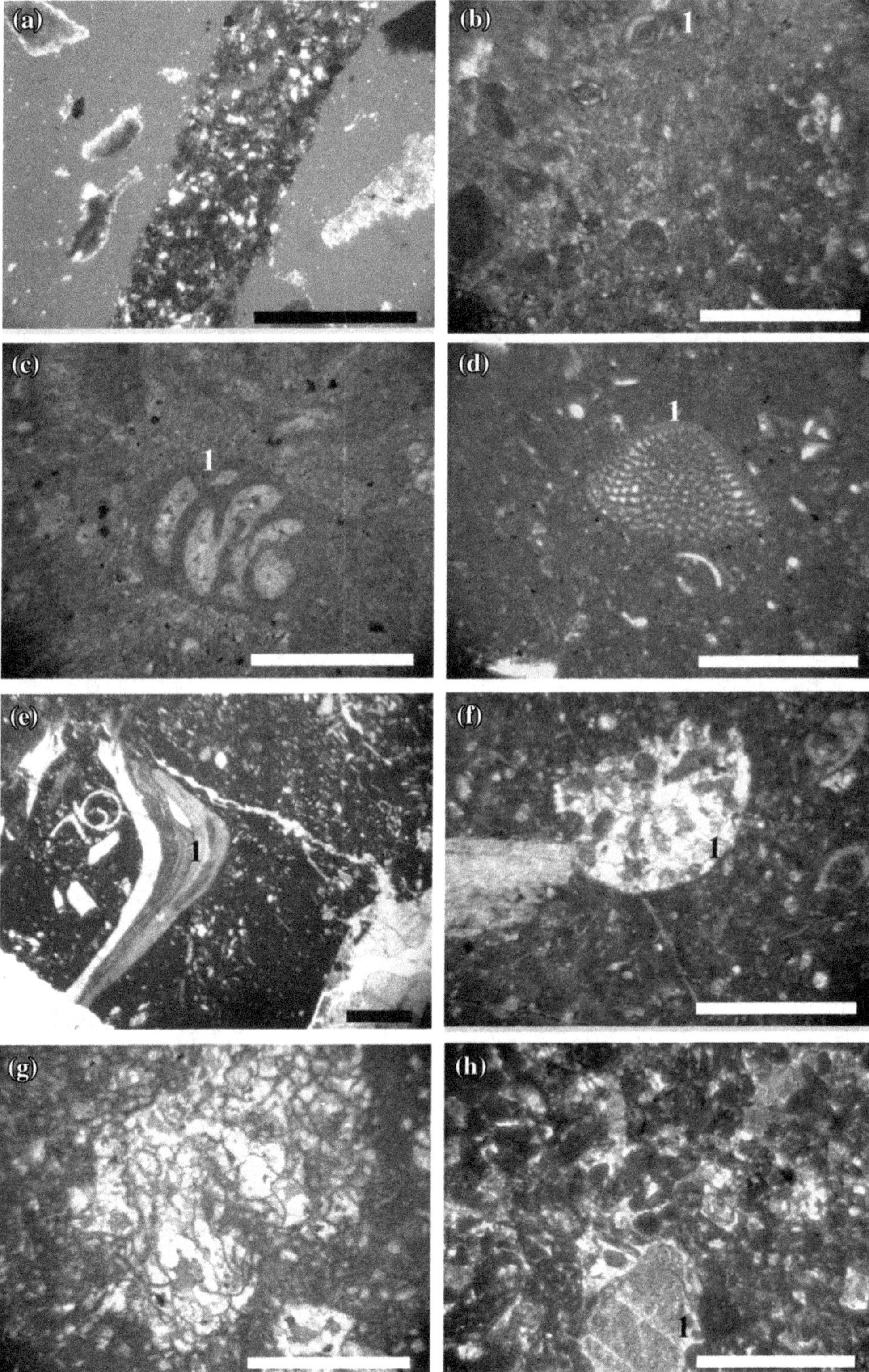

Fig. 3 Dariyan Formation lithofacies in studied wells, **a** sandy mudstone (A1), well 5, depth: 4257 m, PPL; **b** milliolidae wackestone/pellet grainstone (A2), *1* Milliolidae, well 1, 9758 ft, XPL; **c** Benthic foraminifera wackestone/pellet grainstone (A3), *1* Dukhania, well 1, 9747 ft, XPL; **d** conical Orbitolina wackestone/packstone (A4), *1* Conical Orbitolina, well 1, 9645 ft, XPL; **e** rudist rudstone (B1), *1* Rudist debris, well 1, 9711 ft, PPL; **f** coral rudstone (B2), *1* coral fragment, well 1, 9743 ft, XPL; **g** *Lithocodium aggregatum* rudstone/boundstone (B3), well 1, 9750 ft, XPL; **h** echinoid pellet grainstone (B4), *1* Echinoid debris, well 2, 10,890 ft, XPL. Scale bar = 1 mm

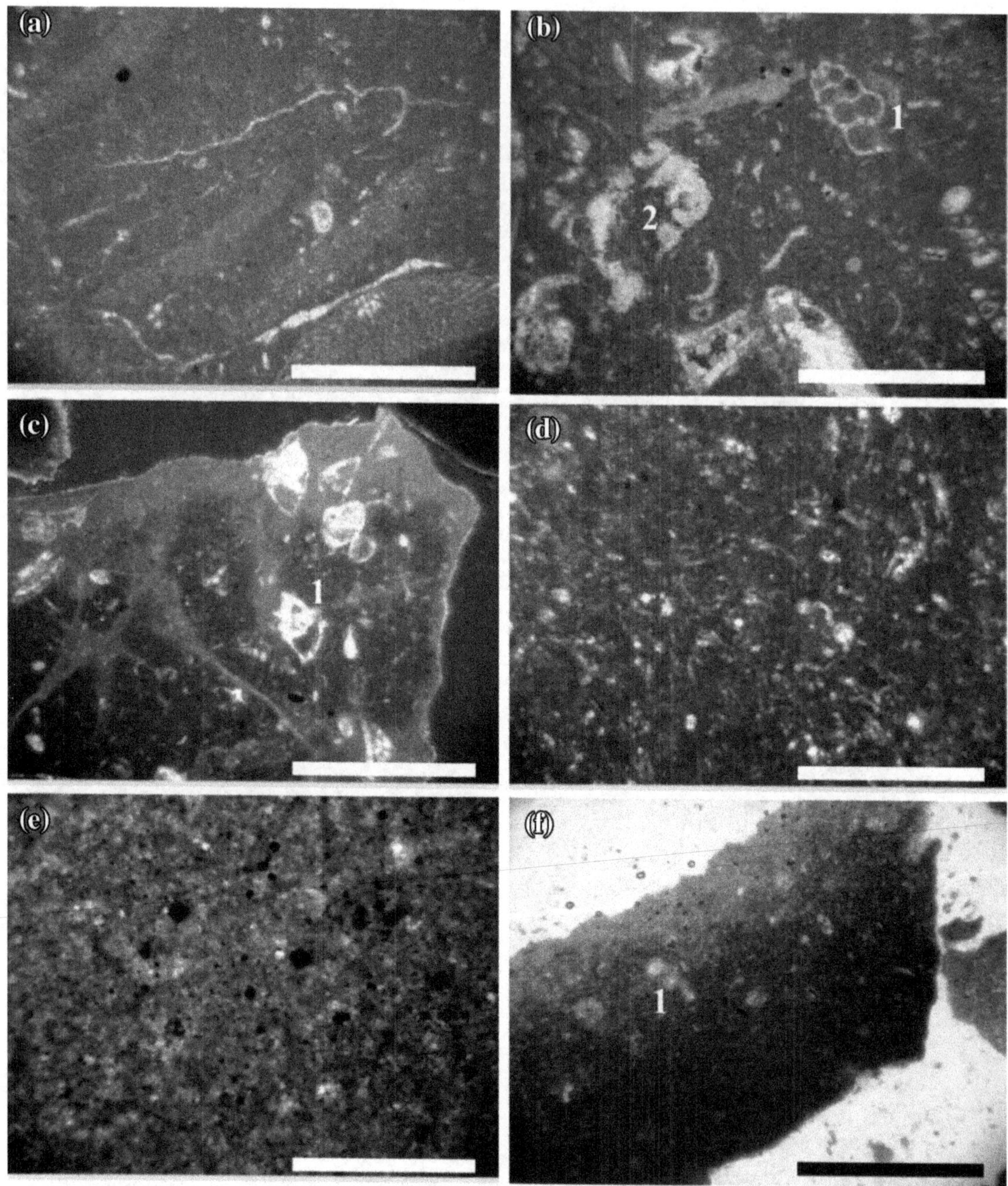

Fig. 4 Dariyan Formation lithofacies in studied wells, **a** discoidal Orbitolina wackestone/packstone, well 1, 9698 ft, XPL; **b** bioclastic wackestone/packstone (B5), well 1, *1* Gastropod, *2* Echinoid, 9692 ft, XPL; **c** lenticulina/epistomina wackestone (C1), well 6, *1* lenticulina, 9950-55 ft, XPL; **d** sponge spicule wackestone/packstone (C2), well 2, 10,730 ft, PPL; **e** pyritized mudstone (C4), well 1, 9633 ft, PPL; **f** Pelagic foraminifera wackestone (C3), well 6, *1* Globigerina, 9885-90 ft, PPL. Scale bar = 1 mm

conditions created grainstone textures in bars and local platform topographic highs. Conical Orbitolina packstone/wackestone consists mainly of conical forms of orbitolinids that are related to shallower water environmental settings (e.g., Vilas et al. 1995; Simmons et al. 2000; Hughes 2000; for further details see Pittet et al. 2002). Benthic foraminifers such as trocholina indicate lagoonal and shallow marine lithofacies (e.g., Meyer 2000; Hughes 2004, 2005). Dasyclad green algae such as salpingoporella dinarica are other common constituents of this lithofacies association that proves a restricted lagoon environment.

5.2 Mid-ramp lithofacies association

This lithofacies association contains rudist rudstone, coral rudstone, lithocodium aggregatum rudstone/boundstone,

Table 1 Lithofacies analysis of Dariyan Formation

Lithofacies association	Lithofacies	Texture	Major components	Minor components	Depositional environment
A	A1	Sandy mudstone	Terrigenous material, pellet	Trocholina, pyrite, milliolidae	Inner ramp-tidal flat
	A2	Milliolidae wackestone/ pellet grainstone	Milliolidae, nezzazata, pellet, intraclast, dasyclad green algae, bivalve, pellet	Trocholina, Orbitolina, textularia, acicularia, dukhania	Inner ramp-restricted lagoon and bar
	A3	Benthic foraminifera wackestone/pellet grainstone	Dukhania, trocholina, milliolidae, nezzazata, pellet, intraclast	Orbitolina, textularia, pyrite, acicularia, gastropod, dasyclad green algae	Inner ramp-restricted lagoon and bar
	A4	Conical Orbitolina wackestone/packstone	Conical Orbitolina, textularia, dukhania, pellet	Echinoid, dasyclad green algae, discoidal Orbitolina, terrigenous material, pyrite, gastropod, rudist debris	Inner ramp-restricted to open lagoon
B	B1	Rudist rudstone	Rudist, lithocodium aggregatum, gastropod, bivalve	Echinoid, conical and discoidal Orbitolina, coral, dasyclad green algae, textularia, pellet	Mid-ramp
	B2	Coral rudstone	Coral, rudist, Orbitolina	Lithocodium aggregatum, benthic foraminifera, dasyclad green algae, pellet	
	B3	Lithocodium aggregatum rudstone/ boundstone	Lithocodium aggregatum, rudist, Orbitolina, coral, spongiomorphid, bivalve	Benthic foraminifera, dasyclad green algae, pellet	
	B4	Echinoid pellet grainstone	Echinoid, pellet, discoidal Orbitolina, bivalve, lithocodium aggregatum	Oyster, gastropod, salpingoporella dinarica	
	B5	Discoidal orbitolina wackestone/packstone	Discoidal Orbitolina	Choffatella decipiens, rudist, echinoid, gastropod, conical Orbitolina, bivalve, benthic and pelagic foraminifera, sponge spicule, terrigenous material	
	B6	Bioclastic wackestone/packstone	Echinoid, bivalve, gastropod	Pelagic foraminifera, sponge spicule, oyster, pellet, discoidal Orbitolina, lenticulina	
C	C1	Lenticulina/epistomina wackestone	Lenticulina, epistomina	Pelagic foraminifera, sponge spicule, choffatella decipiens, bivalve, Orbitolina	Outer ramp
	C2	Sponge spicule wackestone/packstone	Sponge spicule, echinoid, lenticulina, calcispher	Echinoid, debarina, pelagic foraminifera, Orbitolina, radiolaria	Outer ramp-intrashelf
	C3	Pyritized mudstone	Pyrite, terrigenous material, calcispher, pellet	Radiolaria, Orbitolina	Intrashelf
	C4	Pelagic foraminifera wackestone	Globigerina, radiolaria	Sponge spicule, radiolaria, pellet	Intrashelf

echinoid pellet grainstone, discoidal Orbitolina wackestone/packstone, and bioclastic wackestone/packstone (Figs. 3, 4; Table 1). This lithofacies association is characterized by a higher content of open marine bioclasts and existing large rudist shells, corals, and lithocodium aggregatum faunas. Rudist rudstone and lithocodium aggregatum rudstone/boundstone are typical lithofacies of the Dariyan Formation and have been studied and reported from the Shu'aiba Formation in nearby regions (e.g., Van Buchem et al. 2002; Maurer et al. 2010; Pierson et al. 2010). In the studied wells, these two lithofacies sometimes coexisted with each other indicating deposition in similar conditions. Rudists have developed in greater abundance on the southern part of the Arabian Plate (UAE: Wilson 1975; Yose et al. 2006; Oman: Witt and Gökdag 1994; Masse et al. 1997; Van Buchem et al. 2002; Boote and Mou 2003; Qatar: Raven et al. 2010) than southwest Iran (Van Buchem et al. 2010b) and form the main reservoirs in that area (e.g., Amthor et al. 2010). This lithofacies is mainly found in upper parts of the Dariyan Formation and the similar Shu'aiba Formation (e.g.,

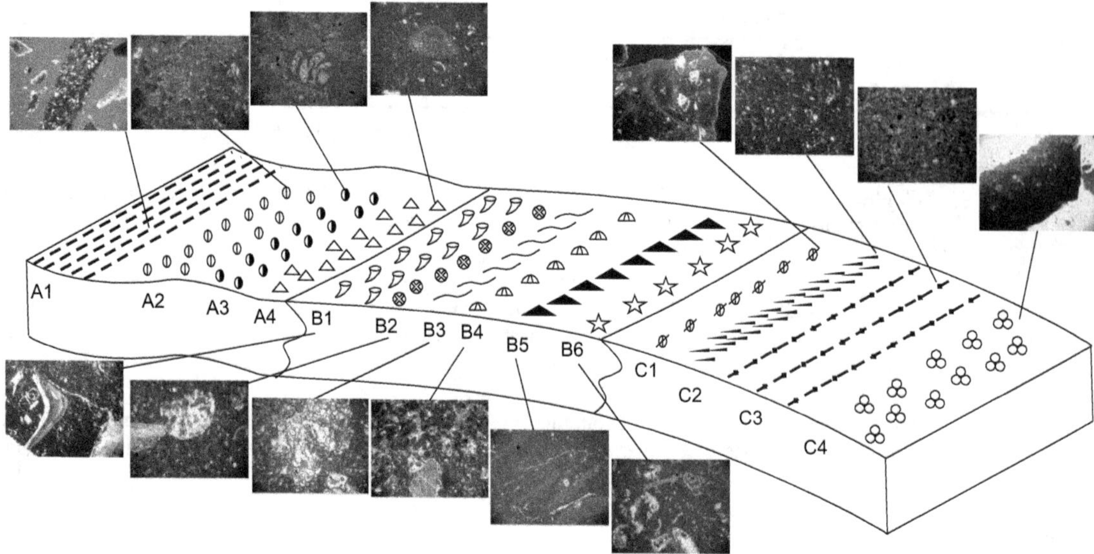

Fig. 5 Depositional model for carbonate sediments of the Dariyan Formation

Vahrenkamp 2010; Yose et al. 2010). It may be formed as a bank and build-ups in the shallow marginal parts of Aptian carbonate platforms with diverse faunas. On the other hand, coral rudstone/boundstone, which sometimes coexists with lithocodium aggregatum rudstone/boundstone and rudist rudstone is another mid-ramp environment lithofacies. Similar to nearby regions (e.g., Granier et al. 2003; Van Buchem et al. 2010a, b), lithocodium aggregatum rudstone/boundstone deposited during the initial construction phase of the Shu'aiba platform and rudists nucleated on them. Lithocodium aggregatum is a very common association in the Early Aptian (Immenhauser et al. 2004, 2005; Hillgärtner et al. 2003) and has a paleogeographic range that is comparable to tropical and subtropical reefs in the Mesozoic (Rameil et al. 2010). This lithofacies quickly colonized and started aggrading following the rise of sea level in the fair-weather wave base area in the shelf margin near rudist rudstones (e.g., Hillgärtner 2010). Echinoid pellet grainstone lithofacies mainly contains open marine echinoid bioclasts and deposited in high-energy conditions of platform margin bars and pellet grains probably transported from shallower areas such as lagoons. The discoidal Orbitolina packstone/wackestone consists mainly of discoidal forms of orbitolinids that are formed in deeper parts of open marine environments in platform margin conditions and probably were deeper than units with abundant lithocodium aggregatum. The flattened form of palorbitolina is probably a response to the relatively low-light setting, therefore indicating their formation in deeper parts of mid-ramp (Immenhauser et al. 1999; Simmons et al. 2000). Finally, bioclastic wackestone/packstone lithofacies consist mainly of diverse bioclasts and were deposited in a low-energy shelf margin to slope conditions.

5.3 Outer ramp/intrashelf basin lithofacies association

This association deposited on the outer ramp or after creation of the intrashelf basin. It consists of lenticulina/epistomina wackestone, sponge spicule wackestone/packstone, pyritized mudstone, and pelagic foraminifera wackestone lithofacies (Fig. 4; Table 1). Sponge spicule wackestone consists mainly of deeper water sponge spicules (Flugel 2010), which were deposited in the outer ramp to intrashelf basin. The occurrence of lenticulina and epistomina along with some pelagic foraminifera in lenticulina/epistomina wackestone also prove deeper marine conditions (Meyer 2000). Pyritized mudstone is another lithofacies that deposited in reducing conditions. Some deep-sea faunas such as radiolaria, sponge spicule, and pelagic foraminifers were identified in this lithofacies and other lithofacies of this association. Pelagic foraminifera wackestone lithofacies is another deep-water lithofacies that contains diverse globigerina and hedbergella faunas within a dark micritic context that partly has been pyritized in reducing conditions.

6 Sequence stratigraphy

Sequence stratigraphy of the Barremian–Aptian strata in the Arabian Plate has been studied in outcrop (Pittet et al. 2002; Van Buchem et al. 2002, 2010b) and subsurface (Sharland et al. 2001; Yose et al. 2010; Maurer et al. 2010; Pierson et al. 2010). The Dariyan Formation (equivalent to the Shu'aiba Formation and Hawar Member) deposited during Aptian supersequence transgressive–regressive cycles and was bounded by unconformities (Yose et al.

2010). Van Buchem et al. (2010a) also proposed a new sequence stratigraphic framework that correlates through the Arabian Plate and Persian Gulf region of this plate. According to this framework, the Arabian Plate Aptian Supersequence was divided into four third-order sequences, named Arabian Plate Aptian Sequences 1 to 4 (abbreviated: Apt 1–Apt 4; Table 2). The Apt 5 third-order sequence of the Late Aptian–Albian supersequence is another sequence that was identified in carbonate sequences of the Dariyan Formation (Van Buchem et al. 2010a; Table 2). Identification of these sequences in the studied area is according to correlation of log data and lithofacies characteristics with Arabian Plate sequences. Table 2 shows some characteristics of these sequences in the Arabian Plate and the studied area, and Fig. 13 shows the correlation of these sequences with each other and Arabian Plate Apt 1–5 sequences in Abu Dhabi. Time ranges of each third-order sequence are according to GTS time scale of Ogg et al. (2004).

6.1 Apt 1 sequence (from 125 to 124.6 Ma)

This sequence comprises the Lower Dariyan Formation (equivalent to the lower Shu'aiba Formation and Hawar Member). The base of this sequence in the Arabian platform shows evidence of exposure and is the lower boundary of a second-order sequence (Vahrenkamp 2010; Van Buchem et al. 2010b) with an abrupt gamma ray increases probably due to uranium concentration by groundwater movement. In offshore Abu Dhabi, Granier (2008) interpreted the Hawar Member as an unconformity-bounded unit and Yose et al. (2010) divided this sequence as Apt 1a (Hawar Member) and Apt 1b (Lower Shu'aiba). According to chronostratigraphic studies of Schroeder et al. (2010) (Fig. 1), there was a hiatus between the Hawar Member and Kharaib Formation (equivalent to the Gadvan Formation) so we can interpret this boundary as a type 1 sequence boundary. In the studied wells, the trend in gamma ray readings is similar to other parts of the Arabian platform and the lower boundary of the Dariyan Formation is thought to be an exposure surface. This sequence consists of shallow inner to mid-ramp lithofacies with similar thicknesses ranging from 12 to 23 m. This sequence in the Arabian platform forms an intraformational seal or dense unit (Van Buchem et al. 2002). This sequence formed TST and HST that are described below:

6.1.1 TST

Transgressive system tract (TST) lithofacies mainly consists of sandy mudstone, milliolidae wackestone/pellet grainstone, benthic foraminifera wackestone/pellet grainstone, echinoid pellet grainstone, discoidal Orbitolina wackestone, and bioclastic wackestone/packstone. Thickness of TST lithofacies ranges from 4 m to 13 m and the shallowest sandy mudstone lithofacies was identified in

Table 2 Characteristics of Arabian Plate Aptian sequences (Van Buchem et al. 2010a) and the studied area

Super sequences	Third-order sequences/age	Characteristics in the Arabian Plate	This study (main characteristics)
Late Aptian–Albian	Apt 5 (LST)/Late Aptian	Deposition of mixed argillaceous/carbonate sediments on exposed sequence boundary	In well 1: TST: B5 HST: A4
Aptian	Apt 4 (Late HST)/latest Early Aptian to early Late Aptian	Progradation of slope-restricted fringing shoals and carbonate mud banks in low-angle clinoforms, development of small rudist shoals and orbitolinid-dominated slopes	In wells 1, 5, and 6: TST: A4, B2, B5, B6, C3, C4 HST: B1, B2, C1, C2, C3, C4
	Apt 3 (Early HST)/latest Early Aptian	Rudist-dominated platform aggradation and progradation, basin starvation, production surpasses accommodation space.	TST: A2, A3, A4, B1, B3, B4, B5, B6, C1, C2, C4 HST: A1, A2, A3, A4, B1, B3, B5, C1, C2, C4
	Apt 2 (Late TST)/late Early Aptian	Development of topographic relief, proliferation of lithocodium–bacinella boundstone lithofacies, which formed mound-like features and eventually aggradational platforms, the time-equivalent starvation of the intrashelf basins where a condensed organic-rich sedimentation took place, time-equivalent to the OAE-1a event	TST: A2, A3, B2, B3, B5, B6, C3, C4 lithofacies HST: A1, A3, B5, B6, C1, C3, C4 lithofacies
	Apt 1 (Early TST)/early Early Aptian	Development of flat, orbitolinid-dominated, argillaceous low-angle ramp system, known as the Hawar member	TST & HST: A1, A2, A3, B4, B5, and B6 lithofacies

wells 5 and 7. The maximum flooding surface lies in low gamma ray lithofacies such as echinoid pellet grainstone, bioclastic wackestone/packstone, and discoidal Orbitolina wackestone/packstone (Figs. 6, 7, 8, 9, 10, 11, 12).

6.1.2 HST

The thickness of highstand system tract (HST) lithofacies ranges between 2 and 10 m and shallowest inner ramp sandy mudstone lithofacies deposited in wells 5 and 7. The upper boundary of this sequence is type 2 and is positioned at the top of sandy mudstone, milliolidae wackestone/pellet grainstone, benthic foraminifera wackestone/pellet grainstone, discoidal Orbitolina wackestone, and bioclastic wackestone/packstone lithofacies (Figs. 6, 7, 8, 9, 10, 11, 12).

6.2 Apt 2 sequence (from 124.6 to 124 Ma)

Apt 2 is an age equivalent to the global oceanic anoxic event (OAE1a) (Van Buchem, 2010a; Yose et al. 2010). This sequence is bounded by type 2 sequence boundaries and is indicated by development of lithocodium boundstone lithofacies along with other inner to outer ramp lithofacies. In wells 1, 2, 5–7, the upper part of this sequence shows higher gamma spikes than lower parts. In the lower Shu'aiba Formation, this trend reflects variations in reducing conditions/organic matter content rather than clay content (Droste 2010). Similar to other studies (e.g., Vahrenkamp 2010), the maximum flooding surface is placed near a high gamma ray spike and represents the maximum flooding surface of the Aptian supersequence. In studied wells, this sequence thickened in well 2 (29 m) and thinned toward wells 6 (12 m) and 1 (15 m). This sequence consists of TST and HST as follows:

6.2.1 TST

The thickness of TST lithofacies ranges between 6 and 24 m and mainly starts with developing lithocodium boundstone lithofacies. Other lithofacies consists of milliolidae wackestone/pellet grainstone, benthic foraminifera wackestone/pellet grainstone, coral rudstone, discoidal Orbitolina wackestone/packstone, bioclastic wackestone/packstone, lenticulina/epistomina wackestone, pyritized mudstone, and pelagic foraminifera wackestone. The deepest lithofacies (pyritized mudstone and pelagic foraminifera wackestone) is identified in wells 5 and 6 and the maximum flooding surface is mainly placed near high spike gamma ray readings (Figs. 6, 7, 8, 9, 10, 11, 12).

6.2.2 HST

Lithofacies of HST consist of sandy mudstone, benthic foraminifera wackestone/pellet grainstone, discoidal Orbitolina wackestone/packstone, bioclastic wackestone/packstone, lenticulina/epistomina wackestone, pyritized mudstone, and pelagic foraminifera wackestone (Figs. 6, 7, 8, 9, 10, 11, 12). The thickness of HST facies changes from 1 to 13 m and the upper part of this sequence is bounded by a type 2 sequence boundary. During deposition of HST, the shallowest lithofacies deposited in well 5 (sandy mudstone) which is probably related to reactivation of older faults.

6.3 Apt 3 (from 124 to 121 Ma)

During this stage, the lithocodium boundstone facies was replaced by colonized rudist and probably formed build-up and flank complexes. According to Al-Husseini and Matthews (2010), Apt 2 and Apt 3 represent the differentiation of the southeastern Arabian Plate into the Shu'aiba Platform and intrashelf Bab Basin. In studied wells, this sequence has a type 2 lower boundary and the upper boundary with the Burgan Formation in wells 2, 3, 4, and 7 is type 1. The thickness of this sequence ranges from 11 to 86 m. This sequence thickened toward well 3 (86 m) and thinned toward SW in wells 1 and 2 and NE in wells 4, 5, and 6. Higher thickness of carbonates in wells 2, 3, 4, and 7 represents good carbonate productivity during platform shallowing and deposition of mid-ramp build-ups toward the location of these wells (Figs. 6, 7, 8, 9, 10, 11, 12). This sequence also consists of TST and HST as follows:

6.3.1 TST

The thickness of TST facies ranges from 7 to 39 m and consists of milliolidae wackestone/pellet grainstone, benthic foraminifera wackestone/pellet grainstone, conical Orbitolina wackestone/packstone, rudist rudstone, lithocodium aggregatum rudstone/boundstone, echinoid pellet grainstone, discoidal Orbitolina wackestone/packstone, bioclastic wackestone/packstone, lenticulina/epistomina wackestone, sponge spicule wackestone/packstone, and pelagic foraminifera wackestone. In wells 1, 5, and 6, the maximum flooding surface is placed at a high gamma ray spike containing discoidal Orbitolina wackestone/packstone, lenticulina/epistomina wackestone, and pelagic foraminifera wackestone lithofacies (Figs. 6, 7, 8, 9, 10, 11, 12).

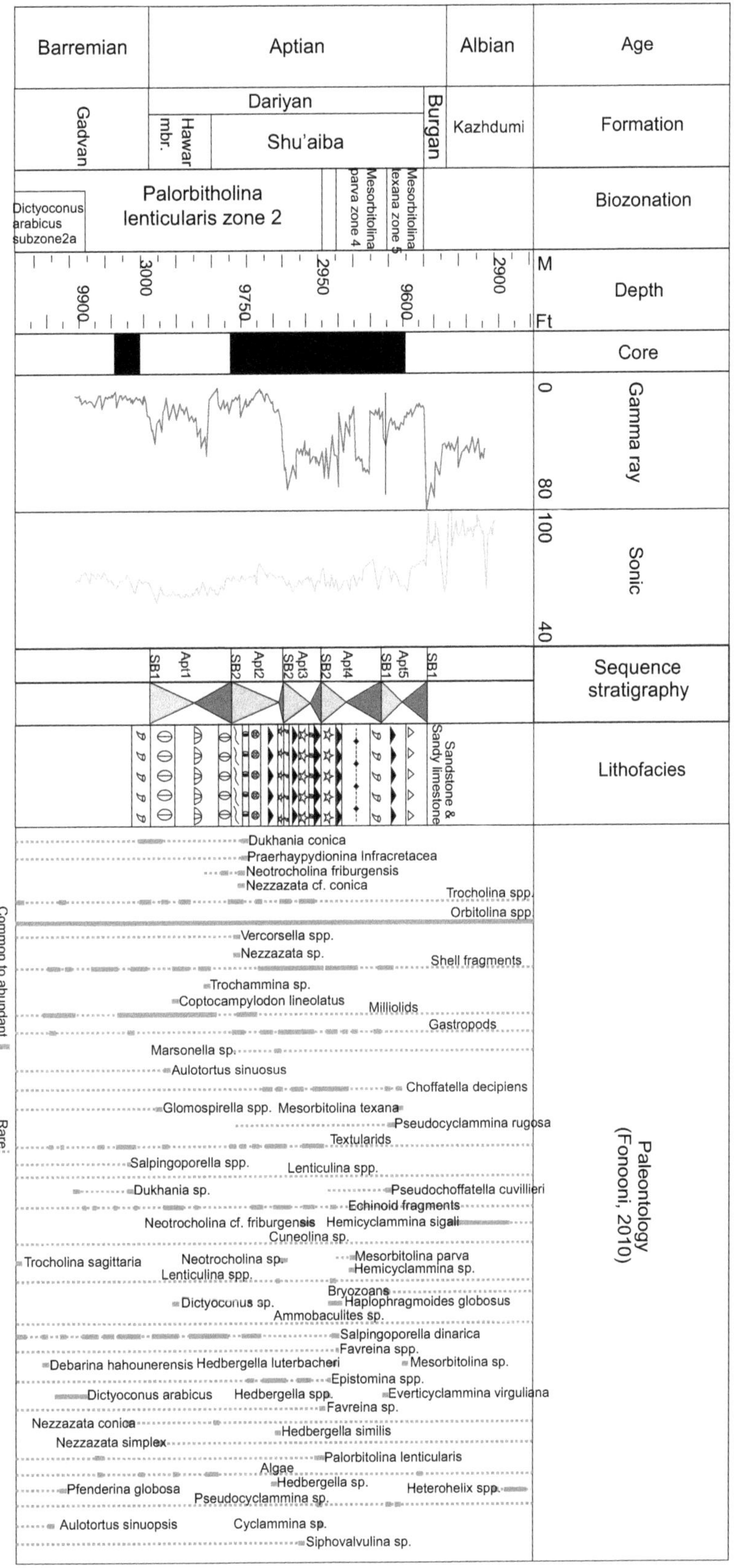

Fig. 6 Sequence stratigraphy, paleontology, gamma ray, and sonic logs in well 1, for lithofacies symbols see Fig. 5

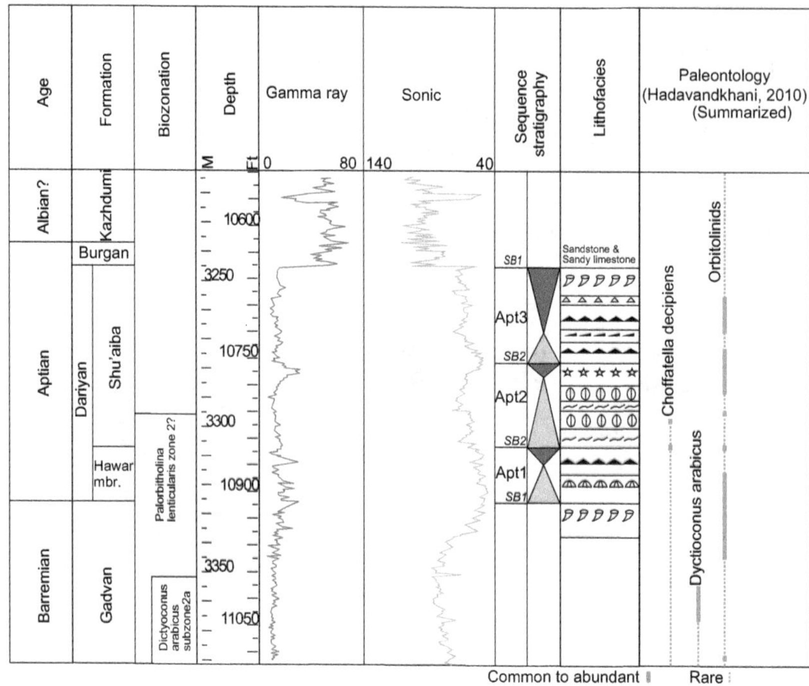

Fig. 7 Sequence stratigraphy, paleontology, gamma ray, and sonic logs in well 2, for lithofacies symbols see Fig. 5

6.3.2 HST

The thickness of HST facies ranges from 3 to 51 m. This phase contains sandy mudstone, milliolidae wackestone/pellet grainstone, benthic foraminifera wackestone/pellet grainstone, conical Orbitolina wackestone/packstone, rudist rudstone, lithocodium aggregatum rudstone/boundstone, discoidal Orbitolina wackestone/packstone, lenticulina/epistomina wackestone, sponge spicule wackestone/packstone, and pelagic foraminifera wackestone. During this phase of deposition, the shallowest sandy mudstone lithofacies were deposited in well 4. The upper boundary is type 2 in wells 1, 5, and 6, and type 1 in wells 2–4 and 7.

6.4 Apt 4 (from 121 to 117.9 Ma)

During this stage, sea level dropped in the Arabian Plate so most parts of the Shu'aiba Platform were exposed and sediments only deposited along the Bab Basin (Al-Husseini and Matthews 2010). Similarly, studied wells 2, 3, 4, and 7 were exposed and Apt 4 did not develop in these wells while deposition continued in wells 1, 5, and 6. Therefore, the lower boundary of Apt 4 in these wells is type 2 and the upper boundary with LST Apt 5 sequence is type 1. Biozonation of well 1 confirms an Early Late Aptian age for Apt 4. Note, in wells 5 and 6, the time range of Apt 4 could not be identified because of the absence of index fossils (Figs. 6, 10, 11). In well 1 (Fig. 6), the thickness of this sequence is about 17 m and contains Mesorbitolina parva with discoidal Orbitolina wackestone/packstone, bioclastic wackestone/packstone, pyritized mudstone, and rudist rudstone lithofacies. The maximum flooding surface is placed at the high gamma ray pyritized mudstone lithofacies. In well 5 (Fig. 10), the thickness of this sequence is about 30 m and contains coral rudstone, conical Orbitolina wackestone/packstone, and pyritized mudstone, and the maximum flooding surface is placed in this well at high gamma ray pyritized mudstone lithofacies. In well 6 (Fig. 11), the thickness of this sequence reaches 30 m and contains pelagic foraminifera wackestone, sponge spicule wackestone/packstone, lenticulina epistomina wackestone, and coral rudstone, and the maximum flooding surface is placed at high gamma ray pelagic foraminifera wackestone.

6.5 Apt 5 (from 117.9 to 112.4 Ma)

This sequence is the lowstand system tract (LST) of the Upper Aptian–Lower Albian Supersequence and deposited during Late Aptian time in mixed argillaceous/carbonate platform. This sequence is depicted in well 1 (Fig. 6) with 13-m thickness and index fossils of Mesorbitolina texana. This sequence consists of conical and discoidal Orbitolina wackestone/packstone lithofacies so the maximum flooding surface is placed at the somewhat high gamma ray discoidal Orbitolina wackestone/packstone lithofacies. As this sequence deposited after exposure of the platform in the Arabian Plate, the lower boundary with Apt 4 is interpreted

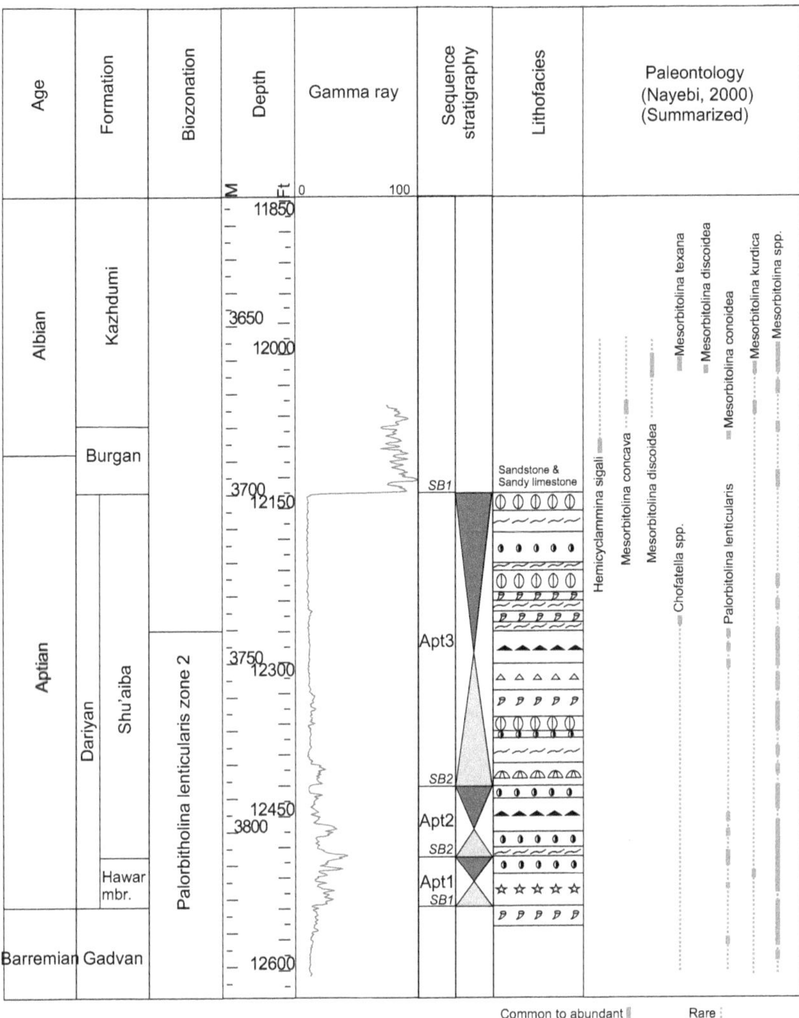

Fig. 8 Sequence stratigraphy, paleontology, and gamma ray log in well 3, for lithofacies symbols see Fig. 5

as type 1 and similarly the upper boundary of this sequence with clastic sediments of the Burgan Formation is also interpreted as a type 1 boundary.

7 Discussion

The sequence stratigraphic correlation of the Dariyan Formation in studied wells and Abu Dhabi wells is illustrated in Fig. 13.

Toward well 1, transition from carbonate platform to intrashelf Bab Basin can be seen, so, well 1 is similar to wells near the intrashelf of the Bab Basin of Vahrenkamp (2010) (wells Y and Z in Fig. 13), while the gamma logs of wells 2–4 are similar to the carbonate platform logs of Vahrenkamp (2010). Toward wells 5 (near Qeshm Island) and 6 (in Qeshm Island), there is a rapid deepening trend into deep-water sediments as seen on the high gamma ray logs; therefore, we believe there might have been an intrashelf basin in this area (Fig. 14). This proposed regional intrashelf basin has not been reported for the Aptian sediments before in the Persian Gulf region and needs more attention. According to Al-Ghamdi and Read (2010), Early Cretaceous intrashelf basins were created because of Neoproterozoic-Early Cambrian Hormuz salt movement that started as early as the Permian time in the Persian Gulf (Motiei 1995). On the other hand, reactivation of Trans-Arabian Bostaneh basement fault (Bahroudi and Koyi 2004) could have played a major role in the creation of this intrashelf basin. According to Glennie (2000), the start of subduction during the Aptian was accompanied by crustal warping of the Arabian continental platform creating erosional highs and intrashelf euxinic basins between them.

Van Buchem et al. (2002) proposed a mechanism for the creation of the Aptian age Bab intrashelf basin. According to Van Buchem et al. (2002), small topographic

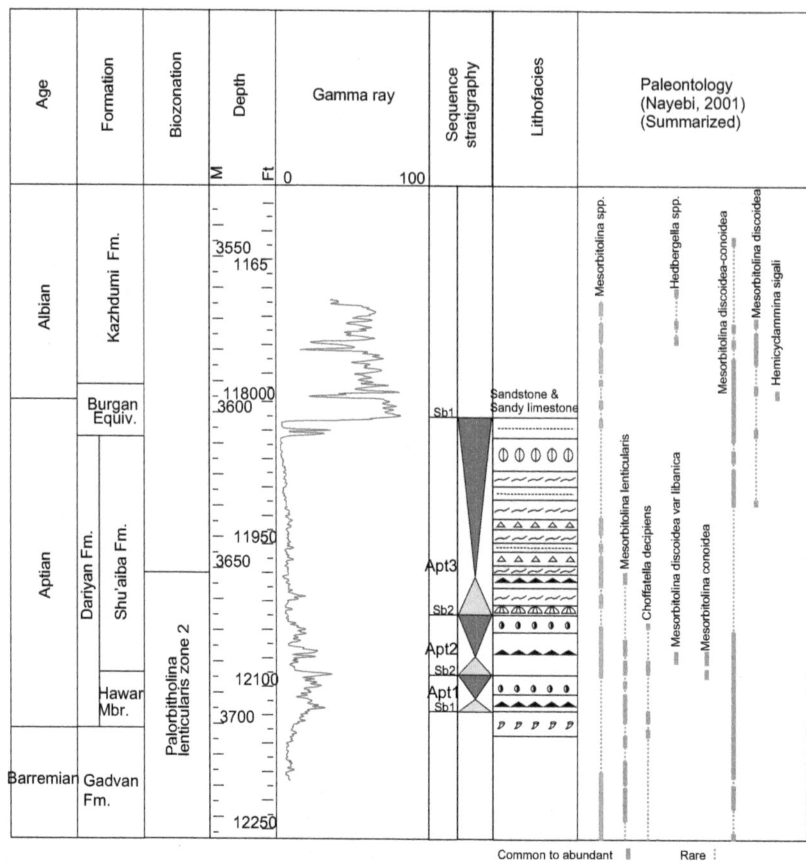

Fig. 9 Sequence stratigraphy, paleontology, and gamma ray log in well 4, for lithofacies symbols see Fig. 5

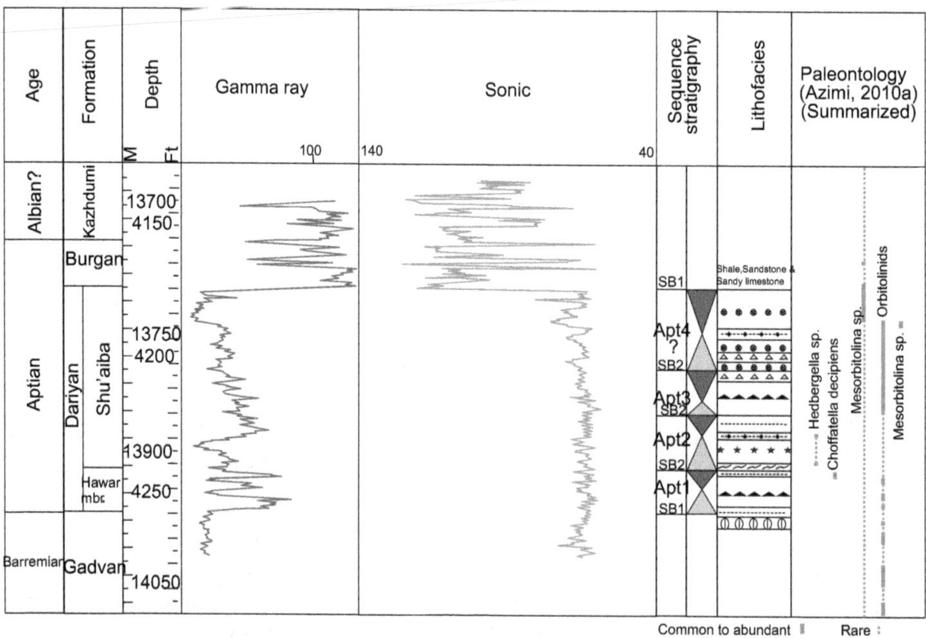

Fig. 10 Sequence stratigraphy, paleontology, gamma ray, and sonic logs in well 5, for lithofacies symbols see Fig. 5

differences are considered to have triggered differential sedimentation rates during an increased rate of relative sea-level rise. We interpret that topographic differences were mainly created by salt movement, crustal warping, and basement fault reactivations (especially the Trans-Arabian Bostaneh fault) in Qeshm Island and adjacent

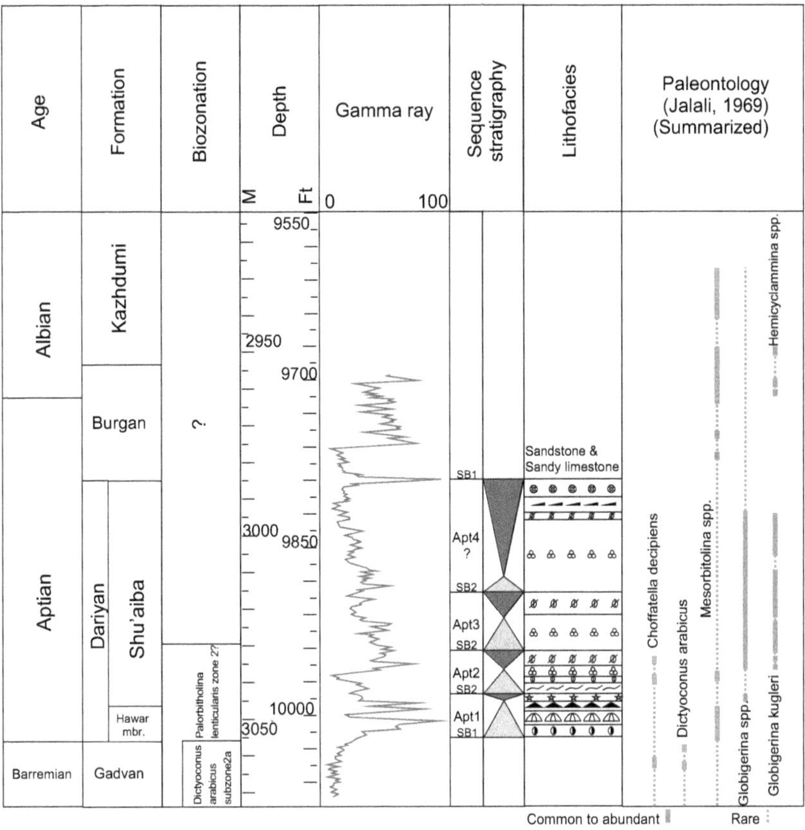

Fig. 11 Sequence stratigraphy, paleontology, and gamma ray log in well 6, for lithofacies symbols see Fig. 5

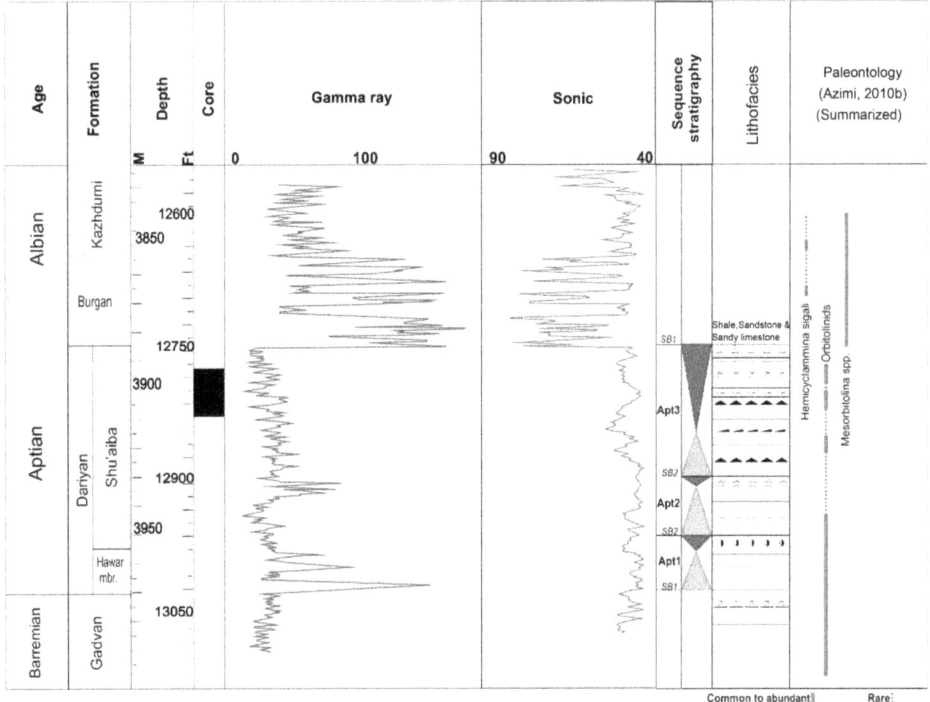

Fig. 12 Sequence stratigraphy, paleontology, gamma ray, and sonic logs in well 7, for lithofacies symbols see Fig. 5

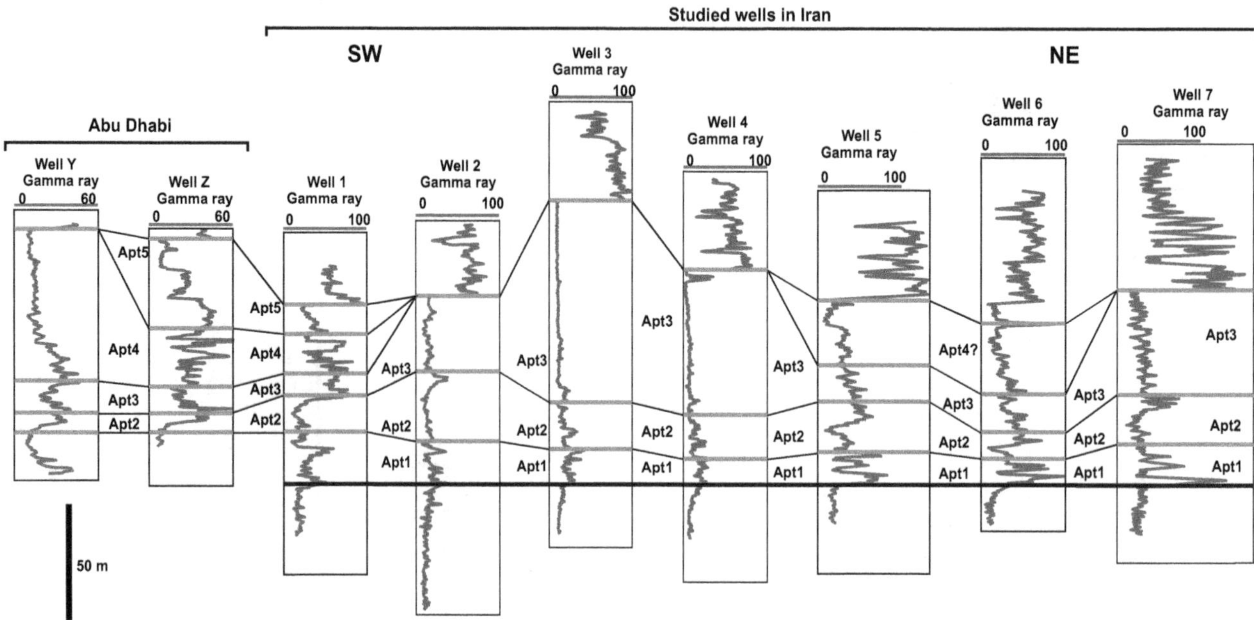

Fig. 13 Sequence stratigraphic correlation between studied wells and comparison of Iranian wells with Abu Dhabi well logs of Vahrenkamp (2010) located in a field near the Bab Basin (Fig. 2)

area. The intrashelf basin, due to conditions such as a stratified water column and slow water circulation, was suitable for the deposition of organic-rich source rock material in the Arabian platform (Van Buchem et al. 2002), so high gamma ray intrashelf deposits in wells 1, 5, and 6 probably have a good source rock capacity. Based on correlation of sequences (Fig. 13) and paleogeographic mapping (Fig. 14), Apt 1 shows similar thicknesses of shallow marine inner to mid-ramp lithofacies. During Apt 1, transgressive onlaps developed toward the eastern exposed area in northern Oman (Droste 2010) (Fig. 14). After the second major transgression (Apt 2), higher differences between sequence thicknesses as well as lithofacies types are due to intrashelf basin topographies. The deepest intrashelf lithofacies of the Apt 2 sequence developed in wells 5 and 6 (Fig. 14). During deposition of Apt 3, the deepest part of the basin is placed in wells 1, 2 (Bab Intrashelf Basin), 6 and 7 (Qeshm Intrashelf Basin) (Fig. 14), while the shallowest part with thickest deposits is located in wells 3–5. Rudist and coral rudstone lithofacies, which probably have good reservoir properties, deposited mid-ramp in wells 2–4, 7, and partly in well 1. After this phase, the Apt 4 sequence developed in wells 1 (Bab Intrashelf Basin), 5, and 6 (Qeshm Intrashelf Basin) (Fig. 14). During this phase, other wells were exposed and Burgan Formation deposited in coastal marine areas (Fig. 14). During the deposition of the Apt 4 interval, rudist and coral rudstone lithofacies deposited in shallow marine environments, probably with good reservoir properties in wells 1, 5, and 6. On the other hand, well 6 contains thicker intrashelf sediments and we interpret that a continuous deepening of the platform existed in well 6. The Apt 5 sequence only developed in well 1 near the Bab Basin and is not present in other wells (Fig. 14). Finally, after deposition of Apt 5, the entire Dariyan carbonate platform was exposed and terrigenous materials of the Burgan Formation deposited after a long break in coastal to shallow marine environment throughout the study area.

Global cooling and warming events (e.g., Herrle and Mutterlose 2003; Mutterlose et al. 2009) played an important role in controlling sea-level changes of the Arabian Plate Cretaceous epicontinental sea and deposition of sequences. The Aptian was a cooler period with some ice sheets at the poles (Al-Ghamdi and Read 2010). According to Van Buchem et al. (2010a), Aptian sedimentation on the Arabian Platform (marginal part of Arabian Plate) was influenced by two cold phases: One short-lived phase in the earliest Aptian and a second longer lived phase in the Late Aptian. These cold phases caused glacio-eustatic large and rapid sea-level fluctuations in the order of 40 m and exposure of the platform before deposition of the Apt 1 and Apt 5 sequences. Extinction of the rudists in Apt 5 of the Arabian Platform and studied area may have been the result of climatic cooling of Aptian seawater (e.g., Strohmenger et al. 2010). Exposure of Aptian sediments in the latest Aptian–Early Albian (upper boundary of Dariyan Formation) also fits with another cooling event that was recognized in sedimentary facies and in the C- and O-isotope records (e.g., Weissert and Lini 1991) and discussed by Raven et al. (2010) in Aptian sediments of Qatar.

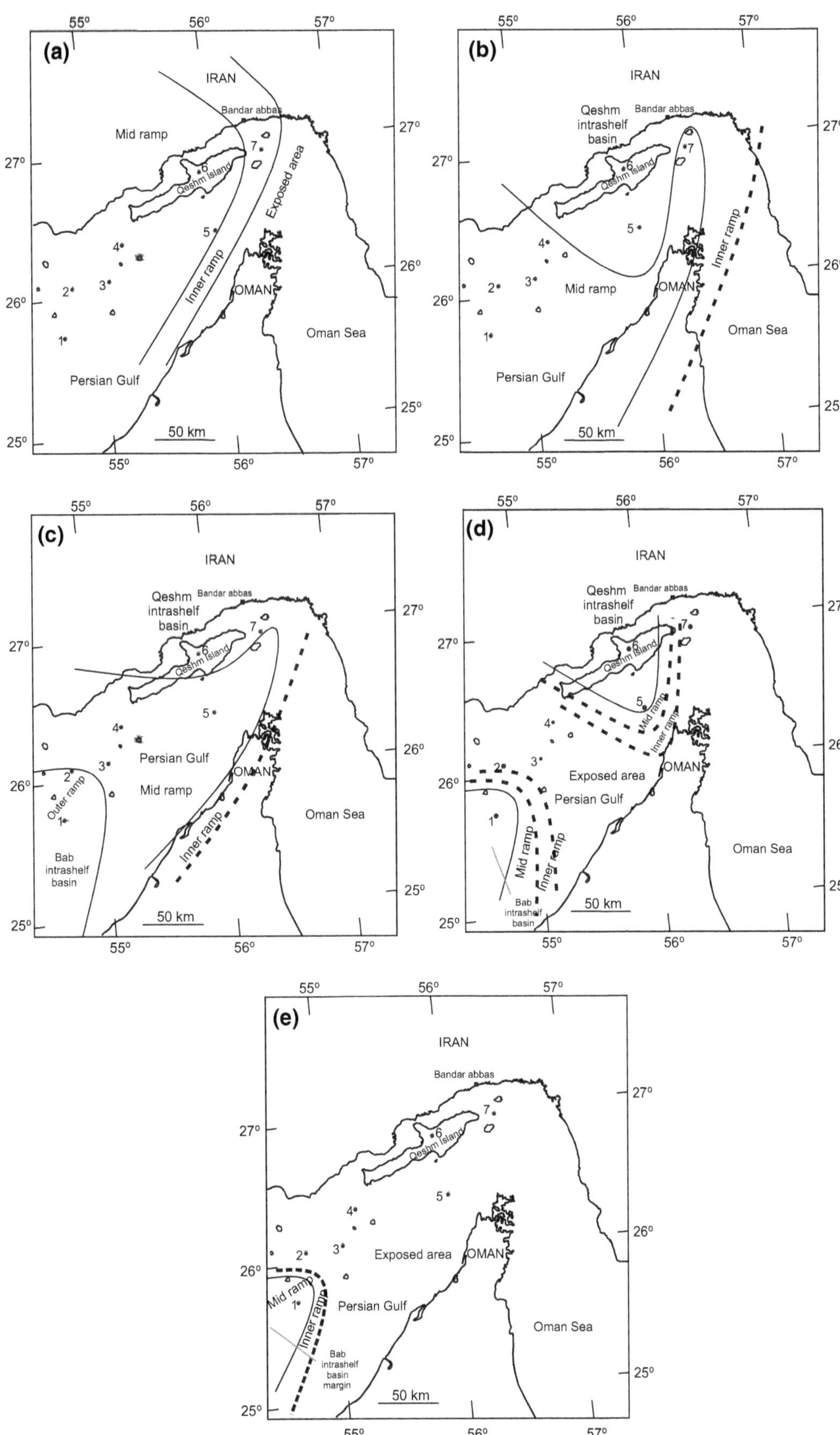

Fig. 14 Paleogeographic map of Aptian time in Qeshm island and offshore during the maximum flooding surface of Apt 1 (**a**), Apt 2 (**b**), Apt 3 (**c**), Apt 4 (**d**), and Apt 5 (**e**) sequences based on studied wells 1–7

8 Conclusion

The Dariyan Formation in Qeshm Island and offshore (southern Iran and the eastern part of Persian Gulf) is equivalent with the Hawar Member and Shu'aiba Formation of the Arabian Plate and contains 14 carbonate lithofacies which deposited on a carbonate ramp system. Sequence stratigraphic studies resulted in recognition of 5 Aptian third-order sequences toward Bab Basin (SW—well 1) and 4 Aptian third-order sequences toward Qeshm Island (NE wells 5 and 6). Toward wells 1, 5, and 6, a typical long-lasting transition from carbonate platform to intrashelf basin has been identified. The present of higher gamma ray peaks as maximum flooding surfaces in Apt 2–Apt 4 indicates that there might be a good source rock similar to the Arabian Platform. We interpret that salt movement, crustal warping, and basement fault reactivations were probably responsible for creation of an intrashelf basin in Qeshm Island and offshore. On the other hand, rudist and coral rudstone lithofacies, which probably have good reservoir properties, were deposited mid-ramp during deposition of Apt 3 (wells 2–4, 7 and partly in well 1) and Apt 4 (wells 1, 5, and 6) sequences. The Dariyan Formation around Qeshm Island mainly deposited in an intrashelf basin and should be considered as a new intrashelf basin in future Aptian paleogeographic maps.

Acknowledgments The authors would like to thank the National Iranian Oil Company, Exploration Directorate, for the support of this research. We also thank the Department of Geology at Ferdowsi University of Mashhad for their support.

References

Ainsworth RB. Sequence stratigraphic-based analysis of reservoir connectivity: influence of sealing faults—a case study from a marginal marine depositional setting. Petrol Geosci. 2006;12(2):127–41.

Ala MA. Salt diapirism in southern Iran. AAPG Bull. 1974;58(9):1758–70.

Alavi M. Regional stratigraphy of the Zagros fold-thrust belt of Iran and its proforeland evolution. Am J Sci. 2004;304:1–20.

Alavi M. Structures of the Zagros fold-thrust belt in Iran. Am J Sci. 2007;307(9):1064–95.

Al-Ghamdi N, Read FJ. Facies-based sequence stratigraphic framework of the Lower Cretaceous rudist platform, Shu'aiba Formation, Saudi Arabia. In: Van Buchem FSP, Al-Husseini MI, Maurer F, Droste HJ, editors. Barremian–Aptian stratigraphy and hydrocarbon habitat of the eastern Arabian Plate, vol. 1. GeoArabia Special Publication 4. Bahrain: Gulf PetroLink; 2010. p. 367–410.

Al-Husseini MI, Matthews RK. Tuning Late Barremian – Aptian Arabian Plate and global sequences with orbital periods. In: Van Buchem FSP, Al-Husseini MI, Maurer F, Droste HJ, editors. Barremian–Aptian stratigraphy and hydrocarbon habitat of the eastern Arabian Plate, vol. 1. GeoArabia Special Publication 4. Bahrain: Gulf PetroLink; 2010. p. 199–228.

Ali MY, Watts AB, Searle MP. Seismic stratigraphy and subsidence history of the United Arab Emirates (UAE) rifted margin and overlying foreland basins. In: Al Hosani K, Roure F, Ellison R, Lokier S, editors. Lithosphere dynamics and sedimentary basins: The Arabian Plate and analogues, frontiers in Earth sciences. Berlin: Springer; 2013. p. 127–43.

Alsharhan AS, Nairn AEM. Sedimentary basins and petroleum geology of the Middle East. Amsterdam: Elsevier; 1997.

Amthor JE, Kerans C, Gauthier P. Reservoir characterisation of a Shu'aiba carbonate ramp-margin field, northern Oman. In: Van Buchem FSP, Al-Husseini MI, Maurer F, Droste HJ, editors. Barremian–Aptian stratigraphy and hydrocarbon habitat of the eastern Arabian Plate, vol. 2. GeoArabia Special Publication 4. Bahrain: Gulf PetroLink; 2010. p. 549–76.

Azimi R. Biostratigraphy, micropalaeontology and micropaleontological studies on the cutting and core samples of drilled sequence of well 5. Report 1801. Tehran: Exploration Directorate of National Iranian Oil Company; 2010a.

Azimi R. Biostratigraphy, micropalaeontology and micropaleontological studies on the cutting and core samples of drilled sequence of well 7. Report 1808. Tehran: Exploration Directorate of National Iranian Oil Company; 2010b.

Bahroudi A, Koyi HA. Tectono-sedimentary framework of the Gachsaran Formation in the Zagros foreland basin. Mar Petrol Geol. 2004;21(10):1295–310.

Bechennec F, Le Metour J, Rabu D, et al. The Hawasina Nappes; stratigraphy, palaeogeography and structural evolution of a fragment of the South-Tethyan passive continental margin. Geol Soc Spec Publ. 1990;49:213–23.

Berberian M, King GCP. Towards a paleogeography and tectonic evolution of Iran. Can J Earth Sci. 1981;18(2):210–65.

Boote DRD, Mou D. Safah field, Oman: retrospective of a new-concept exploration play, 1980 to 2000. GeoArabia. 2003;8(3):367–430.

Droste HJ. Sequence-stratigraphic framework of the Aptian Shu'aiba formation in the Sultanate of Oman. In: Van Buchem FSP, Al-Husseini MI, Maurer F, et al., editors. Barremian-Aptian stratigraphy and hydrocarbon habitat of the eastern Arabian Plate, vol. 1. GeoArabia Special Publication 4. Bahrain: Gulf PetroLink; 2010. p. 229–83.

Dunham RJ. Classification of carbonate rocks according to depositional textures. In: Ham WE, editor. Classification of carbonate rocks, vol. 1. AAPG Memoir; 1962. p. 108–21.

Edgell HS. Salt tectonism in the Persian Gulf basin. Geol Soc London Spec Publ. 1996;100(1):129–51.

Ehrenberg SN, Svånå TA. Use of spectral gamma-ray signature to interpret stratigraphic surfaces in carbonate strata: an example from the finnmark carbonate platform (Carboniferous-Permian), Barents Sea. AAPG Bull. 2001;85(2):295–308.

Embry AF, Klovan JE. A Late Devonian reef tract on north-eastern Banks Island, Northwest Territories. Bull Can Petrol Geol. 1971;19:730–81.

Flugel E. Microfacies of carbonate rocks: analysis, interpretation and application. New York: Springer; 2010.

Fonooni B. Biostratigraphy, micropalaeontology and micropaleontological studies on the cutting and core samples of drilled sequence of well 1. Report 1788. Tehran: Exploration Directorate, National Iranian Oil Company; 2010.

Glennie KW. Cretaceous tectonic evolution of Arabia's eastern plate margin: a tale of two oceans. SEPM Spec Publ. 2000;69:9–20.

Granier B. Holostratigraphy of the Kahmah regional series in Oman. Qatar and the United Arab Emirates: Notebooks of Geology; 2008.

Granier B, Al Suwaidi AS, Busnardo R, et al. New insight on the stratigraphy of the "Upper Thamama" in offshore Abu Dhabi (UAE), vol. 5. Qatar and the United Arab Emirates: Notebooks of Geology; 2003.

Hadavandkhani N. Biostratigraphy, micropalaeontology and micropaleontological studies on the cutting and core samples of drilled sequence of well 2. Report 1807. Tehran: Exploration Directorate, National Iranian Oil Company; 2010.

Halfar J, Ingle JC. Modern warm-temperate and subtropical shallow-water benthic foraminifera of the southern Gulf of California, Mexico. J Foramin Res. 2003;33(4):309–29.

Herrle JO, Mutterlose J. Calcareous nannofossils from the Aptian–Early Albian of SE France: paleoecological and biostratigraphic implications. Cretaceous Res. 2003;24:1–22.

Hillgärtner H. Anatomy of a microbially constructed, high-energy, ocean-facing carbonate platform margin (earliest Aptian, northern Oman Mountains). In: Van Buchem FSP, Al-Husseini MI, Maurer F, et al., editors. Barremian–Aptian stratigraphy and hydrocarbon habitat of the eastern Arabian Plate, vol. 1. GeoArabia Special Publication 4. Bahrain: Gulf PetroLink; 2010. p. 285–300.

Hillgärtner H, Van Buchem F, Gaumet F, et al. The Barremian–Aptian evolution of the eastern Arabian carbonate platform margin (northern Oman). J Sediment Res. 2003;73(5):756–73.

Hughes GW. Bioecostratigraphy of the Shu'aiba Formation, Shaybah field. Saudi Arabia. GeoArabia. 2000;5(4):545–78.

Hughes GW. Middle to Upper Jurassic Saudi Arabian carbonate petroleum reservoirs: biostratigraphy, micropalaeontology and palaeoenvironments. GeoArabia. 2004;9:79–114.

Hughes GW. Complex-walled agglutinated foraminiferal biostratigraphy and palaeoenvironmental significance from the Jurassic supercycle-associated carbonates of Saudi Arabia. In: Abstracts 7th international workshop Agglutinated Foraminifera, Urbino, Italy; 2005. p. 19–20.

Immenhauser A, Hillgärtner H, Sattler U, et al. Barremian-Lower Aptian Qishn Formation, Haushi-Huqf area, Oman: a new outcrop analogue for the Kharaib/Shu'aiba reservoirs. GeoArabia. 2004;9(1):153–94.

Immenhauser A, Hillgärtner H, Van Bentum EC. Microbial-foraminiferal episodes in the Early Aptian of the southern Tethyan margin: ecological significance and possible relation to oceanic anoxic event 1a. Sedimentology. 2005;52:77–99.

Immenhauser A, Schlager W, Burns SJ, et al. Late Aptian to Late Albian sea level fluctuations constrained by geochemical and biological evidence (Nahr Umr Formation, Oman). J Sediment Res. 1999;69:434–66.

Jahani S, Callot JP, Frizon de Lamotte D, et al. The salt diapirs of the Eastern Fars Province (Zagros, Iran): A brief outline of their past and present. In: Lacombe O, Lavé J, Roure F, Vergés, J, editors. Thrust belts and foreland basin: from fold kinematics to hydrocarbon systems. Berlin: Springer; 2007. p. 289–308.

Jalali MR. Paleolog chart of well 6. Tehran: Exploration Directorate, National Iranian Oil Company; 1969.

James GA, Wynd JG. Stratigraphic nomenclature of Iranian oil consortium agreement area. AAPG Bulletin. 1965;49(12):2182–245.

Konyuhov AI, Maleki B. The Persian Gulf Basin: geological history, sedimentary formations, and petroleum potential. Lithol Miner Resour. 2006;41(4):344–61.

Masse JP, Borgomano J, Al-Maskiry S. Stratigraphy and tectonosedimentary evolution of a Late Aptian-Albian carbonate margin: the northeastern Jebel Akhdar (Sultanate of Oman). Sediment Geol. 1997;113:269–80.

Maurer F, Al-Mehsin K, Pierson BJ, et al. Facies characteristics and architecture of Upper Aptian Shu'aiba clinoforms in Abu Dhabi. In: Van Buchem FSP, Al-Husseini MI, Maurer F, et al., editors. Barremian–Aptian stratigraphy and hydrocarbon habitat of the eastern Arabian Plate, vol. 2. GeoArabia Special Publication 4. Bahrain: Gulf PetroLink. 2010. p. 445–68.

Meyer M. Le complexe récifal kimméridgien-tithonien du Jura meridional interne (France), évolution multifactorielle, stratigraphie et tectonique. Terre & Environment; 2000. p. 24.

Motiei H. Stratigraphy of the Zagros. Geological Survey of Iran (in Persian language): Treatise on the Geology of Iran; 1993.

Motiei H. Petroleum Geology of Zagros, 1 and 2. Geological Survey of Iran Publications (in Persian language); 1995.

Motiei H. Simplified table of rock unit in southwest Iran (a map unpublished, KEPS Company); 2001.

Mutterlose J, Bornemann A, Herrle J. The Aptian-Albian cold snap: evidence for "mid" Cretaceous icehouse interludes. Neues Jahrb Geol P-A. 2009;252(2):217–25.

Nayebi Z. Biostratigraphy, micropalaeontology and micropaleontological studies on the cutting and core samples of drilled sequence of well 3. Report 1174, Exploration Directorate, National Iranian Oil Company; 2000.

Nayebi Z. Biostratigraphy, micropalaeontology and micropaleontological studies on the cutting and core samples of drilled sequence of well 4. Report 517, Exploration Directorate, National Iranian Oil Company; 2001.

Ogg JG, Agterberg FP, Gradstein FM. The Cretaceous Period. In: Gradstein F, Ogg J, Smith A, editors. A Geological Time Scale 2004. Cambridge University Press; 2004.

Pierson BJ, Eberli GP, Al-Mehsin K, et al. Seismic stratigraphy and depositional history of the Upper Shu'aiba (Late Aptian) in the UAE and Oman. In: Van Buchem FSP, Al-Husseini MI, Maurer F, et al., editors. Barremian–Aptian stratigraphy and hydrocarbon habitat of the eastern Arabian Plate. GeoArabia Special Publication 4, Gulf PetroLink, Bahrain. 2010. 2. p. 411–4.

Piryaei A, Reijmer JJG, Borgomano J, et al. Late Cretaceous tectonic and sedimentary evolution of the Bandar Abbas area, Fars region. Southern Iran. J Petrol Geol. 2011;34(2):157–80.

Pittet B, Van Buchem FSP, Hillgärtner H, et al. Ecological succession, palaeoenvironmental change, and depositional sequences of Barremian-Aptian shallow water carbonates in northern Oman. Sedimentology. 2002;49(3):555–81.

Rameil N, Immenhauser A, Warrlich G, et al. Morphological patterns of Aptian Lithocodium–Bacinella geobodies: relation to environment and scale. Sedimentology. 2010;57(3):883–911.

Raven MJ, Van Buchem FSP, Larsen PH, et al. Late Aptian incised valleys and siliciclastic infill at the top of the Shu'aiba Formation (Block 5, offshore Qatar). In: Van Buchem FSP, Al-Husseini MI, Maurer F, et al., editors. Barremian–Aptian stratigraphy and hydrocarbon habitat of the eastern Arabian Plate. GeoArabia Special Publication 4, Gulf PetroLink, Bahrain. 2010. 2. p. 469–502.

Ruban DA, Al-Husseini MI, Iwasaki Y. Review of Middle East Paleozoic plate tectonics. GeoArabia. 2007;12:35–55.

Sadooni FN. Stratigraphic sequence, microfacies and petroleum prospectus of the Yamama Formation. Lower Cretaceous, southern Iraq, AAPG Bull. 1993;77(11):1971–88.

Schroeder R, Van Buchem FSP, Cherchi A, et al. Revised orbitolinid biostratigraphic zonation for the Barremian–Aptian of the eastern Arabian Plate and implications for regional stratigraphic correlations. In: Van Buchem FSP, Al-Husseini MI, Maurer F, et al., editors. Barremian–Aptian stratigraphy and hydrocarbon habitat of the eastern Arabian Plate. GeoArabia Special Publication 4, Gulf PetroLink, Bahrain. 2010. 1. p. 49–96.

Scott RW, Simo JAT, Masse JP. Economic resources in Cretaceous carbonate platforms: An Overview: Chapter 2. In: Simo JAT, Scott RW, Masse JP, editors. Cretaceous carbonate platforms, AAPG Memoir. 1993; 56:15–23.

Sharland PR, Archer R, Casey DM, et al. Arabian Plate Sequence Stratigraphy. GeoArabia Special Publication 2, Gulf PetroLink, Bahrain; 2001.

Simmons MD, Whittaker JE, Jones RW. Orbitolinids from Cretaceous sediments of the Middle East—A revision of the Henson, F.R.S. and Associates Collection. In: Hart MB, Kaminski MA, Smart CW, editors. Proceedings of the fifth international workshop on agglutinated foraminifera, vol. 7. Grzybowski Found Special Publication; 2000. p. 411–37.

Strohmenger CJ, Steuber T, Ghani A, et al. Sedimentology and chemostratigraphy of the Hawar and Shu'aiba depositional sequences, Abu Dhabi, United Arab Emirates. In: Van Buchem FSP, Al-Husseini MI, Maurer F, et al., editors. Barremian-Aptian stratigraphy and hydrocarbon habitat of the eastern Arabian Plate, vol. 2. Geo Arabia Special Publication 4. Bahrain: Gulf PetroLink; 2010. p. 341–65.

Vahrenkamp VC. Chemostratigraphy of the Lower Cretaceous Shu'aiba Formation: A $\delta^{13}C$ reference profile for the Aptian Stage from the southern Neo-Tethys Ocean. In: Van Buchem FSP, Al-Husseini MI, Maurer F, et al., editors. Barremian–Aptian stratigraphy and hydrocarbon habitat of the eastern Arabian Plate, vol. 1. GeoArabia Special Publication 4. Bahrain: Gulf PetroLink; 2010. p. 107–37.

Van Buchem FSP, Pittet B, Hillgartner H, et al. High-resolution sequence stratigraphic architecture of Barremian/Aptian carbonate systems in Northern Oman and the United Arab Emirates (Khairab and Shuaiba Formations). GeoArabia. 2002;7:461–500.

Van Buchem FSP, Al-Husseini MI, Maurer F, et al. Sequence stratigraphic synthesis of the Barremian–Aptian of the eastern Arabian Plate and implications for the petroleum habitat, vol. 1. GeoArabia Special Publication 4. Bahrain: Gulf PerrLlink; 2010a. p. 9–48.

Van Buchem FSP, Baghbani D, Bulot LG, et al. Barremian–Lower Albian sequence stratigraphy of southwest Iran (Gadvan, Dariyan and Kazhdumi formations) and its comparison with Oman, Qatar and the United Arab Emirates. In: Van Buchem FSP, Al-Husseini MI, Maurer F, et al., editors. Barremian–Aptian stratigraphy and hydrocarbon habitat of the eastern Arabian Plate, vol. 2. GeoArabia Special Publication 4. Bahrain: Gulf PetroLink; 2010b. p. 503–48.

Van Wagoner JC, Posamentier HW, Mitchum RMJ, et al. An overview of the fundamentals of sequence stratigraphy and key definitions. Sea-level changes: an integrated approach, vol. 42. Tulsa: Society of Economic Paleontologists and Mineralogists Special Publication; 1988, p. 39–45.

Vilas L, Masse JP, Arias C. Orbitolina episodes in carbonate platform evolution: the Early Aptian model from SE Spain. Palaeogeogr Palaeocl. 1995;119:35–45.

Vincent, B., Van Buchem FSP, Bulot LG, et al. Carbon-isotope stratigraphy, biostratigraphy and organic matter distribution in the Aptian–Lower Albian successions of southwest Iran (Dariyan and Kazhdumi formations). In: van Buchem FSP, Al-Husseini MI, Maurer F, et al., editors. Barremian–Aptian stratigraphy and hydrocarbon habitat of the eastern Arabian Plate, vol. 1. GeoArabia Special Publication 4. Bahrain: Gulf PetroLink; 2010. p. 139–97.

Weissert H, Lini A. Ice age interludes during the time of Cretaceous Greenhouse Climate? In: Mueller, DW, McKenzie JA, Weissert H, editors. Controversies in modern geology. London: Academic Press; 1991. p. 173–91.

Wignall PB, Myers KJ. Interpreting benthic oxygen levels in mudrocks: a new approach. Geology. 1988;16:452–5.

Wilson JL. Carbonate facies in geologic history. Berlin: Springer; 1975.

Witt W, Gökdağ H. Orbitolinid biostratigraphy of the Shu'aiba Formation (Aptian), Oman, implications for reservoir development. In: Simmons MD, editor. Micropalaeontology and hydrocarbon exploration in the Middle East. Chapman and Hall: London; 1994. p. 221–42.

Yose LA, Ruf AS, Strohmenger CJ, et al. Three dimensional characterization of a heterogeneous carbonate reservoir, Lower Cretaceous, Abu Dhabi (United Arab Emirates). In: Harris PM, Weber LJ, editors. Giant hydrocarbon reservoirs of the world: from rock to reservoir characterization and modeling. AAPG Memoir 88/Society of Economic Paleontologists and Mineralogists Special Publication; 2006. p. 173–212.

Yose LA, Strohmenger CJ, Al-Hosani I, et al. Sequence stratigraphic evolution of an Aptian carbonate platform (Shu'aiba Formation), eastern Arabian Plate, onshore Abu Dhabi, United Arab Emirates. In: Van Buchem FSP, Al-Husseini MI, Maurer F, et al., editors. Barremian–Aptian stratigraphy and hydrocarbon habitat of the eastern Arabian Plate, vol. 2. GeoArabia Special Publication 4. Bahrain: Gulf PetroLink; 2010. p. 309–40.

2

Velocity calibration for microseismic event location using surface data

Hai-Yu Jiang[1] · Zu-Bin Chen[1] · Xiao-Xian Zeng[1] · Hao Lv[1] · Xin Liu[1]

Abstract Because surface-based monitoring of hydraulic fracturing is not restricted by borehole geometry or the difficulties in maintaining subsurface equipment, it is becoming an increasingly common part of microseismic monitoring. The ability to determine an accurate velocity model for the monitored area directly affects the accuracy of microseismic event locations. However, velocity model calibration for location with surface instruments is difficult for several reasons: well log measurements are often inaccurate or incomplete, yielding intractable models; origin times of perforation shots are not always accurate; and the non-uniqueness of velocity models obtained by inversion becomes especially problematic when only perforation shots are used. In this paper, we propose a new approach to overcome these limitations. We establish an initial velocity model from well logging data, and then use the root mean square (RMS) error of double-difference arrival times as a proxy measure for the misfit between the well log velocity model and the true velocity structure of the medium. Double-difference RMS errors are reduced by using a very fast simulated annealing for model perturbance, and a sample set of double-difference RMS errors is then selected to determine an empirical threshold. This threshold value is set near the minimum RMS of the selected samples, and an appropriate number of travel times within the threshold range are chosen. The corresponding velocity models are then used to relocate the perforation-shot. We use the velocity model with the smallest relative location errors as the basis for microseismic location. Numerical analysis with exact input velocity models shows that although large differences exist between the calculated and true velocity models, perforation shots can still be located to their actual positions with the proposed technique; the location inaccuracy of the perforation is <2 m. Further tests on field data demonstrate the validity of this technique.

Keywords Velocity calibration · Microseismic monitoring · Double-difference RMS error · Very fast simulated annealing · Perforation-shot relocation

✉ Zu-Bin Chen
czb@jlu.edu.cn

Hai-Yu Jiang
joyjiang1987@126.com

[1] Key Laboratory of Geo-Exploration and Instrumentation of Ministry of Education, College of Instrumentation and Electrical Engineering, Jilin University, Changchun 130026, Jilin, China

Edited by Jie Hao

1 Introduction

Hydraulic fracturing of low-permeability reservoirs generates many microseismic events due to pressure increase associated with fluid injection into treatment wells (Warpinski et al. 2005). Fracture development can be characterized by various microseismic monitoring techniques (Liang et al. 2015; Wang et al. 2013). Generally speaking, when the approximate locations of perforation shots can be resolved, we have the confidence to locate nearby microseismic events, and a usable velocity model plays an important role to achieve this goal (Usher et al. 2013). At present, because of the convenience of operation, surface observations are an effective technique when monitoring wells cannot be used. They are one of the main targets for improvement in future microseismic monitoring. Microseismic monitoring with surface observations requires a well-resolved velocity model, yet many factors can interfere with model calibration, as follows. (1) Well logs are influenced by many extraneous factors, such as pore pressure, stress accumulation, and mud invasion; in addition, seismic wave velocities around the reservoir can be altered by prior resource extraction, including mining. Consequently, velocity measurements from well logs are often unsuitable for microseismic event location (Grechka et al. 2011; Pei et al. 2009; Quirein et al. 2006; Zhang et al. 2013a, b). Moreover, log data may be incomplete, which naturally reduces the accuracy of the initial model. Methods based on searching for a local optimal solution (Pei et al. 2008; Tan et al. 2013) are not suitable for this task. (2) A particularly common problem in microseismic monitoring is a combination of little available source information (e.g., perforation shots), few receivers, and poor network coverage, resulting in a poorly constrained velocity model. (3) Perforations are often not precisely timed, so a velocity model cannot always be calibrated using perforation travel times alone. Although seismic tomography is widely used to image earth structure on local to global scales, the above limitations mean that we cannot expect the same high-quality results from microseismic monitoring data (Bardainne and Gaucher 2010). Several papers have proposed methods to construct reservoir velocity models for microseismic event location, most of which are based on the following steps: (1) A simple velocity model, using only a few parameters, is constructed from well logging data. (2) Known positions of perforation shots are iteratively relocated until a suitable velocity model is obtained. Pei et al. (2009) and Bardainne and Gaucher (2010) developed a fast simulated annealing algorithm to invert for a velocity model, which showed little dependence on initial values and outperformed the local optimal solution technique. However, the method still faced the problem that perforation shot origin times are generally inaccurate. Tan et al. (2013) proposed an inversion method based on time differences calculated from picked arrival times, which circumvented the issue of origin time inaccuracies. However, their method was still sensitive to the initial model. Anikiev et al. (2014) described a method in which the initial velocity of each layer was simultaneously increased or decreased using the accuracy of perforation shot relocations as an evaluation standard. They obtained a relatively accurate velocity model by inversion. However, their method still could not satisfy the precision requirements of microseismic event location.

This paper presents a new method to address the problem of velocity model calibration using surface data. A one-dimensional layered model is built, in which the difference between theoretical and expected models is characterized by the root mean square (RMS) errors of time double differences (DDrms) (Concha et al. 2010; Waldhauser and Ellsworth 2000; Zhang et al. 2009a, b; Zhang and Thurber 2003; Zhou et al. 2010). Using the relative differences of the first arrival times of multiple events, DDrms values are minimized using very fast simulated annealing (VFSA) (Pei et al. 2009). In order to obtain an optimal velocity model for perforation relocations, we select a subset of DDrms from the results of simulated annealing. A threshold is set near the minimum value, and velocity models with DDrms values between the threshold and the minimum are chosen for further analysis. These models are then used to relocate the perforation shots. We choose the velocity model with the smallest perforation shot location errors as the model for locating microseismic events.

This paper first introduces the principles of the method, and then conducts tests on synthetic data. We investigate the influences of velocity range constraints and picking errors on the proposed technique. Finally, the proposed technique is applied to data from a perforation shot at a gas shale reservoir as an example of velocity model calculation.

2 Travel time calculation

This study uses ray tracing to obtain travel times for microseismic events and perforation shots. Traditional two-point ray tracing algorithms mainly comprise shooting (e.g., Xu et al. 2004) and ray bending algorithms (e.g., Li et al. 2013). More recent works use wave front extension methods based on the eikonal equation and Huygens' principle (e.g., Zhang et al. 2006a, b); the shortest path algorithm (Wang and Chang 2002; Zhang et al. 2006a, b; Zhao and Zhang 2014); and the LTI method (Zhang et al. 2009a, b), based on graph theory and Fermat's principle. Compared with the above methods, ray tracing based on Snell's law is not restricted by nodes and can provide accurate travel time and azimuth information (Zhang et al. 2013a, b). Traditional shooting methods were improved by Gao and Xu (1996), who proposed a new type of step-by-step iterative ray tracing algorithm that greatly improved computational efficiency. This method can also be used with a slightly more complicated velocity model than other techniques. In this paper, we expand the method to a 3D layered structure for calculating travel times.

2.1 Ray tracing in a layered medium

As shown in Fig. 1, the dichotomy is used to determine the shortest path between two points in difference medium. We set the medium interface to $Z = z_2$, where P_1 is the launch point, P_3 is the receiver, P_2 is the intersection of P_1 and P_3

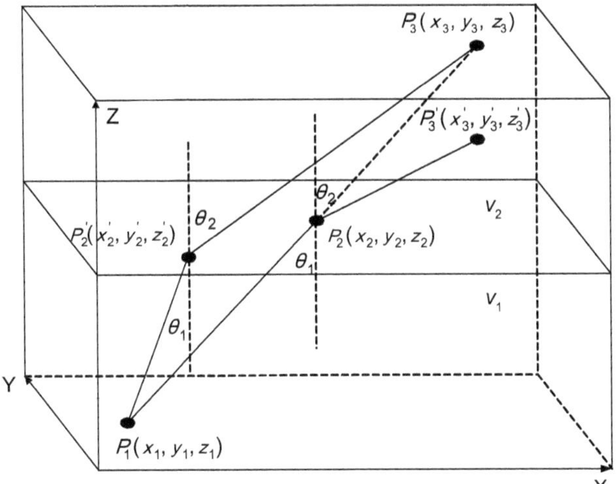

Fig. 1 Solving the refraction points by dichotomy

with the medium interface, and P_3' is the end point of the test ray path.

Beginning with Snell's law of refraction,

$$\frac{\sin\theta_1}{\sin\theta_2} = \frac{v_1}{v_2} \tag{1}$$

and substituting P_1, P_2, P_3' into the equation above, we have

$$\frac{v_1\sqrt{(x_3'-x_2)^2+(y_3'-y_2)^2}}{\sqrt{(x_3'-x_2)^2+(y_3'-y_2)^2+(z_3'-z_2)^2}} = \frac{v_2\sqrt{(x_1-x_2)^2+(y_1-y_2)^2}}{\sqrt{(x_1-x_2)^2+(y_1-y_2)^2+(z_1-z_2)^2}} \tag{2}$$

If $x_3' = x_3$, $y_3' = y_3$, then

$$a = \frac{c(1-b^2)}{b^2} \tag{3}$$

$$\begin{cases} z_3' = z_3 + \sqrt{a}(z_1 < z_2) \\ z_3' = z_3 - \sqrt{a}(z_1 > z_2) \end{cases} \tag{4}$$

If $c = (x_3' - x_2)^2 + (y_3' - y_2)^2$, then

$$b = \frac{v_2\sqrt{(x_1-x_2)^2+(y_1-y_2)^2}}{v_1\sqrt{(x_1-x_2)^2+(y_1-y_2)^2+(z_1-z_2)^2}} \tag{5}$$

We can then solve for P_3'. If the vertical error satisfies $\varepsilon = (z_3' - z_3) < 0$, or if $b > 1$, then

$$\begin{cases} x_2' = (x_1 + x_2)/2 \\ y_2' = y_2 + k_{xy}\times(x_2' - x_2) \\ z_3' = z_3 \end{cases} \tag{6}$$

On the other hand, if $\varepsilon = (z_3' - z_3) > 0$, then

$$\begin{cases} x_2' = (x_3 + x_2)/2 \\ y_2' = y_2 + k_{xy}\times(x_3' - x_3) \\ z_3' = z_3 \end{cases} \tag{7}$$

where $k_{xy} = (y_2 - y_1)/(x_2 - x_1)$ is the slope of the projection of the line segment onto the plane $Z = z_2$. P_2' is then obtained by Eqs. (6) and (7), and P_2 is replaced by P_2'. The steps above are repeated until ε is sufficiently small, which yields an estimate of P_3'.

2.2 Step-by-step iterative ray-tracing method

In this paper, source–receiver paths in a layered medium are modified using a step-by-step iterative method. The specific steps (also shown in Fig. 2) are as follows.

(1) The starting point P_0 and endpoint P_n are connected with a straight line. The intersections of the line with each layer (denoted P_1, P_2, P_3, … P_{n-1}) are calculated.

(2) Y_1 is taken as the first interface. A new intermediate refraction point P_1' is calculated between P_0 and P_2 using the dichotomy method, and P_1 is replaced by P_1'. P_1, P_2, P_3, …, P_{n-1} can be obtained in the same way.

(3) Repeat step (2) until $t - t' < \varepsilon$, where t is the travel time of the previous iteration and t' is the current travel time. A series of intermediate points is obtained. The line that connects these points with the two endpoints is taken as the minimum travel time path.

3 Principle of velocity model perturbance

3.1 Very fast simulated annealing with DDrms

In field data, and particularly for surface observations, the location error of a perforation shot is always very large

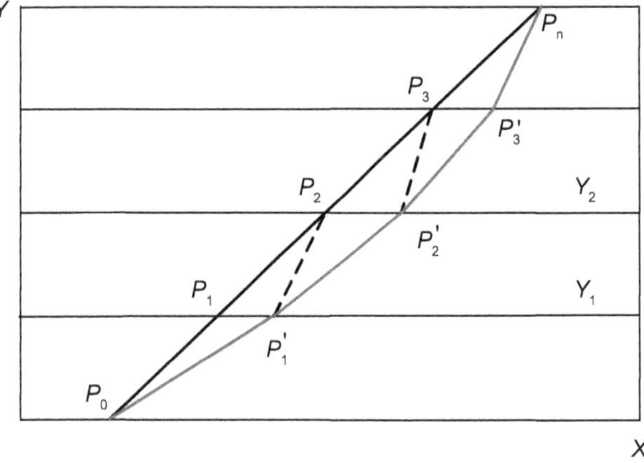

Fig. 2 Iterative node point adjustment (*red solid line* represents the ray path after processing)

when one adopts a velocity model based on a priori well log data. Moreover, part of the data might be missing, which naturally affects the accuracy of the initial model. Therefore, methods of searching for a local optimal velocity model are not applicable. Simulated annealing (SA) is a search algorithm that seeks the global minimum of an objective function in a given model space. There is no need to solve large matrix equations, and constraints can be added easily.

Compared with other techniques, such as the Gaussian–Newton and Levenberg–Marquard methods, SA does not depend on the initial value. As long as the initial annealing temperature is sufficiently high, the method converges stably to the neighborhood of the global minimum. Ingber (1989) presented a very fast simulated annealing (VFSA) algorithm based on iterative calculation of an exponent. Computation was much faster than either the conventional SA algorithm or the standard genetic algorithm (Ingber and Rosen 1992). VFSA has already been used for velocity model estimation based on borehole observations (Pei et al. 2009) by constructing a solution space with six (sets of) parameters:

1. Velocity vector, $\mathbf{V} = (\mathbf{V}_{p1}, \mathbf{V}_{p2}, \mathbf{V}_{p3}, \ldots, \mathbf{V}_{pn})^T$, where $\mathbf{V}_{pi}$ denotes the *P*-wave velocity of layer *i*.
2. Objective function, $E(\mathbf{V})$. Because perforation shot origin times are inaccurate, we use the RMS error of the time double-difference (DDrms) value, which can be computed from first arrival time differences. The procedure to compute the DDrms value is described below.
3. Initial temperature T_0. The initial temperature must satisfy the requirement that all proposed models are acceptable solutions for the next iteration of the calculation. We choose a small positive number at first, then multiply by a constant value $b > 1$, until the probability of acceptance of each proposed model converges to unity.
4. Temperature annealing parameter, T_k, which for VFSA obeys the relationship

$$T_k = T_0 \exp(-ck^{1/2N}) \tag{8}$$

where T_0 is temperature, c is a constant (for this application, $c = 0.5$ is a suitable value), and N is the total number of layers.
5. A random perturbation to the velocity vector, described below.
6. Termination criteria. In this application, we terminate the algorithm the first time one of the following three conditions is satisfied:
   (a) Temperature T_k is reduced to a certain value or close to zero.
   (b) DDrms value decreases below a predetermined threshold.
   (c) DDrms value does not decrease after multiple iterations.

3.1.1 The objective function

We use the following procedure to construct an objective function based on DDrms values:

1. Select the reference trace with the highest signal-to-noise ratio. This is denoted by the subscript M.
2. Compute the differences between the observed first arrival times of all traces and those of trace M:

$$\Delta t^{obs} = [t_1 - t_k, t_2 - t_k, \ldots, t_M - t_k].$$

3. Generate an initial velocity model, $\mathbf{V}' = \left[\mathbf{V}'_{p1}, \mathbf{V}'_{p2}, \mathbf{V}'_{p3}, \ldots, \mathbf{V}'_{pn}\right]$, from sonic log data.
4. Calculate theoretical time differences for the reference trace, $\Delta t^{cal} = [t_1 - t_k, t_2 - t_k, \ldots, t_M - t_k]$, based on $\mathbf{V}'$.
5. Determine the RMS error of the DDrms value using the equations

$$\delta\Delta t = [\Delta t_1^{obs} - \Delta t_1^{cal}, \Delta t_2^{obs} - \Delta t_2^{cal}, \ldots, \Delta t_n^{obs} - \Delta t_n^{cal}] \tag{9}$$

$$E(\mathbf{V}) = \sqrt{\frac{1}{n}\sum_{i=1}^{n} \delta\Delta t_i^2} \tag{10}$$

3.1.2 Velocity perturbation vector

The velocity vector is perturbed using the equation

$$\mathbf{V}_i^{k+1} = \mathbf{V}_i^k + x \times S_{fact} \times \left(\mathbf{V}_i^{max} - \mathbf{V}_i^{min}\right) \tag{11}$$

where $\mathbf{V}_i^{max}$ and $\mathbf{V}_i^{min}$ are the minimum and maximum values of velocity in layer *i*, respectively, subject to the constraint $\mathbf{V}_i \in [\mathbf{V}_i^{min}, \mathbf{V}_i^{max}]$; S_{fact} is a step-size factor that guarantees the DDrms value decreases stably; and $x \in [-1,1]$ is a random number generated from the equation

$$x = \mathrm{sgn}(\mu - 0.5) T_k \left[\left(1 + \frac{1}{T_k}\right)^{|2\mu - 1|} - 1\right] \tag{12}$$

where sgn denotes the signum function. A suitable value for S_{fact} is approximately 0.1.

3.2 Selecting the optimal velocity model

Before selecting the optimal velocity model, a set of DDrms values and the corresponding velocity models must be obtained. In the simulated annealing process, each time we update the DDrms value, both the DDrms value and the corresponding velocity vector $\mathbf{V}$ are preserved. In the simulated annealing algorithm, $\mathbf{V}'$ replaces $\mathbf{V}$ if the condition $E(\mathbf{V}') < E(\mathbf{V})$ is satisfied. On the other hand, if $E(\mathbf{V}') > E(\mathbf{V})$, then $\mathbf{V}$ is updated using the replacement probability.

$$P(\mathbf{V} \to \mathbf{V}') = \exp\left[\alpha \frac{E(\mathbf{V}) - E(\mathbf{V}')}{T}\right] \tag{13}$$

where α is an adjustment parameter. The number of velocity models in the model set is determined by α; the more velocity models are preserved, the greater the likelihood of obtaining reliable results. However, computation time will increase accordingly.

The purpose of establishing a velocity model in this way is to obtain accurate travel times for perforation shots and microseismic events. We mainly focus on the relationships between DDrms value, absolute travel time RMS (i.e., the differences between travel times calculated with the theoretical and synthetic models), and the velocity model RMS for each layer (i.e., the difference between the theoretical and synthetic layer velocities) (Fig. 3).

4 Synthetic examples

In this section, the effects of a hydraulic fracture treatment are simulated to investigate the accuracy of the proposed method. We define a synthetic velocity model with five layers, and exact velocities are given in Table 1. The geophone array geometries and relative perforation shot location are shown in Fig. 4. The exact shot position is $X_{s1} = 830$ m, $Y_{s1} = 840$ m, $Z_{s1} = -1180$ m. This study uses a star-shaped array (6 lines, 96 geophones), as the aim is to place as many geophones as possible in a small area. The initial velocity values for each layer are 950, 1300, 1800, 2800, and 3300 m/s. The simulation requires 2133 s on a notebook computer with a 2.26-GHz Intel® processor. Observed values of first arrival time differences are determined from the true synthetic model using the ray tracing method described above (Fig. 5).

When the minimum DDrms value is determined (here, 2.97e−5 s), the iterative calculation stops and a threshold of 3.97e−5 is set. Ten DDrms values are chosen randomly between the minimum DDrms value and the threshold. The velocity models corresponding to these DDrms values are used to relocate the perforation shot, as shown in Fig. 6. The optimal velocity model is then picked. From the above results, we can see that although there are significant differences between the initial and synthetic velocity models, the perforation shot can be still relocated to its true position; the relocation error is only 1.67 m.

4.1 Sensitivity to the constraints

Here, we consider two main constraints on the viability of adopting a model for use with surface observations. One is the range of P-wave velocities used in the model. Increasing this range will increase the solution space; if we use a simulated annealing algorithm with a larger velocity range and parameters that are otherwise unchanged, then source location accuracy and computational efficiency will both be reduced. If the range of velocity variations is too small, then we probably will not obtain a viable result, as shown in Fig. 7. It is therefore desirable to choose a reasonable range of velocity variations in each layer. Generally, the range of velocities in the objective layers is mainly determined from well logging data and local geology.

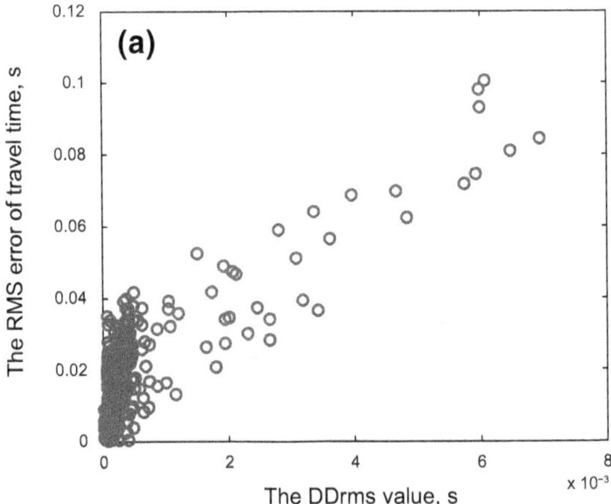

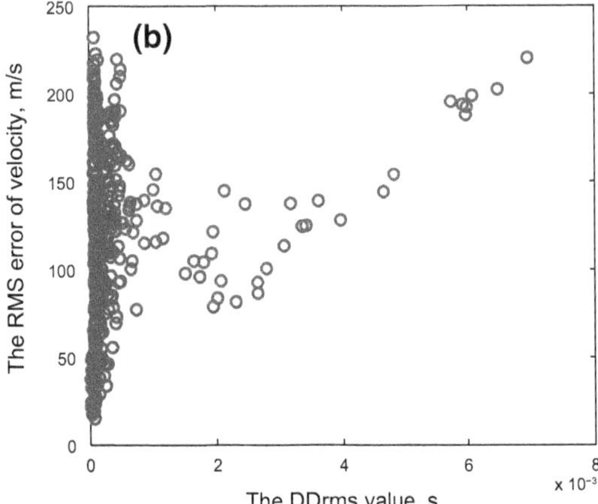

Fig. 3 Sample velocity model distribution. Each *blue circle* represents a velocity model corresponding to one DDrms value. **a** Plot of the relationship between DDrms value and travel time RMS error. The travel time error describes the deviation between the travel time of the perforation shot calculated from the actual medium and that calculated from the velocity model corresponding to the plotted DDrms value. **b** Relationship between DDrms value and velocity model RMS. The model error reflects differences between the velocity model and the actual medium

Table 1 Synthetic velocity model parameters

Layer	Depth, m	Synthetic velocity model, m/s	Velocity constraint range ($\mathbf{V}_{min} - \mathbf{V}_{max}$), m/s
1	0–200	1200	600–1300
2	200–500	1600	1000–1800
3	500–700	2200	1600–2400
4	700–900	3200	2400–3600
5	900–1200	3800	3000–4200

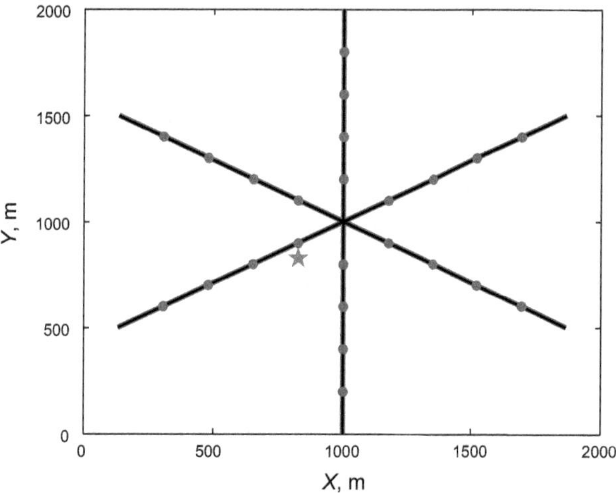

Fig. 4 Geometry of recording stations and perforation shot. Each station has 4 geophones, for a total of 96 sensors. *Black lines* represent the arms of the star-patterned array. *Blue dots* represent geophone positions. The *green star* represents the position of the perforation shot

Another constraint is the number of surface geophones, which determines the number of time double-differences available for the inversion. The same termination conditions are used for the SA algorithm, and adding surface geophones improves the convergence of the DDrms value, as shown in Fig. 8. However, the algorithm requires more computation time to converge. Less accurate travel time information is obtained when DDrms is reduced to a small value, and it can be the case that no acceptable velocity model is obtained at all. Figure 8 compares the velocity calibration results with different numbers of surface geophones; the termination conditions are the same for all simulations. Using the same line pattern as in Fig. 4, the number of geophones increases or decreases uniformly in each line.

The minimum DDrms value of Fig. 8a is 1.3e−6. The minimum DDrms value in Fig. 8b–d is less than 2.5e−6 s. As shown in Fig. 8, reliable location results are more likely when a large number of surface geophones are used.

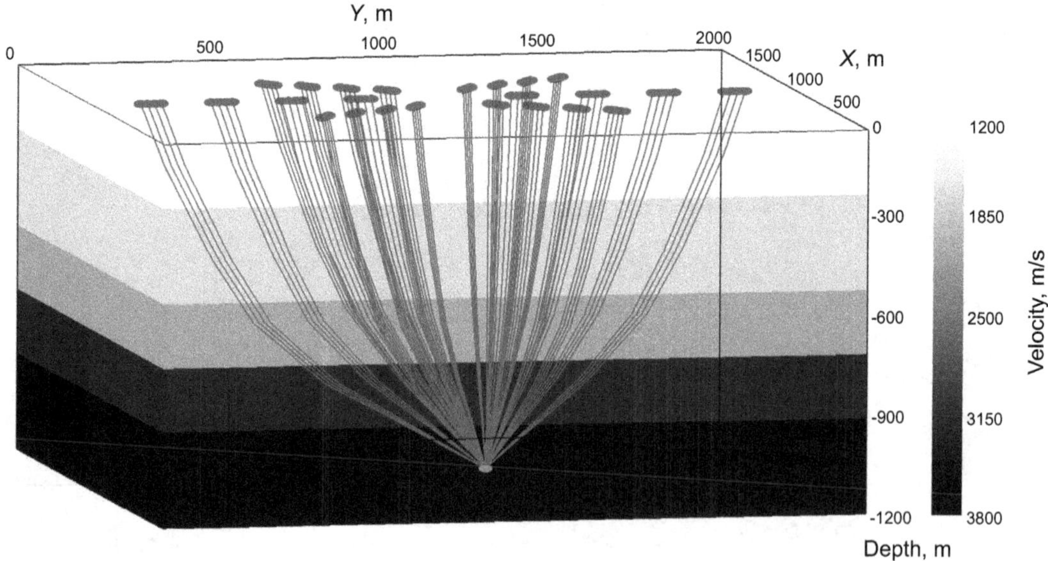

Fig. 5 Velocity model and ray paths. The *green dot* represents the perforation position, *red lines* represent ray paths, and *blue dots* represent receivers

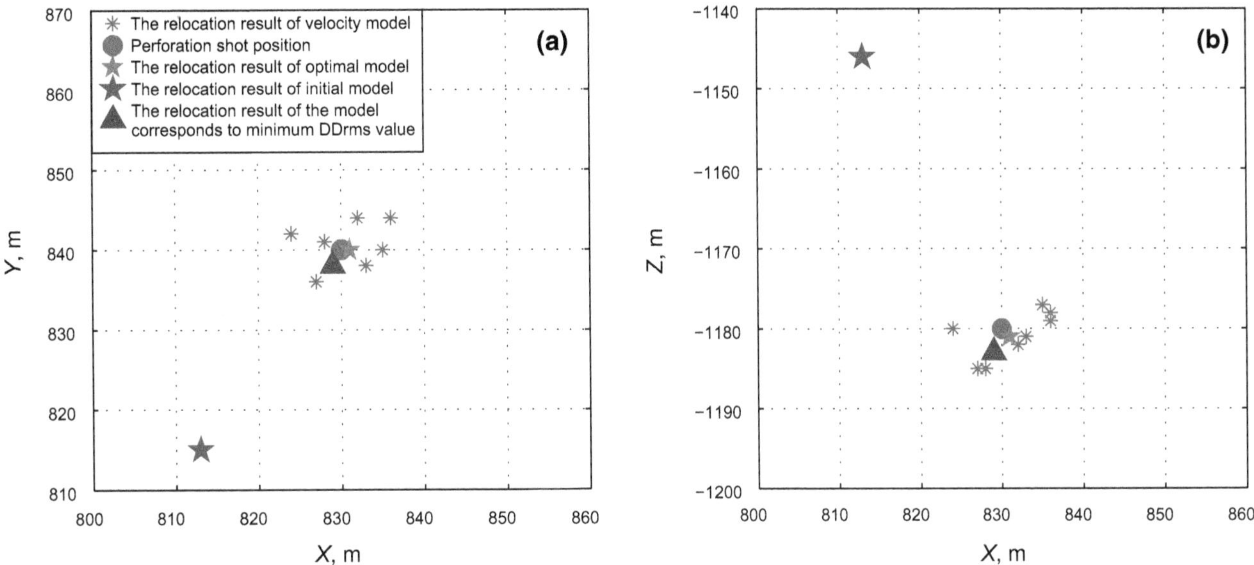

Fig. 6 Calculated perforation shot locations using 10 sample velocity models. The selected models corresponded to DDrms values within a certain range of the minimum DDrms value. **a** The *top* view, **b** the *side* view

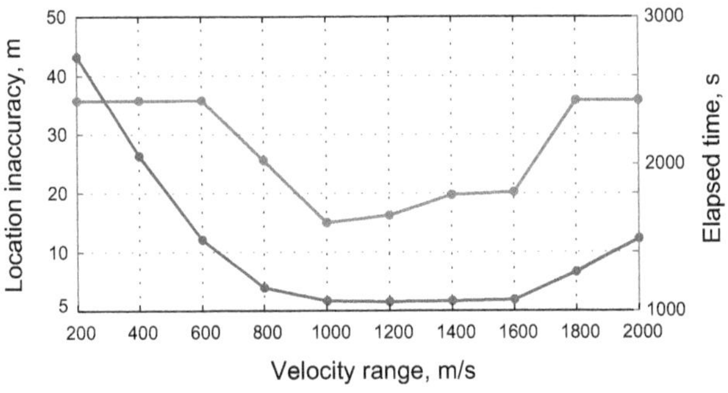

Fig. 7 Influence of velocity range on the accuracy and efficiency of the source location. The *horizontal axis* represents the total range of *P*-wave velocities in the synthetic velocity structures of different tests

However, increasing the number of surface arrays arbitrarily may lead the reduction of DDrms value to be difficult. In this case, more computation time is needed; notably, increasing the number of rays also increases the forward calculation time.

4.2 The sensitivity to picking errors

Calibrating a velocity model for microseismic event location requires accurate information about perforation shots. In actual situations, seismic signals recorded by geophones are usually contaminated by noise, which may cause picking errors (Rodriguez et al. 2012; Song et al. 2010; Tan et al. 2014). Compared with borehole observations, the picking errors of *P*-wave arrivals are relatively large at the surface. This will affect the proposed technique. Therefore, to model our algorithm's sensitivity to picking errors, we add a set of random picking errors to the synthetic arrival times at each receiver, ranging from 0 to 5 % of the calculated travel time. We use the same stop condition as in the numerical experiments above. The algorithm terminates when DDrms value reaches 7.84e−4 s. A threshold of 8.84e−4 s is set; 10 velocity models are selected to relocate the perforation shot, and an optimal velocity model is picked out by the method described above.

Figure 9 shows that the perforation shot can still be located close to its true position, despite picking errors. The location inaccuracy is again within 2 m, and relatively accurate results can still be obtained. The velocity model can be considered an "equivalent" velocity model. Because of the large discrepancies between the recovered velocity model and the true model, large errors are possible when locating microseismic events far from the perforation.

In Fig. 10, the velocity models used in Fig. 9 are used to locate a synthetic microseismic event (true hypocenter 534, 532, −1165). This illustrates the relative accuracy of microseismic event location using these velocity models. Figure 11 illustrates the relationship between first arrival time picking errors and minimum DDrms values, using the same stop conditions as for the previous tests.

Figure 10 confirms that for microseismic events located far from the perforation, the optimal velocity model for the perforation shot introduces an inherent location error. Figure 11 suggests that increasing picking errors will increase location errors; for example, if picking errors reach 20 % of computed travel times, locations of microseismic events will be poorly constrained. The main reason

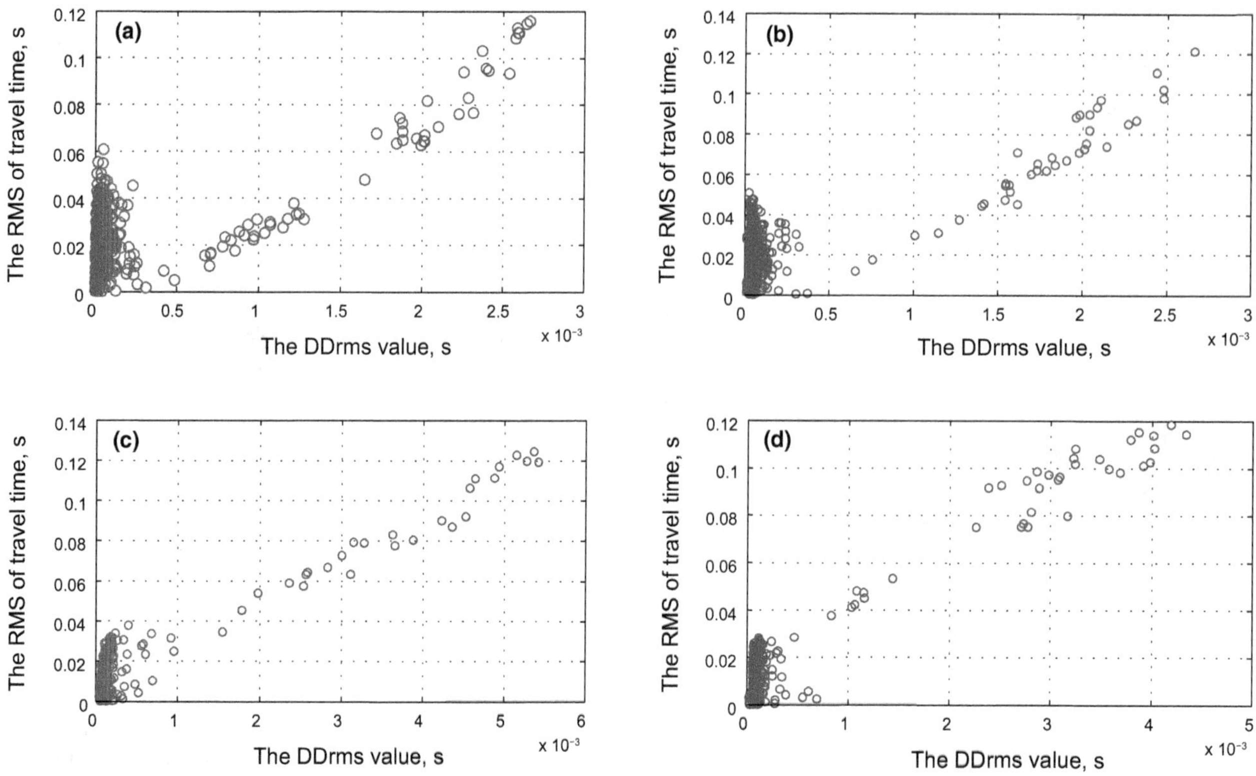

Fig. 8 Comparative plots of tomography inversion results with different constraints. The inversion takes approximately 47 s with 6 geophones in (**a**); 232 s with 24 geophones in (**b**); 813 s with 48 geophones in (**c**); and 3137 s with 96 geophones in (**d**). Legends are the same as for Fig. 3

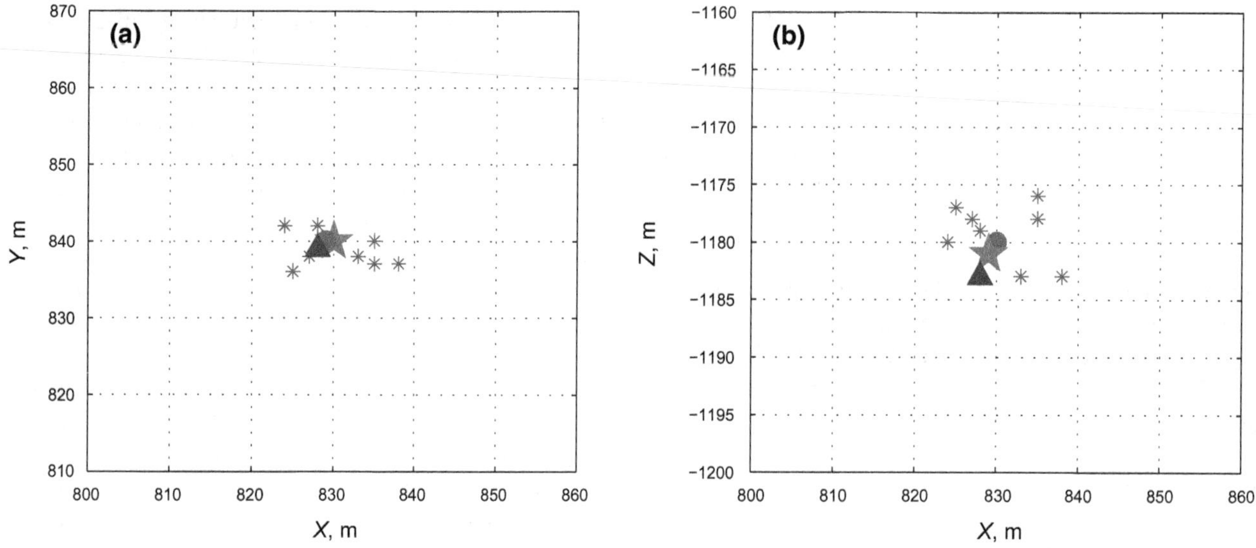

Fig. 9 Relocation of a perforation shot using data with picking errors, legends are the same as for Fig. 6. **a** *Top* view, **b** *Side* view

for this is that the process of reducing the DDrms value is influenced by the picking errors; when picking errors are large, the DDrms value cannot be reduced to a sufficiently small value, and this reduces the chances to obtain a meaningful velocity model.

5 Field data experiments

In this section, we test our algorithm's performance on data recorded by an experiment in Shanxi province, China. As shown in Fig. 12, six survey lines were deployed in this

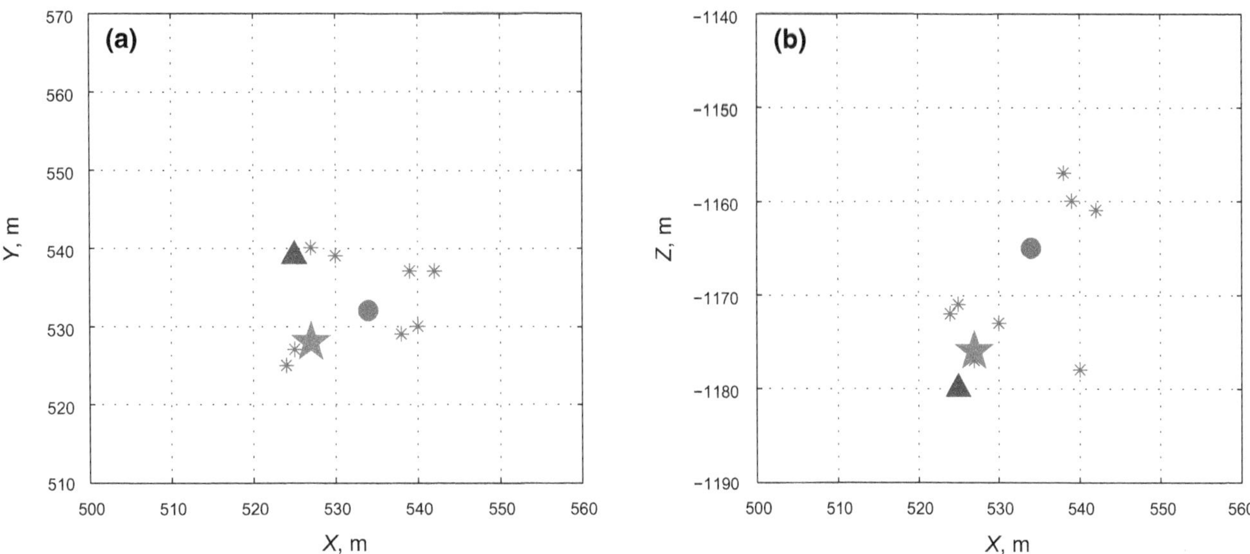

Fig. 10 The location result of microseismic event using the data with picking errors, legends are the same as for Fig. 6. **a** *Top* view, **b** side view

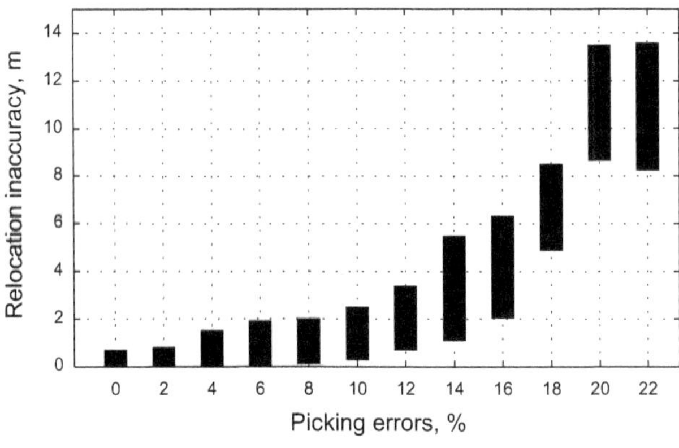

Fig. 11 Relationship between picking errors of first arrival times and relocation inaccuracy. The x-axis represents the maximum values of picking errors added to the synthetic data; the *thick black lines* indicates the range of relocation inaccuracies after 100 trials

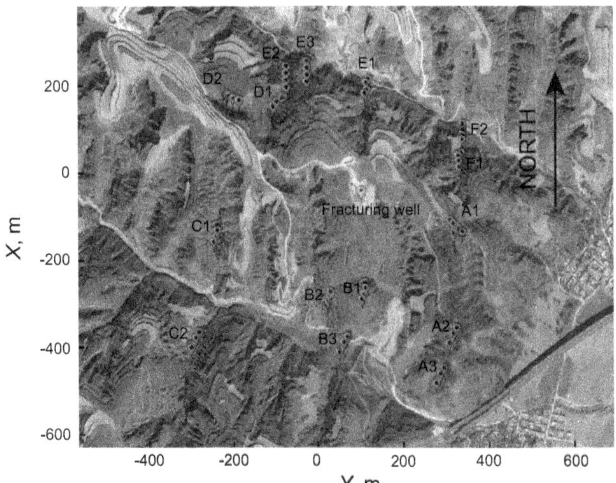

Fig. 12 Microseismic monitoring array geometry. Geophones are arranged in a *star-like* surface array around production wells

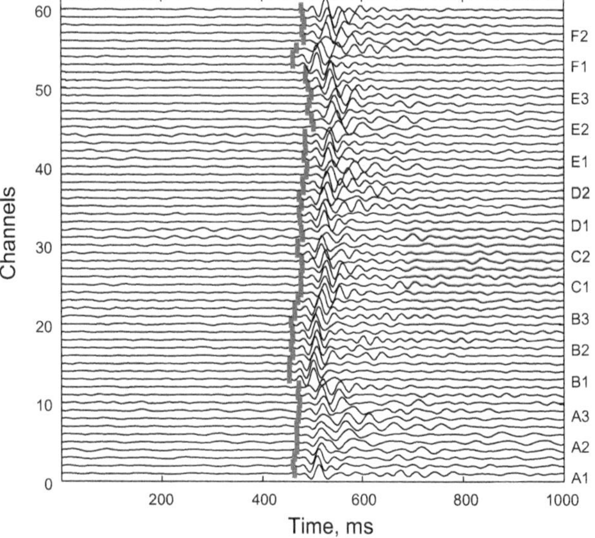

Fig. 13 Perforation shot monitoring records for the surface geophones in Fig. 12. The x-axis represents time since the beginning of the record. The y-axis corresponds to geophone channel numbers

experiment, with two or three data loggers per line. Each data logger was equipped with four vertical-component geophones, with a horizontal sensor spacing of 20 m along the line. The first geophone of each survey line was placed at a fixed distance from the center of the array, to ensure that the data loggers were evenly distributed and all sensors were far enough from the injection well to minimize noise from processes related to injection (e.g., mechanical pump noise). The position of the straight well is at the center of the observation system; the wellhead coordinates were (−68.025, 107.258, −1.34) under the unified GPS

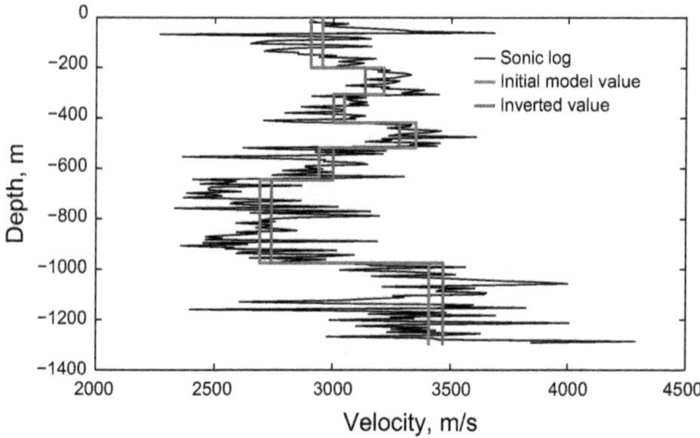

Fig. 14 Inversion results, showing initial and optimal velocity models

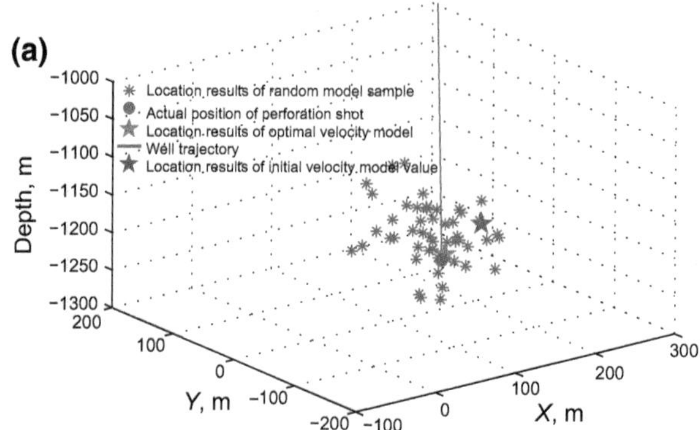

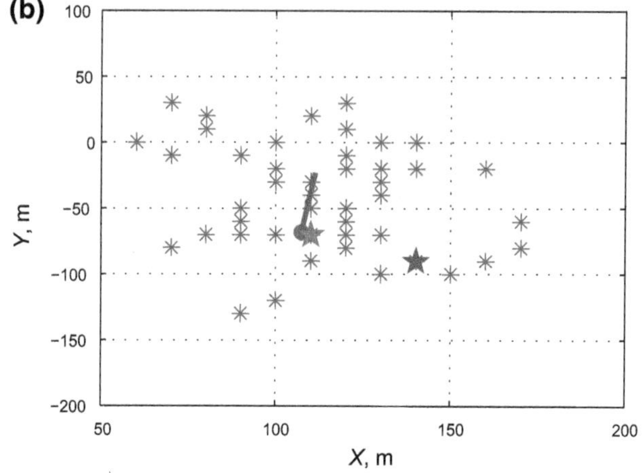

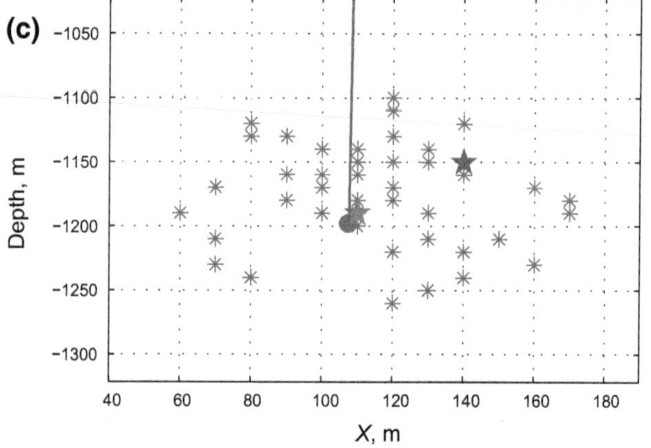

Fig. 15 Location results based on 50 candidate velocity models. Models are selected if their final DDrms value lies within the range of the minimum DDrms value; each model is used to locate the perforation shot independently. **a** Three-dimensional figure. **b** *Top* view. **c** *Side* view

observation system we defined. The perforation (fracturing point) coordinates are (107.258, −68.025, −1197.8). The maximum geophone elevation is −3.87 m and the minimum is −102.73 m. Because the waveforms of Fig. 13 are not in good agreement, we obtain first arrival picks for each geophone manually. Figure 14 shows the initial velocity obtained from well logging data, and the optimal velocity model obtained by the method of this paper. The velocity structure was divided into seven layers, based on sonic logs (Table 2). Layer boundaries corresponded to sudden velocity changes. The number of layers and their respective thicknesses do not need to be a constraint. The perforation positions obtained from the initial velocity model contain significant errors, as shown in Fig. 15. However, the perforation could be located close to its true position using models obtained by inversion. Therefore, we infer that the final velocity model is suitable for microseismic event locations. The VFSA algorithm reduced the DDrms value from 0.0215 s to 4.4e−4 s. To improve the accuracy of the inversion, we set a selection threshold of 6.4e−4 s for candidate velocity models. Fifty models with DDrms values between the threshold and the minimum were selected. These were used to relocate the perforation shot, and an optimal velocity model was picked based on the results. From Fig. 15, compared with the initial velocity model, the perforation shot can be located very close to its actual position; the location inaccuracy is 5.23 m. Therefore, we conclude that the velocity model obtained by our method is suitable for microseismic event location.

6 Conclusion

In this paper, we present a non-linear inversion method for the calculation of velocity models suitable for locating microseismic events with surface sensor data. The proposed technique is based on the RMS error of time double-differences, which are determined from surface records of perforation shots. The DDrms value is minimized using a VFSA algorithm, and velocity model viability is evaluated based on the accuracy of perforation shot relocations. This technique can overcome many of the difficulties caused by monitoring hydraulic fractures with surface instruments alone. Using tests on synthetic and field data, our interpretations and conclusions are as follows.

Table 2 Stratum velocity structure parameters

Layer	Depth, m	Starting velocity model, m/s	Velocity constraint range ($\mathbf{V}_{min} - \mathbf{V}_{max}$), m/s
1	0–200	2954.5	2650–3200
2	200–305	3214.5	3100–3500
3	305–418	3047.8	2800–3200
4	418–517	3348.6	3200–3500
5	517–645	2942.2	2600–3300
6	645–975	2690.1	2400–3000
7	975–1300	3408.2	3000–3800

1. The proposed technique does not strongly depend on the initial velocity model, and can also overcome the problem of inaccurate perforation shot origin times. However, due to complex local geology and a lack of available information, the velocity model inversion is non-unique. Therefore, whether or not a velocity model is suitable for microseismic event location is determined based on the accuracy of perforation-shot relocation.
2. Constraints on velocity structure have an effect on the proposed technique's results. Reasonable velocity constraints should be imposed on each layer; if these are improperly selected, analyst errors will negatively affect the outcome. Second, the DDrms value should be reduced as much as possible; the smaller the DDrms value, the more reliable the optimal velocity model. When using the same simulated annealing termination condition as our examples, more surface geophones are needed.
3. If the arrival times of a perforation shot include significant picking errors, it may be the case that DDrms values do not converge to a reasonable minimum. This can greatly influence the effectiveness of our technique, so it is imperative to minimize picking errors.
4. When processing real field data, after velocity model calibration, a perforation shot could be located to its actual position. Thus, we believe nearby microseismic events can be located with confidence. However, due to many limitations of microseismic monitoring operations, we do not expect to obtain an accurate velocity model from only one perforation shot; this implies a certain risk for microseismic events located far from the perforation. Therefore, it may be necessary to introduce more complex velocity models and source information in our application.

Acknowledgments This study is supported by the National Natural Science Foundation of China (No. 41074074).

References

Anikiev D, Valenta J, Staněk F, et al. Joint location and source mechanism inversion of micro-seismic events: benchmarking on seismicity induced by hydraulic fracturing. Geophys J Int. 2014;198(1):249–58.

Bardainne T, Gaucher E. Constrained tomography of realistic velocity models in microseismic monitoring using calibration shots. Geophys Prospect. 2010;58(5):739–53.

Concha D, Fehler M, Zhang HJ, et al. Imaging of the Soultz enhanced geothermal reservoir using microseismic data. In: 35th Workshop on Geothermal Reservoir Engineering Stanford University. 2010; SGP-TR-188.

Gao EG, Xu GM. A new kind of step by step iterative ray tracing method. Chin J Geophys. 1996;39(Supplement):302–8 **(in Chinese)**.

Grechka V, Singh P, Das I. Estimation of effective anisotropy simultaneously with locations of microseismic events. Geophysics. 2011;76(6):WC143–55.

Ingber L, Rosen B. Genetic algorithms and very fast simulated re-annealing: a comparison. Math Comput Model. 1992;16(11):87–100.

Ingber L. Very fast simulated re-annealing. Math Comput Model. 1989;12(8):967–73.

Li F, Xu T, Wu ZB. Segmentally interactive ray tracing in 3-D heterogeneous geological models. Chin J Geophys. 2013;56(10):3514–22 **(in Chinese)**.

Liang BY, Chen S, Leng CB, et al. Development of microseismic monitoring for hydro-fracturing. Progr Geophys. 2015;30(1):0401–10.

Pei DH, Quirein JA, Cornish BE, et al. A very fast simulated annealing (VFSA) approach for joint-objective optimization. Geophysics. 2009;74(6):WCB47–55.

Pei DH, Quirein JA, Cornish BE, et al. Velocity calibration using microseismic hydraulic fracturing perforation and string shot data. In: 49th Annual Logging Symposium. 2008; SPWLA: 2008-H.

Quirein JA, Grable J, Cornish BE, et al. Microseismic fracture monitoring. In: Annual Logging Symposium. Society of Professional Well Log Analysts. 2006.

Rodriguez IV, Bonar D, Sacchi M. Microseismic data denoising using a 3C group sparsity constrained time-frequency transform. Geophysics. 2012;77(2):21–9.

Song FX, Kuleli SH, Toksöz NM. An improved method for hydrofracture-induced microseismic event detection and phase picking. Geophysics. 2010;75(6):A47–52.

Tan YY, He C, Hou XC, et al. Detection and location of microseismic events with low signal-to-noise ratios. In: International Petroleum Technology Conference. 2014; Kuala Lumpur, Malaysia.

Tan YY, He C, Zhang HL. Time difference-based velocity model inversion for microseismic event location. In: SEG Houston 2013 Annual Meeting. 2013; Segam: 2013-0165.

Usher PJ, Angus DA, Verdon JP. Influence of velocity model and source frequency on microseismic waveforms: some implications for microseismic locations. Geophys Prospect. 2013;61(Suppl. 1):334–45.

Waldhauser F, Ellsworth WL. A double-difference earthquake location algorithm: method and application to the Northern Hayward Fault, California. Bull Seismol Soc Am. 2000;90(6):1353–68.

Wang CL, Cheng JB, Yin C, et al. Microseismic events location of surface and borehole observation with reverse-time focusing using interferometry technique. Chin J Geophys. 2013;56(9):3184–96 **(in Chinese)**.

Wang H, Chang X. 3 D ray tracing method based on graphic structure. Chinese Journal of Geophysics. 2002;43(4):534–41 **(in Chinese)**.

Warpinski NR, Sullivan RB, Uhl JE, et al. Improved microseismic fracture mapping using perforation timing measurements for velocity calibration. SPE J. 2005;10:14–23.

Xu T, Xu GM, Gao EG. Block modeling and shooting ray tracing in complex 3-D media. Chin J Geophys. 2004;47(6):1118–26 **(in Chinese)**.

Zhang D, Fu XR, Yang Y, et al. 3-D Seismic ray tracing algorithm based on LTI and partition of grid interface. Chin J Geophys. 2009a;52(9):2370–6.

Zhang HJ, Sarkar S, Toksöz MN, et al. Passive seismic tomography using induced seismicity at a petroleum field in Oman. Geophysics. 2009b;74(6):WCB57–69.

Zhang HJ, Thurber CH. Double-difference tomography: the method and its application to the Hayward Fault, California. Bull Seismol Soc Am. 2003;93(5):1875–89.

Zhang MG, Cheng BJ, Li XF. A fast algorithm of shortest path ray tracing. Chin J Geophys. 2006a;18(1):146–50 **(in Chinese)**.

Zhang MG, Jia YG, Wang MY, et al. A global minimum travel time ray tracing algorithm of wave front expanding with interface points as secondary sources. Chin J Geophys. 2006b;49(4):1169–75 **(in Chinese)**.

Zhang XL, Zhang F, Li XY. The influence of hydraulic fracturing on velocity and microseismic location. Chin J Geophys. 2013a;56(10):3552–60 **(in Chinese)**.

Zhang XL, Zhang F, Li XY, et al. The influence of hydraulic fracturing on velocity and microseismic location. Chin J Geophys. 2013b;56(10):3552–60.

Zhao HY, Zhang MG. Tracing seismic shortest path rays in anisotropic medium with rolling surface. Chin J Geophys. 2014;18(1):2910–7 **(in Chinese)**.

Zhou RM, Huang LJ, Rutledge J. Microseismic event location for monitoring CO_2 injection using double-difference tomography. Lead Edge. 2010;29(2):208–14.

3

Molecular simulation studies of hydrocarbon and carbon dioxide adsorption on coal

Junfang Zhang[1] · Keyu Liu[1,2] · M. B. Clennell[1] · D. N. Dewhurst[1] · Zhejun Pan[3] · M. Pervukhina[1] · Tongcheng Han[1]

Abstract Sorption isotherms of hydrocarbon and carbon dioxide (CO_2) provide crucial information for designing processes to sequester CO_2 and recover natural gas from unmineable coal beds. Methane (CH_4), ethane (C_2H_6), and CO_2 adsorption isotherms on dry coal and the temperature effect on their maximum sorption capacity have been studied by performing combined Monte Carlo (MC) and molecular dynamics (MD) simulations at temperatures of 308 and 370 K (35 and 97 °C) and at pressures up to 10 MPa. Simulation results demonstrate that absolute sorption (expressed as a mass basis) divided by bulk gas density has negligible temperature effect on CH_4, C_2H_6, and CO_2 sorption on dry coal when pressure is over 6 MPa. CO_2 is more closely packed due to stronger interaction with coal and the stronger interaction between CO_2 molecules compared, respectively, with the interactions between hydrocarbons and coal and between hydrocarbons. The results of this work suggest that the "a" constant (proportional to T_c^2/P_c) in the Peng–Robinson equation of state is an important factor affecting the sorption behavior of hydrocarbons. CO_2 injection pressures of lower than 8 MPa may be desirable for CH_4 recovery and CO_2 sequestration. This study provides a quantitative understanding of the effects of temperature on coal sorption capacity for CH_4, C_2H_6, and CO_2 from a microscopic perspective.

Keywords Molecular simulation · GROMOS force field · Coal bed methane · Sorption isotherm · Bituminous coal · Hydrocarbons · Carbon dioxide

1 Introduction

Enhanced coalbed methane (ECBM) is increasingly important unconventional gas. Carbon dioxide (CO_2) injection in coal seams can replace hydrocarbons and release renewable energy. Understanding the mechanism of CH_4, C_2H_6, and CO_2 sorption on coal is a key factor for CO_2 storage and hydrocarbons recovery. Sorption of gases in coal has been studied for decades (Bae and Bhatia 2006; Billemont et al. 2011, 2013; Brochard et al. 2012a, 2012b; Busch and Gensterblum 2011; Busch et al. 2003, 2004; Day et al. 2008; Fitzgerald et al. 2005; Gensterblum et al. 2014; Goodman et al. 2007; Krooss et al. 2002; Li et al. 2010; Ottiger et al. 2008; Pini et al. 2009, 2010). Theoretical models have been developed and improved to study gas sorption on coal by several authors (Connell et al. 2010; Lu and Connell 2007; Lu et al. 2008; Pan and Connell 2007, 2009, 2012; Sakurovs et al. 2007; Vandamme et al. 2010). They have applied different approaches to describe how sorption capacity, sorption rate, gas diffusion, and permeability are affected. Several groups reported that the sorption of gas on coals at a given pressure increases with decreasing temperature (Azmi et al. 2006; Menon 1968; White et al. 2005; Zhou et al. 2001), but the temperature effect on the maximum sorption capacity was controversial (Sakurovs et al. 2010), nor was the temperature effect quantified. Major challenges for

✉ Junfang Zhang
Junfang.Zhang@csiro.au

[1] CSIRO Energy Flagship, 26 Dick Perry Ave, Kensington, WA 6151, Australia

[2] Research Institute of Petroleum Exploration and Development, PetroChina, Beijing 100083, China

[3] CSIRO Energy Flagship Ian Wark Laboratory, Bayview Avenue, Clayton, VIC 3169, Australia

Edited by Xiu-Qin Zhu

ECBM lie in the fact that the density of the adsorbed phase within varying pores is unknown and fundamental understanding of the sorption mechanism is insufficient. In this study, we simulate sorption of CH_4, C_2H_6, and CO_2 on dry intermediate rank bituminous coal at 308 and 370 K up to a pressure of 10 MPa. The temperature effect on sorption capacity of CH_4, C_2H_6, and CO_2 and the interactions between adsorbate and coal and between adsorbate are investigated in an attempt to provide an insight into the sorption mechanism of coal for those gases, as well as to demonstrate the potential use of this simulation method for ECBM.

The remainder of this article is structured as follows. In Sect. 2, we explain the molecular models selected for coal, CH_4, C_2H_6, and CO_2, the methods used, and the implementation of the simulations. In Sect. 3, we present and discuss the results of the molecular simulations. Finally, we summarize our analysis and draw conclusions in Sect. 4.

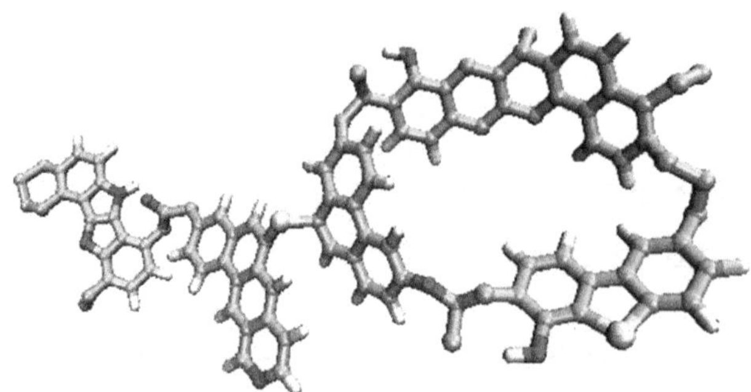

Fig. 1 Building block of intermediate rank coal. Color scheme *O* red; *H* white; *C* cyan; *N* dark blue; *S* yellow

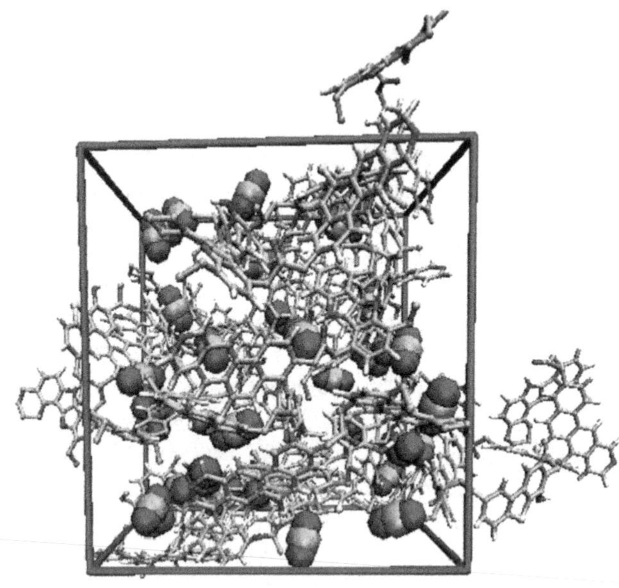

Fig. 2 A snapshot for coal and CO_2. CO_2 molecules are represented by *red* oxygen and *white* hydrogen and the rest is the coal

2 Simulation details

Molecular simulation allows one to describe the interactions between the adsorbate species and between the adsorbate and coal matrix in full detail, without predefined density profiles or sorption patterns. Its advantage lies in its ability to predict micro sorption and to reveal the mechanism of gas adsorption.

2.1 Simulation setup

Coal is characterized by two distinct porosity systems: micropores or matrix and macropores or cleats. The matrix is storage medium where coal seam gas (primarily CH_4 and CO_2) is mainly stored by sorption and moves by molecular diffusion. The cleats constitute a natural fracture network and provide permeability and connectivity to the reservoir but very limited storage volume as free gas.

We simulate gas sorption in the coal matrix. The system studied consists of coal and pure components of CO_2, CH_4, and C_2H_6. In this study, we focus on a model representation of a bituminous coal (Spiro and Kosky 1982; Tambach et al. 2009). A building block of 191 atoms, C100H82O5N2S2, for an intermediate rank coal is shown in Fig. 1 (Zhang et al. 2014, 2015). It was constructed using the Prodrg server (Schuttelkopf and van Aalten 2004). In this model, carbon, hydrogen, oxygen, nitrogen, and sulfur cover about 82.53 %, 5.64 %, 5.5 %, 1.93 %, and 4.4 % of the total mass of the coal, respectively. Constituents and their ratios in this model are similar to that observed in natural coal; therefore, they account for the amorphous and chemically heterogeneous structure of the realistic coal. Generating realistic molecular models of coal is essential for coal simulations applied in coal related research (Mathews et al. 2011).

The initial configuration consists of 12 randomly placed coal building blocks (coal molecules) in an empty space of a simulation box which is large enough to accommodate the coal molecules and has *x*, *y*, *z*-dimensions of $3.2 \times 3.2 \times 3.2$ nm^3. The simulation box with coal and CO_2 is shown in Fig. 2. After the system reaches equilibrium, the volume is around $2.9 \times 2.9 \times 2.9$ nm^3 depending on the pressure applied. Obviously, in our system, the void volume (pores within the coal matrix) is at a nanoscale. We connect our system (coal matrix) with an imaginary gas reservoir to allow gas to exchange between the system and the reservoir. The number of gas molecules in the coal matrix varies depending on the pressure or chemical potential applied.

2.2 Molecular model for coal, CO_2, CH_4, and C_2H_6

Coal, CO_2, CH_4, and C_2H_6 are modeled using the GROMOS force field (Oostenbrink et al. 2004). In this force field, the carbon and the hydrogens that are bonded to it are treated as a single atom, reducing computational effort up to a factor of 9 at the expense of neglecting the slight directional and volume effects of the presence of these hydrogens. Detailed parameter sets can be found in Oostenbrink et al. 2004. In contrast to other biomolecular force fields, this parameterization of the GROMOS force field is based primarily on reproducing the free enthalpies for a range of compounds. The relative free enthalpy is a key property in many biomolecular processes of interest and is why this force field was selected. The non-bonded interactions between atoms which are separated by more than three bonds, or belong to different molecules, are described by pair wise-additive Lennard-Jones (LJ) 12–6 potentials. Cross-interactions between unlike atoms are calculated by the Jorgensen combining rules. The LJ parameters and energy terms and the parameters are taken from Oostenbrink et al. 2004. We use a truncated and shifted potential with a cutoff radius of 14 Å in accordance with the Gromacs force field (Oostenbrink et al. 2004).

2.3 Simulation details

Our simulation procedure consists of MD simulation with a constant number of particles, constant pressure, and constant temperature ensemble (NPT) coupled with MC simulations (Frenkel et al. 1992; Siepmann and Frenkel 1992) in the grand canonical ensemble (GCMC) in which the chemical potentials of the adsorbate, the volume, and the temperature of the system are fixed. The chemical potential (or equivalently the fugacity) is imposed. Instead of setting the chemical potential, it is more intuitive to set the reservoir pressure which is related to the chemical potential by $\mu = \mu^0 + RT\ln\left(\frac{\varphi P}{p^0}\right)$, where μ is the chemical potential, and p^0 and μ^0 are the standard pressure and chemical potential, respectively. P is the reservoir pressure and φ is the fugacity coefficient. The temperature T and the chemical potential of the adsorbate phase μ, which is assumed to be in equilibrium with a gas reservoir, are fixed. MD simulation in the NPT ensemble are carried out using Gromacs software (Berendsen et al. 1995; Lindahl et al. 2001; Van der Spoel et al. 2005), while the GCMC algorithm allows the calculation of the isotherm sorption. Periodic boundary conditions are applied in three directions.

In MC, the energy difference between the new configuration and the old configuration is computed ($\Delta E = E_{new} - E_{old}$). If $\Delta E \leq 0$, the new configuration is accepted. If $\Delta E > 0$, the new configuration is accepted with a Boltzmann-weighted probability of $\exp(-\Delta E/kT)$, where T is temperature and k is the Boltzmann constant. A more detailed description of the GCMC method can be found in Dubbeldam et al. (2004a, b). To update the configuration, several MC moves are involved. They are translation, swap, and orientation-biased insertions. A translation move is to give a particle a random translation, and the move is accepted or rejected based on the energy difference. A swap move is to insert or delete a particle randomly with a probability of 50 % to allow a chemical equilibrium between the system studied and an imaginary gas reservoir. The orientation-biased insertions are commonly used in MC to insert particles to energetically favorable conformations to increase the acceptance ratio of the moves, especially when density is high for the system under high pressure. Equilibrium is attained when the number of successful insertion and deletion attempts balances each other. The MC simulations are performed using the open source package RASPA 1.0 developed by Dubbeldam et al. (2008). In our simulations, the temperature was fixed using the Berendsen thermostat (Berendsen et al. 1984). As with the temperature coupling, the system can also be coupled to a pressure bath. We use the Berendsen pressure coupling scheme to reach the target pressure, and then switch to Parrinello–Rahman coupling (Parrinello and Rahman 1981) for production runs once the system is in equilibrium as Parrinello–Rahman pressure coupling works in a more efficient and sensible way (Berendsen et al. 1995; Lindahl et al. 2001; Van der Spoel et al. 2005). The equations of motion were integrated with a time step of 0.001 ps. A typical MD production run was $\sim$50 ns. We run MPI parallel programming in the Raijin supercomputer in the National Computational Infrastructure (NCI) Australia and the SGI GPU-based system, Fornax, in iVEC.

3 Simulation results

3.1 Absolute sorption and bulk density

The absolute sorption refers to the actual amount of adsorbate present in the simulation box. Absolute sorption isotherms of CH_4, C_2H_6, and CO_2 on dry coal at 308 and 370 K are simulated. Our simulation results for CO_2, CH_4, and C_2H_6 plotted in different units of cm^3 (STP)/g (Fig. 3a), kg/t (Fig. 3b), and mol/kg (Fig. 3c), show that the absolute sorption of CO_2 on dry coal is higher than that of CH_4 and C_2H_6. At 308 K, the initial slope of the absolute sorption of C_2H_6 is higher than that of CH_4 (Fig. 3a and c) due to the stronger affinity between C_2H_6 and coal and the amount of C_2H_6 adsorbed in coal is higher than that of CH_4 in the low-pressure region. Jiang et al. (1994)

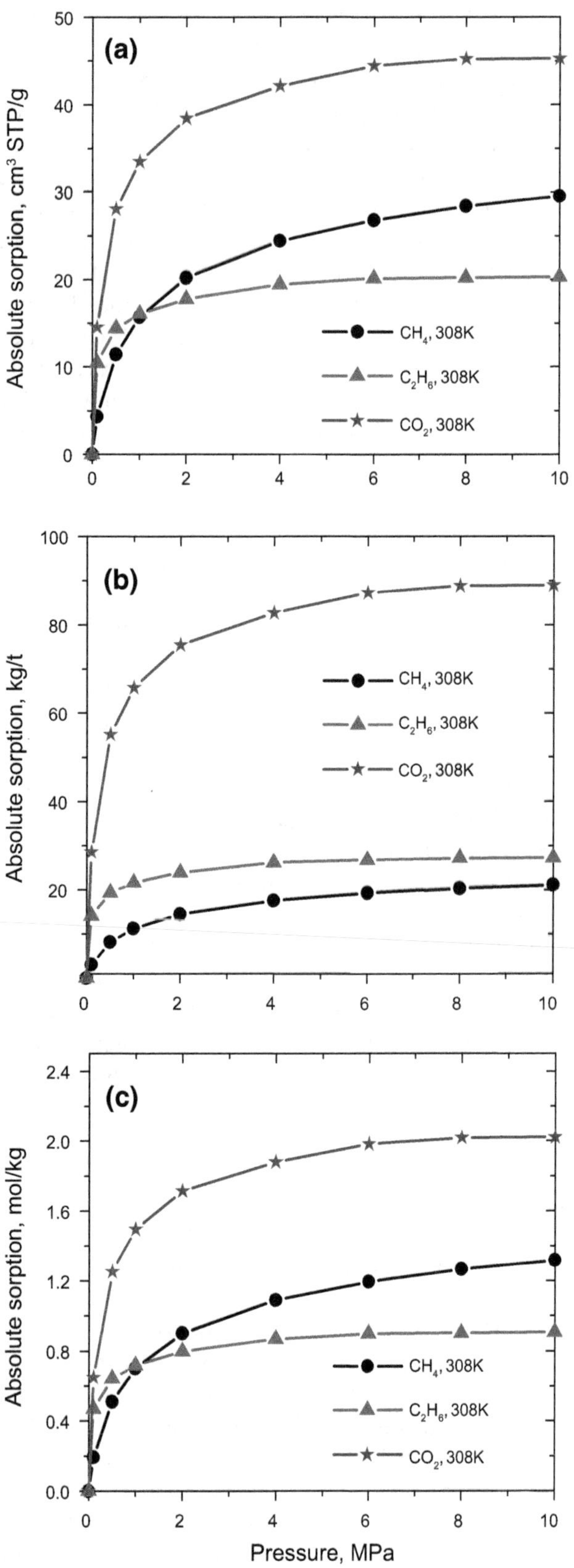

Fig. 3 Absolute sorption isotherms of CH_4, C_2H_6, and CO_2 on coal at a temperature of 308 K in the unit of **a** cm^3 STP/g coal; **b** kg/t coal; **c** mol/kg coal

showed experimentally that C_2H_6 sorption isotherm rises sharply and reaches a maximum at a lower pressure than CH_4. At high pressure, C_2H_6 sorption is lower compared with CH_4. This observation may be explained by comparing the fugacity of C_2H_6 and CH_4, shown in Fig. 4. The calculated bulk C_2H_6 fugacity is lower than that of CH_4 when pressure is higher than 2 MPa.

For high pressure, especially near or above the critical points of the adsorbate, real gas effects must be considered. The Peng–Robinson equation of state (Peng and Robinson 1976) has the general form:

$$P = \frac{RT}{V - b} - \frac{a}{V^2 + 2bV - b^2}, \tag{1}$$

where a and b are constants; P, T, and V are pressure, temperature, and volume, respectively; and R is gas constant. The constants a and b are defined by the following equations:

$$a = 0.45724R^2T_c^2/P_c, \tag{2}$$

$$b = 0.0778RT_c/P_c, \tag{3}$$

where T_c and P_c are critical temperature and pressure, respectively.

Equation (1) defines the compressibility factor (Z):

$$Z = \frac{PV}{RT} = \frac{V}{V - b} - \frac{aV/RT}{V^2 + 2bV - b^2}. \tag{4}$$

The compressibility factor measures the deviation from ideal behavior. Z is related to the fugacity (f):

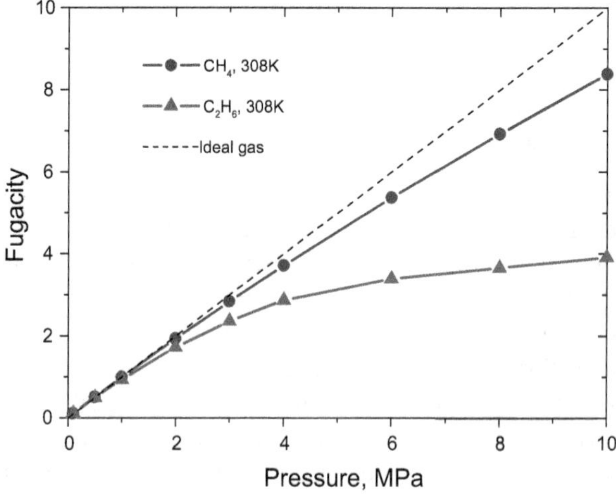

Fig. 4 Bulk gas fugacity as a function of bulk gas pressure at 308 K

$$P\left(\frac{\partial \ln f}{\partial P}\right)_T = \frac{PV}{RT} = Z. \quad (5)$$

According to Eqs. (2) and (3), we obtain the constant a and b for CH_4 and C_2H_6:

$$a_{CH_4} = 0.25\,\mathrm{m^6Pa/mol^2}; \quad b_{CH_4} = 2.68 \times 10^{-5}\,\mathrm{m^3/mol},$$

$$a_{C_2H_6} = 0.60\,\mathrm{m^6Pa/mol^2}; \quad b_{C_2H_6} = 4.05 \times 10^{-5}\,\mathrm{m^3/mol}.$$

From Eq. (4), positive deviations ($Z > 1$) are due to the molecules having finite size and is quantified by the constant b; while negative deviations ($Z < 1$) are due to the molecules having intermolecular forces and are quantified by the constant a. Figure 4 shows negative deviations (below diagonal dash line for ideal gas) for both CH_4 and C_2H_6, indicating that the second term on the right-hand side of Eq. (4) dominates. CH_4 and C_2H_6 deviate from ideal behavior in different ways. Below 2 MPa, they behave similarly. However, above 2 MPa, the deviation of C_2H_6 from ideal behavior is more significant compared with CH_4. This is due to the fact that $a_{C_2H_6}$ is 2.4 times of the value of a_{CH_4}. Therefore, we might infer that for hydrocarbon sorption in coal, the constant "a" is identified as an important factor affecting the performance of gas sorption in coal.

In Fig. 5, we show the absolute sorption results for the temperature of 370 K. We observe the same trend for both 308 and 370 K. The absolute sorption (expressed on a mass basis, Figs. 3b and 5b) increases in the order of CH_4, C_2H_6, and CO_2. In Figs. 3c and 5c, the absolute molar sorption isotherms of CH_4 and C_2H_6 show characteristic features: they intersect and this takes place at higher pressure when the temperature is higher. The pressure where the sorption curves of CH_4 and C_2H_6 cross shifts from around 1.3 to 3.0 MPa, when temperature is increased from 308 to 370 K. In Fig. 6, we illustrate the temperature effect on the absolute molar sorption for CH_4, C_2H_6, and CO_2. It is obvious that the sorption amount decreases with increasing temperature. Other groups also reported that the sorption of gas on coals at a given pressure increases with decreasing temperature (Menon 1968; White et al. 2005; Zhou et al. 2001).

In Fig. 7, we present the bulk density of CH_4, C_2H_6, and CO_2 obtained using the Peng–Robinson equation of state (Peng and Robinson 1976). The bulk density increases with pressure but decreases with temperature. The decreases in the bulk density with increasing temperature are simply reflected by the decrease in the sorption capacity when the temperature is raised, shown in Fig. 6. In the case of 308 K, which is close to the critical temperature of C_2H_6 (305.4 K) and CO_2 (304.13 K), there is a sharp increase in the bulk density for both C_2H_6 and CO_2 around their critical pressures. As shown in Fig. 7b, the critical density of

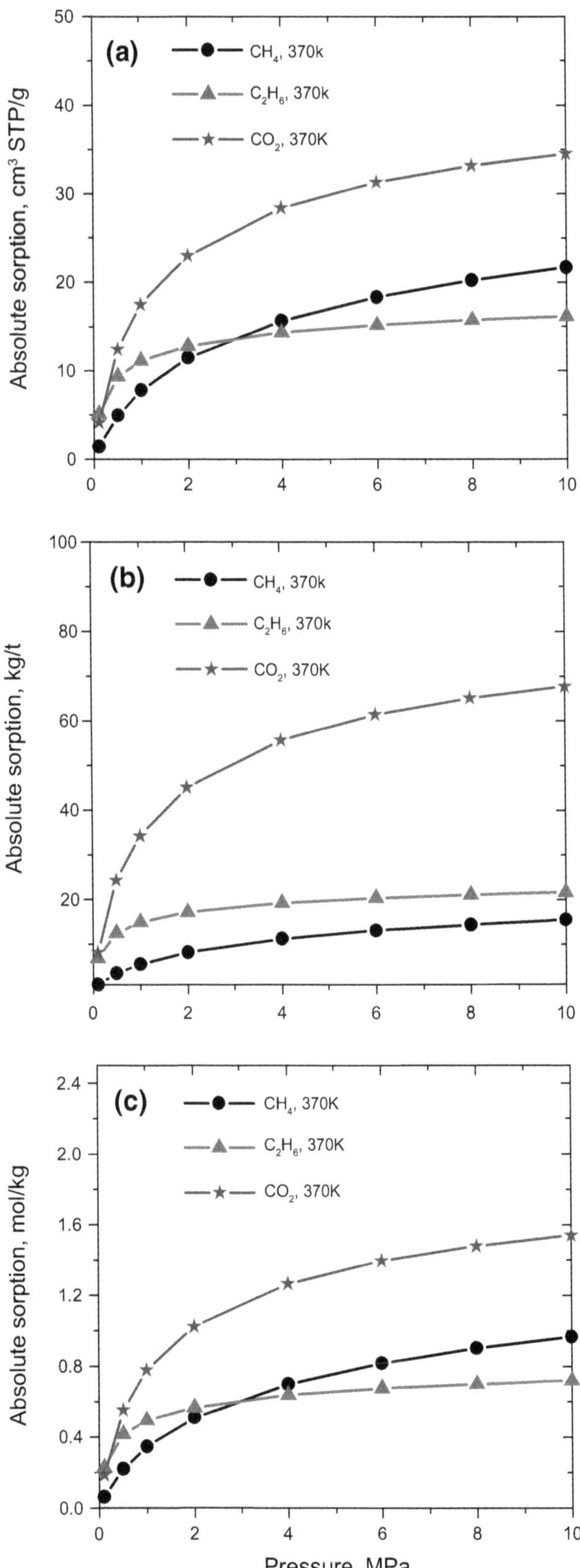

Fig. 5 Absolute sorption isotherms of CH_4, C_2H_6, and CO_2 on coal at a temperature of 370 K in the unit of **a** $\mathrm{cm^3}$ STP/g coal; **b** kg/t coal; **c** mol/kg coal

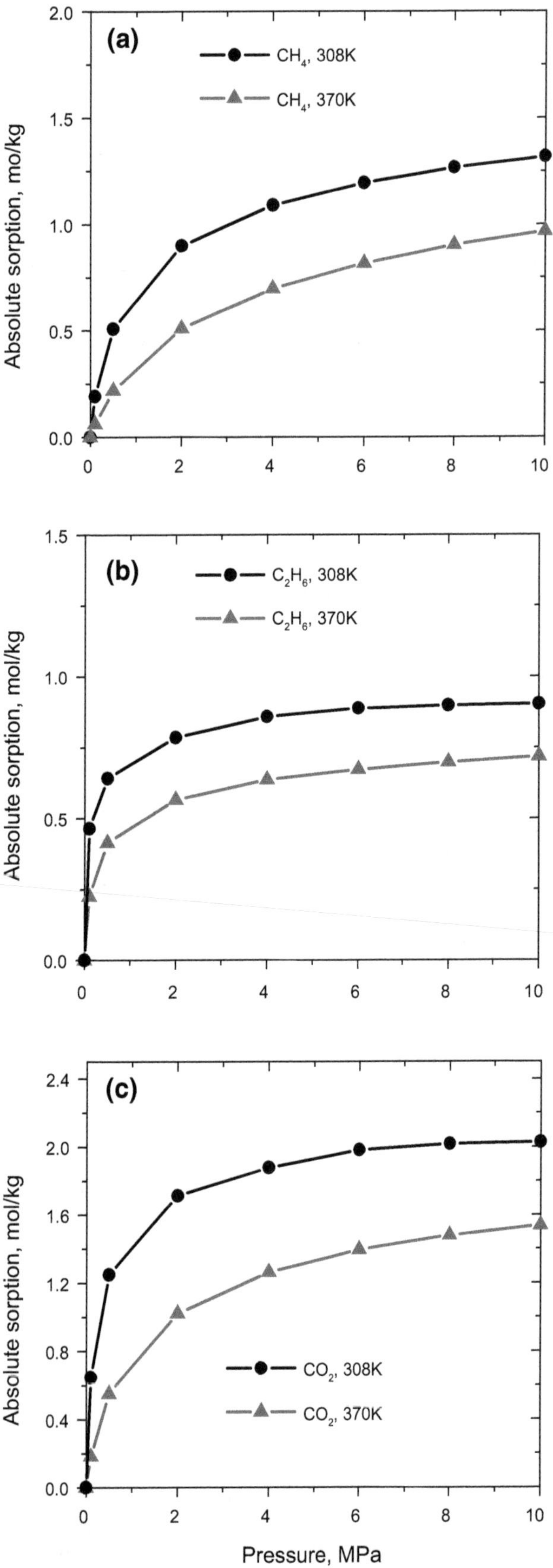

◀Fig. 6 Temperature effect on the absolute sorption isotherms of **a** CH_4; **b** C_2H_6; **c** CO_2

C_2H_6 is 206.18 kg/m^3; a clear inflection point appears at the corresponding density at around 5.2 MPa. After the inflection point, one expects the derivative of the bulk density with respect to pressure to decrease. Similarly, in Fig. 7c, we observe an inflection point for CO_2 at its critical density of 467.6 kg/m^3 at ~8.1 MPa. Bae and Bhatia (2006) have reported an inflection point in the CO_2 bulk phase density at a pressure of 8.93 MPa at 313 K. The pressure corresponding to the inflection point increases with temperature. In Fig. 8, we compare the bulk density at 308 K (Fig. 8a) and 370 K (Fig. 8b) for CH_4, C_2H_6, and CO_2. The bulk density of CH_4 increases linearly with pressure, while the bulk density of C_2H_6 and CO_2 behave non-linearly, especially at 308 K. The linearity of the bulk density with pressure exists up to the pressure of around 4 and 5 MPa for C_2H_6 and CO_2, respectively. Then they go through a significant increase around their inflection point. At the pressure range of 4.9–7.4 MPa, C_2H_6 has the greatest bulk density followed by CO_2 and CH_4.

Interestingly, we found that the absolute sorption expressed on a volume basis (the absolute sorption in terms of kilogram (kg) per ton (t) of coal, shown in Figs. 3b and 5b, divided by bulk density, presented in Fig. 8a, b) for CH_4, C_2H_6, and CO_2 merges with increasing pressure at 308 K (Fig. 9a) and 370 K (Fig. 9b). The Gurvitsch rule (Anderson 1914) predicts that the pore volume occupied by condensable gases and liquids is constant. From our results, it can be inferred that the maximum sorption capacity expressed on volume basis can be described in terms of Gurvitsch's law. In Fig. 10, we show the temperature effect on the absolute sorption on volume basis for CH_4, C_2H_6, and CO_2. As shown in Fig. 10, if pressure is higher than 4 MPa, the temperature dependence of the absolute sorption expressed on a volume basis for CH_4, C_2H_6, and CO_2 is negligible, indicating that for each adsorbate, the maximum sorption capacity, expressed on a volume basis, is independent of temperature.

3.2 Excess sorption

Experiments produce excess sorption. The excess sorption is the difference between the absolute sorption and the amount of gas in the reference system. The reference system has the same volume as the sorption system, but the interaction with the solid surface is neglected. The excess molar sorption is given by

$$n^e = n^a - V^p \rho^b, \quad (6)$$

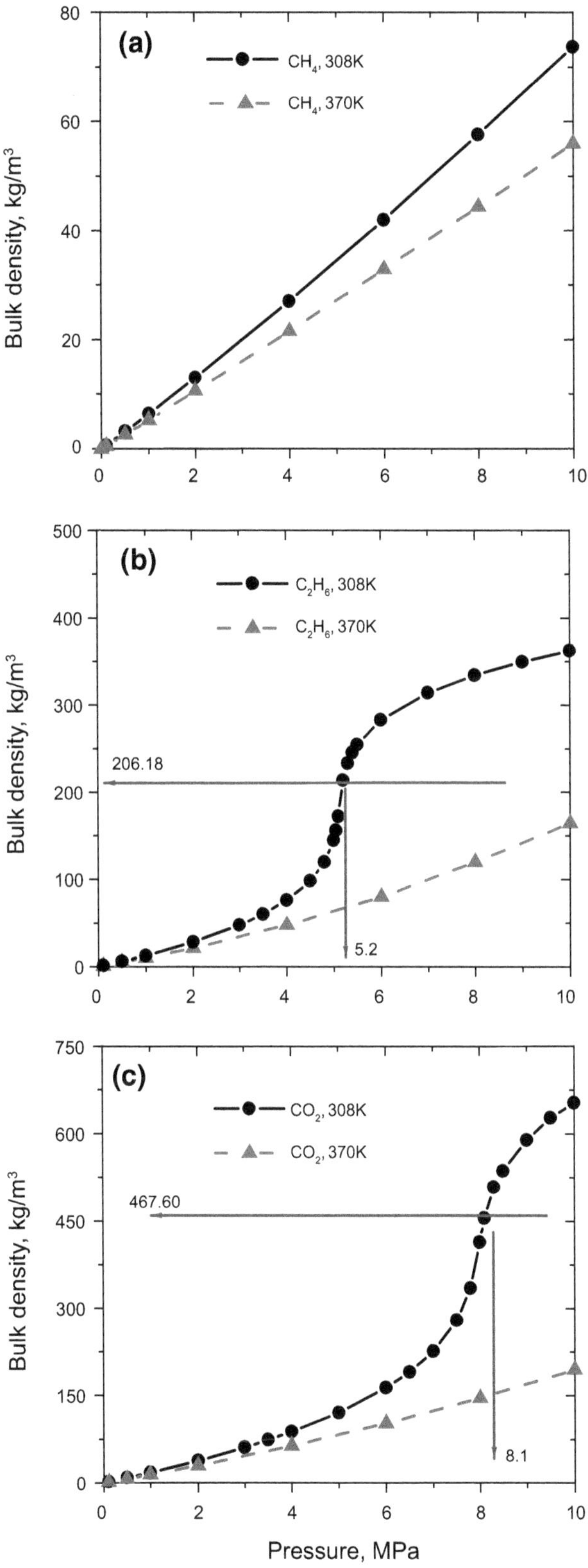

Fig. 7 Bulk densities of CH_4, C_2H_6, and CO_2 as a function of pressure at 308 and 370 K. **a** CH_4; **b** C_2H_6; **c** CO_2

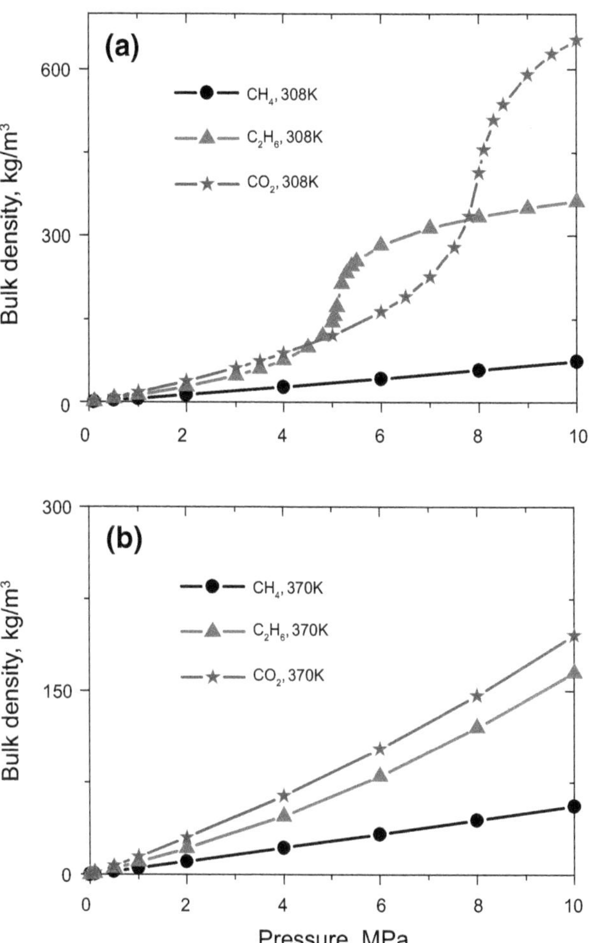

Fig. 8 Comparison of the bulk density for CH_4, C_2H_6, and CO_2 at **a** 308 K; **b** 370 K

where the excess sorption, n^e, is the amount in adsorbed phase in excess of the amount that would be present in the pore volume at the equilibrium density of the bulk gas. n^a is the absolute amount adsorbed; V^p is the pore volume; and ρ^b is the equilibrium density of the bulk gas. The pore volume is the volume fraction free to be occupied by gases in sorption processes. In experiments, it is measured using helium, because helium is hardly adsorbed. The point is that helium is a reference gas for measuring excess sorption of all other gases. Whether or not helium actually "adsorbs" is irrelevant. The requirement is that the procedure for measuring pore volume be identical for theoretical prediction and experimental determination. In molecular simulations, void volume is measured by probing the structure with helium at a room temperature of 25 °C. It was obtained from a separate simulation using the Widom particle insertion method. We probed the coal structure with an LJ helium atom at millions of random points,

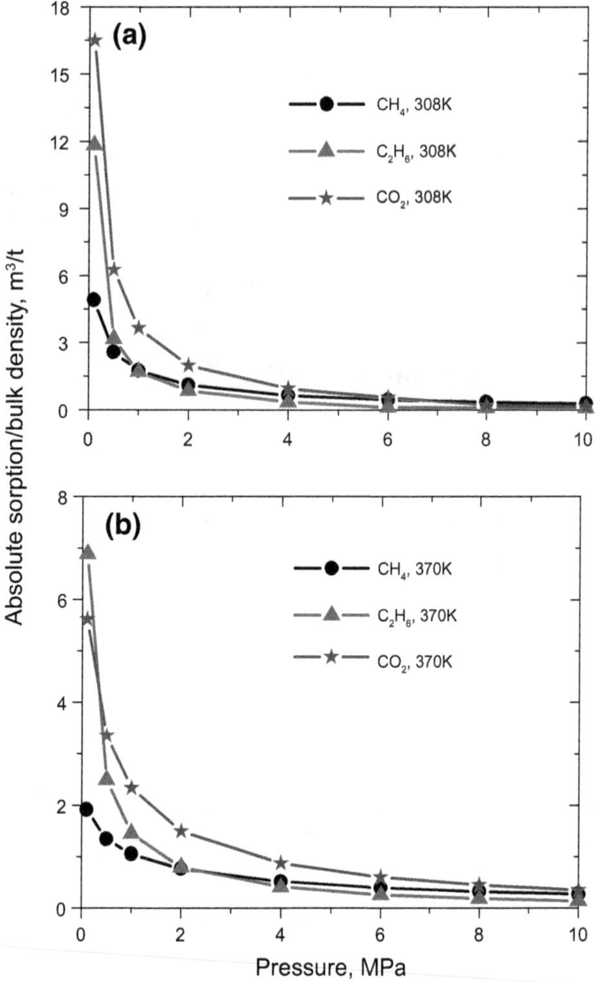

Fig. 9 Absolute sorption expressed on a volume basis (adsorbed amount divided by bulk density) for CH_4, C_2H_6, and CO_2 as a function of pressure at **a** 308 K; **b** 370 K

computed the energy difference with and without the particle, and estimated the average Boltzmann weight which directly corresponds to the void fraction (Talu and Myers 2001). In our simulation, all micropores are accessible because of the inherent feature of the MC method. We obtained an average helium pore fraction of 17 % for the dry intermediate rank bituminous coal. Because of the inherent feature of the MC method, the porosity might be overestimated. Based on our porosity result of 17 % and the absolute sorption, we calculate the excess sorption which is the relevant physical observable in experiments.

In Fig. 11, we show the absolute and excess sorption of CH_4, C_2H_6, and CO_2 at 308 K. The excess and the absolute quantity are indistinguishable at low pressures in the range up to 0.5 MPa. At higher pressure, the absolute sorption and excess sorption are different. The excess quantity reaches a maximum and then declines. This is due to the fact that excess sorption is relative to what would have

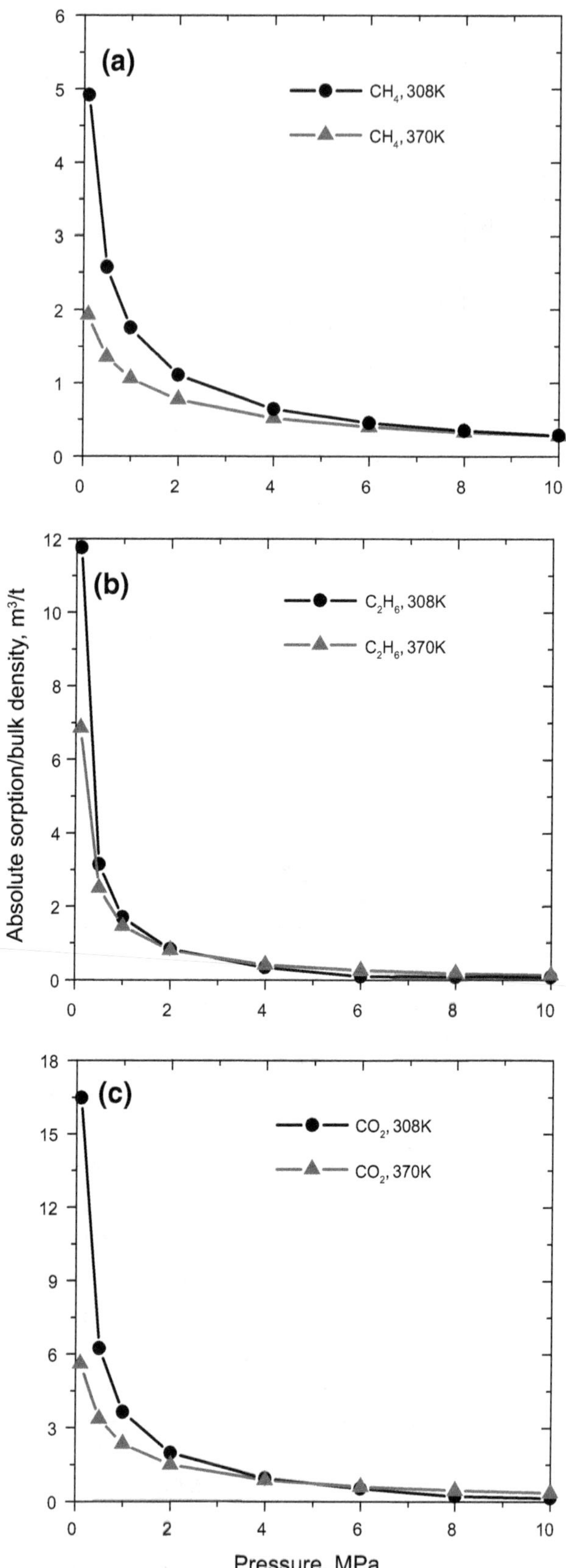

Fig. 10 Temperature effect on the absolute sorption on a volume basis (adsorbed amount divided by bulk density) for **a** CH_4; **b** C_2H_6; **c** CO_2

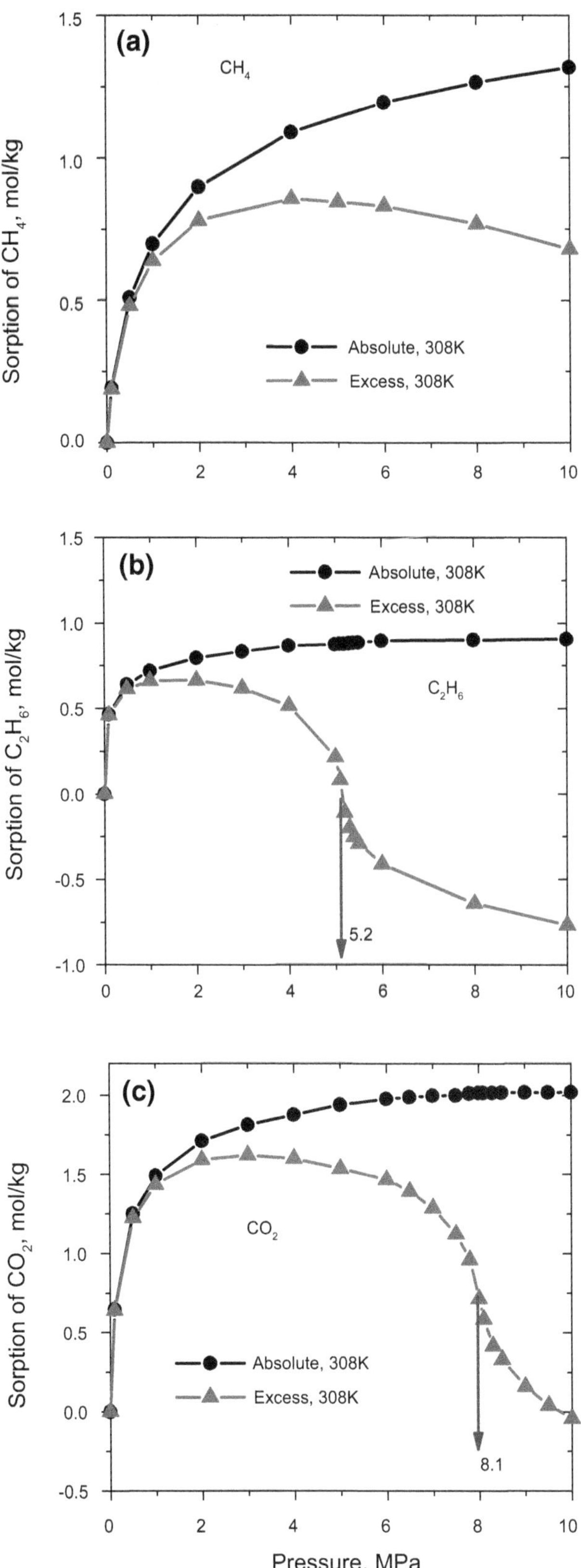

Fig. 11 Absolute and excess sorption of CH_4, C_2H_6, and CO_2 on coal at 308 K. **a** CH_4; **b** C_2H_6; **c** CO_2

been in the pore volume. At high pressures, the bulk phase can still be compressed, but eventually, the pores are filled up and the adsorbed phase density levels off. Once the bulk phase density is higher than the adsorbed phased density, the excess sorption turns negative. The significant increase in the bulk density of C_2H_6 and CO_2 close to their inflection point, shown in Fig. 7, results in a dramatic decrease in the excess sorption of C_2H_6 and CO_2. Inflection points also appear in the excess amount of C_2H_6 and CO_2 at the same pressure as the corresponding inflection point in the bulk density curve, shown in Fig. 7b, c. Li et al. (2010) reported that at 35 °C, the maximum excess sorption capacity of the medium-volatile bituminous coal (dry, ash-free) for CH_4 and CO_2 is around 0.8 and 1.3 mol/kg, respectively. Our molecular simulation results indicate a maximum excess sorption capacity of around 0.82 and 1.57 mol/kg for CH_4 and CO_2 on the bituminous coal (dry, ash-free), respectively. The results of the maximum excess sorption for CH_4 and CO_2 reported by Li et al. (2010) are around 2.4 % and 17.2 % lower than our simulation results. This could be attributed to the fact that the porosity might be overestimated due to the inherent feature of the MC method. A good agreement between the experiment and the molecular simulation indicates that the sorption mainly takes place in the coal matrix.

In Fig. 12, we show the temperature effect on the excess sorption of CH_4, C_2H_6, and CO_2. Similar to those obtained by Li et al. (2010), we also observed that after passing through the maximum, the lower temperature excess sorption isotherms decline more rapidly than the higher temperature isotherms, which results in an intersection of the isotherms at ~4.5 MPa for C_2H_6 and ~7.8 MPa for CO_2. The intersection corresponds to a reversal point of the temperature dependence of the excess sorption isotherms. At the pressure range above the reversal point, in contrast to the low-pressure range, the excess sorption increases with temperature. This effect is clearly related to the bulk density change of the C_2H_6 and CO_2, shown in Fig. 8. It is evident that for high pressures, the bulk density is much higher at lower temperature, and the excess amount reduces more quickly. When the bulk phase density approaches the adsorbed phase density, the excess sorption would become zero.

Figure 13 compares the ratio of absolute and excess adsorbed amounts of CO_2–CH_4 on coal at 308 K. The ratio for both the absolute and excess amount decreases with increasing pressure. Above 8 MPa, the ratio of the excess amounts adsorbed at 308 K is less than one, suggesting that pressures lower than 8 MPa may provide efficient CO_2 sequestration and methane replacement in coal bed. Bae and Bhatia (Bae and Bhatia 2006) suggested an optimum

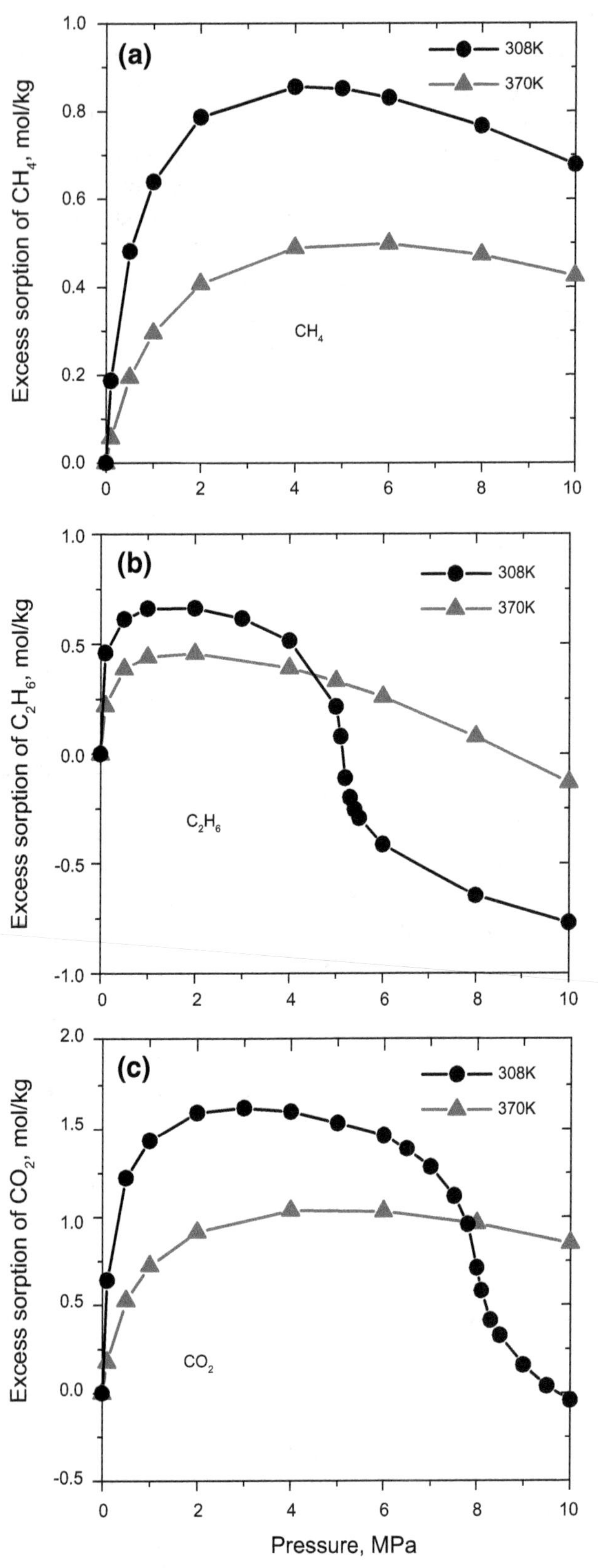

Fig. 12 Temperature effect on the excess sorption of CH_4, C_2H_6, and CO_2 on coal. **a** CH_4; **b** C_2H_6; **c** CO_2

pressure of 10 MPa at a higher temperature of 313 K. Similarly, we present the ratio of the absolute and excess adsorbed amounts of CO_2–C_2H_6 on coal at 308 K in Fig. 14. The ratio of the absolute adsorbed amounts increases when pressure is less than 1 MPa and then levels off around an average value of 2, but the ratio of the excess adsorbed amounts increases sharply when pressure is over 4 MPa, where the excess adsorbed amount of C_2H_6 drops significantly, suggesting an optimum pressure of above 4 MPa for C_2H_6 replacement.

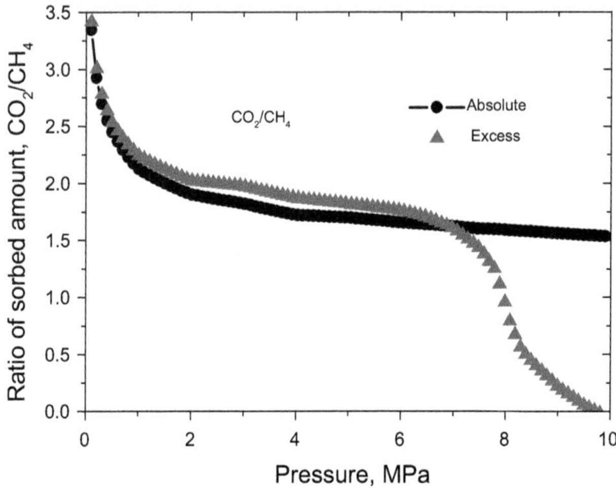

Fig. 13 Ratio of the absolute and excess adsorbed amounts of CO_2–CH_4 on coal at 308 K

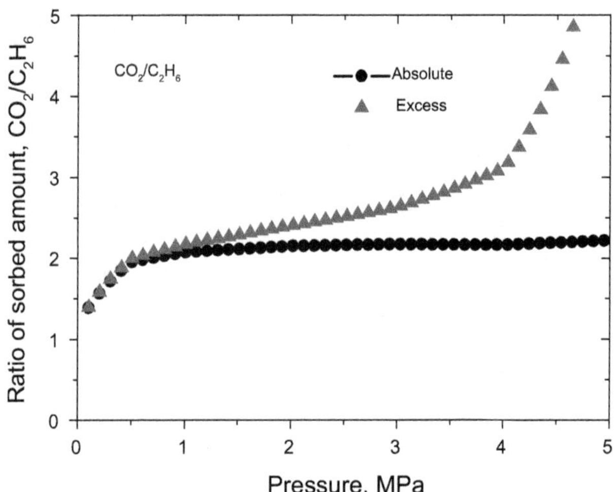

Fig. 14 Ratio of the absolute and excess adsorbed amounts of CO_2–C_2H_6 on coal at 308 K

3.3 Interaction energy and radial distribution functions (RDFs)

We further investigate the sorption mechanism by analyzing the interaction energy between the coal and adsorbate. We compare the interaction energy of adsorbate–adsorbate and adsorbate–coal. The results are given in Fig. 15. For the purpose of clarity, only the results at 308 K are shown. The interaction energy between the coal and CO_2 is much higher than between the coal and hydrocarbons. The interaction energy between the coal and the CO_2 becomes systematically more negative with increasing pressure. The increasingly negative energies signify greater interactions between the coal and CO_2 molecules with an increasing pressure. Based on the observation that the coal-CO_2 interaction is ~2.2–4.5 times of the interaction energy of coal-CH_4 and ~1.3–2.2 times of coal-C_2H_6. We infer that the higher sorption of CO_2 is mainly caused by stronger intermolecular interactions between coal and CO_2.

RDFs are defined as the ratio of the number of atoms at a distance r from a given atom compared with the number of atoms at the same distance in an ideal gas with the same density. The peak and the shape of the RDFs can reflect the density and structure of the system. To further investigate the effect of packing of the adsorbate on the sorption, we compare the RDFs of CH_4–CH_4, C_2H_6–C_2H_6, and CO_2–CO_2 at 308 K and 10 MPa in Fig. 16. As it is shown, the distance of closest contact in the O–O, O–C, and C–C RDFs of CO_2 is ~2.5 Å, but for hydrocarbons, the closest contact appears at 3.2 Å. It indicates that the distance between CO_2 molecules is shorter and CO_2 molecules are more closely packed compared with CH_4 and C_2H_6. We found the first contact peak between CH_4 and CH_4 is more significant than between C_2H_6 and C_2H_6, but at a separation distance of ~4.4–6.9 Å, the curve of the C_2H_6–C_2H_6 RDF is above that of CH_4–CH_4, indicating that the separation between C_2H_6 molecules is larger than the distance between the CH_4 molecules and C_2H_6 is loosely packed compared with CH_4.

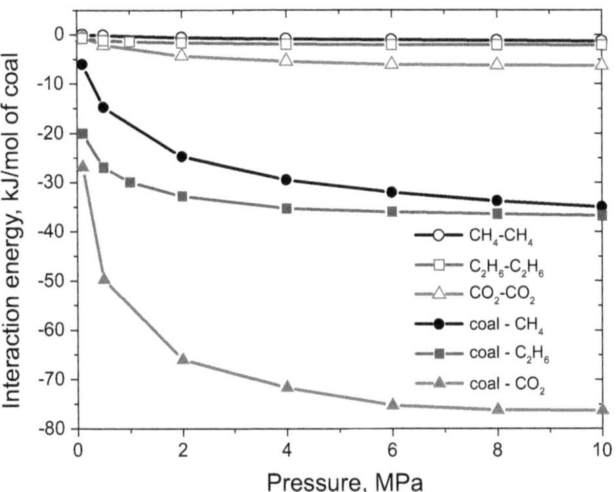

Fig. 15 Interaction energy of coal–CO_2, coal–C_2H_6, coal–CH_4, CO_2–CO_2, C_2H_6–C_2H_6, and CH_4–CH_4 at 308 K

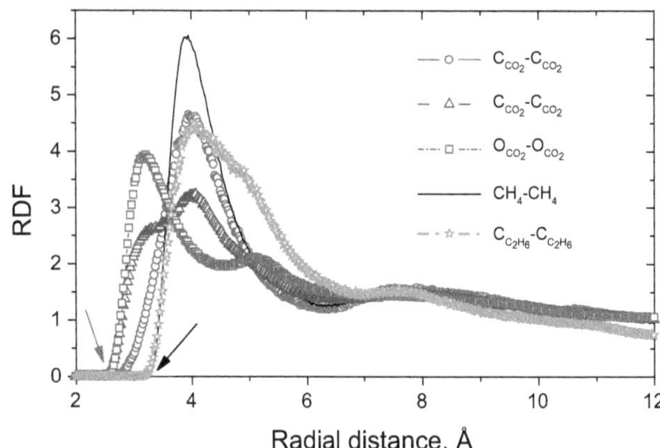

Fig. 16 Radial distribution functions (RDFs) of CH_4, C_2H_6, and CO_2 at 308 K and 10 MPa

4 Conclusion

We have performed the combined MD and MC simulations on CH_4, C_2H_6, and CO_2 sorption on dry intermediate rank coal in a pressure range up to 10 MPa and at temperatures of 308 and 370 K (35 and 97 °C). Our results indicate that absolute sorption (expressed as a mass basis) divided by bulk density is independent of temperature for both hydrocarbons (CH_4 and C_2H_6) and CO_2 when pressure is over 4 MPa. We infer that temperature has negligible effect on the maximum absolute sorption on a volume basis. We also observe that the intermediate rank coal has close maximum sorption capacity expressed on a volume basis for CH_4, C_2H_6, and CO_2. Based on the observation, we could infer that the pore volume occupied by CH_4, C_2H_6, and CO_2 is similar and the sorption capacity expressed on volume basis can be described in terms of Gurvitsch's law. The comparisons of our adsorption isotherm obtained from molecular simulation with published experimental data are satisfactory. It indicates that the CH_4, C_2H_6, and CO_2 are mainly stored in a coal matrix by sorption. Our results reveal that the "a" constant (proportional to T_c^2/P_c) in the Peng–Robinson equation of state is an important factor affecting the sorption behavior of hydrocarbons. This study provides a quantitative understanding of the effect of temperature on CH_4, C_2H_6, and CO_2 sorption capacity from a microscopic perspective. It also offers insights into aspects of the excess sorption and its relationship with bulk phase density. Molecular simulation proves to be a cost-effective and efficient method for directly studying the interactions between coal and gases under various external environments and for predicting gas sorption behavior in complicated and complex systems.

Acknowledgments We thank the National Computing Infrastructure (NCI) national facility and iVEC GPU cluster for a generous allocation of computing time. The second author was supported by the National Basic Research Program of China (2014CB239004) and the "Element and Process Constraint Petroleum System Modeling" project (No. 2011A-0207) under the PetroChina Science Innovation program.

References

Anderson JS. Structure of silicic acid gels. Z Physik Chem. 1914;88: 191–228.

Azmi AS, Yusup S, Muhamad S. The influence of temperature on adsorption capacity of Malaysian coal. Chem Eng Process. 2006; 45(5):392–6.

Bae JS, Bhatia SK. High-pressure adsorption of methane and carbon dioxide on coal. Energy Fuels. 2006;20(6):2599–607.

Berendsen HJC, Postma JPM, Vangunsteren WF, et al. Molecular-dynamics with coupling to an external bath. J Chem Phys. 1984; 81(8):3684–90.

Berendsen HJC, Vanderspoel D, Vandrunen R. Gromacs: a message-passing parallel molecular-dynamics implementation. Comput Phys Commun. 1995;91(1–3):43–56.

Billemont P, Coasne B, De Weireld G. An experimental and molecular simulation study of the adsorption of carbon dioxide and methane in nanoporous carbons in the presence of water. Langmuir. 2011;27(3):1015–24.

Billemont P, Coasne B, De Weireld G. Adsorption of carbon dioxide, methane, and their mixtures in porous carbons: effect of surface chemistry, water content, and pore disorder. Langmuir. 2013; 29(10):3328–38.

Brochard L, Vandamme M, Pelenq RJM, et al. Adsorption-induced deformation of microporous materials: coal swelling induced by CO_2–CH_4 competitive adsorption. Langmuir. 2012a;28(5): 2659–70.

Brochard L, Vandamme M, Pellenq RJM. Poromechanics of microporous media. J Mech Phys Solids. 2012b;60(4):606–22.

Busch A, Gensterblum Y. CBM and CO_2-ECBM related sorption processes in coal: a review. Int J Coal Geol. 2011;87(2):49–71.

Busch A, Gensterblum Y, Krooss BM. Methane and CO_2 sorption and desorption measurements on dry Argonne premium coals: pure components and mixtures. Int J Coal Geol. 2003;55(2–4):205–24.

Busch A, Gensterblum Y, Krooss BM, et al. Methane and carbon dioxide adsorption-diffusion experiments on coal: upscaling and modeling. Int J Coal Geol. 2004;60(2–4):151–68.

Connell LD, Lu M, Pan ZJ. An analytical coal permeability model for tri-axial strain and stress conditions. Int J Coal Geol. 2010;84(2): 103–14.

Day S, Fry R, Sakurovs R. Swelling of Australian coals in supercritical CO2. Int J Coal Geol. 2008;74(1):41–52.

Dubbeldam D, Calero S, Ellis D, et al. RASPA 1.0: Molecular Software Package for adsorption and diffusion in nanoporous materials. Evanston: Northwestern University; 2008.

Dubbeldam D, Calero S, Vlugt TJH, et al. Force field parametrization through fitting on inflection points in isotherms. Phys Rev Lett. 2004a;93(8):088302.

Dubbeldam D, Calero S, Vlugt TJH, et al. United atom force field for alkanes in nanoporous materials. J Phys Chem B. 2004b; 108(33):12301–13.

Fitzgerald JE, Pan Z, Sudibandriyo M, et al. Adsorption of methane, nitrogen, carbon dioxide and their mixtures on wet Tiffany coal. Fuel. 2005;84(18):2351–63.

Frenkel D, Mooij GCAM, Smit B. Novel scheme to study structural and thermal-properties of continuously deformable molecules. J Phys Condens Matter. 1992;4(12):3053–76.

Gensterblum Y, Busch A, Krooss BM. Molecular concept and experimental evidence of competitive adsorption of H_2O, CO_2 and CH_4 on organic material. Fuel. 2014;115:581–8.

Goodman AL, Busch A, Bustin RM, et al. Inter-laboratory comparison II: CO_2 Isotherms measured on moisture-equilibrated Argonne premium coals at 55 °C and up to 15 MPa. Int J Coal Geol. 2007;72(3–4):153–64.

Jiang SY, Zollweg JA, Gubbins KE. High-pressure adsorption of methane and ethane in activated carbon and carbon-fibers. J Phys Chem-Us. 1994;98(22):5709–13.

Krooss BM, van Bergen F, Gensterblum Y, et al. High-pressure methane and carbon dioxide adsorption on dry and moisture-equilibrated Pennsylvanian coals. Int J Coal Geol. 2002;51(2):69–92.

Li DY, Liu Q, Weniger P, et al. High-pressure sorption isotherms and sorption kinetics of CH_4 and CO_2 on coals. Fuel. 2010;89(3): 569–80.

Lindahl E, Hess B, van der Spoel D. GROMACS 3.0: a package for molecular simulation and trajectory analysis. J Mol Model. 2001;7(8):306–17.

Lu M, Connell LD. A model for the flow of gas mixtures in adsorption dominated dual porosity reservoirs incorporating multi-component matrix diffusion - Part I.Theoretical development. J Petrol Sci Eng. 2007;59(1–2):17–26.

Lu M, Connell LD, Pan ZJ. A model for the flow of gas mixtures in adsorption dominated dual-porosity reservoirs incorporating multi-component matrix diffusion-Part II numerical algorithm and application examples. J Petrol Sci Eng. 2008;62(3–4):93–101.

Mathews JP, van Duin ACT, Chaffee AL. The utility of coal molecular models. Fuel Process Technol. 2011;92(4):718–28.

Menon PG. Adsorption at high pressures. Chem Rev. 1968;68(3): 277–94.

Oostenbrink C, Villa A, Mark AE, et al. A biomolecular force field based on the free enthalpy of hydration and solvation: The GROMOS force-field parameter sets 53A5 and 53A6. J Comput Chem. 2004;25(13):1656–76.

Ottiger S, Pini R, Storti G, et al. Competitive adsorption equilibria of CO2 and CH4 on a dry coal. Adsorption. 2008;14(4–5):539–56.

Pan ZJ, Connell LD. A theoretical model for gas adsorption-induced coal swelling. Int J Coal Geol. 2007;69(4):243–52.

Pan ZJ, Connell LD. Comparison of adsorption models in reservoir simulation of enhanced coalbed methane recovery and CO_2 sequestration in coal. Int J Greenhouse Gas Control. 2009;3(1):77–89.

Pan ZJ, Connell LD. Modelling permeability for coal reservoirs: a review of analytical models and testing data. Int J Coal Geol. 2012;92:1–44.

Parrinello M, Rahman A. Polymorphic transitions in single-crystals: a new molecular-dynamics method. J Appl Phys. 1981;52(12): 7182–90.

Peng D, Robinson DB. New 2-constant equation of state. Ind Eng Chem Fund. 1976;15(1):59–64.

Pini R, Ottiger S, Storti G, et al. Pure and competitive adsorption of CO2, CH4 and N2 on coal for ECBM. Greenhouse Gas Control Technologies 9. 2009; 1(1): 1705–10.

Pini R, Ottiger S, Storti G, et al. Prediction of competitive adsorption on coal by a lattice DFT model. Adsorption. 2010;16(1–2):37–46.

Sakurovs R, Day S, Weir S. Relationships between the critical properties of gases and their high pressure sorption behavior on coals. Energy Fuels. 2010;24:1781–7.

Sakurovs R, Day S, Weir S, et al. Application of a modified Dubinin–Radushkevich equation to adsorption of gases by coals under supercritical conditions. Energy Fuels. 2007;21(2):992–7.

Schuttelkopf AW, van Aalten DMF. PRODRG: a tool for high-throughput crystallography of protein-ligand complexes. Acta Crystallogr Sect D-Biol Crystallogr. 2004;60:1355–63.

Siepmann JI, Frenkel D. Configurational Bias Monte-Carlo: a new sampling scheme for flexible chains. Mol Phys. 1992;75(1):59–70.

Spiro CL, Kosky PG. Space-Filling Models for Coal. 2. Extension to Coals of Various Ranks. Fuel. 1982;61(11):1080.

Talu O, Myers AL. Molecular simulation of adsorption: gibbs dividing surface and comparison with experiment. AIChE J. 2001;47(5):1160–8.

Tambach TJ, Mathews JP, van Bergen F. Molecular exchange of CH_4 and CO_2 in coal: enhanced coalbed methane on a nanoscale. Energy Fuels. 2009;23:4845–7.

Van der Spoel D, Lindahl E, Hess B, et al. GROMACS: fast, flexible, and free. J Comput Chem. 2005;26(16):1701–18.

Vandamme M, Brochard L, Lecampion B, et al. Adsorption and strain: the CO_2-induced swelling of coal. J Mech Phys Solids. 2010;58(10):1489–505.

White CM, Smith DH, Jones KL, et al. Sequestration of carbon dioxide in coal with enhanced coalbed methane recovery: a review. Energy Fuels. 2005;19(3):659–724.

Zhang J, Clennell MB, Dewhurst DN, et al. Combined Monte Carlo and molecular dynamics simulation of methane adsorption on dry and moist coal. Fuel. 2014;122:186–97.

Zhang J, Liu K, Clennell MB, et al. Molecular simulation of CO_2–CH_4 competitive adsorption and induced coal swelling. Fuel. 2015;160:309–17.

Zhou L, Zhou YP, Bai SP, et al. Determination of the adsorbed phase volume and its application in isotherm modeling for the adsorption of supercritical nitrogen on activated carbon. J Colloid Interface Sci. 2001;239(1):33–8.

4

The concept and the accumulation characteristics of unconventional hydrocarbon resources

Yan Song[1,2] · Zhuo Li[1,2] · Lin Jiang[3] · Feng Hong[3]

Abstract Unconventional hydrocarbon resources, which are only marginally economically explored and developed by traditional methods and techniques, are different from conventional hydrocarbon resources in their accumulation mechanisms, occurrence states, distribution models, and exploration and development manners. The types of unconventional hydrocarbon are controlled by the evolution of the source rocks and the combinations of different types of unconventional reservoirs. The fundamental distinction between unconventional hydrocarbon resources and conventional hydrocarbon resources is their non-buoyancy-driven migration. The development of the micro- to nano-scale pores results in rather high capillary resistance. The accumulation mechanisms of the unconventional and the conventional hydrocarbon resources are also greatly different. In conventional hydrocarbon resources, oil and gas entrapment is controlled by reservoir-forming factors and geological events, which is a dynamic balance process; while for unconventional hydrocarbon resources, the gas content is affected by the temperature and pressure fields, and their preservation is crucial. Unconventional and conventional hydrocarbons are distributed in an orderly manner in subsurface space, having three distribution models of intra-source rock, basin-centered, and source rock interlayer. These results will be of great significance to unconventional hydrocarbon exploration.

Keywords Unconventional hydrocarbon resources · Non-buoyancy-driven accumulation · Accumulation mechanisms · Distribution model

1 Introduction

Unconventional hydrocarbon resources are becoming increasingly significant in global energy structures. Global petroleum exploration is currently undergoing a strategic shift from conventional to unconventional hydrocarbon resources. Unconventional hydrocarbon resources (including tight oil/gas, shale oil/gas, and coal bed gas) are becoming a significant component of world energy consumption (Jia et al. 2012; Zou 2013). Unconventional hydrocarbon resources are distinct from conventional hydrocarbon resources. The characteristics of the unconventional hydrocarbon resources are as follows: the source and the reservoir coexist; the porosity and the permeability are ultra-low; nano-scale pore throats are widely distributed; there is no obvious trap boundary; buoyancy and hydrodynamics have only a minor effect, Darcy's law does not apply; phase separation is poor; there is no uniform oil–gas–water interface or pressure system; and oil or gas saturation varies (Sun and Jia 2011; Yang et al. 2013). Unconventional hydrocarbons in tight reservoirs show characteristics distinct from those of the hydrocarbon sources hosted in structural and stratigraphic traps. Unconventional petroleum geology differs from traditional petroleum geology in terms of trap conditions, reservoir properties, combination of source and reservoir rocks,

✉ Yan Song
sya@petrochina.com.cn

1 State Key Laboratory of Petroleum Resource and Prospecting, China University of Petroleum, Beijing 102249, China

2 Unconventional Natural Gas Institute, China University of Petroleum, Beijing 102249, China

3 Research Institute of Petroleum Exploration and Development, PetroChina, Beijing 100083, China

Edited by Jie Hao

accumulation features, percolation mechanisms, and occurrence features, so different reservoir conditions and accumulation mechanisms are essential for unconventional hydrocarbon accumulation (Zou et al. 2012). According to the relationship between source rock evolution and reservoir formation, we clarify the relations of various unconventional hydrocarbon resources, propose the identification marks and distribution models for unconventional hydrocarbon resources, and compare the differences between unconventional and conventional hydrocarbon in terms of types, characteristics, distribution models, and accumulation mechanisms, which provide important guidance for unconventional hydrocarbon exploration (Zou et al. 2015).

2 Concept of unconventional hydrocarbon resources

2.1 Generation of unconventional hydrocarbon resources

Unconventional and conventional hydrocarbon resources are both generated during thermal evolution of source rocks. Conventional hydrocarbon is generally defined and classified by generation, migration, trap, and preservation, while the unconventional hydrocarbon is defined by kerogen type, evolution of source rocks, and reservoir types (Song et al. 2013). Hydrocarbon generation and expulsion from type I-II and type III kerogen during thermal maturation are different (Tissot and Welte 1978; Huang et al. 1984; Zhang and Zhang 1981; Martini et al. 2003), and the relationship between reservoir characteristics and hydrocarbon generation and expulsion determines the type of unconventional hydrocarbon reservoirs (Song et al. 2013). For type I–II kerogen, oil is generated from and detected in source rocks at a relatively low maturity stage, and oil shale is formed. During mature stage, source rocks generate and expel a large amount of oil and gas, which accumulates in tight reservoirs close to source rocks to form tight oil, and remains inside source rocks to form shale oil. During the over-mature stage, source rocks mainly generate gas, which accumulates in tight reservoirs adjacent to source rocks to form tight gas, meanwhile a large amount of remaining gas inside source rocks is identified as shale gas (Fig. 1).

Natural gas is generated from type III kerogen during thermal evolution (Dai et al. 1992) and is stored inside the source rocks and adjacent tight reservoirs to form shale gas and tight gas, respectively. Coal bed methane (CBM) is formed in coal beds during thermal maturation of coals (Fig. 2).

As shown in Fig. 3, different types of unconventional hydrocarbons are oil and gas generated during source rock evolution and accumulated in different unconventional reservoirs.

2.2 Identification marks of unconventional hydrocarbons

Non-buoyancy-driven accumulation means that hydrocarbon accumulation is driven by forces excluding buoyancy. Unconventional hydrocarbon resources have the characteristics of coexisting source rocks and reservoirs, no obvious trap boundaries, weak fluid phase differentiation, no uniform water–oil interface, independent pressure system, and oil or gas saturation varying significantly (Zou et al. 2011; Ju et al. 2015). There is a fundamentally important geological distinction between conventional and unconventional hydrocarbon. Conventional gas resources are buoyancy-driven deposits, occurring as discrete accumulations in structural and/or stratigraphic traps, whereas unconventional gas resources are generally non-buoyancy-driven accumulations. Non-buoyancy-driven accumulation means that buoyancy has a weak effect on hydrocarbon migration and cannot overcome resistance.

2.2.1 Key reason of non-buoyancy-driven accumulation

Capillary pressure is the principle resistance for hydrocarbon migration, which is controlled by the radius of pore-throats of reservoirs. The narrower the pore-throats, the higher the capillary pressure. Thus, the key reason of non-buoyancy-driven accumulation of unconventional hydrocarbon can be attributed to small pore-throats of reservoirs. By advanced experimental test methods, it has been proved that the widely developed micro–nano-pore-throats lead to large resistance due to high capillary pressure (Loucks and Ruppel 2007). The statistical analysis of global tight reservoirs' pore-throat diameters shows that the shale reservoirs have the minimum pore-throat diameters, while the tight sandstones have relatively larger pore-throat diameters (Nelson 2009; Zou et al. 2011; Passey et al. 2011). The average pore-throat diameter of the shale gas reservoirs is 5–200 nm (Jarvie et al. 2007), that of the shale oil reservoirs is 30–400 nm (Montgomery et al. 2005), that of the tight gas reservoirs is 40–700 nm, that of the tight sandstone oil reservoirs is 50–900 nm, and that of the tight carbonate oil reservoirs is 40–500 nm (Jia et al. 2012; Du et al. 2014). The development of micro–nano-pores leads to high capillary pressure in the pore structure of reservoirs. If the diameter of pores is 10–50 nm, then the calculated capillary pressure of those pores could be 12–24 MPa (Zhang et al. 2014), indicating that at least under such strength of driving force (buoyancy or abnormal pressure), hydrocarbon could be capable of migrating.

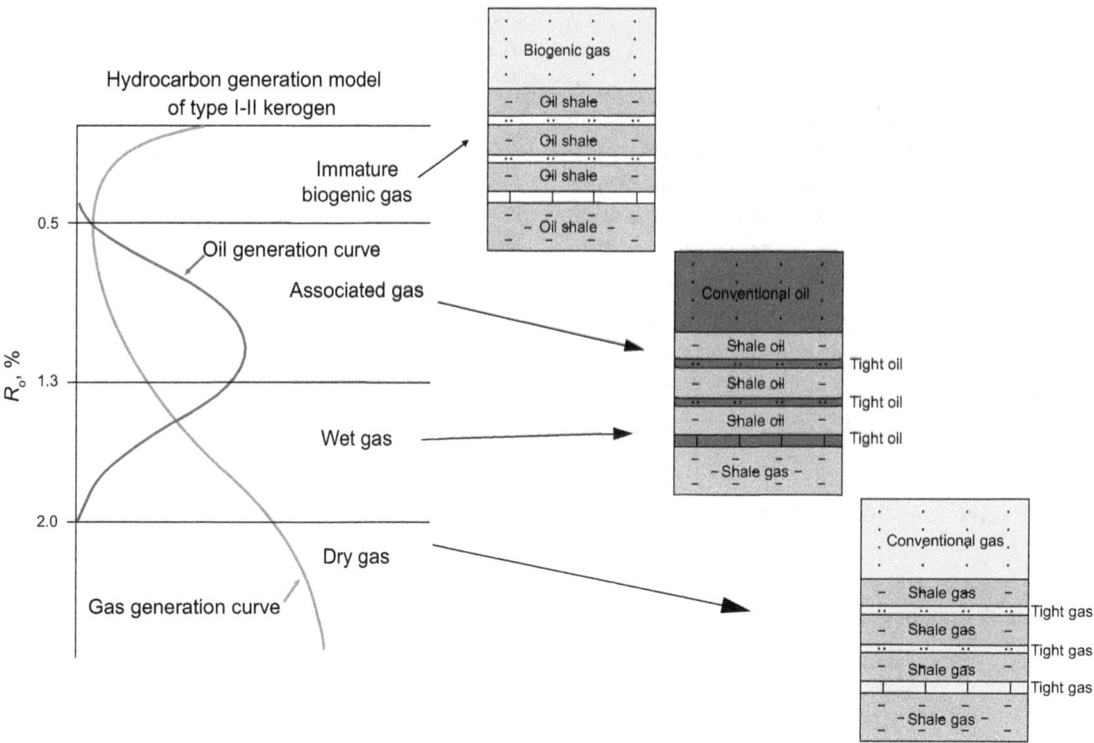

Fig. 1 Relationship between type I–II kerogen maturation and hydrocarbon types

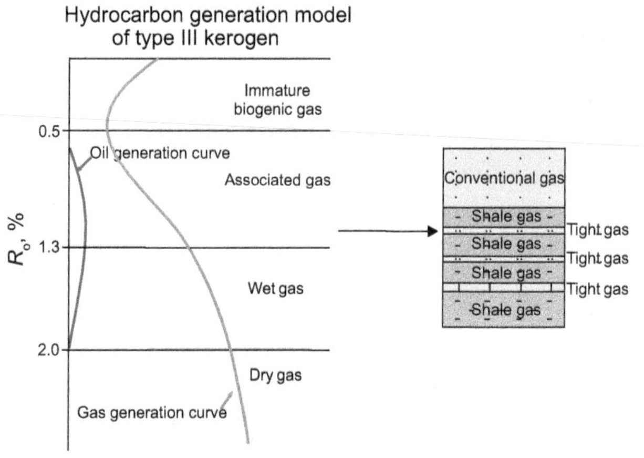

Fig. 2 Relationship between type III kerogen maturation and hydrocarbon types

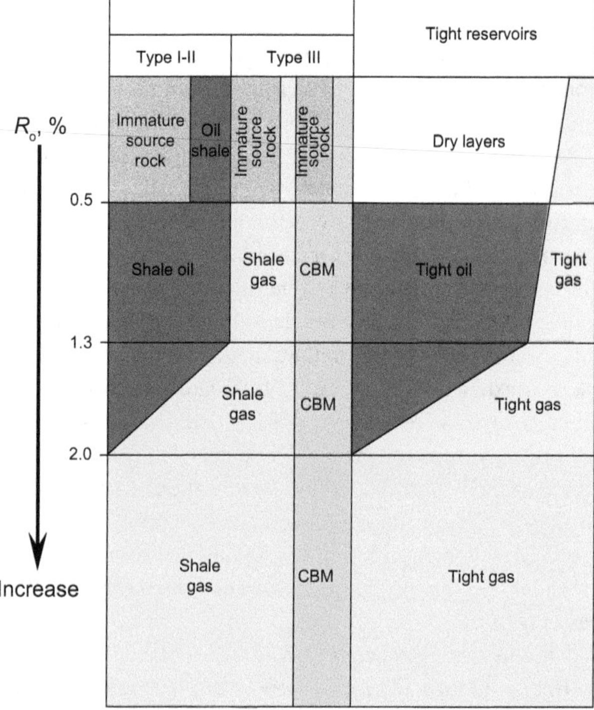

Fig. 3 Unconventional hydrocarbon types under source rock maturation and reservoir type control

2.2.2 Mechanisms of non-buoyancy-driven accumulation

Within conventional petroleum systems, buoyancy is considered to be the driving force, and capillary pressure is the resistance for hydrocarbon migration and accumulation (Davis 1987). According to the equation of buoyancy and capillary pressure (Schowalter 1979), when the radius of pore-throats decreases by 10 %, the capillary pressure would increase tenfold. If buoyancy is still considered to be the driving force, then hydrocarbon migration would happen only when buoyancy correspondingly increases tenfold. Taking one gas column with a height of 3 m and density of 0.2 g/cm^3 for an example, the buoyancy can be 0.024 MPa, but gas cannot enter the pore-throats with a radius of 2 μm. The pore-throat diameter of tight sandstones is mostly less than 1 μm, the capillary pressure is at least more than 0.08 MPa. However, migration of gas with a density of 0.2 g/cm^3 needs a buoyancy of 0.07 MPa, and the height of the gas column required would be over 10 m. Based on the research of outcrops, thickness measurements, and profile interpretation, the fluvial sandbodies with a vertical thickness over 10 m are scarce (Shanley 2004). Therefore, no favorable geological conditions for gas columns can form enough buoyancy, and buoyancy could not be the dominant driving force for unconventional oil and gas accumulation.

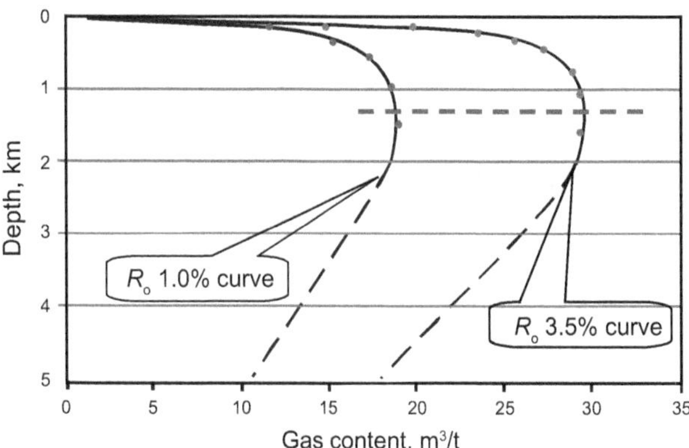

Fig. 4 Relation of gas content with depth in the no. 3 coal bed in the Qinshui Basin, China

3 Characteristics of unconventional hydrocarbon accumulation

The differences between unconventional and conventional hydrocarbons in occurrence and accumulation processes determine the differences in accumulation mechanisms. In order to better understand the characteristics of unconventional hydrocarbon accumulation, unconventional gas reservoirs characterized by adsorbed gas are taken as examples to compare with conventional gas reservoirs.

3.1 The unconventional gas content is affected by temperature and pressure fields while the conventional gas content is controlled by dynamic balance

Conventional gas accumulation can be divided into two processes: natural gas generated and expelled from source rocks migrates and accumulates in reservoirs, and then it is continuously lost by diffusion and seepage. Conventional gas accumulation is the consequence of the balance of gas charge and loss, namely the dynamic balance. Thus, the intensity and time of gas charge and sealing conditions are the key factors to natural gas accumulation.

The unconventional gas with the most common occurrence of adsorbed gas is associated with adsorption capacity which is controlled primarily by temperature and pressure. The higher the pressure and the lower the temperature are, the higher the adsorption capacity (Wang and Reed 2009; Liu et al. 2013; Guo et al. 2014). However, under actual geological conditions, the adsorbed gas content in unconventional reservoirs is controlled by the combination of the temperature and pressure changes. Figure 4 illustrates the relationship between adsorption capacity and depth of two different rank coal samples with R_o of 1.0 % and 3.5 %. At depth shallower than approximate 1000 m, the adsorption capacity is principally controlled by pressure, and the gas content tends to increase with the burial depth increasing; whereas at depths deeper than 1000 m, the adsorption capacity is mainly controlled by temperature, and the gas content tends to decrease with the burial depth increasing.

The diffusion of the unconventional hydrocarbon can be attributed to temperature–pressure fields. Temperature and pressure changes lead to the conversion of adsorbed gas to free gas, and free gas diffuses through caprocks or formation water. Therefore, unlike the conventional gas only needing top caprocks, the preservation of CBM needs not only top caprocks, but also bottom caprocks. Coal beds should be in an enclosed system in order to store a large amount of gas (Fig. 5a). First, an enclosed system can be overpressure to elevate the adsorption capacity of coal beds. Second, free gas can be preserved well from diffusion and hydrodynamic destruction. However, in most cases, an enclosed system can be destroyed by permeable layers at the bottom of coal beds (Fig. 5b) or on top of the coal beds (Fig. 5c), leading to gas loss through diffusion and formation water washing.

3.2 Unconventional gas accumulation is controlled by preservation, while the conventional hydrocarbon accumulation is controlled by the best match of petroleum systems

Conventional gas accumulation generally experiences processes of gas generation, migration, concentration, and preservation. The best match of static factors such as source rocks, reservoirs, and caprocks and the dynamic factors such as natural gas generation, migration, entrapment, and accumulation controls the hydrocarbon accumulation periods.

Unconventional gas accumulation generally undergoes three distinct stages: (a) gas generation and adsorption, (b) increasing adsorption and desorption, and (c) diffusion

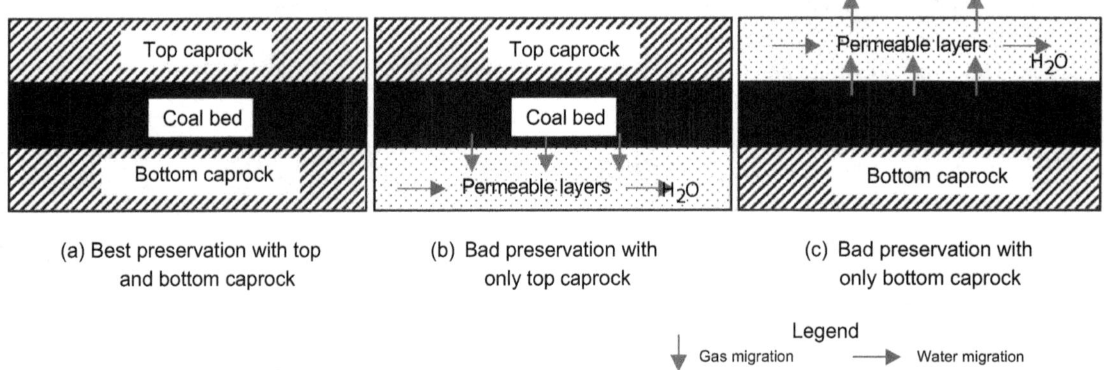

Fig. 5 Relationship between CBM preservation and caprocks

and preservation (Fig. 6). It has been confirmed that most shale gas and coal bed methane reservoirs discovered in China have experienced intense uplift. During the basin evolution, the pressure and temperature increase with time. There were two phases of gas generation and adsorption in most basins with gas generated and stored primarily as an adsorbed phase in the coal seams.

CBM loss is primarily due to tectonic uplift and pressure–temperature changes, which result in desorption of gas. There are three diffusion paths for reservoir gas. First, free gas diffuses by overcoming capillary pressure of sealing rocks (Song et al. 2007). Second, dissolved gas in water diffuses because of a concentration difference. Third, gas is flushed directly by flowing water (Qin et al. 2005). Thus, tectonic evolution, hydrodynamics, and sealing conditions are three major controlling factors for CBM accumulation and enrichment (Song et al. 2012).

CBM reservoir accumulation depends on the preservation conditions resulting from tectonic uplift. The higher the coal seam is uplifted, the poorer the preservation conditions will be. During tectonic uplifting when gas generation ceased, if the coal seam was uplifted to a depth still below the present weathering zone, CBM would be preserved through enhanced adsorption capacity (Song et al.

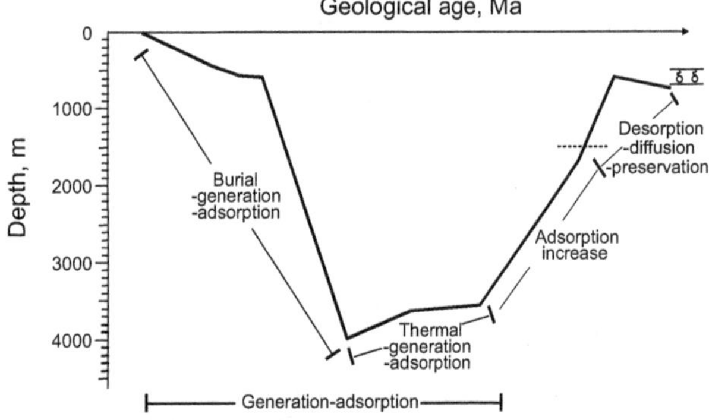

Fig. 6 Accumulation mechanisms of CBM entrapment

2005). The CBM abundance is then dependent on the thickness of the overlying strata. The thicker the overlying strata are, the higher the CBM abundance will be (Fig. 6). The formation of unconventional gas reservoirs is controlled by the key time of structural evolution, which is different from the charge time of the conventional gas.

3.3 Synclinal accumulation of unconventional gas is controlled by water potential and pressure and conventional gas is distributed in structural highs under control of gas potential

Conventional gas is featured by accumulation in structural highs under control of gas potential. Regionally, unconventional gas is characterized by synclinal accumulation mainly controlled by water potential and pressure field. A low potential area enclosed by high potential layers is located in reservoirs. The low potential area with high porosity and permeability is a favorable area for hydrocarbon accumulation and preservation, indicating that the oil and gas potential controls the accumulation of conventional hydrocarbon. Low potential is generally located at structural highs and is the migration direction, so the conventional hydrocarbon mainly accumulates in the anticline structures.

Synclinal accumulation of the unconventional hydrocarbon is a combined result of favorable tectonic evolution, hydrodynamic, and sealing conditions. The synclinal CBM pooling model from the Qinshui Basin in China is illustrated in Fig. 7. For a regional syncline, the surface water may permeate through outcrops near the elevated margins of the syncline and flow downwards to the axis direction due to gravity, forming water seals on both limbs of the syncline by downward water flow, and thus resulting in excellent preservation conditions for CBM. In addition, the central axial area with thick and stable caprocks above is deeply buried and structurally stable and less susceptible to fracturing, which is favorable for preservation of

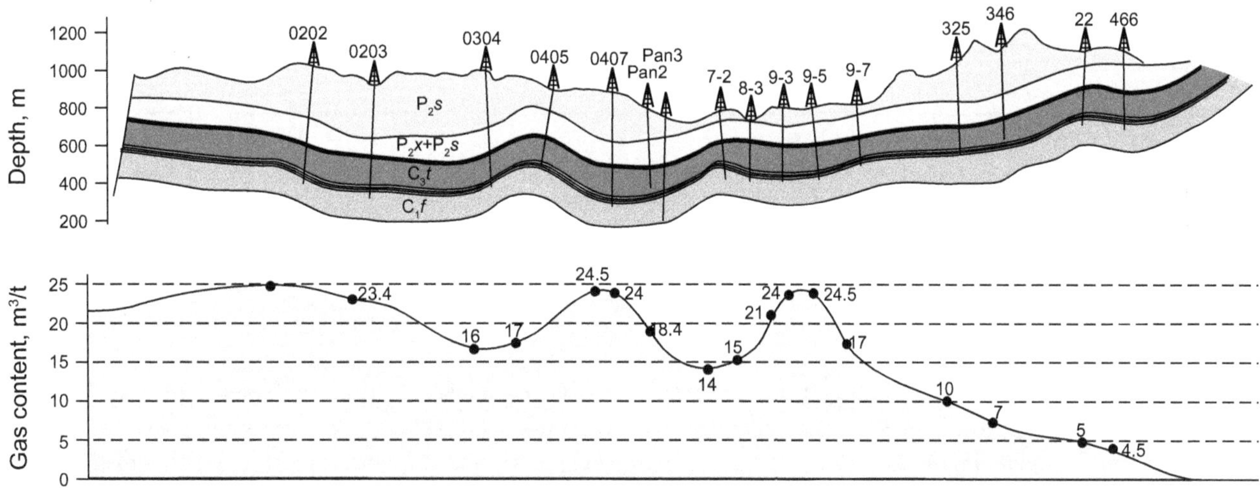

Fig. 7 Structure and gas contents of coal seams in the Jincheng area, Qinshui Basin

overpressure. CBM accumulation is dually controlled by flow potential and pressure potential in the reservoir system. Therefore, the central axial area of a syncline is the most favorable place for CBM accumulation and preservation, usually with the highest abundance and saturation.

4 Distribution characteristics of unconventional hydrocarbon

4.1 Coexistence of unconventional and conventional hydrocarbon

Unconventional hydrocarbon can be generated in different maturation stages of source rocks (reservoirs), and the genetic types of unconventional hydrocarbon are controlled by the evolution process of source rocks and the characteristics of reservoirs. Therefore, different unconventional hydrocarbons can be distributed in an orderly manner with conventional hydrocarbon reservoirs (Zou et al. 2014). Down slope into the basin center, sandstone reservoirs may change to mudstone reservoirs. Vertically, as the burial depth increases, the source rocks become mature or overmature, and generate oil and gas. Meanwhile, the reservoirs also become tight during complex diagenesis processes. Thus, in self-source systems, unconventional hydrocarbon (shale gas, tight gas, shale oil, tight oil and oil shale) and conventional hydrocarbon are always spatially distributed from the deep formations to the shallow formations, characterized by spatial integration and continuity (Fig. 8).

An introduction to an unconventional hydrocarbon accumulation mechanism is provided by comparing its characteristics of pore diameter and the relationship with source rocks (Table 1). Conventional hydrocarbon

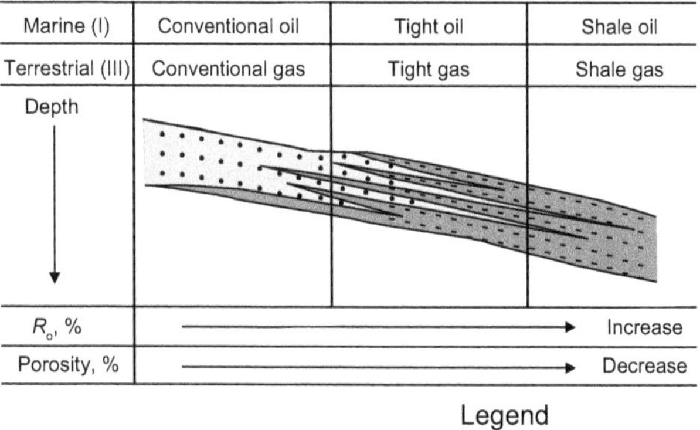

Fig. 8 Coexistence of conventional and unconventional hydrocarbons

accumulation usually refers to an individual hydrocarbon accumulation in a single trap with a uniform pressure system and oil–water contact. In conventional hydrocarbon accumulations, hydrocarbon migration is attributed to effects of gravitational segregation and buoyancy, and fluid flow follows Darcy's law. Conventional hydrocarbon is entrapped individually or sealed in a low potential zone or in a structural trap under impermeable rocks. Tight oil and gas accumulates close to source rocks under control of a pressure difference between source rocks and reservoirs, experiencing primary migration or short-distance secondary migration with the occurrence of free gas (Li et al. 2015; Sun et al. 2014). Shale gas refers to unexpelled gas in shale generated in the mature stage, occurring as adsorbed gas and free gas, and shale gas often changes between the adsorbed state and free state during accumulation, i.e., as the temperature–pressure conditions change, after the fulfillment of self adsorption of shale, free shale gas occurs

Table 1 Accumulation mechanisms of different unconventional hydrocarbons

Type	Conventional oil and gas	Tight oil and gas	Shale oil and gas	Coal bed methane
Pore diameter	$d > 2\ \mu m$	$2\ \mu m > d > 0.03\ \mu m$	$0.1\ \mu m > d > 0.0005\ \mu m$	$d > 2\ \mu m$
Accumulation mechanisms	Long distance migration through preferential pathways, secondary migration	Driven by pressure difference, short-distance migration	Adsorption and free gas, primary migration	Adsorption
Relationship with source rocks	Distant from source rocks	Near source rocks	In source rocks	In source rocks

and migrates inside shales (Curtis 2002; Bowker 2007; Ross and Bustin 2009). CBM is mainly adsorbed gas also characterized by accumulating in the source, and the generated gas is directly adsorbed by coal on the surfaces of pores.

4.2 Distribution models of unconventional hydrocarbon

Three models of unconventional hydrocarbon distribution can be determined in petroliferous basins, namely the intra-source rock model, the basin-centered gas model, and the source rock interlayer model.

4.2.1 The intra-source rock model

Oil shale, shale oil, shale gas, and CBM are accumulated in mudstones, shales, and coal beds, characterized by "self-source and self-reservoir" (Fig. 9). Shale gas generated at mature and overmature stages exists in three forms: (1) free gas in pores and fractures, (2) adsorbed gas in organic matter and on inorganic minerals, and (3) dissolved gas in oil and water (Curtis 2002; Martini et al. 2003; Bowker 2007; Kinley et al. 2008). A strong positive correlation between total organic carbon (TOC) and total gas content shows that the total organic matter content is primarily responsible for shale gas yield. CBM generated during different maturation processes is primary adsorbed in coal beds. The methane adsorption content linearly increases with the increase of TOC and micropore surface area (micropore size < 2 nm), indicating microporosity associated with the organic fraction has a primary control on CBM accumulation. Mudstones with relatively higher capillary pressure on the top and bottom of coal seams are not only advantageous to provide favorable sealing conditions for the free CBM in coal beds, but also favorable for overpressure and adsorbed CBM preservation.

4.2.2 The source rock interlayer model

Oil and gas is expelled from source rocks, migrates within short-distances into the coexisting tight sandstone and carbonate interlayers of the source rocks, and forms tight oil and tight gas, such as the interlayer tight sandstone gas in the Triassic Xujiahe Formation in the west Sichuan foreland basin (Li et al. 2010; Zou et al. 2013) (Fig. 10). The formation of the Xujiahe gas reservoirs is primarily attributed to the pressure gradient from source rocks to interlayer tight sandstone reservoirs. After short-distance migration from source rocks to reservoirs, oil and gas mainly charges the sheet-like interlayer tight sandstone reservoirs, and the source rock interlayer distribution model develops in a large area.

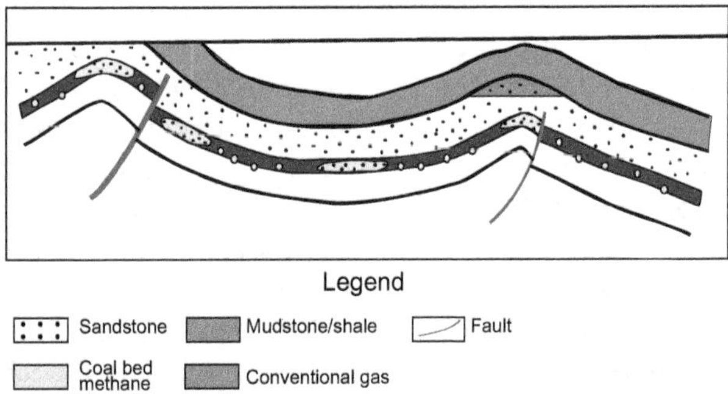

Fig. 9 The intra-source rock distribution model of coal bed methane

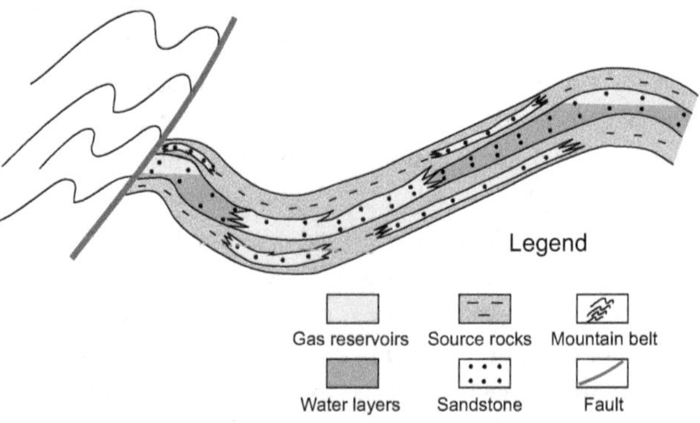

Fig. 10 Distribution model of the tight sandstone gas in the Triassic Xujiahe Formation in Sichuan

4.2.3 The basin-centered gas model

The basin-centered distribution model of tight sandstone gas reservoirs is characterized by regionally pervasive accumulation, abnormal pressure (high or low), an inverse or ill-defined gas–water contact and low-permeability reservoirs. For instance, the Elmworth gas field in the Alberta Basin and the Mesa Verde tight sandstone gas field in the Piceance Basin, overpressure is the primary driving force for hydrocarbon migration from source rocks upwards into tight sandstone reservoirs. Basin-centered tight sandstone gas reservoirs always have low porosity and low permeability, so buoyancy is not the driving force for gas accumulation. Unlike conventional gas accumulation, basin-centered gas reservoirs always show a characteristic of gas–water inversion. Tight sandstone gas reservoirs are widely distributed regionally, covering several thousand square kilometers, and consist of single or isolated reservoirs a few meters thick or vertically stacked reservoirs several thousand meters thick, controlled by structural traps, stratigraphic traps and lithological traps (Fig. 11). Tight sandstone gas reservoirs are gas-saturated with little or no producible water, do not have an obvious trap boundary or intact caprocks, and are downdip from water-bearing reservoirs, widely distributed in deep depressions, central synclines and downdip of structural slopes.

5 Conclusions

(1) The types of unconventional hydrocarbon resources include oil shale, tight oil/gas, shale oil/gas, and CBM. These are controlled by the evolution of source rocks and the combinations of different unconventional reservoirs.

(2) The fundamental differences of unconventional hydrocarbon from conventional hydrocarbon resources are tight reservoir properties, non-buoyancy-driven migration, and no obvious trap boundary. The essential reasons for non-buoyancy-driven accumulation are widespread micro- and nano-scale pores, the resistance of high capillary pressure in tight reservoirs and lack of formation conditions providing strong buoyancy.

(3) The differences in occurrence and accumulation processes between unconventional and conventional hydrocarbon result from the great differences in accumulation mechanisms. For unconventional hydrocarbon, subsurface temperature–pressure fields control the gas content, preservation conditions affect the critical time for hydrocarbon accumulation, and water potential and pressure result in accumulation in synclines. For the conventional hydrocarbon resources, dynamic balance processes

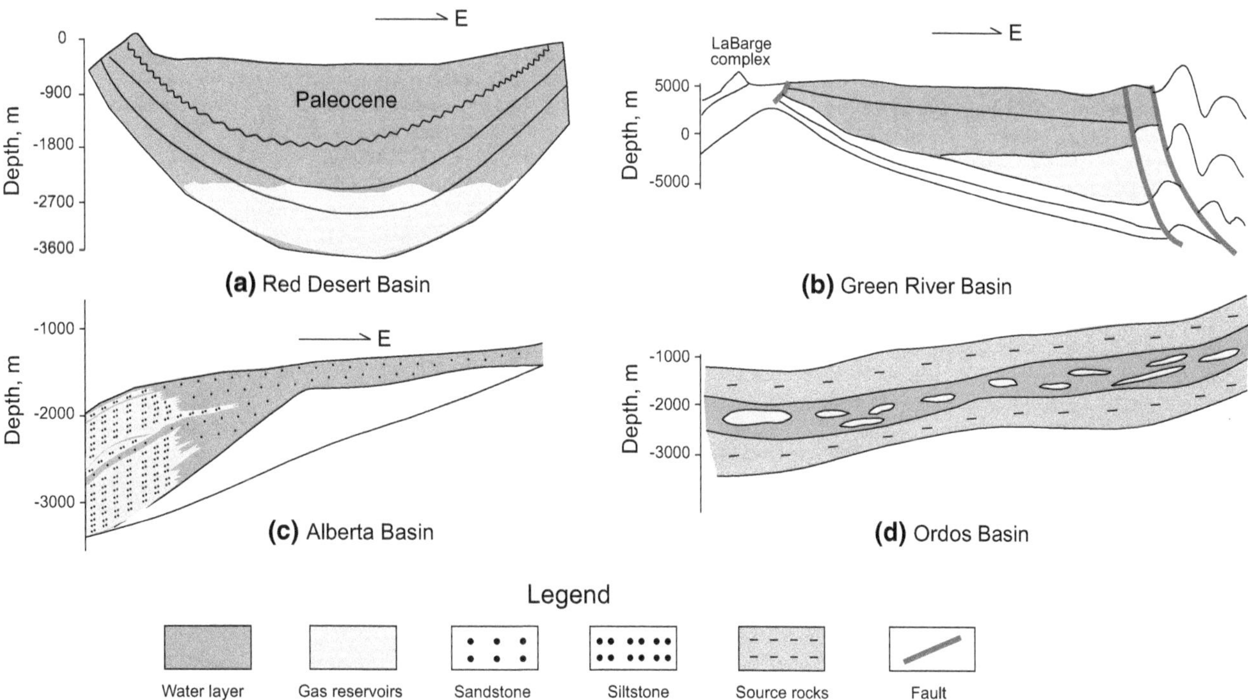

Fig. 11 Basin-centered gas model of unconventional hydrocarbon

control the hydrocarbon accumulation, the best match of reservoir-forming factors and geological events controls the entrapment time, and gas potential controls the accumulation in structural highs.

(4) Unconventional and conventional hydrocarbons coexist and are distributed in an orderly manner in sedimentary basins. The unconventional hydrocarbon has three distribution models, namely the intra-source rock model, the basin-centered gas model, and the source rock interlayer model.

Acknowledgments This research was supported by Major Projects of Oil and Gas of China (No. 2011ZX05018-002). We thank Profs. Zou Caineng, Jiang Zhenxue, and anonymous reviewers for their critical and constructive comments. We also thank Ji Wenming and Xiong Fengyang for improving the English of the manuscript.

References

Bowker KA. Barnett shale gas production, Fort Worth Basin: issues and discussion. AAPG Bull. 2007;91(4):523–33.

Curtis JB. Fractured shale-gas systems. AAPG Bull. 2002;86(11): 1921–38.

Dai JX, Pei XG, Qi HF. Natural gas geology in China, vol. Vol. 1. Beijing: Petroleum Industry Press; 1992. p. 118–31 **(in Chinese)**.

Davis RW. Analysis of hydrodynamic factors in petroleum migration and entrapment. AAPG Bull. 1987;71(6):643–9.

Du JH, He HQ, Yang T. Progress in China's tight oil exploration and challenges. China Pet Explor. 2014;19(1):1–9 **(in Chinese)**.

Guo XS, Hu DF, Li YP, et al. Geological features and reservoiring mode of shale gas reservoirs in Longmaxi Formation of the Jiaoshiba Area. Acta Geol Sinica. 2014;88(6):1811–21.

Huang DF, Li JC, Zhou ZH, et al. Evolution and hydrocarbon generation mechanism of continental organic matter. Beijing: Petroleum Industry Press; 1984 **(in Chinese)**.

Jarvie DM, Hill RJ, Ruble TE, et al. Unconventional shale-gas systems: the Mississippian Barnett shale of north-central Texas as one model for thermogenic shale-gas assessment. AAPG Bull. 2007;91(4):475–99.

Jia CZ, Zou CN, Li JZ, et al. Assessment criteria, main types, basic features and resource prospects of the tight oil in China. Acta Pet Sin. 2012;33(3):344–50 **(in Chinese)**.

Ju YW, Lu SF, Sun Y, et al. Nano-geology and unconventional oil and gas. Acta Geol Sin. 2015;89(s1):192–3.

Kinley TJ, Cook LW, Breyer JA, et al. Hydrocarbon potential of the Barnett Shale (Mississippian), Delaware Basin, west Texas and southeastern New Mexico. AAPG Bull. 2008;92(8):967–91.

Li HB, Guo HK, Yang ZM, et al. Tight oil occurrence space of Triassic Chang 7 Member in Northern Shaanxi Area, Ordos Basin,NW China. Pet Explor Dev. 2015;42(3):434–8.

Li W, Zou CN, Yang JL, et al. Types and controlling factors of accumulation and high productivity in the upper Triassic Xujiahe Formation gas reservoirs, Sichuan Basin. Acta Sedimentol Sin. 2010;18(5):1037–45 **(in Chinese)**.

Liu GD, Sun ML, Zhao ZY, et al. Characteristics and accumulation mechanism of tight sandstone gas reservoirs in the Upper Paleozoic, northern Ordos Basin, China. Pet Sci. 2013;10(4): 442–9.

Loucks RG, Ruppel SC. Mississippian Barnett Shale: lithofacies and depositional setting of a deep-water shale-gas succession in the Fort Worth Basin, Texas. AAPG Bull. 2007;91(4):579–601.

Martini AM, Walter LM, Ku TCW, et al. Microbial production and modification of gases in sedimentary basins: a geochemical case study from a Devonian shale gas play, Michigan Basin. AAPG Bull. 2003;87(8):1355–75.

Montgomery SL, Jarvie DM, Bowker KA, et al. Mississippian Barnett Shale, Fort Worth Basin, north-central Texas: gas-shale play with multi-trillion cubic foot potential. AAPG Bull. 2005;89(2): 155–75.

Nelson PH. Pore-throat sizes in sandstones, tight sandstones, and shales. AAPG Bull. 2009;93(1):329–40.

Passey QR, Bohacs KM, Klimentidis RE, et al. My source rock is now my shale-gas reservoir—Characterization of organic-rich rocks. AAPG Annual Convention. Houston, Texas; 2011.

Qin SF, Song Y, Tang XY, et al. The destructive mechanism of the groundwater flow on gas containing of coal seams. Chin Sci Bull. 2005;50(S1):99–104 **(in Chinese)**.

Ross DJK, Bustin RM. The importance of shale composition and pore structure upon gas storage potential of shale gas reservoirs. Mar Pet Geol. 2009;26(6):916–27.

Schowalter TT. Mechanics of secondary hydrocarbon migration and entrapment. AAPG Bull. 1979;63(5):723–60.

Shanley K W. Fluvial reservoir description for a giant low permeability gas field: Jonah field, Green River Basin, Wyoming, USA. AAPG studies in geology 52 and Rocky Mountain Association of Geologists 2004 Guidebook. 2004; 52:159–82.

Song Y, Jiang L, Ma XZ. Formation and distribution characteristics of unconventional oil and gas reservoirs. J Palaeogeogr. 2013; 15(5):677–86 **(in Chinese)**.

Song Y, Liu H, Hong F, et al. Syncline reservoir pooling as a general model for coalbed methane (CBM) accumulations: mechanisms and case studies. J Pet Sci Eng. 2012;88–89:5–12.

Song Y, Qin SF, Zhao MJ. Two key geological factors controlling the coalbed methane reservoirs in China. Nat Gas Geosci. 2007; 18(4):545–52 **(in Chinese)**.

Song Y, Zhao MJ, Liu SB, et al. The influence of tectonic evolution on the accumulation and enrichment of coalbed methane (CBM). Chin Sci Bull. 2005;50(S1):1–5.

Sun L, Zou CN, Liu XL, et al. A static resistance model and the discontinuous pattern of hydrocarbon accumulation in tight oil reservoirs. Pet Sci. 2014;11(4):469–80.

Sun ZD, Jia CZ. Exploration and development of unconventional hydrocarbon. Beijing: Petroleum Industry Press; 2011. p. 67–70 **(in Chinese)**.

Tissot BP, Welte DH. Petroleum formation and occurrence: a new approach to oil and gas exploration. Berlin: Springer; 1978. p. 23–33.

Wang FP, Reed RM. Pore networks and fluid flow in gas shales. Paper SPE124253 presented at the SPE annual technical conference and exhibition. New Orleans, Louisiana, 4–7 October 2009.

Yang H, Li SX, Liu XY. Characteristics and resource prospects of tight oil and shale oil in Ordos Basin. Acta Pet Sin. 2013;34(1): 1–6 **(in Chinese)**.

Zhang HF, Zhang WX. Petroleum geology. Beijing: Petroleum Industry Press; 1981 **(in Chinese)**.

Zhang H, Zhang SC, Liu SB, et al. A theoretical discussion and case study on the oil-charging throat threshold for tight reservoirs. Pet Explor Dev. 2014;41(3):408–16.

Zou CN. Unconventional hydrocarbon geology. 2nd ed. Beijing: Geological Publishing House; 2013. p. 1–10 **(in Chinese)**.

Zou CN, Gong YJ, Tao SZ, et al. Geological characteristics and accumulation mechanisms of the "continuous" tight gas reservoirs of the Xu2 Member in the middle-south transition region, Sichuan Basin, China. Pet Sci. 2013;10(2):171–82.

Zou CN, Tao SZ, Hou LH. Unconventional hydrocarbon geology. Beijing: Geological Publishing House; 2011. p. 33–56 **(in Chinese)**.

Zou CN, Yang Z, Zhang GS, et al. Conventional and unconventional petroleum "orderly accumulation": concept and practical significance. Pet Explor Dev. 2014;41(1):14–30.

Zou CN, Zhai GM, Zhang GY, et al. Formation, distribution, potential and prediction of global conventional and unconventional hydrocarbon resources. Pet Explor Dev. 2015;42(1):14–28.

Zou CN, Zhu RK, Wu ST, et al. Types, characteristics, genesis and prospects of conventional and unconventional hydrocarbon accumulations: taking tight oil and tight gas in China as an instance. Acta Pet Sin. 2012;33(2):173–87 **(in Chinese)**.

5

A scenario analysis of oil and gas consumption in China to 2030 considering the peak CO_2 emission constraint

Xi Yang[1] · Hong Wan[2] · Qi Zhang[1] · Jing-Cheng Zhou[2] · Si-Yuan Chen[1]

Abstract China is now beginning its 13th five-year guideline. As the top CO_2 emitter, China has recently submitted the intended nationally determined contributions and made the commitment to start reducing its total carbon emissions in or before 2030. In this study, a bottom-up energy system model is built and applied to analyze the energy (mainly coal, oil, and gas) consumption and carbon emissions in China up to 2030. The results show that, the total energy consumption will reach a peak of 58.1 billion tonnes of standard coal and the CO_2 emissions will get to 105.8 billion tonnes. Moreover, in the mitigation scenario, proportion of natural gas consumption will increase by 7 % in 2020 and 10 % in 2030, respectively. In the transportation sector, gasoline and diesel consumption will gradually decrease, while the consumption of natural gas in 2030 will increase by 2.7 times compared to the reference scenario. Moreover, with the promotion of electric cars, the transport electricity consumption will increase 3.1 times in 2030 compared to the reference scenario. In order to fulfill the emission peaking target, efforts should be made from both the final demand sectors and oil and gas production industries, to help adjust the energy structure and ensure the oil and gas supply in future.

✉ Qi Zhang
zhangqi@cup.edu.cn; zhangqi56@tsinghua.org.cn

[1] Academy of Chinese Energy Strategy, China University of Petroleum (Beijing), Changping, Beijing 102249, China

[2] Policy Research Office, China National Petroleum Corporation, Beijing 100007, China

Edited by Xiu-Qin Zhu

Keywords Energy system planning · Oil and gas consumption · China-ESPT model · Carbon mitigation · Transportation sector

1 Introduction

With the targets of the 12th five-year plan being successfully fulfilled, China is now starting its 13th five-year plan. In April 2014, the National Development and Reform Commission of China claimed that it is crucial to correctly handle the relationship between the government and market, also scientific planning objectives should be set in the 13th five-year period. At present, China is facing pressures from both economy development and carbon mitigation. On one hand, China is experiencing a transition to the new normal economy. The economic growth is slowing down, with the GDP growth rate decreasing from 10.4 % in 2010 to 7.4 % in 2014. On the other hand, parallel challenges from carbon mitigation and environmental protection are emerging and drawing the attention of China's government. The emission of sulfur dioxide and carbon dioxide has been strictly controlled since last year, by multiple related regulations. In addition, recently in 2015, China has submitted the intended nationally determined contributions (INDCs), and made the commitment that by 2030, the carbon dioxide emissions per GDP should decrease by 60 %–65 % compared to the level of 2005; China's carbon dioxide emissions should reach a peak in or before 2030. The non-fossil fuel consumption should achieve 20 % of the total primary energy consumption (State Council of China 2015).

According to the BP Statistical Review of World Energy (2015a), China is the world's largest energy consumer, with the total energy consumption of about 4.3 billion tonnes of coal equivalent, which accounts for about 23 %

of the world's total energy consumption. Although coal consumption still accounts for an absolutely dominant proportion, about 66 % of the total energy consumption, oil consumption accounts for 17.5 %, and natural gas accounts for 5.6 % in 2014 (BP 2015a). Plans aiming to adjust China's energy structure are already on schedule. The State Council of China has claimed that by 2020, natural gas consumption is planned to account for 10 % of the primary energy consumption, the proportion of non-fossil fuel consumption would be more than 15 %, and the proportion of coal consumption should be controlled to below 62 % (State Council of China 2014). To successfully achieve the goal, it is urgent to take effective actions to reduce carbon emissions, to fully explore the potential of natural gas and non-fossil energy, and to have clean energy sources as feasible alternatives.

Among all the final demand sectors, the transportation sector, covering road, air, railway, and waterways, is the main sector for oil consumption, responsible for 59 % of the total oil consumption in 2011. According to the prediction of the Organization for Economic Co-operation and Development (OECD) and International Energy Agency (IEA), the oil consumption of the transportation sector will continue to increase, with a predictable rise to 63 % of all oil demand by 2040. For developing countries, the proportion of oil consumption from transportation sector is around 65 % of all the sectors, as shown in Fig. 1. Given the oil consumption for transportation sector, it can be found that less than 40 % of oil consumed in 2040 will be from the other sectors. The petrochemical and other industries account for approximately one-quarter of the total oil consumption in 2040. Residential and agricultural consumption, together with some consumption in the commercial sector, contributes only to 10 % of the total consumption. Thus, the transportation sector is the most critical final demand sector, when we are analyzing future oil and gas consumption in China. Besides, in China's road transport long-term planning, it is clearly claimed that the number of electric vehicles, as well as plug-in hybrid electric vehicles, should reach at least 5 million in 2020. Technological progress plays an important role in promoting the optimization of energy consumption structure, especially fuel standard upgrading and the improvement of natural gas consumption. Thus, in this paper, the effect of technological improvement on the energy consumption of the transportation sector is studied, according to the transportation long-term planning.

During the last decade, the traditional methods for predicting the final energy consumption, especially oil and gas consumption, mainly include time series, regression, econometric modeling, autoregressive integrated moving average model, and soft computing methods, such as fuzzy logic, genetic algorithm, and neural networks, are also being extensively used for final demand side management. In recent years, the support vector regression, ant colony, and particle swarm optimization have been new methods adopted for energy demand studies. Besides, bottom-up models such as MARKet ALlocation (MARKAL; energy technology systems analysis program, ETSAP 2013) and the Long-range Energy Alternatives Planning System (Stockholm Environment Institute, SEI 2013) are also being used at the national and regional levels for final energy demand analysis. There have been studies on the consumption of conventional energy in India, and one of them is carried out based on three time series models, namely the Grey–Markov model, the Grey model with a rolling mechanism, and singular spectrum analysis (Kumar and Jain 2010). Besides, a logistic function is used to characterize the peak and ultimate production of global crude oil and petroleum-derived liquid fuels (Gallagher 2011). Using a curve-fitting approach, a population-growth logistic function was applied to complete the cumulative production curve. An idealized Hubbert curve is defined as having properties of production data resulting from a constant growth rate under fixed resource limits.

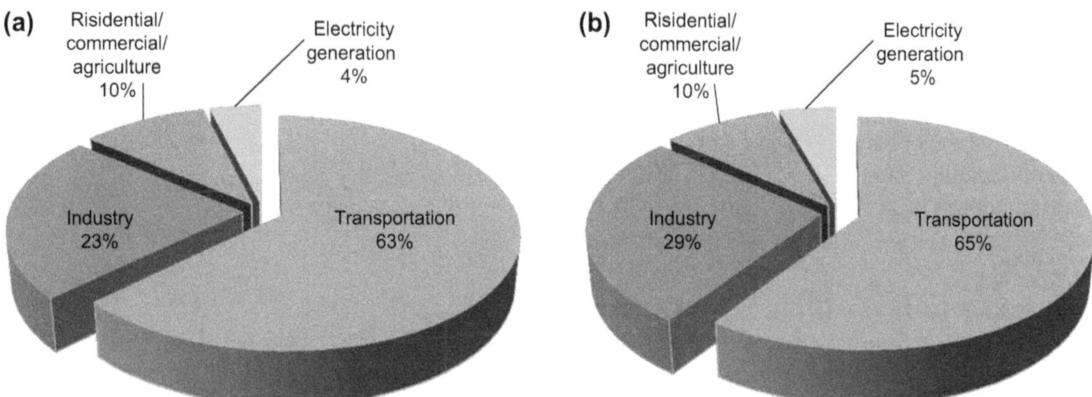

Fig. 1 Percentage of oil demand by sector in 2040, for the world and developing countries (%). *Data source* OECD/IEA Energy Balances of OECD/Non-OECD Countries (2013) and OPEC Secretariat calculations

A bottom-up model can forecast future energy supply and demand, and analyze the impacts of the energy system on the environment, using various engineering technology models and by simulating the modes of energy consumption and production. The typical representatives of the bottom-up model are the MARKAL, EFOM, and the integrated MARKAL–EFOM system (TIMES; ETSAP 2013). Many studies have been undertaken in recent years to evaluate energy system planning, and these works offer good references for the possible future development pathways (Adams and Shachmurove 2008; Kypreos and Bahn 2003; Cox et al.1985; Lund 2010; Manne et al. 1995; Krewitt and Nitsch 2003; Bollen et al. 2009; Budzianowski 2012; Gillingham et al. 2008; da Graça Carvalho 2012; Messner and Schrattenholzer 2000; Klaassen and Riahi 2007; Klaassen et al. 2004; Nordhaus 1993; Paltsev et al. 2005; Tu et al. 2007; Rafaj and Kypreos 2007). For example, the US Energy Information Administration (EIA) presents long-term annual projections of energy supply, demand, and prices through 2040 based on results from EIA's national energy modeling system (2015). According to the BP Energy Outlook 2035 (2015b) and Joint IEA–IEF–OPEC report (OPEC 2015), global demand for energy is expected to rise by 37 % from 2013 to 2035, with an average increase of 1.4 % per year. Demand for the world's natural gas will grow faster among the fossil fuels over the period up to 2035, with a rate of increase of 1.9 % per year, driven mainly by the high demand from Asia. Meanwhile, the IEA declares that the energy use worldwide is expected to grow by one-third to 2040 in their central scenario, and the change of China's energy demand is estimated to be over 900 million tonnes of oil equivalent by 2040, with a total energy demand of nearly 4000 million tonnes of oil equivalent worldwide. Moreover, the natural gas demand of Asia is claimed to increase by approximately 700 billion m^3 by 2040 (IEA 2015). As well, the Grantham Institute for Climate Change at Imperial College London analyzed China's energy system based on the sector-specific and energy consumption bottom-up model for the transport, buildings, and industrial sectors based on the IIASA mix scenario and the IIASA HCB scenario (Gambhir et al. 2013). Besides, studies about China's energy system optimization have been carried out based on enenrgy model analysis (Jiang et al. 2010; Yin and Chen 2013; Cai et al. 2008; Dai et al. 2011; Zhang et al. 2011; Zhou et al. 2013a,b; De Laquil et al. 2003; Chen et al. 2007).

The paper mainly focuses on analysis of oil and gas consumption in future, the effect of road transport technology improvement on fuel consumption, and the optimization of China's energy system. Quantitative analysis has been carried out based on the China energy system planning and technology (China-ESPT) evaluation model. In Sect. 1, the background of the 13th five-year plan and the critical role of the transportation sector for oil and gas consumption are stated, with a literature study. In Sect. 2, the China-ESPT model is introduced for the structure, the main parameters, and the scenario design. Section 3 shows the main results based on scenario analysis, mainly focusing on the change of the energy system and the transportation sector. Section 4 concludes the analysis and further discusses relevant recommended policy.

2 Methodology

In this study, we introduce the China-ESPT evaluation model, an energy optimization model based on TIMES-VEDA (ETSAP 2013), and review a large amount of literature to quantitatively analyze and predict energy consumption up to 2030. Furthermore, the China-ESPT model is based on the reference energy system (RES) to describe the process of energy extraction, energy conversion, and energy distribution. Also RES is helpful for better describing the main characteristics, complex internal relations, and external constraints of the energy system. The model is driven by the future energy demand, considering the capacity of the related equipment, the characteristics of the future optional technology, and the constraints of the current and future energy supply. The model is based on the simulation of investment and operation, primary energy supply, and greenhouse gas emission reduction constraints. Scenario analysis is an important and widely used tool for the long-term simulation of energy and economic systems.

The objective function of this model is to minimize the total cost of the energy system while meeting the energy demand and constraints. The investment and dismantling costs are transformed into streams of annual payments, computed for each year of the horizon. Using China-ESPT, a total net present value (NPV) of the stream of annual costs is then computed, discounted to a user-selected reference year. The objective function of the model is listed in Eqs. (1)–(2):

$$\mathrm{OBJ}(z) = \sum_{r \in \mathrm{REG}} \mathrm{NPV}(z, r), \tag{1}$$

$$\mathrm{NPV}(z, r) = \sum_{y \in \mathrm{YEARS}} (1 + d_{r,y})^{\mathrm{REFYR}-y} \times \mathrm{COST}(r, y) - \mathrm{SAL}(z, r), \tag{2}$$

where NPV is the net present value of the total cost for all regions, $d_{r,y}$ is the general discount rate, REFYR is the reference year for discounting, YEARS is the set of years for which there are costs, R is the set of regions in the area of study, COST(r, y) is the total annual cost in region r and year y, and SAL is the residual value of the assets. The total cost includes several elements: the capital costs incurred for investment, fixed and variable annual operation and maintenance costs, the fuel cost for domestic production and import, transport cost, tax and subsidies, revenues from export, and so on.

The main constraints of the model mainly consist of the constraints of span-time capacity transfer, technical activities, flow balance, etc. First, the equation of capacity transfer balance is shown in Eq. (3):

$$\mathrm{CAPT}(r,\ t,\ p) = \sum_{t_n} \mathrm{NCAP}(r,\ t_n,\ p) + \mathrm{RESID}(r,\ t,\ p), \tag{3}$$

$$t - t_n < \mathrm{LIFE}(r,\ t_n,\ p),$$

where CAPT(r, t, p) is the total installed capacity of technology p, in region r and period t, NCAP(r, t, p) stands for new capacity addition (investment) for technology p, in period t and region r, and RESID(r, t, p) is the (exogenously provided) capacity of technology p due to investments that are made prior to the initial model period and that still exists in region r, at time t. The total available capacity for each technology p, in region r, in period t (all vintages), is equal to the sum of investments made by the model in the past and current periods, and whose physical life has not yet ended, plus capacity in place prior to the modeling horizon that is still available.

The equation of technical activities constraint is given in Eq. (4):

$$\mathrm{ACT}(r,\ v,\ t,\ p,\ s) = \sum_{c} \mathrm{FLOW}(r,\ v,\ t,\ p,\ c,\ s) / \mathrm{ACTFLO}(r,\ v,\ p,\ c), \tag{4}$$

where ACT(r, v, t, p, s) is the activity level of technology p, in region r and period t (optionally vintage and time-slice s), FLOW(r, v, t, p, c, s) is the quantity of commodity c consumed or produced by process p, in region r and period t (optionally with vintage v and time-slice s), and ACTFLO(r, v, p, c) is a conversion factor (often equal to 1) from the activity of the process to the flow of a particular commodity. As shown in Eq. (2), the quantity of commodity c consumed or produced equals the activity level of technology p times a conversion factor.

The equation of use of capacity is given in Eq. (5):

$$\mathrm{ACT}(r,\ v,\ t,\ p,\ s) = \mathrm{AF}(r,\ v,\ t,\ p,\ s) \times \mathrm{CAPUNIT}(r,\ p) \times \mathrm{FR}(r,\ s) \times \mathrm{CAP}(r,\ v,\ t,\ p), \tag{5}$$

where AF(r, v, t, p, s) is the availability factor, which serves to indicate the nature of the constraint as an inequality or equality. CAPUNIT(r, p) is the conversion factor between units of capacity and activity (often equal to 1, except for power plants). FR(r, s) parameter is equal to the duration of time-slice s, such as the proportion of daytime and nighttime. CAP(r, v, t, p) constraints mainly describe the relationship between the capacities of technology p with its activity level.

For the energy system analysis, first, the production of oil and gas must meet the demand for energy in future, and the development of the future energy system needs to meet the requirements of the low-carbon development path. Second, from the point of view of the supply of oil and natural gas, there are many factors influencing the energy production, such as exploitation of China's energy reserves, import–export ratio, unconventional gas development, etc. In addition, two factors determining the energy supply curve, which are average production and transportation costs, should be taken into account. Third, it is significant to set different scenarios according to different energy plans.

The model (China-ESPT model) is based on full economy sectors, including resource supply, electricity generation sector, transportation sector, industry sector, building sector, and other sectors. The electricity generation sector in this paper mainly focuses on power generation technology, nuclear generation technology, and renewable-energy generation technology. Coal and gas power generation are the main constituents of the power generation technology. In this model, the coal generation includes the traditional subcritical generation technologies, circulating fluidized bed furnace, and supercritical and ultra-supercritical technologies. In addition, the integrated gasification combined cycle (IGCC) technology and cogeneration technologies are also considered.

The improvement of gas power generation technology is mainly based on gas-boiler and gas-steam combined cycle units. The gas-based power generation technologies are of higher efficiency and more environmental friendly than coal-based power generation technologies. The investment cost of gas boiler and combined cycle is much lower than that of the coal-fired generation unit. The cost of a gas boiler plant is around 1350–1780 RMB/kW, and the investment cost of natural gas-fired combined cycles (NGCCs) is around 3700–4230 RMB/kW on average, which is far below the cost of coal-fired units. Besides, the power generation based on biomass sources mainly includes biomass direct combustion power generation and biomass gasification power generation technology. The average efficiency and cost of power generation technologies are summarized in Table 1.

As the main oil product consumption sector, and also the main end-use sector of oil and gas products, the transportation sector is further studied in this paper. For the transportation sector, the final demand is related to the passenger or freight task and the fuel efficiency of vehicles.

Table 1 Efficiency and cost of power generation

Power generation	Technology	Efficiency, %	Investment per unit, RMB/kW
Coal-based	Ultra-supercritical (U-SUC)	45–47	3600–3800
	Supercritical (SUC)	41–42	3700–3850
	Subcritical	38–39	4400–4600
	Coal circulating fluidized bed	35–40	4500–6000
	Integrated gasification combined cycle (IGCC)	40–43	8000–10,000
Gas-based	Gas boiler	Below 35	3100–3400
	Natural gas combined cycle (NGCC)	55–67	3282–3350
Oil-based	Oil boiler	Below 25	3300–3500
	Oil circulating fluidized bed	37	4500–6000
Biomass	Biomass-fired	23	6500–8500
	Biomass gasification	36	6500–12,000

In the model structure, the output is calculated in Eqs. (6) and (8), with the total demand of the transportation sector shown in Eqs. (7) and (9):

$$\mathrm{TUR\_}P_i = \sum_m \mathrm{AR\_}P_{i,m} \times \mathrm{STOCK\_}P_{i,m} \times \mathrm{LOAD\_}P_m, \tag{6}$$

$$\mathrm{DEM\_}P_i = \sum_m \mathrm{FEUL}_{i,m} \times \mathrm{TUR\_}P_{i,m}, \tag{7}$$

$$\mathrm{TUR\_}F_i = \sum_m \mathrm{AR\_}F_{i,m} \times \mathrm{STOCK\_}F_{i,m} \times \mathrm{LOAD\_}F_m, \tag{8}$$

$$\mathrm{DEM\_}F_i = \sum_m \mathrm{FEUL}_{i,m} \times \mathrm{TUR\_}F_{i,m}, \tag{9}$$

where $\mathrm{TUR\_}P_{i,m}$ is the passenger task of the transport mode m in year i and $\mathrm{TUR\_}F_{i,m}$ stands for the freight task of the transport mode m in year i. $\mathrm{AR\_}P_{i,m}$ and $\mathrm{AR\_}F_{i,m}$ are, respectively, the average annual transport distance of transport mode m in year i for the passenger and freight transport. $\mathrm{STOCK\_}P_{i,m}$ and $\mathrm{STOCK\_}F_{i,m}$ are the stock of vehicles in year i for different modes m. $\mathrm{LOAD\_}P_m$ and $\mathrm{LOAD\_}F_m$ are, respectively, the carrying capacity for different kinds of vehicles m, for the passenger transport, where the carrying capacity is the number of persons, and for the freight transport, where the carrying capacity is expressed in tonnes of freight. $\mathrm{FEUL}_{i,m}$ is the fuel efficiency of the regular vehicles per kilometer, with the unit of L/km, and $\mathrm{DEM\_}P_i$ and $\mathrm{DEM\_}F_i$ are the final demand of the transportation sector, with the unit of million tonnes of coal equivalent. Among them, the data of the traffic department are mainly from the China Statistical Yearbook on Transportation (China Transportation Association 2011), the data of carrying capacity of the vehicle are obtained from Compilation of National Highway and Waterway Transportation Survey Data of China (China Ministry of Transport 2010), and the data of the average mileage of vehicles are from the setting of the GREET model (Argonne National Laboratory 2015).

According to different transport modes m, the transport sector includes passenger transport and freight transport, further divided into road transport, railway transport, water transport, air transport, and other transport. The road transport includes two- and three-wheeled small vehicles, ordinary vehicles, buses, trucks, and other transport modes. Furthermore, in this model, the road transportation technologies are defined referring to both different fuel consumption and the national emission standards. The main modes and technologies of transportation sector are listed in Table 2.

On the premise of meeting energy demands, the China-ESPT model aims at system cost optimization, the future development of oil and gas, as well as the proportion changes of application in some typical terminal sectors. The main economic assumptions for the model are shown in Table 3.

3 Results and analysis

3.1 Energy consumption and effect of key policies in the mitigation scenario

First of all, we set the mitigation scenario referring to the objectives of the "Thirteenth Five-Year Plan," including total targets and branch targets, and the long-term development plan for different sectors. The main constraints include the following: (1) the CO_2 emission intensity per GDP in 2020 should decrease 40 %–45 % compared to the year 2005; (2) the proportion of the non-fossil energy accounted in primary energy consumption should rise to 15 %; (3) according to the latest submitted INDC, the carbon dioxide emissions per GDP should decrease by 60 %–65 % and also start to decrease in 2030. Based on

Table 2 Main transport modes and technologies of the transportation sector

Transport modes	Modes and technologies	Final demand	Units
Road transport	Two- and three-wheeled small vehicles (oil product/electricity)	Passenger/freight task	MPkm/MTkm
	Vehicles (gasoline/diesel/hybrid/plug-in hybrid/electricity/fuel cell/others)	Passenger task	MPkm
	Buses (gasoline/diesel/natural gas/hybrid/electricity/others)	Passenger task	MPkm
	Trucks (mini truck/small truck/medium truck/heavy truck)	Freight task	MTkm
Railway transport	Railway (passenger/freight)	Passenger/freight task	MPkm/MTkm
Water transport	Shipping (passenger/freight)	Passenger/freight task	MPkm/MTkm
Air transport	Airplane (passenger/freight) aviation kerosene	Passenger/freight task	MPkm/MTkm
Others	Pipeline and others	Freight task	MTkm

The gasoline- and diesel-based transport modes are further grouped by the national emission standard of China (GB18352.3-2005: limits and measurement methods for emissions from light-duty vehicles)

Table 3 Assumptions of economy development for China-ESPT model

	2010	2020	2030
Population, millions	1360	1440	1470
GDP per capita (10^4 RMB/capita, 2010 price level)	2.95	5.74	9.88
Urbanization ratio, %	51.1	58.2	67.1

the above targets, we compare carbon dioxide emission reduction between the reference scenario and the mitigation scenario as shown in Table 5.

In the mitigation scenario, carbon dioxide emissions will peak in 2030 and then decrease to achieve a significant reduction in total emission. Compared to the reference scenario, the total reduction of carbon dioxide emissions will reach 13 million tonnes, which is a decrease from 118.8 million tonnes of emissions for the reference scenario to 105.8 million tonnes, with an average decrease rate of around 2.4 %. The average annual growth rate of carbon emissions before 2030 is about 1.5 %, which is 0.6 % lower than the reference scenario (Table 4).

Moreover, the emissions per GDP in 2020 is significantly lower than that in 2005, decreasing by 46 %–50 % in 2020, which is consistent with the target of carbon emission intensity in 2009; and the emissions per GDP will continue to decline by 30 %–40 % until 2030 on the basis of 2020. Compared to 2005, emission intensity per GDP will decrease to about 60 %–68 %, and the decline of the carbon emissions in 2020 is between 40 % and 45 %. The reduction of carbon emissions in 2030 is in accordance with the targets in China's latest submitted INDC documents. Based on the mitigation scenario analysis, the commitments are achievable in 2030 for China.

The main primary energy consumption is shown in Figs. 2 and 3 which show that China's total energy consumption will reach 4.95 billion tonnes of standard coal and 5.81 billion tonnes of standard coal, respectively, in 2020 and 2030.

In Fig. 3, the primary energy consumption structure of the mitigation scenario is compared with that of the reference scenario. After 2020, the coal-dominant energy structure in the process of electric power production and industry sectors will be improved gradually, but the consumption of fossil fuels will still account for a significant proportion. The oil consumption accounts for about 23 % in 2020 and about 25 % in 2030, on average 3 % lower than that in the reference scenario. Non-fossil energy accounts for more than 15 % after 2015 and will reach more than 20 % in 2030.

As a relatively clean energy, the future consumption of natural gas will significantly increase. It is estimated that the gas consumption will grow up to 304.4 billion m^3 by 2020, and increase to 483.3 billion m^3 by 2030, due to the structural adjustment of the power generation sector and industry sector. In the mitigation scenario, the proportion of natural gas will increase to 7 % in 2020 and to 10 % in 2030. However, the growth rate of the use of natural gas will slow down after 2030. The main reasons include the

Table 4 Comparison of carbon dioxide emissions between the reference scenario and the mitigation scenario (million tonnes)

Years	2010	2015	2020	2025	2030	2035	2040
Reference scenario	78.36	89.32	108.77	112.15	118.84	122.59	128.74
Mitigation scenario	78.36	89.46	104.37	105.42	105.77	99.65	99.63

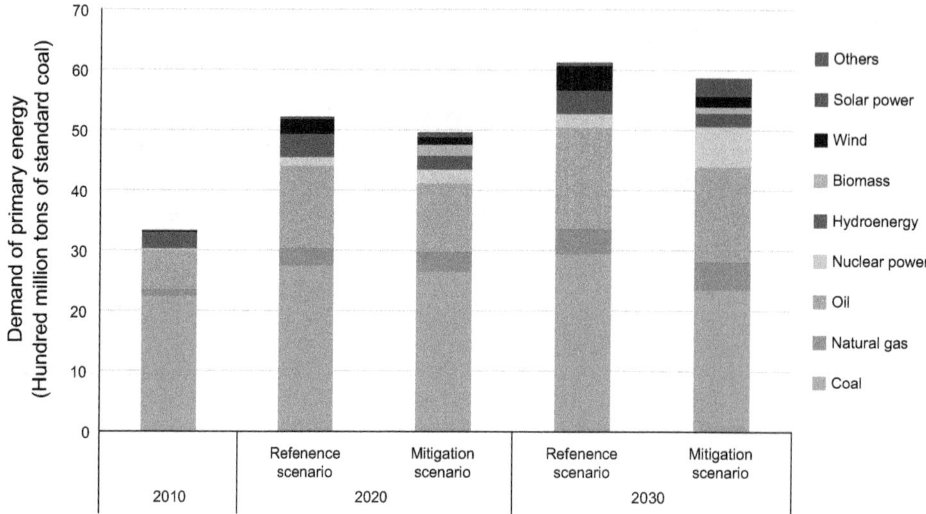

Fig. 2 Comparison of the primary energy consumption between the reference scenario and the mitigation scenario

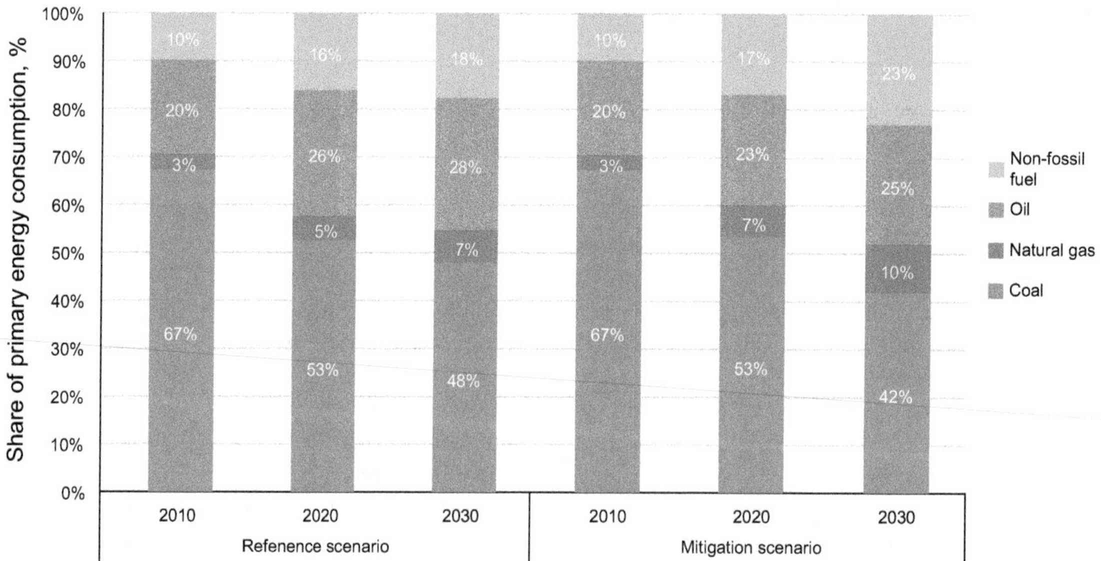

Fig. 3 Comparison of structure of the primary energy between the reference scenario and the mitigation scenario

following: first, the amount of China's natural gas imports continues to increase, while the growth rate is slowing down; second, the new pipelines being put into operation are limited by the resources and efficiency; third, the use of gas in some industries is especially sensitive to the fluctuation of gas prices; last but not least, the energy demand increase is also gradually slowing down.

In order to analyze the effect of key policies, we take the power generation sector as an example. In the power generation sector, several technology options are adopted in the mitigation scenario. Firstly, the cleaner use of fossil fuels will be promoted. All the newly installed coal-based power plants after 2020 will be supercritical, ultra-supercritical, or IGCC power generation technologies. Natural gas power generation technologies should be applied in an appropriate place and act as an important back-up technology for intermittent generation technologies. About 90 % of coal power plants and 80 % of natural gas power plants will be equipped with carbon capture and storage facilities. Secondly, the renewable power generation technologies will become more cost competitive in the future and enjoy a rapid growth. The generation cost of wind power and solar photovoltaic technologies will decrease gradually and will be lower than that of fossil fuel power generation technologies in the future. The biomass power generation will not see an apparent growth after 2020 due to limited resources. Next, the nuclear power will see a large increase and all new nuclear plants should be built in accordance with the third-generation standard.

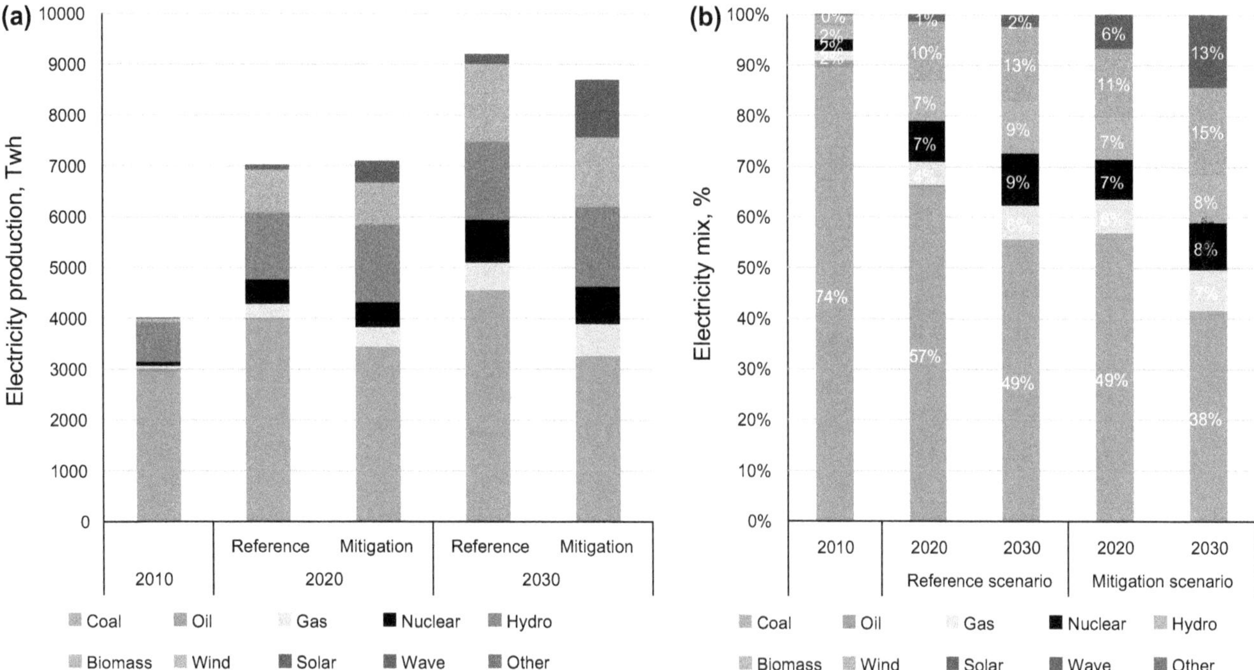

Fig. 4 Electricity production (TWh) and electricity mix (%) during 2010–2030

The most important measure for the carbon emission reduction will be the extension of the use of natural gas, renewable, and nuclear energy, replacing a large amount of coal-fired power generation. Also, the high-percentage combination of carbon capture facilities with coal-fired and gas-fired power generation units also has an effect on the carbon mitigation. The electricity generation and structure are shown in Fig. 4. First, due to energy conservation measures in the final demand sectors, the total electricity needed for the final demand in 2030 is about 8692 TWh, which is 2.17 times of that in 2010, and a 6.78 % reduction compared to the reference scenario in 2030. Second, for the structure of electricity production, the proportion of power generated by fossil energy will reach to 45 % in 2030, with a 31 % reduction compared to that in 2010, and a 10 % reduction compared with the reference scenario in 2030. Third, although the total power generation ratio is decreasing, the relevant clean gas-fired power generation shows an increasing trend. Gas-fired power generation will reach 6 % in 2020 and continue to increase to 7 % in 2030 in the mitigation scenario, compared to 2 % in 2010, 4 % in 2020, and 6 % in 2030 of the reference scenario.

3.2 Impact of technological improvement on the transport sector

According to the model, the main fuel consumption of the transportation sector is still oil products. The consumption of gasoline and diesel accounts for 74 % of the total fuel consumption in 2030, decreasing from 79 % in 2010. The total fuel consumption of transport sector will increase to 397 million tonnes of standard coal in 2030, while it is 272 million tonnes in 2010.

Based on the increase of electric vehicles and hybrid electric vehicle, which is estimated to reach 5 million in 2020, and the improvement of fuel economy, the change of main oil products consumption in the transport sector is studied using the model. Also, the contribution of different transport modes to passenger and freight service is calculated with the model. At the same time, the model takes into consideration the full life cycle of the main transport mode of China. The average life expectancy of the vehicles is 16.5 years, while a truck's average life expectancy is 11 and that of a train is 19. In addition, the annual inventories of vehicles must meet the demand for passenger and freight service, which can be proved by the average carrying rate and annual average driving distance of all transport modes.

Apart from the common hybrid technology, pure electric vehicles, plug-in hybrid vehicles, and fuel-cell vehicles are also considered. Due to the limitations of technical maturity, costs, infrastructure, and other factors, the future development of these technologies is full of uncertainty. From the perspective of technology costs, the model assumes that the annual costs of conventional vehicles will remain constant during the calculation period, and costs of hybrid electric vehicles will annually reduce by 2.5 % on average, while costs of electric vehicles and fuel-cell vehicles will reduce by 4.5 %.

First, for the road transportation, this paper mainly considers three different modes, namely light-duty vehicles

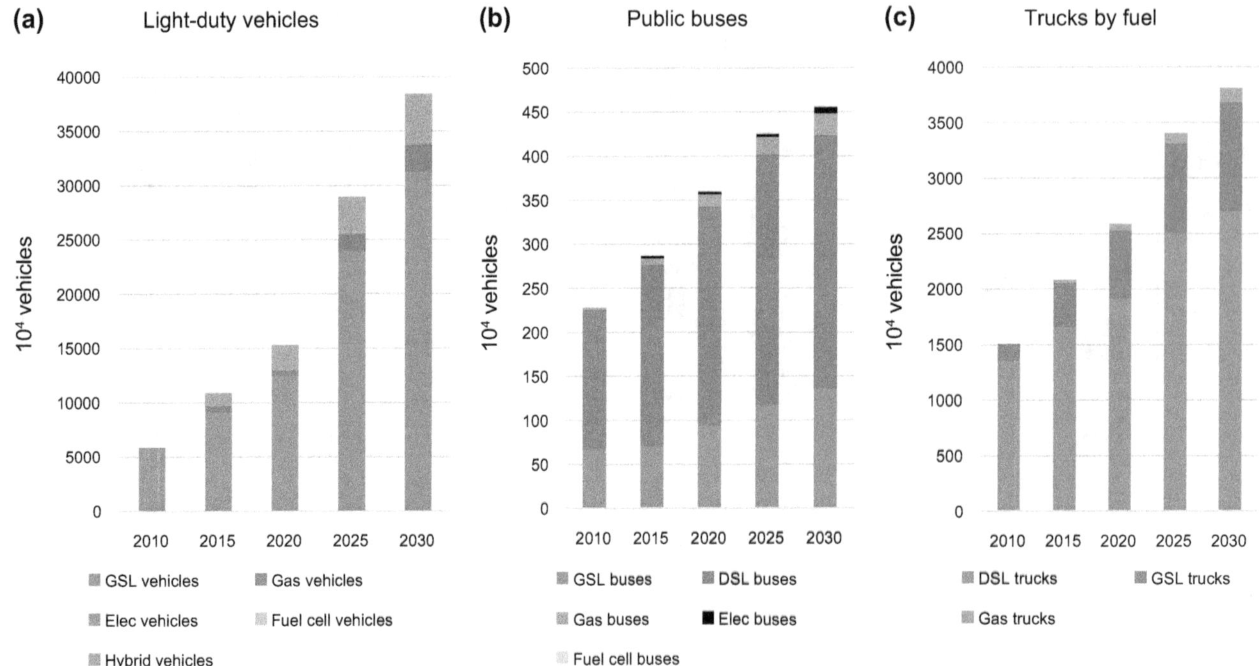

Fig. 5 Numbers of light-duty vehicles, public buses, and trucks during 2010–2030 (10^4 vehicles) (*DSL* diesel, *GSL* gasoline)

(LDVs), public transportation for passengers (mainly buses), and freight transportation (which is mainly by truck). In the reference scenario, when focusing on road transportation, the main vehicle stocks are shown in Fig. 5. Among the private cars, gasoline cars are still the dominant vehicles with 58.1 million vehicles in 2010 and will increase to 311.7 million vehicles in 2030. The gas-fuel vehicles also will increase from 0.19 million in 2010 to 24.5 million in 2030. The electric vehicles in the reference scenario will increase to 3.5 million in 2030, while it is 5.1 million in the technological improvement scenario. For the public buses, generally, the gasoline buses account for the dominant proportion and will increase from 0.67 million in 2020 and to 1.35 million in 2030. The expansion of gas buses and electric buses in recent years cannot be ignored, and the number will increase to 253,000 and 70,000, respectively, in 2030 according to the current policies. The trucks are mainly diesel based, with the diesel trucks increased from 13.6 million in 2010 to 27.1 million in 2030, while the proportion of diesel is decreasing from 90.3 % in 2010 to 70.9 % in 2030. At the same time, the number of gas trucks will increase to 1.30 million with the proportion of 3.7 % in 2030.

When focusing on different modes of road transport, besides the vehicle stocks, the contribution of different vehicles to the road passenger turnover task or freight turnover task is much more important when analyzing the final demand of the transportation sector. In the technological improvement scenario, the fuel efficiency is supposed to improve according to the China national standard, and the electric vehicles stock is encouraged to be more than 5 million with China national subsidy policies. Besides, the elimination of old vehicles and the technological improvement of new vehicles according to their life expectancy are also considered.

For LDVs, the main technology structure and the share rate of passenger transport of the passenger car are shown in Fig. 6. It can be seen that the proportion of the hybrid vehicles takes second place (gasoline car takes first place), with the contribution to the passenger service being about 17.3 % in 2020, and further expanding to 18.9 % in 2030. On the other hand, the electric vehicles and fuel-cell vehicles only account for 2.1 % in 2020 for the reference scenario and 4.3 % for the technological improvement scenario.

From the point of view of road passenger transport (mainly buses), the main technologies used and the contribution of different fuel-based buses are shown in Fig. 7. The main fuels of road passenger transport include diesel and gasoline. The proportion of buses based on natural gas is gradually increasing as the technology improves, accounting for approximately 8.7 % in 2020 and further increasing to 11.2 % in 2030. In addition, the proportion of electric buses is also gradually growing, reaching 3.5 % in 2020 and further rising to 4.6 % in 2030.

For the road freight transport, the main technologies and the contribution of various trucks to freight transport are shown in Fig. 8. The road freight mode is divided into mini trucks, small trucks, medium trucks, and heavy trucks, and the main fuel consumed is diesel, especially for the

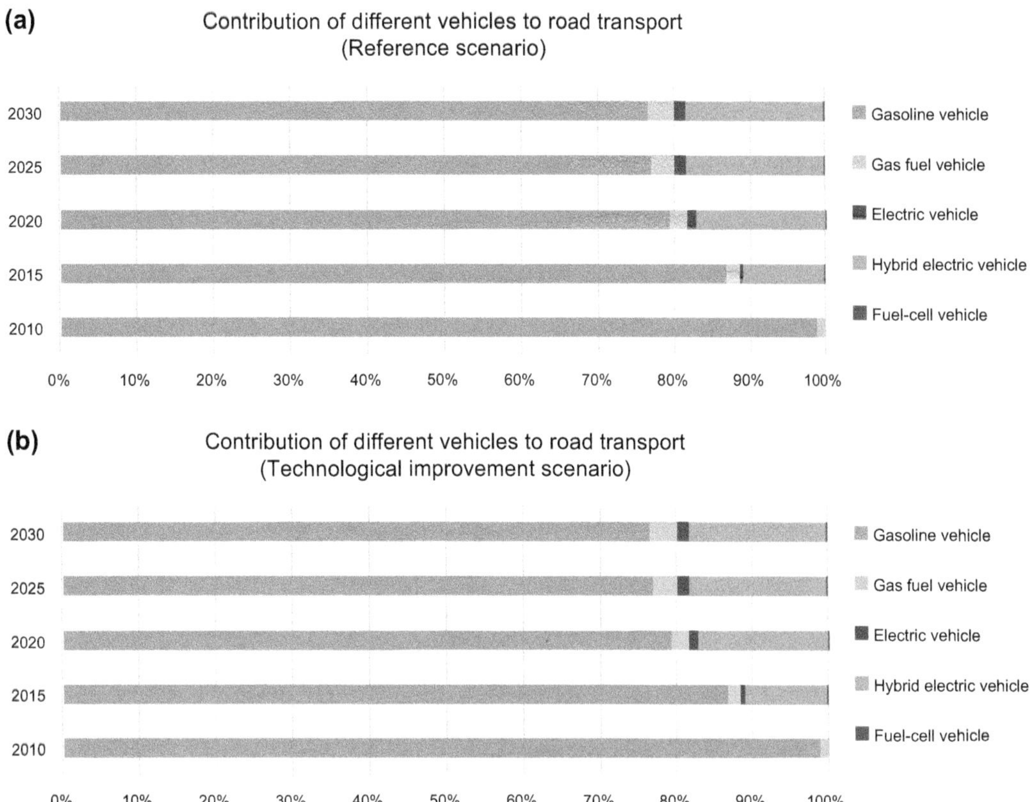

Fig. 6 Contribution of different vehicles to road passenger transport

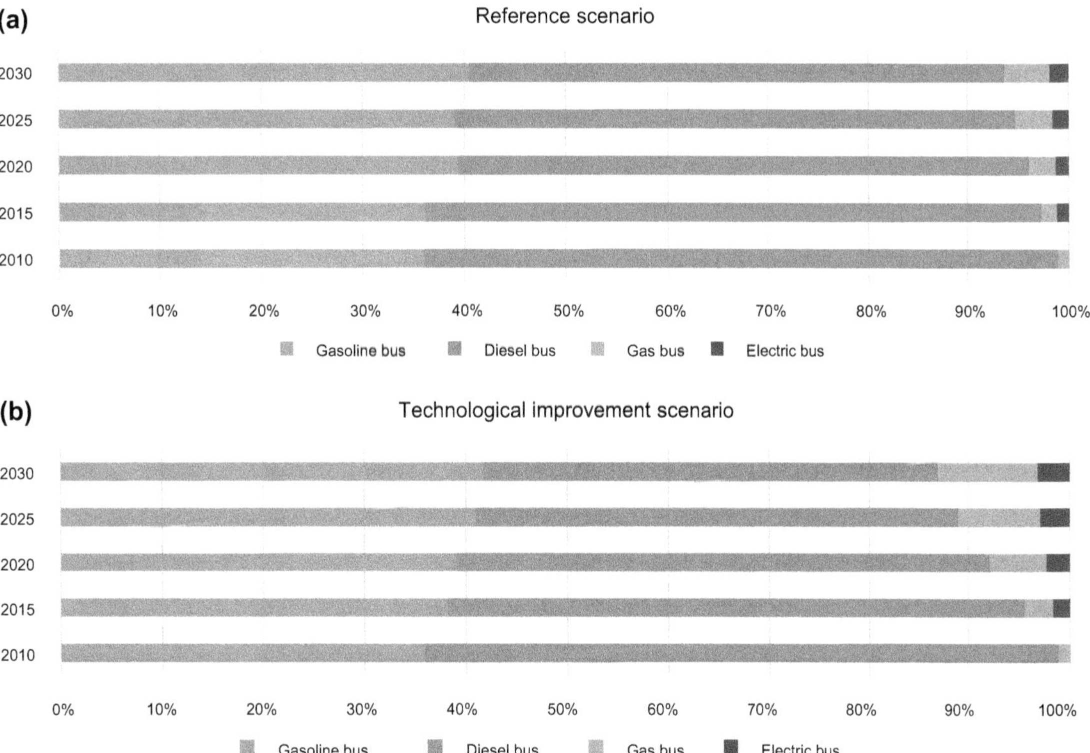

Fig. 7 Contribution of different buses to road passenger transport

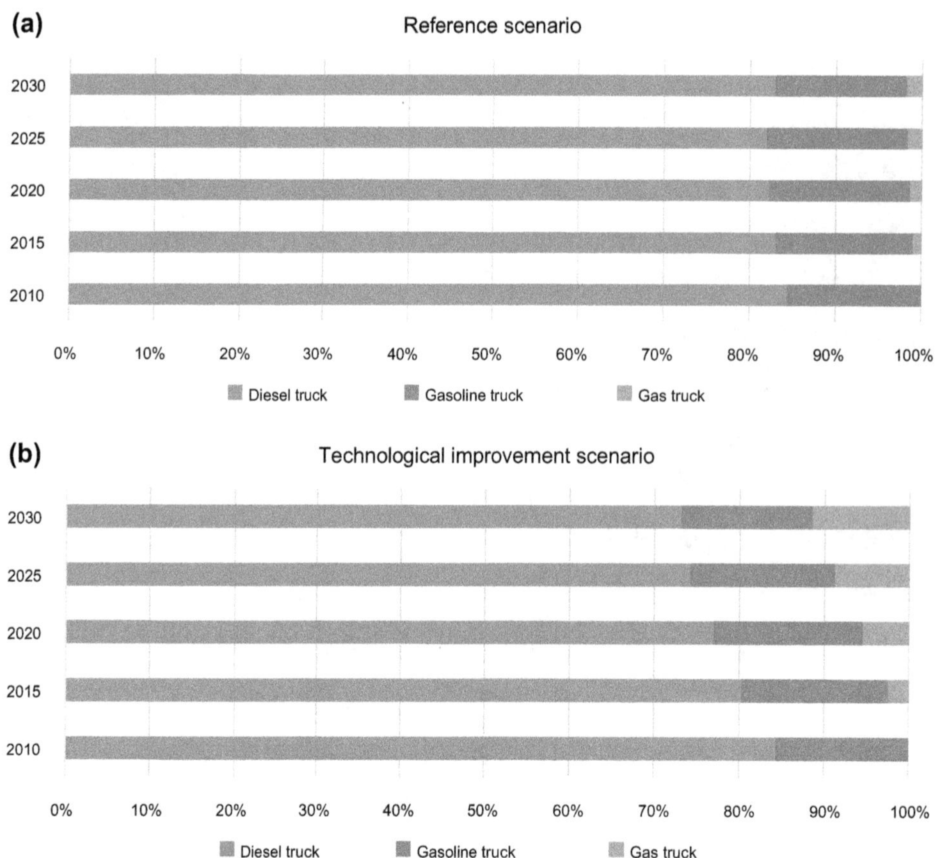

Fig. 8 Contribution of different trucks to the road freight task

medium truck transportation. Also gasoline cars account for a small proportion. In the technological improvement scenario, the contribution of gasoline trucks to freight transport will be more than 5.5 % in 2020 and over 11.7 % in 2030.

The total energy consumption of transport sector is shown in Fig. 9. The growth rate of the total energy consumption in the reference scenario is gradually slowing down due to improvement of fuel economy. The proportion of compressed natural gas, liquefied petroleum gas, and fuel ethanol will increase to 5.1 % in 2030. Overall, although the proportion of oil fuel will decrease, it is still the main component of future energy consumption in the transportation sector, with large requirements for the future oil supply of China.

As shown in Table 5, the calculation results of China-ESPT are consistent with other models, like MESSAGE V.4 (IIASA 2013) and IMAGE 2.4 (PBL 2010). The total energy consumed by freight transport is about 55.7 % of the total energy consumption, and passenger transport accounts for 44.3 % of the total energy consumption. In addition, passenger vehicles account for 26.9 % of all modes of passenger transport, with an energy consumption of approximately 1/4 of total energy consumption.

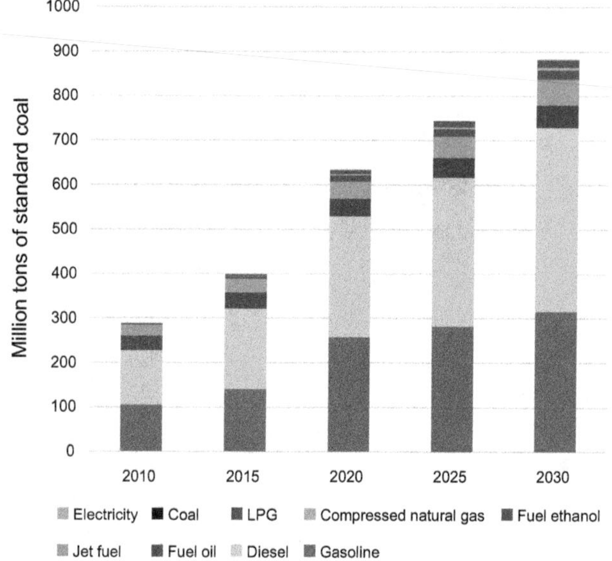

Fig. 9 Energy consumption of transport sector before 2030 (the reference scenario)

As shown in Fig. 10, in the technological improvement scenario, the total fuel consumption of the transportation sector will reduce by about 7.9 % in 2020, in which the

Table 5 Oil consumption of transportation (EJ)

	2010	2020	2030
China-ESPT	8.33	17.91	24.75
IMAGE 2.4 (PBL 2010)	6.58	15.70	24.59
MESSAGE V.4 (International Institute for Applied System Analysis, IIASA 2013)	7.99	18.40	28.34
REMIND 1.5 (PIK 2013)	7.47	15.74	23.29
ChinaTimes (Yin et al. 2015)	8.68	17.17	24.33

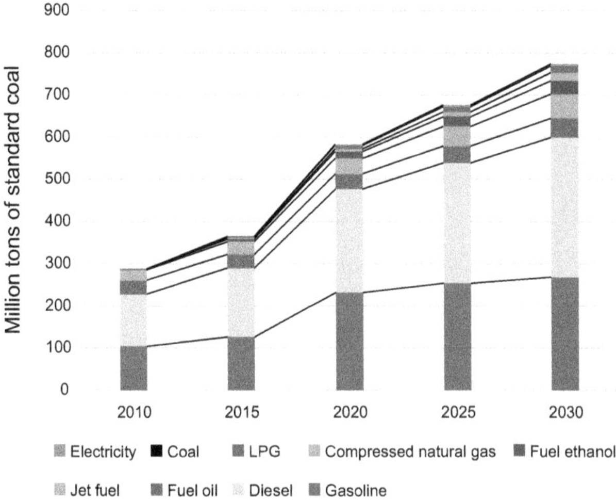

Fig. 10 Energy consumption of transport sector (technological improvement scenario)

consumption of diesel, gasoline, and fuel oil will reduce by 10.4 % compared to the reference scenario. In addition, it is estimated that the total fuel consumption will decline by 12.3 % in 2030, and diesel, gasoline, and fuel oil will, respectively, decrease by 19.1 %, 15.4 %, and 10.2 %. Moreover, in terms of fuel substitution, the consumption of natural gas will increase 1.8 times that of the reference scenario in 2020, and 2.7 times that of the reference scenario in 2030 as a result of technology improvement. Because of the increase of electric vehicles and fuel-cell vehicles, electricity consumption in transport will be 2.5 times and 3.1 times that of the reference scenario, respectively, in 2020 and 2030.

4 Discussion

According to the model analysis, in the mitigation scenario, the primary energy consumption will reach 4.95 billion tonnes of standard coal in 2020 and 5.81 billion tonnes in 2030, with CO_2 emissions limits of 105 tonnes in 2030. The consumption of fossil energy will still occupy a significant proportion. In the mitigation scenario, non-fossil energy consumption will reach more than 15 % in 2020 and more than 20 % in 2030.

The consumption of natural gas will significantly increase to 304.4 billion m^3 in 2020 and 483.3 billion m^3 in 2030. This is mainly due to an increase in final energy demand, especially in industry sector and transportation sector. Besides, China has launched a plan to control the total amount of coal consumption to ensure that carbon emissions will peak around 2030. The gas resource is cleaner than coal, and less expensive than renewables, thus has large development space. In addition, the development of nature gas is highly encouraged by the government. The oil consumption will keep increasing, but with the rate of increase decreasing. The main reason is the driving of the increasing demands from transportation sector. Thus, in this paper the transportation sector is further studied.

As a main consumer of oil products, the transportation sector is studied based on the reference scenario and technology improvement scenario. The design of the scenario mainly considers the fuel economy updates and technological improvement of hybrid cars, electric vehicles, and fuel-cell car. This paper analyzes changes in fuel, oil, and natural gas consumption, as well as changes in contributions of all types of vehicles to future passenger and freight service. Because of the technical maturity, technical cost, and infrastructure, the future development of these technologies has some uncertainty. In this study, we can predict the contribution rate of different types of vehicles to passenger and freight service, by assuming the service life of the vehicle, average carrying rate, annual distance, and changes in costs. The total energy consumption of the transportation sector is also analyzed.

Focusing on the oil and gas consumption, the consumption of diesel, gasoline, and fuel oil of the technology improvement scenario will reduce by 10.4 % compared to reference scenario. In addition, it is estimated that the total fuel consumption will decline by 12.3 % in 2030. The diesel, gasoline, and fuel oil will, respectively, decrease by 19.1 %, 15.4 %, and 10.2 %. In terms of fuel substitution, the consumption of natural gas will increase 1.8 times that of the reference scenario in 2020, and 2.7 times that of the reference scenario in 2030 due to technological improvement. With the increase of hybrid electric vehicles, electric vehicles, and fuel-cell vehicles, the transport electricity consumption in 2020 will increase 2.5 times that of the reference scenario, and 3.1 times that of the reference scenario in 2030.

According to the consumption of oil and gas in the transportation sector, on one hand, the consumption of refined oil is still the main component of the transportation sector. It is important to improve the fuel economy, under the premise of ensuring oil supply and energy security. Moreover, the development of electric vehicles, plug-in hybrid vehicles, and fuel-cell vehicles can effectively promote the achievement of goals due to energy structure adjustment and carbon dioxide emission reduction, and it is important to push forward the upgrading of oil standards and natural gas consumption.

Based on the results above, the oil and gas industry should also strengthen the implementation of relevant policies to achieve the target of peaking carbon emissions in 2030. First, it is important to promote the update of oil quality, especially the emission standard. Up to now, China's diesel product can achieve the national III standard and the gasoline product would be able to achieve national V standard. Second, it is necessary to ensure the clean energy, especially the natural gas supply. Based on the analysis, compared to the reference scenario, the proportion of natural gas consumption will increase by 7 % in 2020 and 10 % in 2030. Besides intensifying the domestic natural gas exploration, the importation of natural gas from Kazakhstan and Burma should also be encouraged, under the premise of ensuring energy security. Moreover, encouraging fuel substitution in transportation sector will help adjust the energy consumption structure. The promotion of electric or hybrid vehicles and natural gas buses will definitely improve the energy structure of the transportation sector, with an increase of electricity consumption by 3.1 times in 2030 compared to reference scenario. Thus, efforts should be made from both oil and gas supply industry and final demand sectors, to fulfill the emission peaking target in 2030.

5 Conclusions

In the present study, the energy consumption during the period of the "Thirteenth Five-Year Plan" and by 2030 was analyzed based on the developed bottom-up energy model. Three important constraints were considered in the analysis: (i) total CO_2 emissions will reach a peak value by 2030 as presented in INDC; (ii) CO_2 emission per GDP in 2020 should decrease 40 %–45 % from those in 2005; and (iii) the proportion of non-fossil energy in total energy consumption should reach 15 %. China's total carbon dioxide emission reductions and energy (mainly coal, oil, and gas) consumptions were calculated and analyzed subject to the constraints. Especially, the scenario analysis of oil and gas consumptions in transportation sector was conducted, and the importation roles of oil and gas were emphasized. Finally, in Sect. 4, the analysis results were reviewed and the improvement of fuel economy, quality of oil product, and fuel substitution in transportation sector were analyzed and discussed.

Acknowledgments This study is funded by the Project supported by China National Petroleum Corporation (Project Name: The development trend of oil and gas industry till 2030; Project Number: 20150114).

References

Adams FG, Shachmurove Y. Modeling and forecasting energy consumption in China: implications for Chinese energy demand and imports in 2020. Energy Econ. 2008;30:1263–78.

Argonne National Laboratory. Greenhouse gases, regulated emissions, and energy use in transportation model document. https://greet.es.anl.gov/. Accessed 10 Dec 2015.

Bollen J, Zwaan B, Brink C, et al. Local air pollution and global climate change: a combined cost–benefit analysis. Resour Energy Econ. 2009;31:161–81.

BP statistical review of world energy 2015. http://www.bp.com/en/global/corporate/ (2015a). Accessed 18 Jan 2016.

BP. BP energy outlook 2035. http://www.bp.com/content/dam/bp/pdf/ (2015b). Accessed 18 Oct 2016.

Budzianowski WM. Target for national carbon intensity of energy by 2050: a case study of Poland's energy system. Energy. 2012;46:575–81.

Cai W, Wang C, Chen J, et al. Comparison of CO_2 emission scenarios and mitigation opportunities in China's five sectors in 2020. Energy Policy. 2008;36:1181–94.

Chen W, Wu Z, He J, et al. Carbon emission control strategies for China: a comparative study with partial and general equilibrium versions of the China MARKAL model. Energy. 2007;32:59–72.

China Ministry of Transport. Compilation of national highway and waterway transportation survey data of China, 2010.

China Transportation Association. China transportation yearbook, 2011–2014.

Cox JC, Ingersoll JE, Ross S. An intertemporal general equilibrium model of asset prices. Econometrica. 1985;53:363–84.

da Graça Carvalho M. EU energy and climate change strategy. Energy. 2012;40:19–22.

Dai H, Masui T, Matsuoka Y, et al. Assessment of China's climate commitment and non-fossil energy plan towards 2020 using hybrid AIM/CGE model. Energy Policy. 2011;39:2875–87.

De Laquil P, Wenying C, Larson ED. Modeling China's energy future. Energy Sustain Dev. 2003;7:40–56.

EIA. Annual energy outlook 2015. Washington, DC: US Energy Information Administration; 2015.

ETSAP. Energy technology systems analysis program 2013. http://www.iea-etsap.org/. Accessed 10 Dec 2015.

Gallagher B. Peak oil analyzed with a logistic function and idealized Hubbert curve. Energy Policy. 2011;39(2):790–802.

Gambhir A, Schulz N, Napp T, et al. A hybrid modelling approach to develop scenarios for China's carbon dioxide emissions to 2050. Energy Policy. 2013;59:614–32.

Gillingham K, Newell RG, Pizer WA. Modeling endogenous technological change for climate policy analysis. Energy Econ. 2008;30:2734–53.

IEA. World energy outlook 2015. http://www.worldenergyoutlook.org/ (2015). Accessed 20 Jan 2016.

International Institute for Applied System Analysis (IIASA). http://www.iiasa.ac.at/web/home/research/modelsData/MESSAGE/MESSAGE.en.html (2013). Accessed 20 Jan 2016.

Jiang B, Sun Z, Liu M. China's energy development strategy under the low-carbon economy. Energy. 2010;35:4257–64.

Klaassen G, Riahi K. Internalizing externalities of electricity generation: an analysis with MESSAGE–MACRO. Energy Policy. 2007;35:815–27.

Klaassen, G, Amann M, Berglund C, et al. The extension of the RAINS model to greenhouse gases. An interim report describing the state of work as of April 2004. IIASA IR-04-015; 2004.

Krewitt W, Nitsch J. The German Renewable Energy Sources Act—an investment into the future pays off already today. Renew Energy. 2003;28:533–42.

Kumar U, Jain VK. Time series models (Grey–Markov, Grey Model with rolling mechanism and singular spectrum analysis) to forecast energy consumption in India. Energy. 2010;35(4):1709–16.

Kypreos S, Bahn O. A MERGE model with endogenous technological progress. Environ Model Assess. 2003;8:249–59.

Lund H. The implementation of renewable energy systems. Lessons learned from the Danish case. Energy. 2010;35:4003–9.

Manne A, Mendelsohn V, Richels R. MERGE: a model for evaluating regional and global effects of GHG reduction policies. Energy Policy. 1995;23:17–34.

Messner S, Schrattenholzer L. MESSAGE–MACRO: linking an energy supply model with a macroeconomic module and solving it iteratively. Energy. 2000;25:267–82.

Nordhaus WD. Rolling the 'DICE': an optimal transition path for controlling greenhouse gases. Resour Energy Econ. 1993;15:27–50.

OECD. Energy balances of non-OECD countries 2013. http://www.oecd-ilibrary.org/energy/energy-balances-of-non-oecd-countries-2013_energy_bal_non-oecd-2013-en. Accessed 10 Sept 2015.

OPEC. Joint IEA–IEF–OPEC report 2015. http://www.opec.org/opec_web/en/publications. Accessed 17 Jan 2016.

Paltsev S, Reilly JM, Jacoby HD, et al. The MIT emissions prediction and policy analysis (EPPA) model, in, joint program on the science and policy of global change. Cambridge: Massachusetts Institute of Technology; 2005.

PBL Netherlands Environmental Assessment Agency. http://themasites.pbl.nl/models/ (2010). Accessed 20 Oct 2015.

Potsdam Institute for Climate Impact Research (PIK). https://www.pik-potsdam.de/research/sustainable-solutions/models/remind (2013). Accessed 20 Oct 2015.

Rafaj P, Kypreos S. Internalization of external cost in the power generation sector: analysis with Global Multi-regional MARKAL model. Energy Policy. 2007;35:828–43.

Stockholm Environment Institute (SEI). Long range energy alternatives planning system 2014. Joint IEA–IEF–OPEC report. http://www.opec.org/opec_web/en/publications. Accessed 20 Oct 2015.

The State Council of China. Enhanced actions on climate change: China's intended nationally determined contributions [EB/OL]. http://www.gov.cn/xinwen/2015-06/30/content_2887330.htm/. Accessed 21 Oct 2015.

The State Council of China. Strategic action plan for energy development (2014–2020). http://www.gov.cn/zhengce/content/2014-11/19/content_9222.htm. Accessed 21 Oct 2015.

Tu J, Jaccard M, Nyboer J. The application of a hybrid energy–economy model to a key developing country—China. Energy Sustain Dev. 2007;11:35–47.

Wicke L. Beyond Kyoto—a new global climate certificate system. Heidelberg: Springer; 2005.

Yin X, Chen W. Trends and development of steel demand in China: a bottom-up analysis. Resour Policy. 2013;38:407–15.

Yin X, Chen W, Eom J, et al. China's transportation energy consumption and CO_2 emissions from a global perspective. Energy Policy. 2015;82:233–48.

Zhang N, Lior N, Jin H. The energy situation and its sustainable development strategy in China. Energy. 2011;36:3639–49.

Zhou N, Fridley D, Khanna NZ, et al. China's energy and emissions outlook to 2050: perspectives from bottom-up energy end-use model. Energy Policy. 2013a;53:51–62.

Zhou S, Kyle GP, Yu S, et al. Energy use and CO_2 emissions of China's industrial sector from a global perspective. Energy Policy. 2013b;58:284–94.

6

Thermal decomposition characteristics and kinetics of methyl linoleate under nitrogen and oxygen atmospheres

Xue-Chun Wang[1] · Jian-Hua Fang[1] · Bo-Shui Chen[1] · Jiu Wang[1] · Jiang Wu[1] · Di Xia[1]

Abstract The thermal decomposition characteristics of methyl linoleate (ML) under nitrogen and oxygen atmospheres were investigated, using a thermogravimetric analyzer at a heating rate of 10 °C/min from room temperature to 600 °C. Furthermore, the pyrolytic and kinetic characteristics of ML at different heating rates were studied. The results showed that the thermal decomposition characteristics of ML under nitrogen and oxygen atmospheres were macroscopically similar, although ML exhibited relatively lower thermal stability under an oxygen atmosphere than under a nitrogen atmosphere. The initial decomposition temperature, the maximum weight loss temperature, the peak decomposition temperature, and the rate of maximum weight loss of ML under an oxygen atmosphere were much lower than those under a nitrogen atmosphere and increased with increasing heating rates under either oxygen or nitrogen atmosphere. In addition, the kinetic characteristics of thermal decomposition of ML were elucidated based on the experimental results and by the multiple linear regression method. The activation energy, pre-exponential factor, reaction order, and the kinetic equation for thermal decomposition of ML were obtained. The comparison of experimental and calculated data and the analysis of statistical errors of pyrolysis ratios demonstrated that the kinetic model was reliable for pyrolysis of ML with relative errors of about 1 %. Finally, the kinetic compensation effect between the pre-exponential factors and the activation energy in the pyrolysis of ML was also confirmed.

Keywords Methyl linoleate · Pyrolysis characteristics · Kinetics · Thermogravimetric analysis · Biodiesel

✉ Jian-Hua Fang
fangjianhua71225@sina.com

[1] Department of Military Oil Application & Administration Engineering, Logistical Engineering University, Chongqing 401311, China

Edited by Xiu-Qin Zhu

1 Introduction

With the decrease of petroleum reserves and the increase of environmental pollution brought by the extensive use of petroleum, alternative fuels and renewable resources have attracted increasing attention worldwide from the perspective of environmental protection and resource strategy (Lin et al. 2011; Demirbas 2009; Huang et al. 2012; Sharma and Singh 2009; Nigam and Singh 2011). Biodiesel, referred to as mixtures of fatty acid mono-alkyl esters with relatively high contents of long-chain, mono- and poly-unsaturated compounds, is produced from vegetable oils and animal fats by transesterification with alcohols of low molecular weights over catalysts (Candeia et al. 2009; Leung et al. 2010; Moser and Vaughn 2010). Fatty acid methyl esters (FAME) such as soybean methyl ester (SME) are typically mixtures of esters with 16-18 carbon atoms, where 80 %–85 % (w/w) of the mixture is unsaturated (Knothe et al. 2005). However, the presence of such mono- and poly-unsaturated compounds makes biodiesel extremely liable to thermal decomposition at elevated temperatures.

As we know, during the operation of a biodiesel-powered engine, small amounts of biodiesel will leak into the crankcase by oil seepage flow or gas entrainment. The leakage of biodiesel into the engine crankcase markedly lowers the quality of the engine oils due to thermal instability of biodiesel. At present, some studies have been made on biodiesel-induced deterioration of engine oil for facilitating development and application of biodiesel as a clean and renewable petro-diesel substitute (Gili et al. 2011; Watson and Wong 2008; Wang et al. 2009). However, the thermal decomposition characteristics and kinetics of biodiesels, as well as their influences on deterioration of engine oils, have so far not been investigated intensively, partly because of the complexity of their pyrolytic chemical behavior and mechanisms. In fact, thermal instability of biodiesel is governed by its chemical nature, especially the structure and composition of unsaturated fatty acid methyl esters. It is therefore essential to investigate the thermal decomposition behavior of unsaturated FAME so as to better understand the nature of biodiesel-induced deterioration of engine oil. In this paper, the thermal decomposition characteristics and kinetics of unsaturated methyl linoleate were studied by thermogravimetric (TG) analysis. The present investigation is of importance for further understanding the pyrolytic behaviors of biodiesel and, in the nature of things, its influence on engine oil deterioration.

2 Experimental

2.1 Materials and apparatus

(1) Methyl linoleate (ML): an analytically purified chemical supplied by Xiya Reagent Research Center (China).
(2) TG analyzer: SDT-Q600 model, TA Instruments, USA.

2.2 TG analysis

The pyrolysis characteristics of ML were tested on the TG analyzer. In each test run, approximately 7 mg of ML was put uniformly on the bottom of an alumina crucible and the crucible was placed at the same position of the beam platform of the analyzer. Subsequently, the sample was progressively heated from room temperature to 600 °C at different heating rates of 10, 15, 20, and 30 °C/min under nitrogen or oxygen gas carrier flowing at 50 mL/min. The weight loss and heat flow changes in response to temperature were recorded. Finally, the TG and differential thermogravimetric (DTG) curves were plotted and the pyrolysis kinetics were studied.

3 Results and discussion

3.1 Pyrolysis characteristics of ML

Figure 1 shows the TG–DTG curves of ML at a heating rate of 10 °C/min. Table 1 shows the pyrolysis parameters of ML. From Fig. 1 and Table 1, it can be clearly observed that the thermal decomposition characteristics of ML under nitrogen and oxygen atmospheres were macroscopically similar. With increasing temperatures, the pyrolysis process in TG curves can be divided into three stages viz., the drying stage ($<T_s$), the main pyrolysis stage (T_s–T_f), and the residue decomposition stage ($>T_f$). The first stage occurred as the temperature increased from room temperature to the initial decomposition temperature T_s. In this

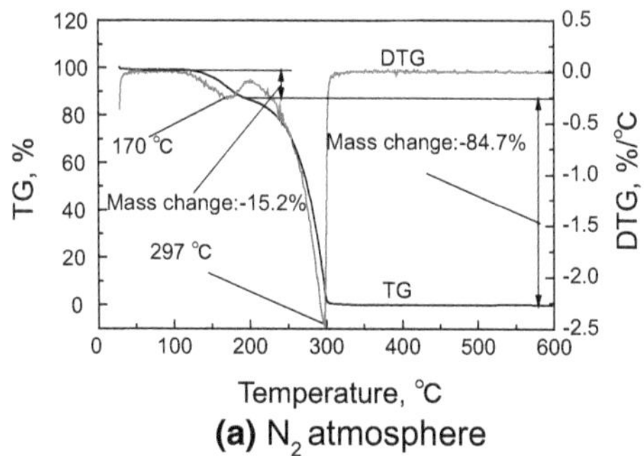

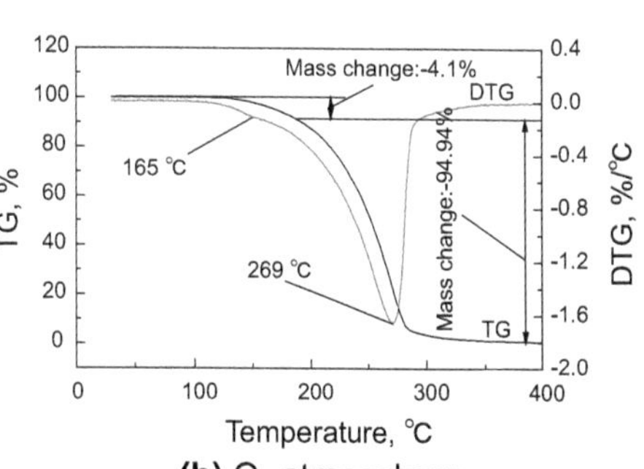

Fig. 1 TG/DTG curves of ML at different atmospheres ($\beta = 10$ °C/min)

Table 1 Pyrolysis parameters of ML at different atmospheres

Pyrolysis stage	N_2 atmosphere				O_2 atmosphere			
	Temperature range, °C	T_{max}, °C	$(dM/d\tau)_{max}$, %/°C	Mass loss, %	Temperature range, °C	T_{max}, °C	$(dM/d\tau)_{max}$, %/°C	Mass loss, %
Stage I	105–210	170	0.265	15.2	110–145	165	0.165	4.1
Stage II	210–300	297	2.47	84.7	145–285	269	1.65	94.9

T_{max}—peak temperature of DTG curve, °C; $(dM/d\tau)_{max}$—the maximum weight loss rate %/°C

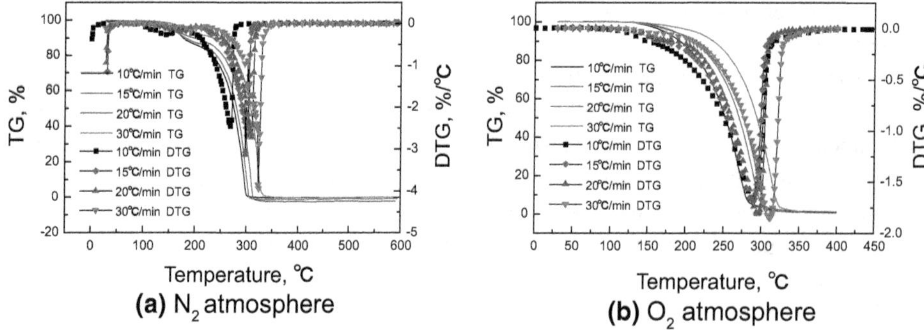

Fig. 2 TG/DTG curves of ML at different heating rates

Table 2 Pyrolysis parameters of ML at different heating rates

Pyrolysis stage	β, °C/min	N_2 atmosphere				O_2 atmosphere			
		Temperature range, °C	T_{max}, °C	$(dM/d\tau)_{max}$, %/°C	Mass loss, %	Temperature range, °C	T_{max}, °C	$(dM/d\tau)_{max}$, %/°C	Mass loss, %
Stage I	10	105–210	170	0.265	15.2	110–145	165	0.165	4.1
	15	115–225	192	0.279	13.5	115–160	167	0.169	3.9
	20	120–230	229	0.280	11.4	120–170	172	0.207	3.6
	30	120–230	238	0.286	9. 5	125–175	177	0.212	2.9
Stage II	10	210–300	297	2.47	84.7	145–285	269	1.65	94.9
	15	225–305	302	2.68	87.3	160–310	283	1.70	96.7
	20	230–310	310	3.14	90.9	170–315	287	1.81	96.7
	30	230–330	323	3.94	90.5	175–330	299	1.85	97.1

stage, a small weight loss could be observed due to evaporation of volatiles. The second stage occurred as the temperature increased from T_s to the maximum weight loss temperature T_f. The great weight loss in this stage was attributed to significant thermal decomposition of ML at elevated temperatures. The third stage occurred as the temperature increased from T_f to 600 °C; during the third stage almost no further weight loss could be detected. Figure 1 indicates that there were only two main weight loss stages in the TG curve, corresponding to two peaks in the DTG curve viz., the first stage ($<T_s$) and the second stage (T_s–T_f). However, it can also be observed from Fig. 1 and Table 1 that the pyrolysis characteristics of ML under nitrogen and oxygen atmospheres were slightly different. The peak temperature (T_{max}) and the initial decomposition temperature (T_s) of ML under an oxygen atmosphere were lower than those under nitrogen during the whole process, indicating that the thermal decomposition was easier under an oxygen atmosphere. Lower thermal stability of ML under oxygen might be attributed to a unique molecular structure containing two double bonds, thus enhancing decomposition of ML and promoting its thermal oxidation degradation at lower temperature (Wongsiriamnuay and Tippayawong 2010).

To further understand the pyrolysis characteristics of ML, thermogravimetric tests at different heating rates, i.e., 10, 15, 20, and 30 °C/min, were also conducted. The pyrolysis curves of ML at different heating rates are shown in Fig. 2. Also the main pyrolysis characteristics parameters of ML are shown in Table 2. It can be observed from the TG curves in Fig. 2 that the decomposition temperatures at different heating rates were slightly different. When the heating rate increased from 10 to 30 °C/min, the initial temperature of the main decomposition shifted to a higher one. Figure 2 shows the DTG profiles obtained from the thermal decomposition of ML at different heating rates. The maximum weight loss rate increased, and the corresponding peak temperature at maximum weight loss rate shifted to higher temperatures, with increasing heating rate. When the heating rate increased from 10 to 30 °C/min, the maximum weight loss rate sharply increased, and the peak of weight loss in the DTG curves became higher and broader. The increase of the initial decomposition temperature, the maximum weight loss temperature, the peak decomposition temperature, and of the maximum weight loss rate with increasing heating rates may be resulted from the reduction of the activation energy, as well as from the delay of heat transfer during ML thermal decomposition (Park et al. 2009; Chen et al. 2011; Liang et al. 2014).

3.2 Kinetics of thermal decomposition of ML

3.2.1 Kinetic modeling

The kinetic parameters obtained from TG and DTG analysis are exceedingly crucial for efficient evaluation and calculation of the thermal decomposition process of ML. Assuming that the thermal decomposition of ML was a non-isothermal process, the decomposition rate equation can be given as follows:

$$d\alpha/dt = kf(\alpha)^n, \tag{1}$$

where α is the conversion rate and is defined as $\alpha = (m_0 - m_t)/(m_0 - m_\infty)$; m_0 is the initial weight of the test sample; m_t is the weight after a specified decomposition duration t; m_∞ is the weight of the indecomposable residue; n is the reaction order; and k is the reaction rate constant.

Generally, the Arrhenius equation is applicable in thermal decomposition reactions. Based on the Arrhenius equation, the reaction rate constant, k, for thermal decomposition of ML is given below:

$$k = A\exp(-E/RT), \tag{2}$$

where A is the pre-exponential factor, E is the activation energy (kJ/mol), T is the reaction temperature (K) and R is the ideal gas constant.

Since the mechanism function $f(\alpha)$ in Eq. (1) is dependent on the reaction model and reaction mechanism during pyrolysis process, for a simple reaction, $f(\alpha)$ is suggested as

$$f(\alpha) = (1 - \alpha)^n. \tag{3}$$

Then, Eqs. (1), (2), and (3) can be combined to give

$$d\alpha/dt = A\ \exp(-E/RT)f(\alpha) = A\ \exp(-E/RT)(1 - \alpha)^n. \tag{4}$$

Furthermore, substituting the heating rate, β, into Eq. (4) gives

$$d\alpha/dT = \frac{A}{\beta}\exp(-E/RT)(1 - \alpha)^n, \tag{5}$$

where $\beta = dT/dt$ and $d\alpha/dT$ is the ratio of weight changes with temperature.

3.2.2 Determination of kinetic parameters

According to the kinetic model given above, the kinetic parameters such as pre-exponential factor, activation energy, and reaction order, and the most probable mechanism function for thermal decomposition of ML were determined by the multiple linear regression method.

Taking the natural logarithm of Eq. (5) gives

$$\ln(d\alpha/dT) = \ln\frac{A}{\beta} - E/RT + n\ln(1 - \alpha). \tag{6}$$

Equation (6) may be further expressed in the linear form as given below:

$$Y = B + CX + DZ, \tag{7}$$

where $Y = \ln(d\alpha/dT)$, $X = 1/T$, $Z = \ln(1 - \alpha)$, $B = \ln(A/\beta)$, and $C = -E/R$, $D = n$.

The constants B, C, and D were estimated by the multiple linear regression method with the TG–DTG data for pyrolysis of ML using Origin 8.0 software. The kinetic parameters viz., pre-exponential factor, activation energy, and reaction order, for each test run are presented in Table 3. The high correlation coefficients, R^2, shown in Table 3, demonstrated that the kinetic parameters calculated by the multiple linear regression method were reliable.

Table 3 Pyrolysis kinetic parameters of ML at different heating rates

Pyrolysis stage	β, °C/min	N_2 atmosphere					O_2 atmosphere				
		Temperature range, °C	Reaction order (n)	E, kJ/mol	$\ln A$	R^2	Temperature range, °C	Reaction order (n)	E, kJ/mol	$\ln A$	R^2
Stage I	10	105–210	36.8	105.25	33.08	0.970	110–145	28.7	103.14	25.44	0.999
	15	115–225	38.9	100.57	30.18	0.906	115–160	38.7	101.49	24.75	0.999
	20	120–230	47.3	98.62	28.53	0.924	120–170	39.5	95.22	23.53	0.999
	30	120–230	50.6	88.62	20.90	0.883	125–175	38.1	62.14	13.41	0.999
Stage II	10	210–300	0.160	83.14	21.31	0.992	145–285	0.437	49.87	9.54	0.970
	15	225–305	0.310	94.42	26.59	0.974	160–310	0.291	50.56	9.60	0.980
	20	230–310	0.200	95.08	26.90	0.982	170–315	0.278	54.26	10.59	0.975
	30	230–330	0.040	97.36	28.36	0.988	175–330	0.394	59.67	11.84	0.995

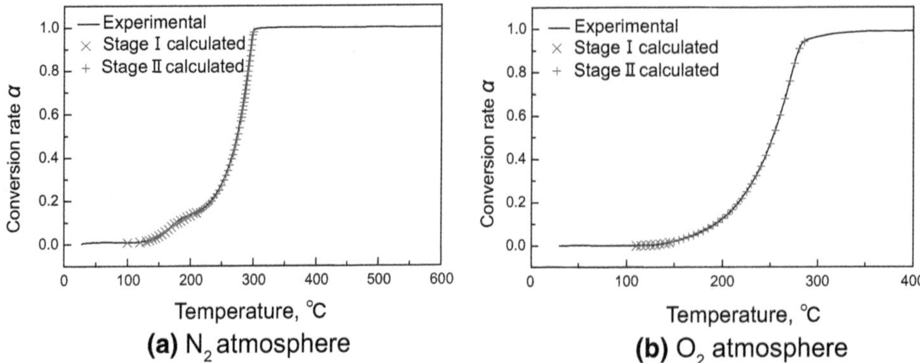

Fig. 3 Comparison of conversion rate between the experimental data and the calculated data by multiple linear regression

Table 4 Numerical statistical errors of conversion rate between the experimental data and the calculated data by multiple linear regression

Atmosphere	Reaction order (n)	Error, %	B, °C/min			
			10	15	20	30
N_2	–	RMSE	1.8793	2.4414	2.3718	2.5014
	–	MAPE	1.6402	1.9774	2.1897	2.0611
O_2	–	RMSE	3.1244	2.7745	2.7561	1.8252
	–	MAPE	2.2188	3.0143	3.1178	2.9112

3.2.3 Kinetic reliability analysis

The relationship and the numerical statistical errors of conversion rate between the experimental data and the calculated data by the multiple linear regression method at the heating rate of 10 °C/min are shown in Fig. 3 and Table 4, respectively. The numerical statistical errors of root mean square error (RMSE) and mean absolute percentage (MAPE) were calculated by the following equations:

$$\mathrm{RMSE} = \sqrt{\frac{1}{p}\sum_{i=1}^{p}(E_i - C_i)^2}$$

$$\mathrm{MAPE} = \sum_{i=1}^{p}\left|\frac{E_i - C_i}{E_i}\right| \times \frac{100}{p},$$

where p is the total number of observations considered in the test; E_i and C_i correspond to the experimental data and the calculated data of conversion rate, respectively.

It can be clearly observed from Fig. 3 and Table 4 that the experimental data and the calculated data of conversion rates obtained from the kinetic model by the multiple linear regression method are in good agreement, and the numerical statistical errors are not notable. This indicated that on one hand the suggested kinetic mechanism function, $f(\alpha) = (1-\alpha)^n$, fitted well with the pyrolysis kinetic regularity of ML and the kinetics modeling was thus reliable and on the other hand the multiple linear regression method was more available in the analysis of ML pyrolysis kinetics model.

3.2.4 Kinetic compensation effect in pyrolysis of ML

It has been found that for thermal decomposition reactions, the kinetic parameters, such as the pre-exponential factor A and the activation energy E, exhibit the following relationship (Cai and Bi 2009; Li et al. 2011):

$$\ln A = aE + b,$$

where a and b are constant coefficients. This relationship is referred to as the kinetic compensation effect, meaning that the pre-exponential factor of a thermal decomposition reaction is not a constant but changes with the variation of activation energy. Therefore, based on the kinetic parameters listed in Table 3, the relationship between $\ln A$ and E for pyrolysis of ML was plotted, as shown in Fig. 4. The kinetic parameters E and A were found to satisfy the kinetic compensation effect relationship, and their linear relationship is listed in Table 5. By comparison of the correlation coefficients ranging from 0.996 to 0.997, the good linear relationship between $\ln A$ and E demonstrated that the pre-exponential factor for thermal decomposition of ML was kinetically well compensated by activation energy.

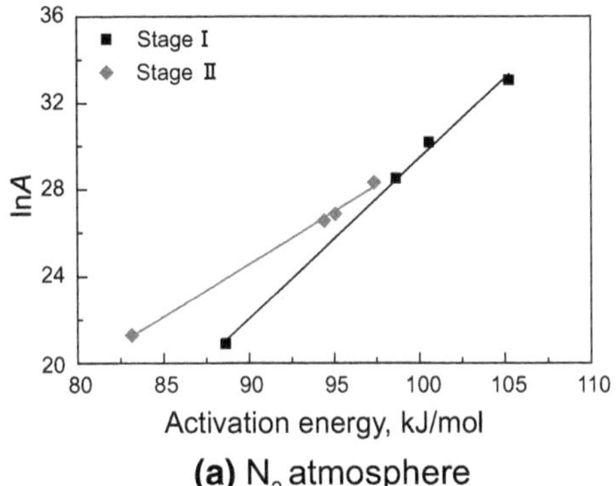

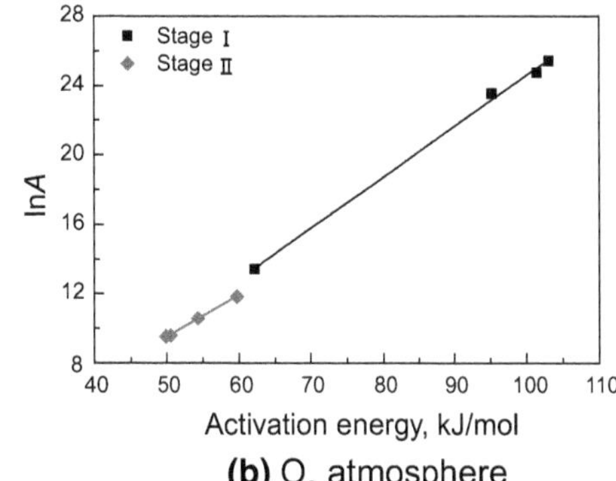

Fig. 4 Natural logarithm of pre-exponential factor against activation energy by MLR

Table 5 Relationship of kinetic compensation effects for ML at different atmospheres

Pyrolysis stage	N_2 atmosphere		O_2 atmosphere	
	Regression equation	R^2	Regression equation	R^2
Stage I	$\ln A_1 = 0.74116E_1 - 44.65774$	0.997	$\ln A_1 = 0.29309E_1 - 4.74191$	0.997
Stage II	$\ln A_2 = 0.48372E_2 - 18.95425$	0.996	$\ln A_2 = 0.23996E_2 - 2.47087$	0.997

4 Conclusions

The thermal decomposition characteristics of methyl linoleate under nitrogen and oxygen atmospheres were macroscopically similar. The initial decomposition temperature, the maximum weight loss temperature, and the peak decomposition temperature of methyl linoleate increased with increasing heating rates due to the increase of the activation energy and to the delay of heat transfer. However, the above-mentioned temperatures were much lower under an oxygen atmosphere, thus indicating that under oxygen methyl linoleate exhibited relatively poorer thermal stability. In addition, a kinetic model for thermal decomposition of methyl linoleate was built up, and the kinetic parameters such as activation energy, pre-exponential factor, and reaction order were obtained. The kinetic model was reliable in predicting the pyrolysis characteristics of methyl linoleate and provided a theoretical basis for expanding the experiment and anti-decomposition in biodiesel applications. Finally, the kinetic compensation effect between the pre-exponential factors and the activation energy in the pyrolysis of methyl linoleate was confirmed.

Acknowledgments The authors gratefully acknowledge the financial support provided by National Natural Science Foundation of China (Project No. 51375491) and the Natural Science Foundation of Chongqing (Project No. CSTC, 2014JCYJAA50021).

References

Cai JM, Bi LS. Kinetic analysis of wheat straw pyrolysis using isoconversional methods. J Therm Anal Calorim. 2009;98(1):325–30.

Candeia RA, Silva MCD, Carvalho Filho JR, et al. Influence of soybean biodiesel content on basic properties of biodiesel–diesel blends. Fuel. 2009;88(4):738–43.

Chen M, Qi X, Wang J, et al. Catalytic pyrolysis characteristics and kinetics of cotton stalk. J Fuel Chem Technol. 2011;39(8):585–9 (in Chinese).

Demirbas A. Progress and recent trends in biodiesel fuels. Energy Convers Manag. 2009;50(1):14–34.

Gili F, Igartua A, Luther R, et al. The impact of biofuels on engine oil's performance. Lubr Sci. 2011;23(7):313–30.

Huang D, Zhou H, Lin L. Biodiesel: an alternative to conventional fuel. Energy Procedia. 2012;16:1874–85.

Knothe G, Van Gerpen J, Krahl J. The biodiesel handbook. Champaign: AOCS Press; 2005.

Leung DYC, Wu X, Leung MKH. A review on biodiesel production using catalyzed transesterification. Appl Energy. 2010;87(4):1083–95.

Li D, Chen L, Zhang X, et al. Pyrolytic characteristics and kinetic studies of three kinds of red algae. Biomass Bioenergy. 2011;35(5):1765–72.

Liang Y, Cheng B, Si Y, et al. Thermal decomposition kinetics and characteristics of *Spartina alterniflora* via thermogravimetric analysis. Renew Energy. 2014;68:111–7.

Lin L, Cunshan Z, Vittayapadung S, et al. Opportunities and challenges for biodiesel fuel. Appl Energy. 2011;88(4):1020–31.

Moser BR, Vaughn SF. Coriander seed oil methyl esters as biodiesel fuel: unique fatty acid composition and excellent oxidative stability. Biomass Bioenergy. 2010;34(4):550–8.

Nigam PS, Singh A. Production of liquid biofuels from renewable resources. Prog Energy Combust Sci. 2011;37(1):52–68.

Park YH, Kim J, Kim SS, et al. Pyrolysis characteristics and kinetics of oak trees using thermogravimetric analyzer and micro-tubing reactor. Bioresour Technol. 2009;100(1):400–5.

Sharma YC, Singh B. Development of biodiesel: current scenario. Renew Sustain Energy Rev. 2009;13(6–7):1646–51.

Wang Z, Xu G, Huang H, et al. Reliability test of diesel engine fueled with biodiesel. Trans Chin Soc Agric Eng. 2009;25(11):169–72.

Watson SAG, VW Wong. The effects of fuel dilution with biodiesel and low sulfur diesel on lubricant acidity, oxidation and corrosion: a bench scale study with CJ-4 and CI-4+ lubricants[C]//STLE/ASME 2008 International Joint Tribology Conference. Am Soc Mech Eng. 2008;2008:233–5.

Wongsiriamnuay T, Tippayawong N. Thermogravimetric analysis of giant sensitive plants under air atmosphere. Bioresour Technol. 2010;101(23):9314–20.

7

Numerical estimation of choice of the regularization parameter for NMR T_2 inversion

You-Long Zou[1] · Ran-Hong Xie[1] · Alon Arad[2]

Abstract Nuclear Magnetic Resonance (NMR) T_2 inversion is the basis of NMR logging interpretation. The regularization parameter selection of the penalty term directly influences the NMR T_2 inversion result. We implemented both norm smoothing and curvature smoothing methods for NMR T_2 inversion, and compared the inversion results with respect to the optimal regularization parameters (α_{opt}) which were selected by the discrepancy principle (DP), generalized cross-validation (GCV), S-curve, L-curve, and the slope of L-curve methods, respectively. The numerical results indicate that the DP method can lead to an oscillating or oversmoothed solution which is caused by an inaccurately estimated noise level. The α_{opt} selected by the L-curve method is occasionally small or large which causes an undersmoothed or oversmoothed T_2 distribution. The inversion results from GCV, S-curve and the slope of L-curve methods show satisfying inversion results. The slope of the L-curve method with less computation is more suitable for NMR T_2 inversion. The inverted T_2 distribution from norm smoothing is better than that from curvature smoothing when the noise level is high.

✉ You-Long Zou
zoyolo_ok@126.com

✉ Ran-Hong Xie
xieranhong@cup.edu.cn

1 State Key Laboratory of Petroleum Resources and Prospecting, China University of Petroleum, Beijing 102249, China

2 Shell International Exploration and Production Inc., Houston 77079, TX, USA

Edited by Jie Hao

Keywords NMR T_2 inversion · Tikhonov regularization · Variable substitution · Levenberg–Marquardt method · Regularization parameter selection

1 Introduction

NMR logging directly measures the signal from protons in the fluid in formation of pores. Its applications include fluid typing, porosity calculation, permeability estimation, fluid saturation determination, and bound water estimation. NMR logging interpretation is based on the inverted T_2 distributions from acquired echo trains. NMR T_2 inversion is an ill-posed problem, so it is critical to choose a robust and efficient inversion method to obtain credible NMR spectra. For NMR T_2 inversion, scholars have proposed many kinds of inversion methods. Butler, Reeds, and Dawson (BRD) proposed a method to solve norm smoothing with a non-negative constraint of solution (Butler et al. 1981). Dunn et al. (1994) proposed another method for solving norm smoothing with a non-negative constraint of solution. Prammer (1994) used the singular value decomposition (SVD) method for NMR T_2 inversion, and adopted a series of measures to improve the inversion speed for the purpose of real-time processing. Borgia et al. (1998) put forward a complex curvature smoothing method which is called uniform-penalty (UPEN) method, and then made further modifications to the method which allows the regularization parameter to be a variable in the iterative process (Borgia et al. 2000). The SVD method implements the non-negative constraint of solution by singular value truncation which decreases the accuracy of the solution (Prammer 1994; Ge et al. 2016). For the SVD method, the low signal-to-noise ratio of NMR logging data leads to a large cutoff of singular value, which seriously decreases the accuracy of the solution. Norm smoothing and curvature smoothing can usually obtain more satisfactory solutions than the SVD method.

The critical issue of norm smoothing and curvature smoothing methods (Dunn et al. 1994) is to determine the optimal regularization parameter. The different optimal regularization parameters selected by different regularization parameter selection methods will cause slightly different inversion results. The published literature mainly used the BRD (Butler et al. 1981) and S-curve (Sezginer 1994; Song et al. 2002) methods to select the regularization parameter for NMR T_2 inversion. Compared with BRD method, the S-curve method does not need to know the noise level. Except for the above two methods, the generalized cross-validation (GCV) (Golub et al. 1979) and L-curve (Hansen 1992) methods are often widely used to select the regularization parameter for data inversion in many fields. But every regularization parameter selection method has its own advantages and disadvantages. For different inversion problems, we need to comprehensively account for both the amount of calculation and the accuracy of inversion result to determine the most satisfactory parameter selection method for the studied inverse problem. For NMR logging T_2 inversion, we implement the inversion procedure at each well-logging depth point, so inversion speed should also be an important consideration. This paper implemented norm smoothing and curvature smoothing methods for NMR T_2 inversion, and compared the inversion results with respect to the optimal regularization parameters which were selected by the DP, GCV, S-curve, L-curve, and the slope of L-curve methods, respectively.

2 NMR T_2 inversion

The measured echo amplitude of NMR logging using a Carr–Purcell–Meiboom–Gill (CPMG) pulse sequence with sufficient polarization time has the following equation:

$$b(t) = \int f(\mathrm{T}_2)\exp(-t/\mathrm{T}_2)\mathrm{d}\mathrm{T}_2 + \varepsilon, \tag{1}$$

where $b(t)$ is the echo amplitude at time t, T_2 is the transverse relaxation time, $\exp(-t/\mathrm{T}_2)$ is the kernel function, $f(\mathrm{T}_2)$ is the amplitude of T_2 distribution, and ε is noise.

The discrete form of Eq. (1) is

$$A_{m\times n}f_{n\times 1} = b_{m\times 1} + \varepsilon_{m\times 1}, \tag{2}$$

where $A_{m\times n}$ is kernel matrix, $f_{n\times 1} = [f(\mathrm{T}_{2,1}), \ldots, f(\mathrm{T}_{2,n})]^{\mathrm{T}}$, $b_{m\times 1} = [b(t_1), \ldots, b(t_m)]^{\mathrm{T}}$.

As is known, NMR T_2 inversion is an ill-posed problem, so regularization terms are needed to be added. The most common form of regularization is the Tikhonov regularization, which has the following objective function:

$$\min\left\{\phi(f) = \frac{1}{2}\|W(Af-b)\|^2 + \frac{\alpha}{2}\|Lf\|^2\right\}, \tag{3}$$

where $\|\cdot\|$ means Euclidean norm and W is a weighted matrix whose diagonal elements equal to the reciprocal of the noise level. If the noise level of data is a constant, W can be an identity matrix. L is the regularization matrix, can be a zero-, or first-, or second-derivative operator which corresponds to norm smoothing, slope smoothing, and curvature smoothing (Dunn et al. 1994). α is the regularization parameter.

To obtain the non-negative constraint of solution, the iterative solution is commonly made by eliminating the columns of kernel matrix corresponding to the negative components in the solution or replacing them with large constants. Unlike the above-mentioned methods, we use a variable substitution method to obtain a non-negative constraint of solution, in which the solution is substituted by a non-negative expression. For example, set $f = \exp(x)$ or x^2, the above objective function of Eq. (3) can be rewritten as

$$\min\left\{\phi(x) = \frac{1}{2}\|W(A\exp(x)-b)\|^2 + \frac{\alpha}{2}\|L\exp(x)\|^2\right\}, \tag{4a}$$

or

$$\min\left\{\phi(x) = \frac{1}{2}\|W(Ax^2-b)\|^2 + \frac{\alpha}{2}\|Lx^2\|^2\right\}. \tag{4b}$$

The objective functions of Eqs. (4a) and (4b) without constraint conditions can be solved by the Levenberg–Marquardt method, an iterative method. The new solution x_{new} is updated by the following equations (Madsen and Nielsen 2010):

$$(\phi'' + \mu I)\Delta x = -\phi', \tag{5}$$

$$x_{\mathrm{new}} = x_{\mathrm{old}} + \Delta x, \tag{6}$$

where I is an identity matrix, ϕ' is the gradient of objective function, ϕ'' is the Hessian matrix of objective function, μ is a parameter that can be updated in every iteration by the updating strategy of Madsen and Nielsen (Madsen and Nielsen 2010).

By calculating the partial derivatives of the objective functions of Eqs. (4a) and (4b) with respect to x, we can obtain the gradients ϕ' and approximate symmetric positive definite Hessian matrices ϕ'' of the objective functions of Eqs. (4a) and (4b), respectively.

For $f = \exp(x)$,

$$\phi' = (W\cdot A\cdot \mathrm{diag}(\exp(x)))^{\mathrm{T}}(W\cdot A\cdot \exp(x) - W\cdot b) + \alpha(L\cdot \mathrm{diag}(\exp(x)))^{\mathrm{T}}(L\cdot \exp(x)), \tag{7a}$$

$$\phi'' \approx (W\cdot A\cdot \mathrm{diag}(\exp(x)))^{\mathrm{T}}(W\cdot A\cdot \mathrm{diag}(\exp(x))) + \alpha(L\cdot \mathrm{diag}(\exp(x)))^{\mathrm{T}}(L\cdot \mathrm{diag}(\exp(x))). \tag{7b}$$

For $f = x^2$,

$$\phi' = 2.0 \cdot (W \cdot A \cdot \mathrm{diag}(x))^{\mathrm{T}}\left(W \cdot A \cdot x^2 - W \cdot b\right) + 2.0 \cdot \alpha(L \cdot \mathrm{diag}(x))^{\mathrm{T}}(L \cdot x^2), \tag{8a}$$

$$\phi'' \approx 4.0 \cdot (W \cdot A \cdot \mathrm{diag}(x))^{\mathrm{T}}(W \cdot A \cdot \mathrm{diag}(x)) + 4.0 \cdot \alpha(L \cdot \mathrm{diag}(x))^{\mathrm{T}}(L \cdot \mathrm{diag}(x)), \tag{8b}$$

where diag(exp(x)) and diag(x) are diagonal matrices which are generated by vectors exp(x) and x, respectively.

3 Regularization parameter selection

Since a too small or too large regularization parameter can result in an undersmoothed or oversmoothed solution, it is critical to choose an optimal regularization parameter (α_{opt}). The commonly used regularization parameter selection methods include the DP, GCV, S-curve, and L-curve methods (Morozov 1966; Golub et al. 1979; Sezginer 1994; Hansen 1992).

3.1 Discrepancy principle (DP)

If the noise level σ is known, the DP (Morozov 1966) suggests that the α_{opt} should be chosen to satisfy the following equation:

$$\zeta(\alpha) = \|Af - b\|^2 = \tau m \sigma^2, \tag{9}$$

where m is the number of the echoes of the echo train b. $\tau \geq 1$ is a predetermined real number, typically $\tau = 1$.

Figure 1 shows the typical "S" shape curves of the variation of regularization parameter (α) with residual norm ($\zeta(\alpha)$) for norm smoothing and curvature smoothing. This method needs to know the noise level σ, but sometimes it is difficult to estimate an accurate σ.

3.2 Generalized cross-validation (GCV)

The GCV method was proposed by Golub et al. (1979) to find the α_{opt} that minimizes the GCV function. For Tikhonov regularization, the GCV function is

$$G(\alpha) = \frac{\|b - Af_\alpha\|^2}{\mathrm{trace}(I - AA^{\#})^2}, \tag{10}$$

where I is an identity matrix, $A^{\#}$ denotes the regularized pseudo-inverse of A, $f_\alpha = A^{\#}b$. For Tikhonov regularization, $A^{\#} = (A^{\mathrm{T}}A + \alpha L^{\mathrm{T}}L)^{-1}A^{\mathrm{T}}$.

Figure 2 shows the variation of the GCV function value with a regularization parameter (α) for norm smoothing and curvature smoothing. As shown in Fig. 2, as α increases, the GCV function value first decreases and then increases.

3.3 S-curve

The S-curve method (Sezginer 1994; Song et al. 2002) finds the minimum α as α_{opt} that satisfies

$$\frac{\mathrm{d}\log\zeta}{2\mathrm{d}\log\alpha} = tol, \tag{11}$$

where $0 < tol < 1$ is a predetermined constant, typically $tol = 0.1$.

This method uses the slope of the S-curve as the criterion of α_{opt} selection. Eq. (21) in Appendix shows the specific formula of the slope of S-curve. The heel of the S-curve is selected as α_{opt}, which balances the residual and known noise variance (Song et al. 2002).

Figure 3 shows the variation of slope of S-curve with regularization parameter (α). Typically, the slope of the S-curve gradually increases at first, then remains at a large value, and finally gradually decreases. We choose the smallest α that satisfies the Eq. (11) as α_{opt}.

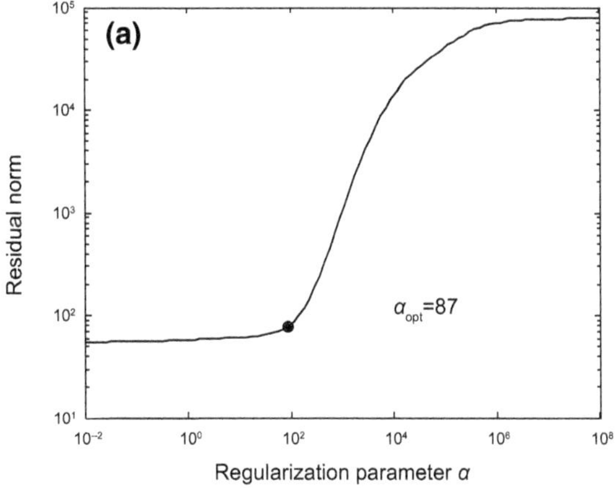

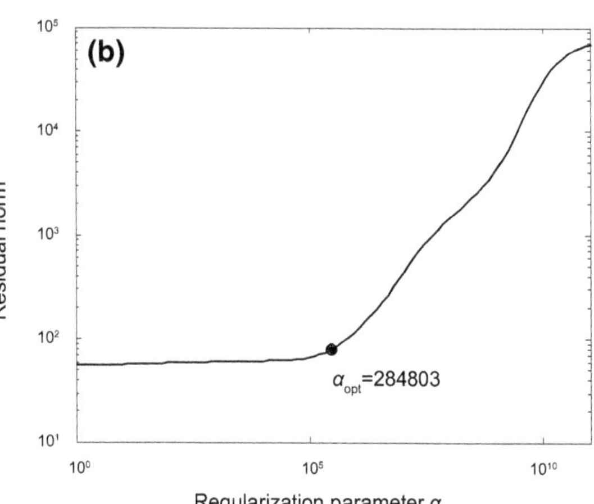

Fig. 1 The variation of residual norm with regularization parameter. **a** Norm smoothing, **b** curvature smoothing

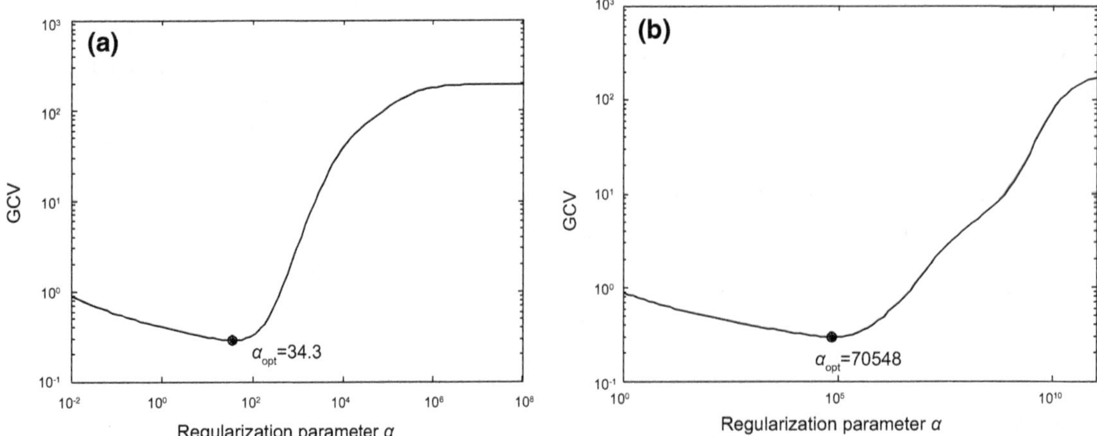

Fig. 2 The variation of GCV function value with regularization parameter. **a** Norm smoothing, **b** curvature smoothing

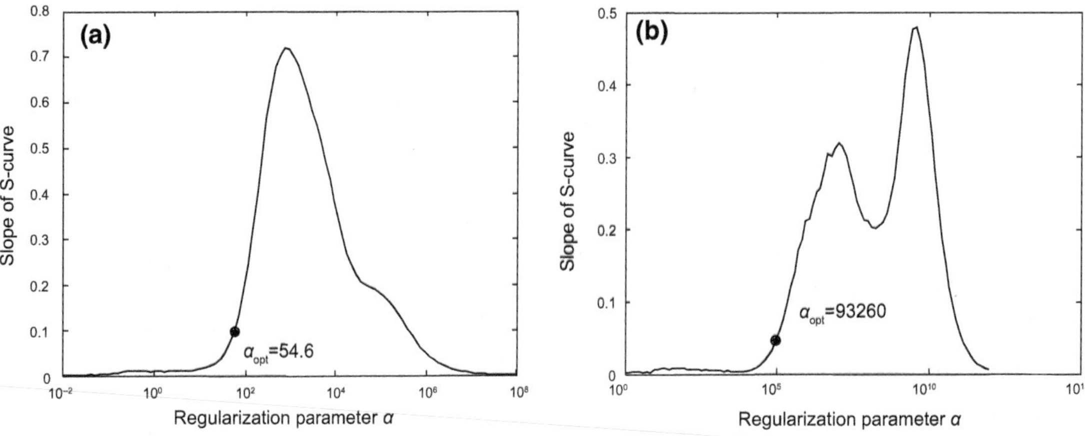

Fig. 3 The variation of slope of the S-curve with regularization parameter. **a** Norm smoothing, **b** curvature smoothing

3.4 L-curve

As α increases, solution norm $\eta(\alpha) = \|Lf\|^2$ decreases, while residual norm $\zeta(\alpha)$ increases. In the log–log scale, the curve formed by $\eta(\alpha)$ versus $\zeta(\alpha)$ for each of a set of α has an "L" shape, so it is called the L-curve. The method was proposed by Lawson and popularized by Hansen (Hansen 1992).

Figure 4 shows the typical L-curves of norm smoothing and curvature smoothing. Intuitively, α_{opt} should lie on the "corner" of the L-curve, for values higher than this "corner", $\zeta(\alpha)$ increases without reducing $\eta(\alpha)$ too much, while for values smaller than this "corner", $\zeta(\alpha)$ decreases little but with a rapid increase of $\eta(\alpha)$.

People have proposed many kinds of methods to locate the "corner" (the point of maximum curvature) of the L-curve. Castellanos et al. (2002) analyzed the drawbacks of three methods for finding the corner of L-curve (Kaufman and Neumaier 1996; Hansen 1998; Guerra and Hernandez 2001), and proposed a robust triangle method. Hansen et al. (2007) proposed an adaptive pruning algorithm, which first calculates the corner candidates at different scales or resolutions and then selects the overall optimal corner from the candidates. The above-mentioned methods are indirect methods to calculate the corner of L-curve. However, indirect methods cannot guarantee correct results in all cases. This article gives an Eq. (27) for directly calculating the curvature of the L-curve of single-parameter Tikhonov regularization, and the specific derivation is shown in Appendix. Figure 5 shows the variation of the curvature of L-curve with regularization parameter (α).

3.5 The slope of the L-curve

Comparing the S-curve (Fig. 1) with the L-curve (Fig. 4), we can see that the two types of curves have similar shapes. If we exchange the horizontal and vertical coordinates of the L-curve, the resulting curve is the mirror image of the S-curve. It naturally refers us to the S-curve method, so we

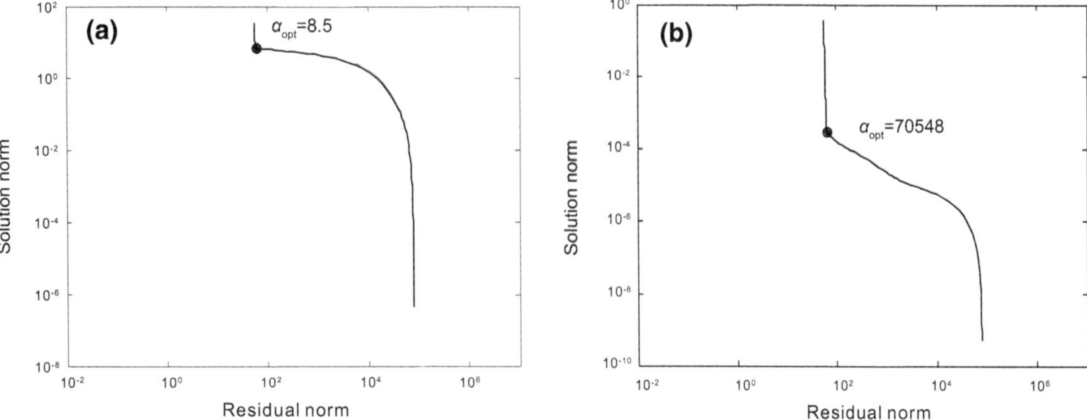

Fig. 4 L-curve. **a** Norm smoothing, **b** curvature smoothing

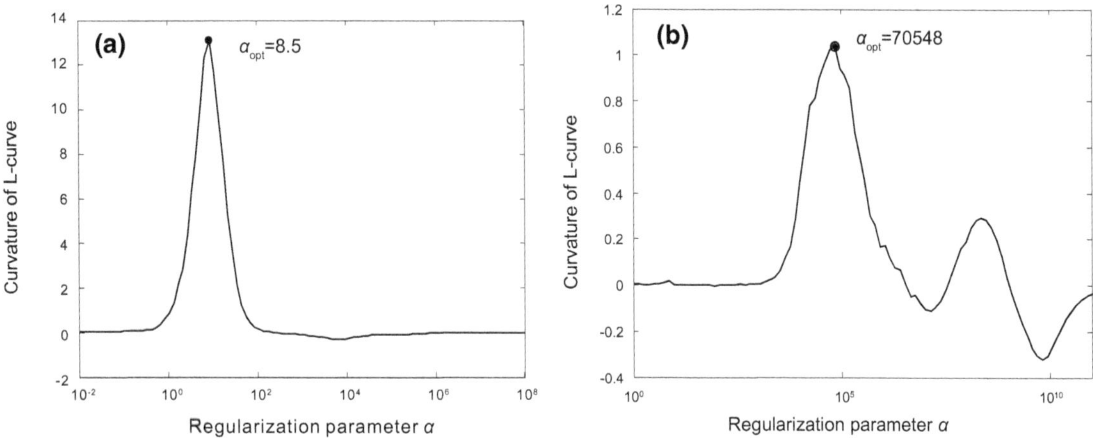

Fig. 5 The variation of the curvature of L-curve with regularization parameter. **a** Norm smoothing, **b** curvature smoothing

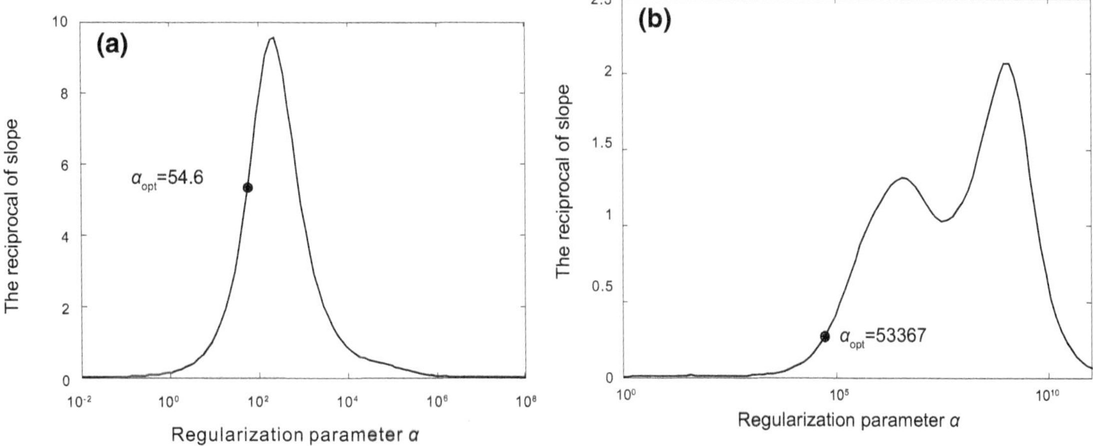

Fig. 6 The variation of the reciprocal of the absolute value of the slope of L-curve with regularization parameter. **a** Norm smoothing, **b** curvature smoothing

can use the slope of the L-curve to select the regularization parameter. According to Eq. (24), the slope of the L-curve has a simple formula and can avoid the matrix inversion of the S-curve method (see Eq. (21)). Figure 6 shows the variation of the reciprocal of the absolute value of the slope of the L-curve with regularization parameter (α).

Comparing Fig. 6 with Fig. 3, it can be found that the two types of curves have striking similarities. So, we attempt to use the slope of the L-curve criterion to select the regularization parameter, and compare its results with those of the other methods. The threshold of the reciprocal of the absolute value of the slope of the L-curve can be selected in the interval [0.1, 10]. Here we choose the threshold values of 5 and 0.25 for norm smoothing and curvature smoothing methods, respectively.

4 Numerical results

A bimodal T_2 distribution model is constructed as shown in Fig. 7, where 64 T_2 components are preselected between 0.1 and 10,000 ms. Using this T_2 distribution model, we generated echo trains with 500 echoes, where echo spacing is 0.9 ms. Different level Gaussian random noise was applied, as shown in Fig. 8. The red line represents the echo train without noise, and the green, blue, magenta, and black lines show the echo trains with noise levels of 0.25, 0.5, 1.0, and 2.0 porosity unit (pu), respectively.

To improve the inversion speed of NMR echo data, the echo trains are usually compressed before inversion (Sezginer 1994; Dunn and LaTorraca 1999; Venkataramanan et al. 2002; Zou and Xie 2015). Here, we compress the NMR echo data with the SVD method (Sezginer 1994; Zou and Xie 2015). After compression, echoes in each echo train shown in Fig. 8 are all compressed to 20 data points for NMR T_2 inversion.

We select optimal regularization parameters (α_{opt}) by the DP, GCV, S-curve, L-curve, and the slope of L-curve methods, respectively, and compare the NMR T_2 inversion results from different regularization parameter selection methods. Figure 9 shows the NMR T_2 inversion results of

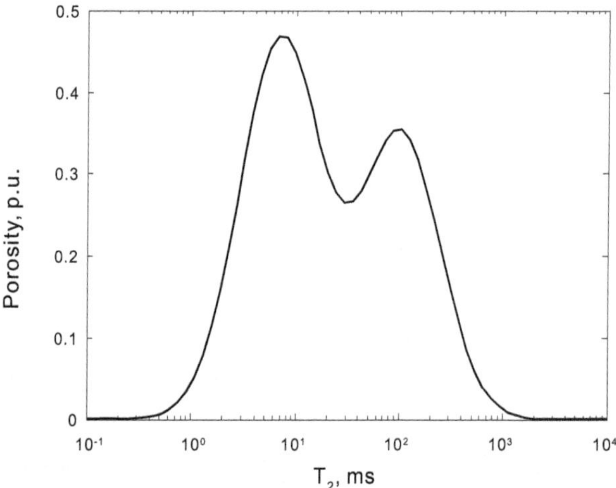

Fig. 7 The T_2 distribution model

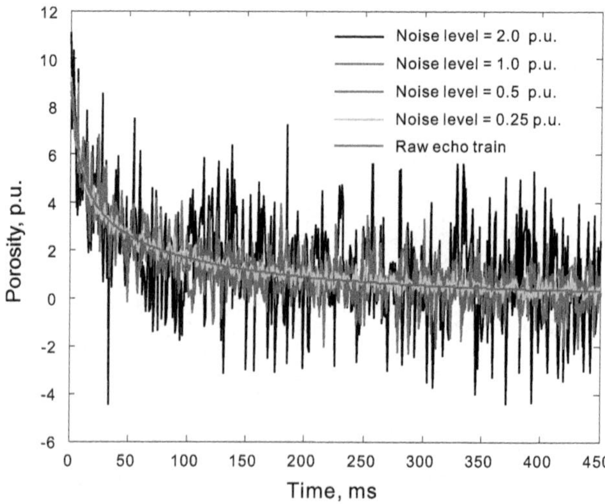

Fig. 8 The computed echo trains from T_2 distribution model with different noise levels

norm smoothing and curvature smoothing methods with respect to noise levels of 0.25, 0.5, 1.0, and 2.0 pu. The black line is the T_2 distribution model, and the green, blue, magenta, red, and cyan lines are the T_2 distribution inversion results according to the DP, GCV, S-curve, L-curve, and the slope of L-curve methods, respectively. As shown in Fig. 9, the selected α_{opt} from the DP method is sometimes small or large (because of the underestimated or overestimated echo data noise level) that leads to an undersmoothed or oversmoothed solution. The α_{opt} selected by L-curve method is occasionally small or large which leads to an undersmoothed or oversmoothed T_2 distribution. The inversion results from the GCV, S-curve, and the slope of L-curve methods are close and satisfactory. Curvature smoothing can better suppress the oscillation caused by noise than norm smoothing, and can obtain a smoother solution than norm smoothing. The curvature smoothing makes the inverted T_2 distribution prone to show single peak shape than norm smoothing when the noise level is high. Table 1 shows the porosity errors of different regularization parameter selection methods, and finds that the DP method occasionally obtains a large porosity error, and the porosity errors of the GCV, S-curve, L-curve, and slope of L-curve methods are close.

5 Well data processing results

Well A is in a tight sandstone reservoir with low porosity. The signal-to-noise ratio (SNR) of the NMR logging data is low. Figure 10 shows the inverted T_2 distributions of norm smoothing, where the fourth track represents the inverted T_2 distributions of the SVD method, the fifth to ninth tracks represent the inversion results of the DP, GCV, S-curve,

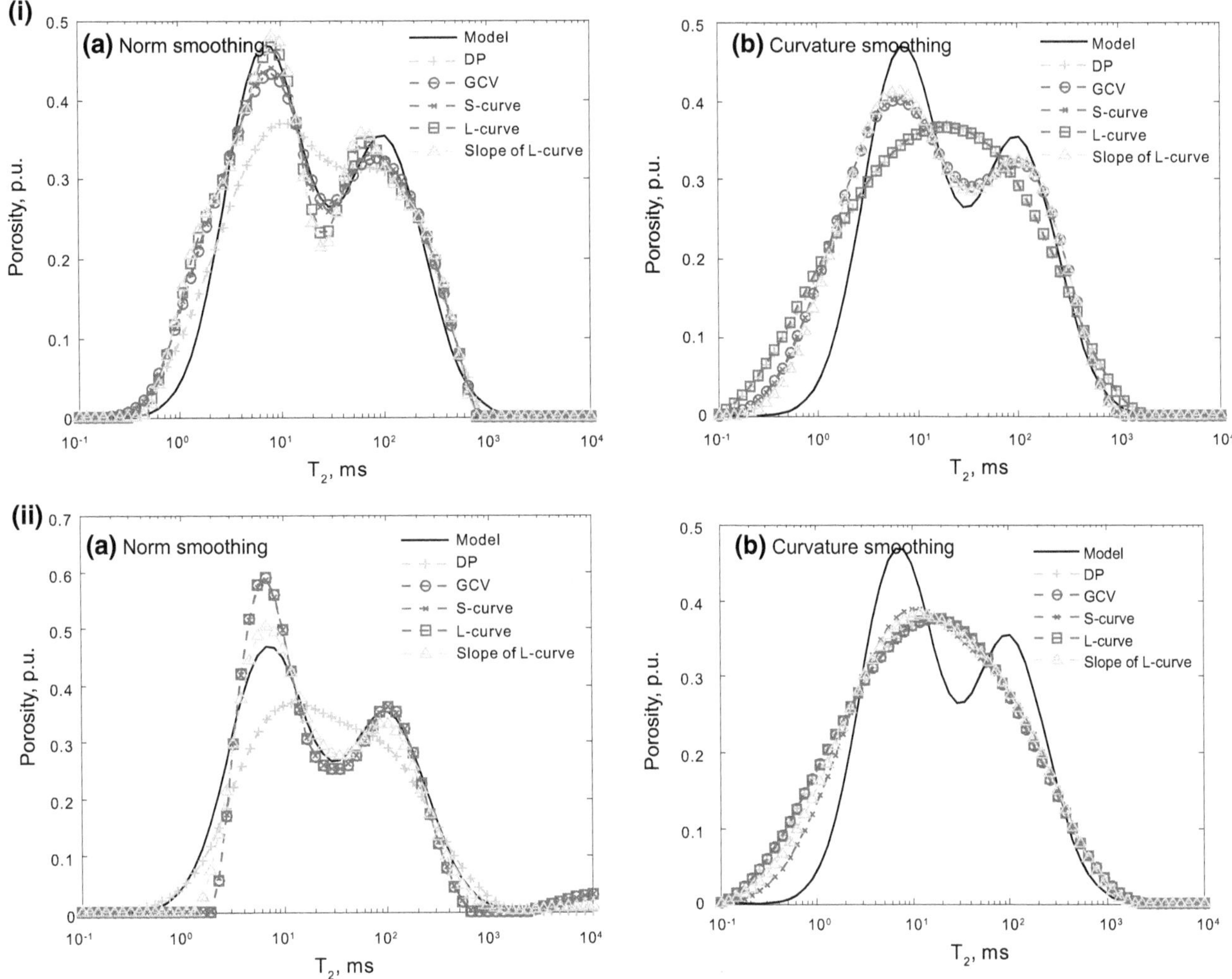

Fig. 9 The inversion results from different regularization parameter selection methods. **i** Noise level is 0.25 pu. **a** norm smoothing; **b** curvature smoothing. **ii** Noise level is 0.5 pu. **a** norm smoothing; **b** curvature smoothing. **iii** Noise level is 1.0 pu. **a** norm smoothing; **b** curvature smoothing. **iv** Noise level is 2.0 pu. **a** norm smoothing; **b** curvature smoothing

L-curve, and slope of L-curve methods, respectively. As shown in Fig. 10, the inverted T_2 distributions of the SVD method, usually with single peak shapes, are oversmoothed. The inverted T_2 distributions of norm smoothing are satisfactory. The inverted T_2 distributions from the DP and L-curve methods are slightly oversmoothed. However, the inverted T_2 distributions from the GCV and S-curve methods are slightly undersmoothed. The inverted T_2 distributions from the slope of L-curve method are relatively more satisfactory than those from other methods.

6 Conclusions

This paper uses the Tikhonov regularization with a non-negative constraint of solution for NMR T_2 inversion. The non-negative constraint of solution is implemented by variable substitution, and then the modified objective function is solved by the Levenberg–Marquardt method. The optimal regularization parameters (α_{opt}) from norm smoothing and curvature smoothing methods are selected by the DP, GCV, S-curve, L-curve, and the slope of L-curve methods, respectively. The following conclusions are obtained.

(1) The inverted NMR T_2 distributions from the DP method depend on the estimated noise level which is difficult to estimate accurately. The inversion results from the GCV, S-curve, and the slope of L-curve methods are satisfactory. The small or large α_{opt} selected by the L-curve method leads to an oscillation or oversmoothed T_2 distribution. When the noise level is high, norm smoothing can more effectively than curvature smoothing avoid the bimodal T_2 distribution being converted to a single peak distribution. The inverted T_2 distribution of

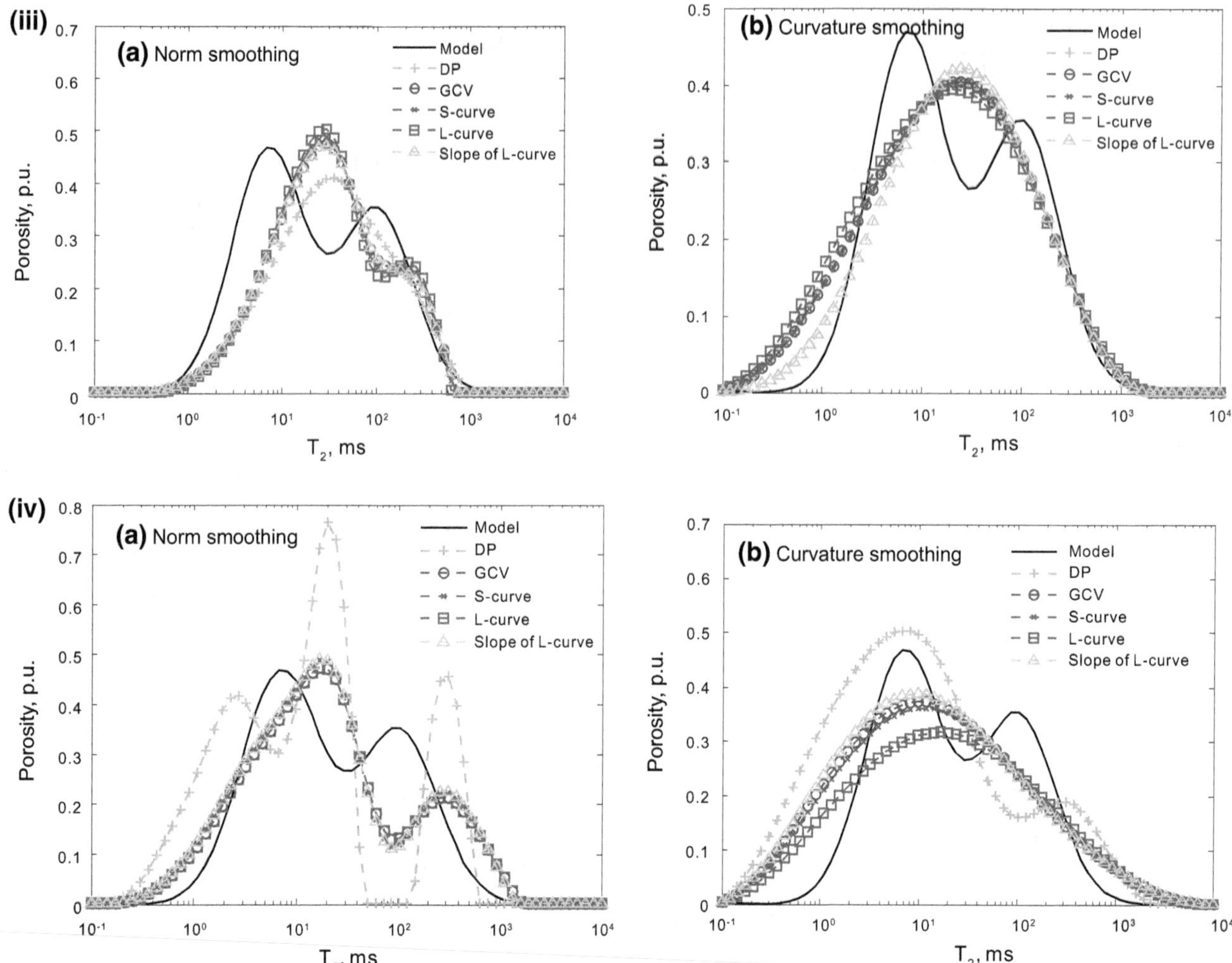

Fig. 9 continued

Table 1 Porosity errors obtained from different regularization parameter selection methods (unit: pu)

Method	DP	GCV	S-curve	L-curve	Slope of L-curve
Norm smoothing					
Noise = 0.25	0.1835	−0.5536	−0.5489	−0.5857	−0.5858
Noise = 0.5	0.9424	0.5039	0.5047	0.5017	0.5295
Noise = 1.0	1.7092	1.1347	1.1220	1.0929	1.1759
Noise = 2.0	−2.2772	−0.1316	−0.3117	0.2869	−0.3521
Curvature smoothing					
Noise = 0.25	−1.0283	−1.1421	−1.0202	−1.0802	−0.8769
Noise = 0.5	−0.8131	−0.8078	−0.4299	−0.8134	−0.6150
Noise = 1.0	−0.6772	−0.5187	−0.6139	−0.9139	0.0128
Noise = 2.0	−3.6340	−1.1841	−1.0550	0.0955	−1.6763

norm smoothing is better than that of curvature smoothing.

(2) The GCV and L-curve methods need to calculate the solution of the regularization parameter over a wide range, which needs a large amount of calculation. The S-curve and the slope of L-curve methods can quickly find the α_{opt} by iteration, suitable for norm smoothing and curvature smoothing. The slope of L-curve method needs less calculation than the S-curve method does, but the T_2 inversion results

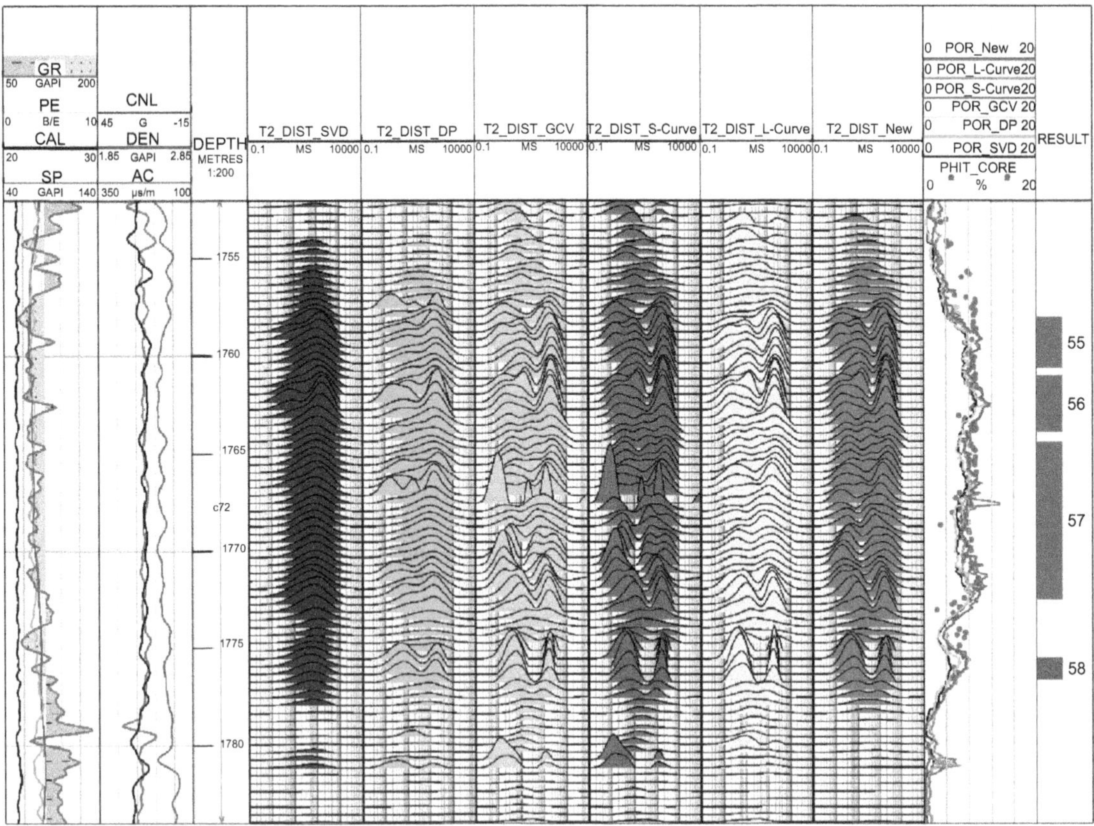

Fig. 10 The inversion results of NMR logging data

from these two methods are close. So, the slope of the L-curve method can be an efficient alternative to the S-curve method.

Acknowledgments This work was funded by Shell International Exploration and Production Inc. (PT45371), the National Natural Science Foundation of China—China National Petroleum Corporation Petrochemical Engineering United Fund (U1262114), and the National Natural Science Foundation of China (41272163).

Appendix

The objective function of single-parameter Tikhonov regularization is given by

$$\min\left\{\frac{1}{2}\|W(Af-b)\|^2+\frac{\alpha}{2}\|Lf\|^2\right\}. \tag{12}$$

The derivative of the above equation with respect to f is written as

$$\left((WA)^{\mathrm{T}}(WA)+\alpha L^{\mathrm{T}}L\right)f=(WA)^{\mathrm{T}}Wb. \tag{13}$$

Let $\zeta=\|W(Af-b)\|^2$, $\eta=\|Lf\|^2$. The curvature κ of L-curve ($\log\zeta$, $\log\eta$) is

$$\kappa=\frac{\frac{\mathrm{d}^2\log\eta}{\mathrm{d}(\log\zeta)^2}}{\left(1+\left(\frac{\mathrm{d}\log\eta}{\mathrm{d}\log\zeta}\right)^2\right)^{3/2}}. \tag{14}$$

To calculate the κ, we need to compute $\frac{\mathrm{d}^2\log\eta}{\mathrm{d}(\log\zeta)^2}$ and $\frac{\mathrm{d}\log\eta}{\mathrm{d}\log\zeta}$. Since

$$\frac{\mathrm{d}\log\eta}{\mathrm{d}\log\zeta}=\frac{\zeta}{\eta}\frac{\mathrm{d}\eta}{\mathrm{d}\zeta}, \tag{15}$$

$$\frac{\mathrm{d}^2\log\eta}{\mathrm{d}(\log\zeta)^2}=\frac{\zeta}{\eta}\frac{\mathrm{d}\eta}{\mathrm{d}\zeta}-\frac{\zeta^2}{\eta^2}\left(\frac{\mathrm{d}\eta}{\mathrm{d}\zeta}\right)^2+\frac{\zeta^2}{\eta}\frac{\mathrm{d}^2\eta}{\mathrm{d}\zeta^2}, \tag{16}$$

the computation of Eq. (14) is converted to compute $\frac{\mathrm{d}\eta}{\mathrm{d}\zeta}$ and $\frac{\mathrm{d}^2\eta}{\mathrm{d}\zeta^2}$.

The derivative of Eq. (13) with respect to α is

$$L^{\mathrm{T}}Lf + \left((WA)^{\mathrm{T}}(WA) + \alpha L^{\mathrm{T}}L\right)\frac{\mathrm{d}f}{\mathrm{d}\alpha} = 0. \tag{17}$$

So,

$$\frac{\mathrm{d}f}{\mathrm{d}\alpha} = -\left((WA)^{\mathrm{T}}(WA) + \alpha L^{\mathrm{T}}L\right)^{-1}L^{\mathrm{T}}Lf. \tag{18}$$

We should note that according to the non-negative constraint, only the rows and columns of the matrices $(WA)^{\mathrm{T}}(WA)$ and $L^{\mathrm{T}}L$ corresponding to $f>0$ will be involved in the calculation.

So, the derivative of η with respect to α is

$$\frac{\mathrm{d}\eta}{\mathrm{d}\alpha} = -2f^{\mathrm{T}}L^{\mathrm{T}}L\left((WA)^{\mathrm{T}}(WA) + \alpha L^{\mathrm{T}}L\right)^{-1}L^{\mathrm{T}}Lf. \tag{19}$$

The derivative of ζ with respect to α is

$$\frac{\mathrm{d}\zeta}{\mathrm{d}\alpha} = 2\alpha f^{\mathrm{T}}L^{\mathrm{T}}L\left((WA)^{\mathrm{T}}(WA) + \alpha L^{\mathrm{T}}L\right)^{-1}L^{\mathrm{T}}Lf. \tag{20}$$

So,

$$\frac{\mathrm{d}\log\zeta}{\mathrm{d}\log\alpha} = \frac{2\alpha^2 f^{\mathrm{T}}L^{\mathrm{T}}L\left((WA)^{\mathrm{T}}(WA) + \alpha L^{\mathrm{T}}L\right)^{-1}L^{\mathrm{T}}Lf}{\zeta}. \tag{21}$$

And,

$$\frac{\mathrm{d}\eta}{\mathrm{d}\zeta} = \frac{\mathrm{d}\eta}{\mathrm{d}\alpha}\frac{\mathrm{d}\alpha}{\mathrm{d}\zeta} = \frac{-1}{\alpha}, \tag{22}$$

$$\frac{\mathrm{d}^2\eta}{\mathrm{d}\zeta^2} = \frac{\mathrm{d}\left(\frac{\mathrm{d}\eta}{\mathrm{d}\zeta}\right)}{\mathrm{d}\zeta} = \frac{\mathrm{d}\left(-\frac{1}{\alpha}\right)}{\mathrm{d}\zeta} = \frac{1}{\alpha^2}\frac{\mathrm{d}\alpha}{\mathrm{d}\zeta}. \tag{23}$$

Substituting Eqs. (22) and (23) into Eqs. (15) and (16), then

$$\frac{\mathrm{d}\log\eta}{\mathrm{d}\log\zeta} = \frac{-\zeta}{\alpha\eta}, \tag{24}$$

$$\frac{\mathrm{d}^2\log\eta}{\mathrm{d}(\log\zeta)^2} = -\frac{\zeta}{\alpha\eta} - \frac{\zeta^2}{\alpha^2\eta^2} + \frac{\zeta^2}{2\alpha^3\eta}\frac{1}{f^{\mathrm{T}}L^{\mathrm{T}}L\left((WA)^{\mathrm{T}}(WA) + \alpha L^{\mathrm{T}}L\right)^{-1}L^{\mathrm{T}}Lf}. \tag{25}$$

Substituting Eqs. (21) and (24) into Eq. (25), then

$$\frac{\mathrm{d}^2\log\eta}{\mathrm{d}(\log\zeta)^2} = \frac{\mathrm{d}\log\eta}{\mathrm{d}\log\zeta} - \left(\frac{\mathrm{d}\log\eta}{\mathrm{d}\log\zeta}\right)^2 - \frac{\mathrm{dlog}\alpha}{\mathrm{dlog}\zeta}\frac{\mathrm{d}\log\eta}{\mathrm{d}\log\zeta}. \tag{26}$$

Substituting Eq. (26) into Eq. (14), then

$$\kappa = \frac{\frac{\mathrm{d}\log\eta}{\mathrm{d}\log\zeta} - \left(\frac{\mathrm{d}\log\eta}{\mathrm{d}\log\zeta}\right)^2 - \frac{\mathrm{dlog}\alpha}{\mathrm{dlog}\zeta}\frac{\mathrm{d}\log\eta}{\mathrm{d}\log\zeta}}{\left(1 + \left(\frac{\mathrm{d}\log\eta}{\mathrm{d}\log\zeta}\right)^2\right)^{3/2}}. \tag{27}$$

References

Borgia GC, Brown RJS, Fantazzini P. Uniform-penalty inversion of multiexponential decay data. J Magn Reson. 1998;132(1):65–77.

Borgia GC, Brown RJS, Fantazzini P. Uniform-penalty inversion of multiexponential decay data II. Data spacing, T_2 data, systematic data errors, and diagnostics. J Magn Reson. 2000;147(2):273–85.

Butler JP, Reeds JA, Dawson SV. Estimating solutions of first kind integral equations with nonnegative constraints and optimal smoothing. SIAM J Numer Anal. 1981;18(3):381–97.

Castellanos JL, Gómez S, Guerra V. The triangle method for finding the corner of the L-curve. Appl Numer Math. 2002;43(4):359–73.

Dunn KJ, LaTorraca GA. The inversion of NMR log data sets with different measurement errors. J Magn Reson. 1999;140(1):153–61.

Dunn KJ, LaTorraca GA, Warner JL, et al. On the calculation and interpretation of NMR relaxation time distributions. In: SPE annual technical conference and exhibition. Society of Petroleum Engineers. 1994.

Ge X, Wang H, Fan Y, et al. Joint inversion of T_1–T_2 spectrum combining the iterative truncated singular value decomposition and the parallel particle swarm optimization algorithms. Comput Phys Commun. 2016;198:59–70.

Golub GH, Heath M, Wahba G. Generalized cross-validation as a method for choosing a good ridge parameter. Technometrics. 1979;21(2):215–23.

Guerra V, Hernandez V. Numerical aspects in locating the corner of the L-curve. In: Approximation, optimization and mathematical economics. Heidelberg: Physica-Verlag; 2001. p. 121–31.

Hansen PC. Analysis of discrete ill-posed problems by means of the L-curve. SIAM Rev. 1992;34(4):561–80.

Hansen PC. Rank-deficient and discrete ill-posed problems: numerical aspects of linear inversion. Philadelphia: SIAM; 1998.

Hansen PC, Jensen TK, Rodriguez G. An adaptive pruning algorithm for the discrete L-curve criterion. J Comput Appl Math. 2007;198(2):483–92.

Kaufman L, Neumaier A. PET regularization by envelope guided conjugate gradients. IEEE Trans Med Imaging. 1996;15(3):385–9.

Madsen K, Nielsen HB. Introduction to optimization and data fitting. Lyngby: Technical University of Denmark; 2010.

Morozov VA. On the solution of functional equations by the method of regularization. Sov Math Dokl. 1966;7(1):414–7.

Prammer MG. NMR pore size distributions and permeability at the well site. In: SPE annual technical conference and exhibition. 1994.

Sezginer A. Determining bound and unbound fluid volumes using nuclear magnetic resonance pulse sequences. US Patent 5,363,041, 1994.

Song YQ, Venkataramanan L, Hürlimann MD, et al. T_1–T_2 correlation spectra obtained using a fast two-dimensional Laplace inversion. J Magn Reson. 2002;154(2):261–8.

Venkataramanan L, Song YQ, Hürlimann MD. Solving Fredholm integrals of the first kind with tensor product structure in 2 and 2.5 dimensions. IEEE Trans Signal Process. 2002;50(5):1017–26.

Zou YL, Xie RH. A novel method for NMR data compression. Comput Geosci. 2015;19(2):389–401.

8

Tight sandstone gas accumulation mechanism and development models

Zhen-Xue Jiang[1,2] · Zhuo Li[1,2] · Feng Li[1,2] · Xiong-Qi Pang[1,3] · Wei Yang[1,2] · Luo-Fu Liu[1,3] · Fu-Jie Jiang[1,3]

Abstract Tight sandstone gas serves as an important unconventional hydrocarbon resource, and outstanding results have been obtained through its discovery both in China and abroad given its great resource potential. However, heated debates and gaps still remain regarding classification standards of tight sandstone gas, and critical controlling factors, accumulation mechanisms, and development modes of tight sandstone reservoirs are not determined. Tight sandstone gas reservoirs in China are generally characterized by tight strata, widespread distribution areas, coal strata supplying gas, complex gas–water relations, and abnormally low gas reservoir pressure. Water and gas reversal patterns have been detected via glass tube and quartz sand modeling, and the presence of critical geological conditions without buoyancy-driven mechanisms can thus be assumed. According to the timing of gas charging and reservoir tightening phases, the following three tight sandstone gas reservoir types have been identified: (a) "accumulation–densification" (AD), or the conventional tight type, (b) "densification–accumulation" (DA), or the deep tight type, and (c) the composite tight type. For the AD type, gas charging occurs prior to reservoir densification, accumulating in higher positions under buoyancy-controlled mechanisms with critical controlling factors such as source kitchens (S), regional overlaying cap rocks (C), gas reservoirs, (D) and low fluid potential areas (P). For the DA type, reservoir densification prior to the gas charging period (GCP) leads to accumulation in depressions and slopes largely due to hydrocarbon expansive forces without buoyancy, and critical controlling factors are effective source rocks (S), widely distributed reservoirs (D), stable tectonic settings (W) and universal densification of reservoirs (L). The composite type includes features of the AD type and DA type, and before and after reservoir densification period (RDP), gas charging and accumulation is controlled by early buoyancy and later molecular expansive force respectively. It is widely distributed in anticlinal zones, deep sag areas and slopes, and is controlled by source kitchens (S), reservoirs (D), cap rocks (C), stable tectonic settings (W), low fluid potential areas (P), and universal reservoir densification (L). Tight gas resources with great resource potential are widely distributed worldwide, and tight gas in China that presents advantageous reservoir-forming conditions is primarily found in the Ordos, Sichuan, Tarim, Junggar, and Turpan-Hami basins of central-western China. Tight gas has served as the primary impetus for global unconventional natural gas exploration and production under existing technical conditions.

Keywords Tight sandstone gas · Reservoir features · Accumulation mechanism · Type classification · Development mode

✉ Zhen-Xue Jiang
jiangzx@cup.edu.cn

Zhuo Li
zhuo.li@cup.edu.cn

1 State Key Laboratory of Petroleum Resources and Prospecting, China University of Petroleum, Beijing 102249, China

2 Unconventional Natural Gas Institute, China University of Petroleum, Beijing 102249, China

3 College of Geosciences, China University of Petroleum, Beijing 102249, China

Edited by Jie Hao

1 Introduction

The field of tight sandstone gas exploration has witnessed global breakthroughs since the resource was first discovered in the San Juan Basin of the USA in 1927. Tight sandstone gas, belonging to unconventional gas reservoirs, is mainly found in North America, the Asia–Pacific region, Europe, and the Middle East, with total proven reserves of around 210×10^{12} m^3 (IEA 2009). Recently, it has served as the main source of global natural gas reserve and production growth (Zou et al. 2011a; Dai et al. 2012; Pang et al. 2013).

Given the considerable resource potential of tight sandstone gas, a series of studies have focused on development conditions, accumulation mechanisms, and type classifications. Accumulation mechanisms involving relative permeability sealing, diagenesis sealing, force balance sealing, and lateral fault sealing have been proposed (Masters 1979; Gies 1984; Jiang et al. 2000, 2006; Pang et al. 2003; Jin et al. 2003). However, accumulation controlling factors, modes, and mechanisms of tight sandstone gas are poorly understood. Based on a review of previous studies of tight sandstone gas and through a detailed case study of typical tight sandstone gas reservoirs found in China, this paper discusses accumulation mechanisms, type classifications, accumulation controlling factors, development modes, and resource potential of tight sandstone gas. Such efforts will play a significant role in enriching natural gas geological theories and in advancing the exploration and development of tight sandstone gas.

2 Concepts and exploration of tight sandstone gas

2.1 Tight sandstone gas concepts

Tight sandstone gas is natural gas contained in tight sandstone reservoirs with porosity of <10 % and in situ permeability of $<0.1 \times 10^{-3}$ μm^2, which belongs to unconventional gas reservoirs. Tight sandstone gas is mainly found in densified reservoirs with micro-nano pores and throats, having generally limited or no natural productivity that is typically less than the lower bound of industrial gas flows. The industrial gas production can be obtained only under specific economic and technical conditions (hydraulic fracturing reform measures or horizontal and multi-lateral wells) (Zou et al. 2011a). The tight sandstone gas has become an important field of natural gas exploration and development in recent years owing to its great resource potential (Dai et al. 2012; Zou et al. 2013).

Concepts of tight sandstone gas have varied under different technical and economic conditions at different times and in different countries. Masters (1979) first presented a definition of deep basin gas, and several other researchers have attempted to refine its description in the following terms: "tight sandstone gas reservoir," "flip-type syncline gas reservoir," "basin-centered gas reservoir," "continuous gas reservoir," "source-contacting gas," etc. (Masters 1979; Walls 1982; Dai 1983; Rose et al. 1984; Law and Dickinson 1985; Schmoker 1995; Jiang et al. 2000, 2006; Jin et al. 2003; Zhang 2006; Zou et al. 2011a) (Table 1). However, these discrimination criteria are derived from geological features without concern for essential issues pertaining to tight sandstone gas genesis.

A unified classification of tight sandstone gas reservoirs has not been created. Definitions of tight sandstone gas reservoirs are continuously improved (Table 2). In 1978, the US Federal Energy Regulatory Commission (FERC) first created the now prevailing criteria for tight sandstone gas reservoirs with the in situ permeability of less than 0.1×10^{-3} μm^2. Spencer also offered a criterion of reservoir permeability of less than 0.1×10^{-3} μm^2 (Spencer 1985). A consensus in terms of geological evaluation criteria for tight sandstone gas has been reached gradually with intense debate. In 2010, the China National Petroleum Corporation (CNPC) proposed a trade standard for China as reservoir porosity <10 %, in situ permeability

Table 1 Definitions of tight sandstone gas

	Main evidence	Scholars
Deep tight reservoirs	Geological conditions and features	Masters (1979), Gies (1984), Jiang (2000, 2006), Jin (2003), Ma (2008), Pang et al. (2003)
Flip-type syncline reservoirs	Hydrocarbon-water distribution relationship	Dai (1983), Chen (1998), Wu et al. (2007)
Tight sandstone gas reservoirs	Reservoir physical properties	Spencer (1989), Surdam (1997), Yang and Pang (2012), Yang et al. (2013)
Continuous gas reservoirs	Reservoir continuity	Schmoker (1995, 2002, 2005)
Basin-centered gas reservoirs	Distribution zones in basins	Law (2002), Rose et al. (1984), Chen et al. (2003)
Source-contacting gas reservoirs	Gas source characteristics	Zhang et al. (2000)

Table 2 Tight sandstone gas classification criteria

Scholars or organizations	Classification criteria
FERC (1978)	Original reservoir permeability $\leq 0.1 \times 10^{-3}$ μm^2
Wyman (1985)	Porosity <10 %, permeability $\leq 0.1 \times 10^{-3}$ μm^2
Spencer (1985, 1989)	In situ permeability $\leq 0.1 \times 10^{-3}$ μm^2
Surdam (1997)	Permeability $\leq 1 \times 10^{-3}$ μm^2
Guan and Niu (1995)	Porosity ≤12 %, permeability $\leq 1 \times 10^{-3}$ μm^2, gas saturation ≤60 %, water saturation > 40 %
Dai et al. (1996)	Porosity <10 %, permeability $\leq 0.5 \times 10^{-3}$ μm^2
Yuan et al. (1996)	Porosity <12 %, surface permeability $\leq 1 \times 10^{-3}$ μm^2, strata permeability $<1 \times 10^{-3}$ μm^2
Wang et al. (2004)	Porosity 2 %–8 %, permeability: (0.1–0.001) $\times 10^{-3}$ μm^2
Yang et al. (2005)	Porosity 7 %–12 %, air permeability $\leq 1.0 \times 10^{-3}$ μm^2, pore throat radius <0.5 μm
Holditch (2006)	Permeability $\leq 0.1 \times 10^{-3}$ μm^2
Nehring (2008)	Permeability $<1 \times 10^{-3}$ μm^2
USGS	Pore throat diameter 0.03–2 μm
IEA (2009)	Permeability $\leq 0.1 \times 10^{-3}$ μm^2
CNPC	Overburden matrix permeability $<0.1 \times 10^{-3}$ μm^2, pore throat radius <1 μm, porosity <10 %, gas saturation <60 %
NEA (2011)	Overburden matrix permeability $<0.1 \times 10^{-3}$ μm^2
Zou et al. (2011a)	Porosity <10 %, permeability $\leq 1 \times 10^{-3}$ μm^2
Pang et al. (2013)	Porosity ≤12 %, effective permeability $\leq 0.1 \times 10^{-3}$ μm^2 (absolute permeability $\leq 1 \times 10^{-3}$ μm^2)

$<0.1 \times 10^{-3}$ μm^2 (air permeability $<1.0 \times 10^{-3}$ μm^2), and gas saturation <60 % (SY/T6832-2011), and these criteria play a key role in guiding geological evaluation and exploration of tight sandstone gas in China.

2.2 Overview of tight sandstone gas exploration

2.2.1 Overview of world tight sandstone gas exploration activities

The world's tight gas resources are mainly found in North America, the Asia–Pacific region, Europe, and the Middle East (Law 2002; Schmoker 2005; Zou et al. 2011a; Dai et al. 2012; Zhao et al. 2012; Yang et al. 2012). Tight gas reservoirs in North America, distributed across roughly 20 basins in the Rocky Mountain Basin Group and the Gulf Coast, are regarded as some of the most typical reservoirs (Law 2002; Lei et al. 2010). The San Juan Basin includes a large gas-bearing area of deep tight gas of approximately 9325 km^2, with porosity ranging from 5.8 % to 7.6 %, permeability varying from 0.01×10^{-3} to 0.15×10^{-3} μm^2, and with geological reserves of some 0.90×10^{12} m^3 (Fig. 1a). The Red Desert Basin covers an area of 2400 km^2, with porosity ranging from 3.0 % to 7.0 %, with permeability falling below 1.0×10^{-3} μm^2, and with geological reserves of 1.4×10^{12} m^3 (Fig. 1b). The Greater Green River Basin covers an area of 51,022 km^2, with porosity ranging from 8.0 % to 12.0 %, with permeability varying from 0.1×10^{-3} to 0.9×10^{-3} μm^2, and with geological reserves of around 0.26×10^{12} m^3 (Fig. 1c). The Elmworth deep tight gas in the western depression of the Alberta Basin is the largest natural gas reservoir in North America. It covers a gas-bearing area of 13,000 km^2 with a Cretaceous reservoir thickness of 3000 m, an average porosity of 8.0 %, permeability less than 1.0×10^{-3} μm^2, and with geological reserves of 4.8×10^{12} m^3 (Fig. 1d).

Tight sandstone gas has become a major source of global natural gas reserves and production growth (Schmoker 2005; IEA 2009; Zou et al. 2010). Tight sandstone gas has been commercially exploited on a large scale in more than 10 countries including the United States, Canada, and China, with the United States being the first to successfully develop and exploit the resource, and now it is the world leader in this field. Tight sandstone gas exploration and development began in the late 1970s, and by 2010, American research teams had identified roughly 900 tight gas fields across 23 basins with more than 10×10^4 production wells and with gas production of 1754×10^8 m^3, accounting for roughly 26 % of total natural gas production in the United States. By 2013, tight sandstone gas production accounted for a third of total US natural gas production (IEA 2013).

2.2.2 Overview of tight sandstone gas exploration in China

The discovery of the Zhongba gas field in the western Sichuan Basin in 1971 is considered to have initiated tight gas exploration and research in China, with tight sandstone

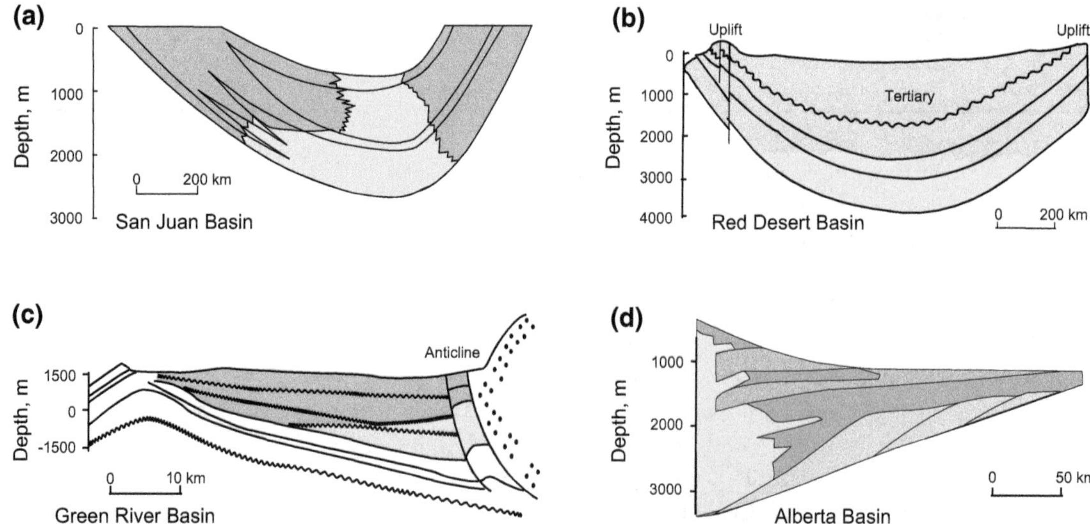

Fig. 1 Typical tight sandstone gas reservoirs in the USA (Modified from Masters 1979; Law 2002). **a** Cretaceous gas reservoirs in the San Juan Basin; **b** Cretaceous gas reservoirs in the Red Desert Basin; **c** Cretaceous, Paleogene, and Neogene gas reservoirs in the Greater Green River Basin; **d** Cretaceous gas reservoirs in the Alberta Basin

gas reserves later being found in the Ordos Basin. However, tight sandstone gas exploration efforts progressed slowly due to a lack of effective evaluation criteria and engineering technologies. In recent years, tight sandstone gas exploration and production have obtained considerable breakthroughs through the application of fracturing technologies. The geological reserves of tight gas reached 3.3×10^{12} m^3 in 2011, accounting for roughly 39 % of total proven reserves of natural gas. The production of tight gas in 2012 and 2013 yielded 300×10^8 and 340×10^8 m^3, respectively, accounting for roughly 28 % of total natural gas production. Tight gas has in turn emerged as the most feasible unconventional gas resource (Dai et al. 2012; Pang et al. 2014).

Ideal geological conditions for tight sandstone gas development (wide distribution and various types) are found in China. Five major confirmed gas-bearing areas include the Ordos, western Sichuan, Tarim, southern Junggar, and Songliao faulted basins. The Ordos and Sichuan Basins have been identified as the tight sandstone gas-bearing areas with the greatest resource potential (Dai et al. 2012; Zou et al. 2012; Zhao et al. 2012).

3 Tight sandstone gas reservoir features

3.1 Reservoir densification and wide distribution

Reservoir densification is the most essential feature of tight sandstone gas. The tight sandstone gas reservoir of the Green River Basin in the United States has porosity ranging from 4.7 % to 11.7 % and permeability ranging from 0.001×10^{-3} to 0.05×10^{-3} μm^2. The tight sandstone gas reservoir of the Alberta Basin in Canada has porosity ranging from 3 % to 13 % and permeability ranging from 0.005×10^{-3} to 0.015×10^{-3} μm^2. The tight sandstone gas reservoir of the Ordos Basin has porosity ranging from 1 % to 12 % and permeability ranging from 0.01×10^{-3} to 1×10^{-3} μm^2. The tight sandstone gas reservoir in the Xujiahe Formation of the central Sichuan Basin has porosity ranging from 4 % to 10 %, and samples with in situ permeability $<0.1 \times 10^{-3}$ μm^2 account for roughly 80 %–92 % (Zou et al. 2011a) (Table 3). Compared with conventional reservoirs, heterogeneous tight reservoirs are characterized by nano pores, with milli-micro pores developing locally. Mercury injection data for typical samples show that 77.2 % of samples have nano pores with a radius of <1 μm, and for tight sandstone gas reservoirs in the Sulige area of the Ordos Basin, pores and throats with a radius of <1 μm account for 83.6 %. Most reservoir spaces consist of intergranular pores, intragranular pores, intercrystal pores, and intergranular cracks, which provide the majority of space for natural gas (Zou et al. 2011b).

In China, the presence of tight sandstone gas is often related to the development of coal strata, as acidic sedimentary, diagenetic and extruded anticline structure environments serve as the main reasons of sandstone densification, where diagenesis processes have the greatest influence and mainly involve compaction, cementation, replacement, dissolution, and clay mineral transformation. Compaction (mechanical and chemical) and cementation serve as the main driving forces behind reservoir densification (Li et al. 2012).

Table 3 Physical properties of typical tight sandstone gas reservoirs in China

	Ordos Basin	Sichuan Basin	Southern Songliao Basin	Turpan-Hami Basin	Junggar Basin	Tarim Basin	
Strata	C–P	T_3x	K	J_1	J_1b	S	J
Depth, m	2000–5200	2000–5200	2200–3500	3000–3650	4200–4800	4800–6500	3800–4900
Porosity							
Mean, %	6.695	4.200	3.200	5.012	9.100	6.513	2.780
Average, %	6.930	5.65	3.350	5.160	9.040	6.980	6.490
Sample number	6015	39,999	61	25	51	1019	4720
Permeability							
Mean, $\times 10^{-3}$ μm^2	0.229	0.057	0.034	0.047	0.455	0.205	0.393
Average, $\times 10^{-3}$ μm^2	0.604	0.351	0.224	0.106	1.250	3.572	1.126
Sample number	5849	32,351	52	25	43	988	4531

Tight sandstone gas reservoirs are also distributed across large areas and are positioned vertically adjacent to source rocks, which is a key to tight sandstone gas reservoir accumulation. For example, the upper Paleozoic strata of the Ordos Basin are composed of a series of sediments of transitional facies and fluvial-delta facies, resulting in poor sandstone physical properties but horizontally wide distribution that covers nearly the entire basin and vertically multi-layered development (found in the Benxi, Taiyuan, Shanxi, and Xiashihezi Formations). In brief, tight sandstone gas reservoirs present features of vertically multi-layered and horizontally large area superimposed distribution (Li et al. 2012).

3.2 Source rocks mainly from coal and adjacent to reservoirs

For tight sandstone gas reservoirs, coal source rocks are characterized by wide distribution, high TOC, and high hydrocarbon-generating intensity. The results of hydrocarbon generation simulation experiments show that coal can continuously generate and expel hydrocarbon while charging reservoirs with no gas generation peak level, even during highly thermal evolution stages. The Sulige gas field of the Ordos Basin, gas fields in the western Sichuan Depression of the Sichuan Basin and tight sandstone gas reservoirs in the Kuqa Depression of the Tarim Basin are primarily associated with coal strata, which can indeed generate continuous and abundant natural gas. This period of large-scale gas generation has been relatively late and considerably long (it still continues), thus facilitating the formation of tight sandstone gas reservoirs (Zou et al. 2011a; Li et al. 2012).

Source rocks of tight sandstone gas reservoirs are distributed widely and are located either within reservoirs or adjacent to them, and this can result in a considerable increase in expulsion efficiency from source rocks to reservoirs for large contacting areas and short distances, presenting sheet-like generating and diffuse hydrocarbon charging properties. For instance, for the Dibei tight gas reservoir in the eastern Kuqa Depression of the Tarim Basin, the source rocks (Triassic coal strata of the Taliqike Formation and lacustrine mudstone of the Huangshanjie Formation) make a close contact with the lower Jurassic tight sandstone reservoirs of the Ahe and Yangxia Formations. This facilitates the large-scale accumulation of tight sandstone gas (Zhou 2002; Liang et al. 2004; Wang 2014).

Tight sandstone gas reservoirs are also found adjacent to widely distributed source rocks and are characterized by short-distance seepage diffusion and non-Darcy seepage migration processes, which can improve accumulation efficiency (Zou et al. 2012). For example, the upper Paleozoic sandstone strata of the Ordos Basin present horizontally wide distribution and poor physical properties, indicating that gas largely migrates only short distances laterally within the tight sandstone. A lack of faults also results in an absence of long-distance vertical migration. Source-proximal hydrocarbon accumulation thus dominates without long-distance vertical and lateral migration.

3.3 Complex gas and water distribution and pressure anomalies

Intricate relationships between gas and water in tight sandstone gas reservoirs are attributed to the fact that the relationship between gas and water is not controlled by structural contours, and thus no unified gas–water interface or even gas–water inversion processes are present (where gas accumulates below water, rather than above it as usual). The Upper Paleozoic tight gas reservoirs, for instance, show complicated gas–water relationships with the presence of water during tight gas operation, no clear regional gas–water inversion processes or apparent edge or bottom water features. Furthermore, gas–water relationship complications are related to highly heterogeneous and well-developed micro-nano pores in tight sandstone reservoirs and show an affinity with gas charging and migration processes. Buoyancy no longer acts as the driving force for natural gas migration and accumulation. Gas rarely migrates long distances within tight sandstone reservoirs and accumulates over short migration distances, typically resulting in poor gas–water differentiation, tangled distribution, and a large number of small water bodies.

Tight sandstone gas reservoirs have no uniform pressure systems and often present pressure anomalies (Wang 2002). For instance, most tight sandstone gas reservoirs of the Alberta, San Juan, and Denver Basins present abnormally low pressure, while those of the Green River, Piceance, and Utah Basins largely show abnormally high pressure (Fig. 2). In China, tight sandstone gas reservoirs in the western Sichuan Depression of the Sichuan Basin and Kuqa Depression of the Tarim Basin also show abnormally high pressure, and those in the Xiaocaohu area of the Turpan-Hami Basin show abnormally low pressure. In comparison with those cases described above, the Upper Paleozoic tight gas reservoirs of the Ordos Basin show complex pressure systems with pressure coefficients varying greatly across all pressure range (i.e., negative, atmospheric, and overpressure). Furthermore, even within the same gas field or the same horizon, pressure varies considerably, denoting the presence of multiple pressure systems and of poor connectivity between them (Zhao et al. 2012).

3.4 Complex distribution and the high resource potential of tight gas reservoirs

Sandstone is found in large quantities in tight gas reservoirs and is typically superimposed in large continuous sheets in the plane with no definite traps, and tight sandstone gas reservoirs are developed in deep concave central areas or in downdip areas of structural slope belts, and structural highlands. Tight sandstone gas reservoirs in the United States are commonly found in downdip areas of foreland basins, in central areas of frontal uplifts, and in deep structural basin synclines. Tight sandstone gas reservoirs in China are mainly developed in downdip areas of basin slopes and in deep structural basin synclines and anticlines. Small reservoirs are also found in central basin areas or in deep depressions, and this distribution may possibly be related to the current low level of exploration.

Furthermore, even though tight sandstone gas reservoirs present large gas-bearing areas, their enrichment is controlled by sweet spots and fractures, and they show signs of partial accumulation (Zhao et al. 2004; Yang et al. 2007). Sweet spots are considered to be central to tight reservoirs. The USGS in 1999 first defined sweet spots as resource blocks that are capable of offering continuous stable production for 30 years. Law (2002) proposed that sweet spots are composed of tight sandstone zones with relatively high porosity and permeability. Sweet spots are now commonly used to study unconventional resources and are defined as local zones of higher porosity and permeability that are capable of offering relatively high daily gas production and continuous economic production within areas with poor physical properties. Sweet spots can be divided into two types: "Pore" sweet spots and "Fracture" sweet spots. "Pore" sweet spots are mainly controlled by sedimentation and diagenesis processes, and "Fracture" sweet spots are controlled by fracture distribution, with structural fractures serving as the most critical factor (Yang et al. 2013).

Fractures control the tight gas reservoir distribution, as they can greatly improve physical properties of reservoirs, in particular permeability. In addition, fractures serve as storage space for reservoir fluid and as the main pathway for reservoir fluid migration, which in turn determine the distribution of seepage systems in tight gas reservoirs, resulting in the stable and significant production of tight gas reservoirs. Meanwhile, fractures also determine natural gas migration and accumulation, as buoyancy prevails within wide fractures, while molecular expansion forces dominate within matrix pores. In serving as the main forcing mechanism of natural gas accumulation and production, fracture systems can further improve the potential value of tight reservoirs.

4 Tight sandstone gas reservoir accumulation mechanisms and classifications

4.1 Model testing and mechanism interpretation of tight sandstone gas

Whether buoyancy serves as the main driving force for the accumulation of tight sandstone gas reservoirs is still debated. Buoyancy conditions have been detected in reservoir settings in numerous physical simulation works (Gies 1984; Zeng 2000; Pang et al. 2003; Xiao et al. 2008; Pang et al. 2014), which include capillary simulation and quartz sand simulation experiments.

4.1.1 Capillary simulation experiments

Buoyancy accumulation processes have been examined through capillary physical modeling in previous studies (Fig. 3).

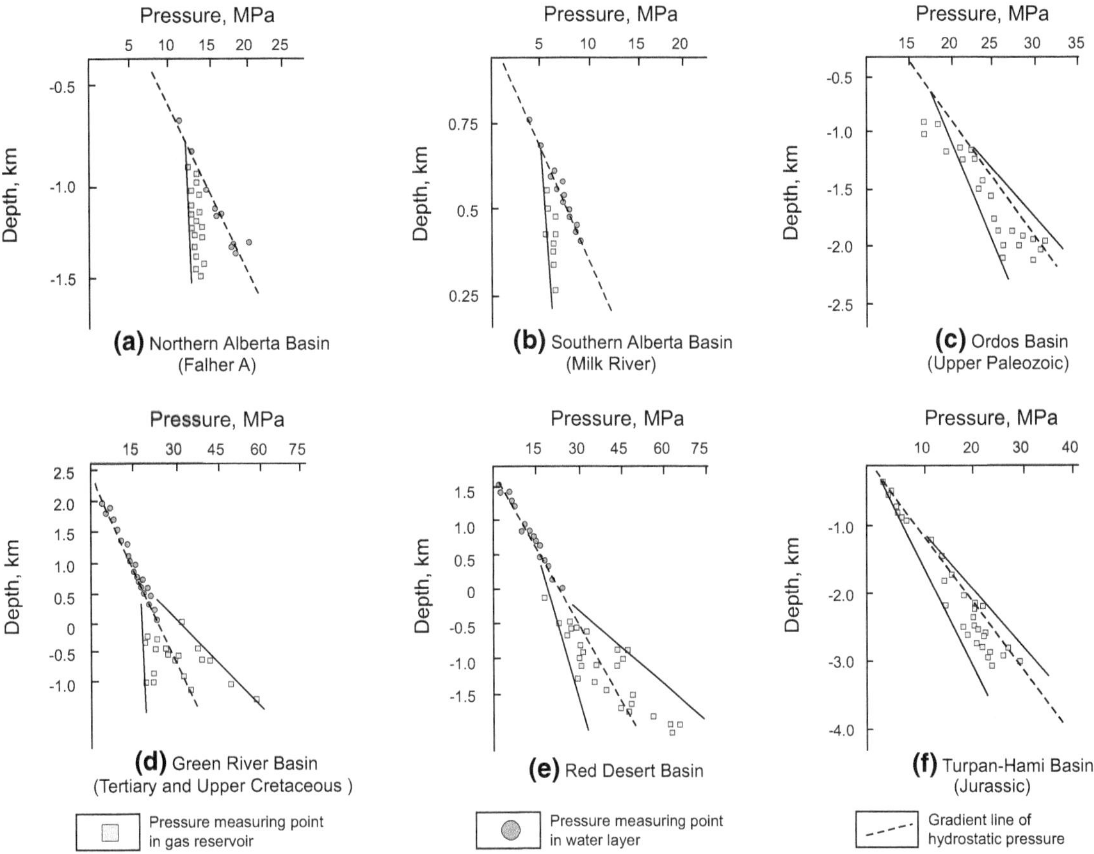

Fig. 2 Pressure distribution of typical tight sandstone gas reservoirs in China and abroad (Wang 2002)

A single homogeneous capillary physical model for hydrosealed threshold measurements has shown that oil and gas injected into the bottom of a glass tube and accumulating in the lower segment without upward migration processes is not affected by buoyancy when the diameter of a glass tube filled with water is less than 4.5 mm, thus proving the lower limit of buoyancy accumulation (Pang et al. 2003).

A capillary physical model was also used in an experiment involving a cone-shaped glass tube, and the critical diameter of pores and throats for hydrosealing was measured to be between 0.10 and 0.36 cm (Pang et al. 2003).

A capillary physical model with a smaller glass tube characterized by changeable cone shapes showed that pore and throat radii corresponding to the lower limit of buoyancy accumulation decrease with an increase in hydrostatic pressure and oil injection pressure (Pang et al. 2013).

In addition, reservoir crack widths were modeled in a physical model based on gap distances between glass slides, which were conducted by using 0.02-mm-thick tinfoil. Gas drainage induced by gas injection into the experimental installation filled with red ink was observed, and gas–water inversion only occurred when crack widths were less than 0.02 mm.

4.1.2 Quartz sand simulation experiments

Recent studies have also studied the importance of buoyancy as a critical geological factor with quartz sand simulation experiments (Fig. 4).

Pressure models with glass pillars developed by Gies (1984) show that gas–water inversion processes can be observed in fine sand without buoyancy control but not in coarse sand.

Physical models of oil–gas migration processes with sand-filled glass tubes show that gas and oil, respectively, injected upwards into a glass tube always accumulates in sand of the lower segment without engaging in upward migration affected by buoyancy when the sand grain diameter is less than 0.1 or 0.2 mm (glass tube filled with water). These diameters were defined as the lower boundaries of buoyancy accumulation for gas and oil, respectively (Pang et al. 2003).

During physical modeling of hydrosealed thresholds using sand-filled glass pillars, gas–water inversion processes were observed in sand grains with diameters ranging from 0.05 to 0.1 mm (Pang et al. 2003).

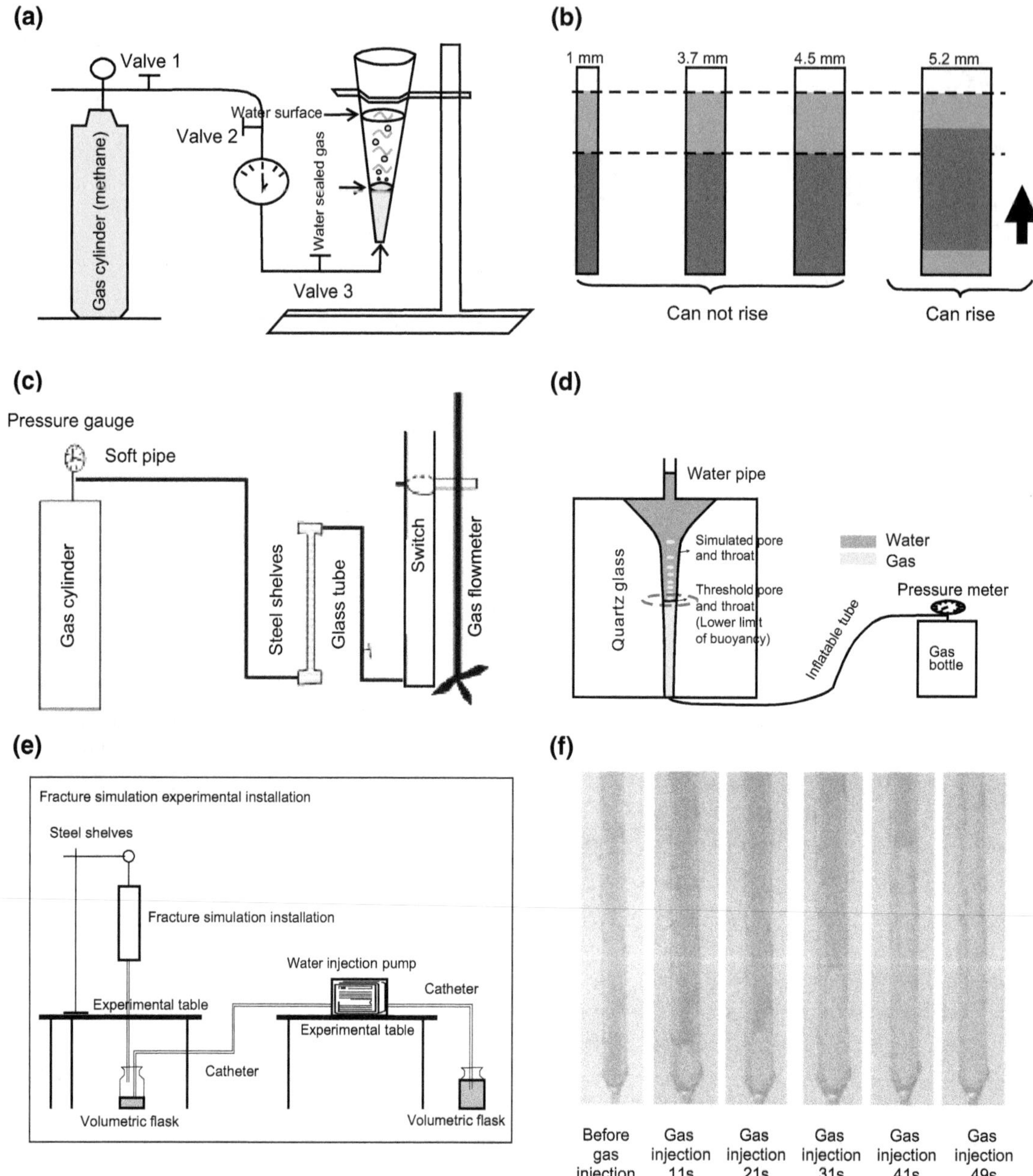

Fig. 3 Physical modeling of critical accumulation conditions. **a** Physical modeling of the gas-sealing threshold using a funnel-shaped capillary (Pang et al. 2003); **b** Zeng (2000); **c** Physical modeling of critical pore throat diameters using a funnel-shaped capillary (Pang et al. 2003); **d** Physical modeling using a cone glass tube (Pang et al. 2014); **e** Fracture simulation using glass slides; **f** Observations of fracture simulation with glass slides. The figure illustrates gas–water contact changes before and after gas injection into the glass slides

Gas injection experiments involving glass tubes have shown that pore water is displaced by natural gas injected upward that accumulates in sand grains in the bottom layer (Xiao et al. 2008).

Physical models involving larger sand-filled glass tubes placed under changing pressure conditions have shown that both pore and throat radii and sand grain diameters corresponding to the lower limit of buoyancy accumulation decrease with an increase in hydrostatic pressure and oil injection pressure in a single tube and sand-filled glass pillar, respectively (Pang et al. 2014).

Despite significant differences in temperature–pressure conditions between the physical models and real geological settings, the simulation experiments noted above are still of great importance to future research. According to the capillary and quartz sand models, capillary forces increase with a decrease in sand grain and glass tube diameters, resulting in the invalidation of buoyancy as the most

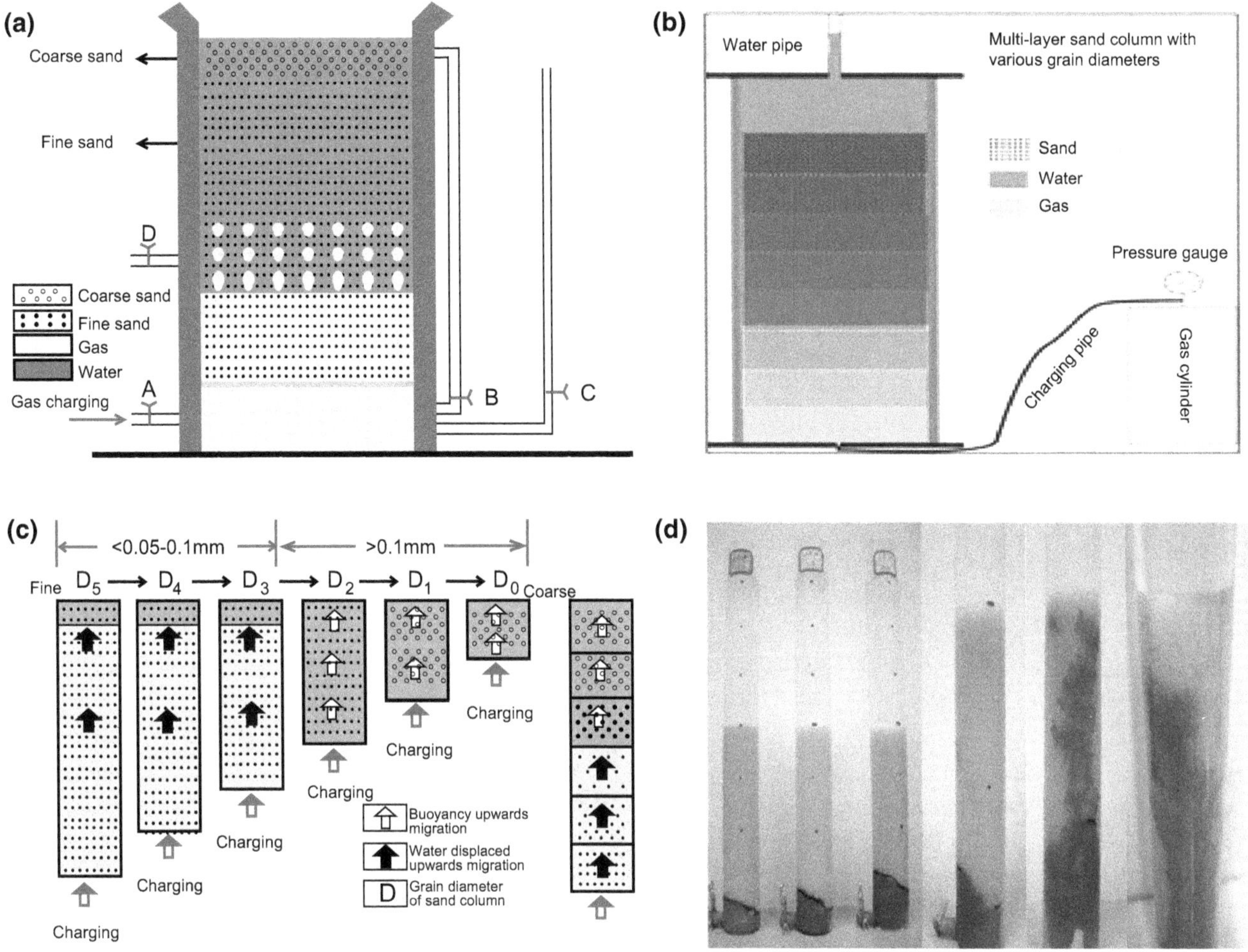

Fig. 4 Quartz sand simulation experiments. **a** Pressure simulation involving a glass column (Gies 1984); **b** Physical modeling involving a sand-filled glass tube under pressure-changing conditions (Pang et al. 2014); **c** Physical modeling of gas-sealing thresholds involving a thick glass tube (Pang et al. 2001); **d** Gas injection experiments involving glass tubes have shown that pore water is displaced by natural gas injected upward that accumulates in sand grains in the bottom layer (Xiao et al. 2008)

important controlling factor. A critical geological condition without buoyancy-driven mechanism that is characterized by gas–water inversion can thus be assumed.

4.2 Accumulation mechanisms of tight sandstone gas reservoirs

4.2.1 Dynamic mechanisms of "accumulation–densification" tight gas

"Accumulation–densification" tight gas reservoirs are formed from conventional gas reservoirs that subsequently undergo densification due to compaction and diagenesis. Their formation mechanism and distribution characteristics resemble those of conventional gas reservoirs, because when gas accumulates, reservoirs are not yet densified (i.e., the completion of gas charging and accumulation occurs in conventional gas reservoirs). Densification occurs as a result of subsequent burial compaction and structural compression. Thus, buoyancy serves as the main driving mechanism for natural gas migration and accumulation, and "accumulation–densification" tight gas reservoirs are always found in structurally high positions associated with anticlines, faults, lithological lenses, and pinch out areas (Fig. 5). Sandstone reservoir densification is induced by compaction, cementation, and structural compression with compaction (mechanical and chemical) serving as major driving forces (Shou et al. 2003; Liu et al. 2006; Zhang et al. 2011).

4.2.2 Dynamic mechanisms of "densification–accumulation" tight gas

"Densification–accumulation" tight gas forms after reservoir densification. A large amount of gas expelled by effective source rocks migrates directly into adjacent tight

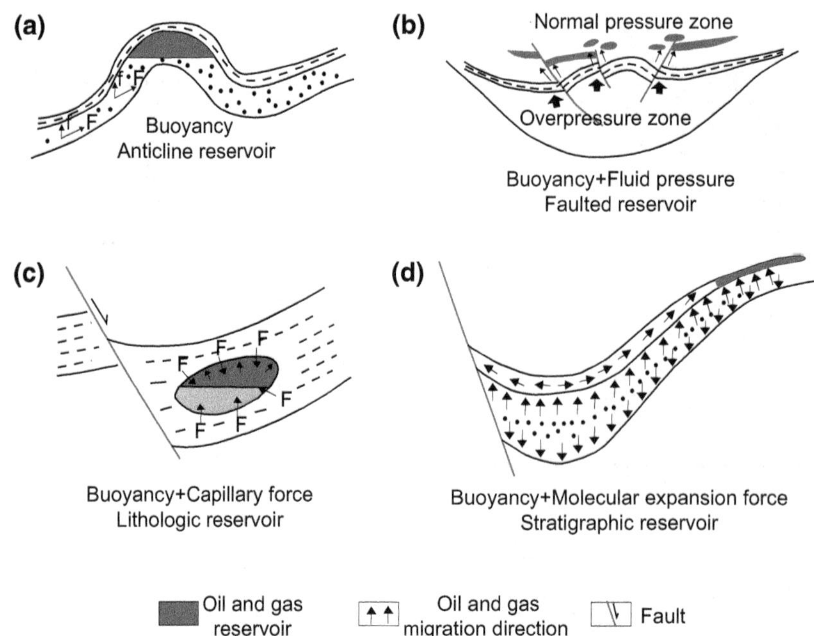

Fig. 5 Classification of tight conventional gas reservoirs (modified from Pang et al. 2014). **a** Anticline tight gas reservoir; **b** Fault-related tight gas reservoir; **c** Lithological lens tight gas reservoir; **d** Stratigraphic pinching out tight gas reservoir

reservoirs under decisive capillary forces and hydrocarbon molecular expansion force (not buoyancy). Pore water is drained to extend the gas distribution scale which is then aggregated to form continuous, large-scale deep tight gas reservoirs. "Densification–accumulation" tight gas reservoirs are always found in low structural positions (e.g., deep depression slopes or basins).

Conventional and unconventional oil and gas accumulations differ in that whether buoyancy serves as the primary driving force for hydrocarbon accumulation (Song et al. 2013). In non-buoyancy accumulation, buoyancy processes cannot overcome gas accumulation resistance and thus cannot act as the main driving force for gas accumulation. Non-buoyancy accumulation is mainly controlled by the difference between resistance and buoyancy, and strong capillary forces in reservoirs are normally induced by developed micro-nano pores.

The distribution area of deep tight gas reservoirs depends on the force balance (Fig. 6), as in critical conditions, hydrocarbon expansive forces are equal to the sum of capillary force and overburden hydrostatic pressure (Pang et al. 2003; Xie et al. 2004). Buoyancy becomes insignificant in densified reservoirs, and the potential maximum distribution area (trap area) of tight gas and minimum burial depth can be obtained using the force balance equation. In addition, the material balance of tight gas reservoirs can be described as follows: gas storage amount is equal to the gas supply amount minus the sum of gas loss from cap rocks, from the gas–water interface, and from the trap spill point. The material balance determines the distribution areas of deep basin gas traps, gas-bearing areas inside traps, and favorable exploration zone borders (Pang et al. 2003).

4.3 Type classifications of tight sandstone gas reservoirs

4.3.1 Previous classifications of tight sandstone gas reservoirs and outstanding uncertainties

Numerous scholars in China and abroad have carried out studies on accumulation mechanisms and type classifications of tight sandstone gas reservoirs. Based on reservoir characteristics, Guan and Niu (1995) divided tight sandstone gas reservoirs into three types: good (dense), moderate (overly dense), and poor (extremely dense). Law (2002) classified tight sandstone gas reservoirs in central basin areas into two types, direct and indirect, in light of organic matter types of source rocks. Jiang et al. (2006) divided tight sandstone gas reservoirs into three types, the "densification–accumulation" (DA) deep tight type and the "accumulation–densification" (AD) tight conventional type, based on hydrocarbon expulsion peak timing and reservoir densification evolution. Dong et al. (2007) classified tight sandstone gas reservoirs into reformed and original types based on relationships between tight gas accumulation and structural evolution and different accumulation principles. Zou et al. (2011a) classified tight

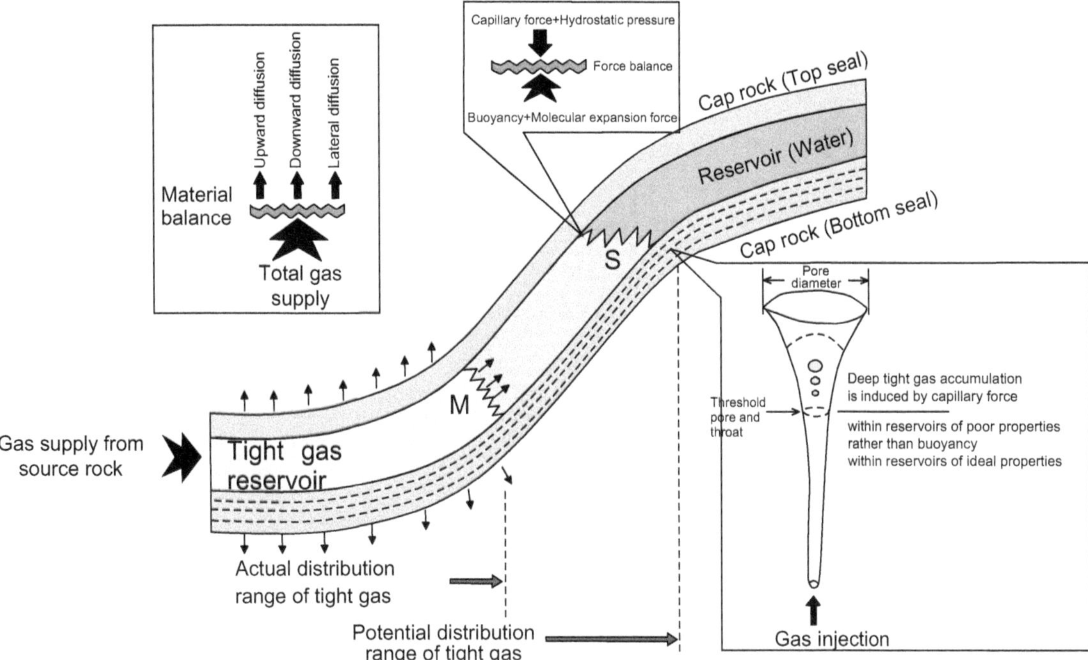

Fig. 6 Accumulation mechanisms of force and material balance for tight gas reservoirs (Pang et al. 2003)

sandstone gas reservoirs into slope lithological and deep structural types based on formation conditions, distribution characteristics, trap types, exploration practices, etc. Guo et al. (2012), based on accumulation modes, divided accumulation zones of tight sandstone gas reservoirs into continuous and transitional types. Dai et al. (2012) classified tight sandstone gas reservoirs into continuous and trap types based on reservoir characteristics, reserves, and structural locations. Li et al. (2012) classified tight sandstone gas reservoirs into slope, anticline structural, and deep sag types based on structural locations, accumulation mechanisms, and evolution principles. Zhao et al. (2012), according to trap types and distribution, proposed continuous, quasi-continuous, and non-continuous type classifications. However, all of the classifications noted above pertain to geological features of gas reservoirs, without classifying tight gas reservoirs based on accumulation dynamics.

4.3.2 Tight gas reservoir classification based on accumulation dynamics

On the basis of gas charging periods (GCP) and reservoir densification periods (RDP) and in consideration of dynamic features and distribution characteristics, we propose the classification of tight sandstone gas reservoirs into three types: conventional tight gas reservoirs, deep tight gas reservoirs, and composite tight gas reservoirs (Fig. 7). During the RDP, buoyancy dynamics cannot act, as critical physical properties, which vary with actual geological settings for different basins.

Conventional tight reservoirs are tight gas reservoirs in which the GCP precedes the RDP. They are formed via accumulation and subsequent densification, with buoyancy serving as the main accumulation force that also determines the distribution of tight gas reservoirs (Fig. 7a).

Deep tight gas reservoirs are defined as tight gas reservoirs in which the RDP precedes the GCP. Formation mechanisms involve densification followed by charging and accumulation under hydrocarbon generation expansive forces without buoyancy. Deep tight gas reservoirs are characterized by close contact between source rocks and reservoirs and by the continuous widespread distribution and the common presence of relatively deep depression areas or sag and slope belts (Fig. 7b).

Composite tight gas reservoirs exhibiting composite features of conventional tight gas reservoirs and deep tight gas reservoirs are characterized by accumulation and trapping in low porosity zones of structurally high areas and by distribution in structurally low areas. Gas charging occurred during both early and late stages of accumulation that involved initial accumulation followed by densification and re-accumulation. These are mainly driven by early buoyancy and a combination of hydrocarbon expansion forces and capillary forces. Composite tight gas reservoirs are distributed widely in anticline belts, deep depressions, and slope zones (Fig. 7c).

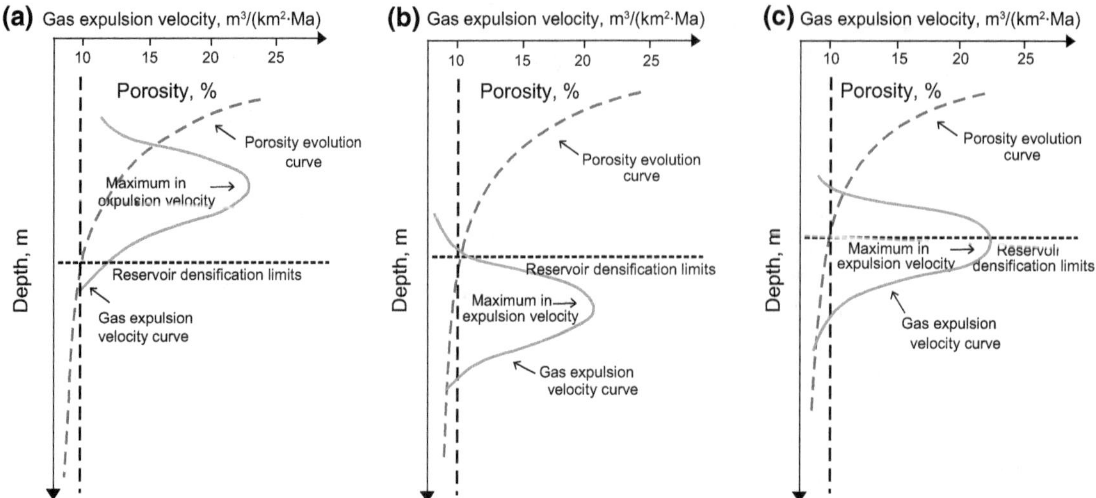

Fig. 7 Genetic type classification of tight sandstone gas reservoirs. **a** Conventional tight gas reservoir; **b** Deep tight gas reservoir; **c** Composite tight gas reservoir

4.3.3 Characteristics of various tight sandstone gas reservoirs

4.3.3.1 Characteristics of "AD" tight gas reservoirs Conventional tight reservoirs form through the compaction and densification of conventional reservoirs, and they often show accumulation in zones of relatively high porosity and permeability. With the exception of relatively low porosity and permeability, there is no obvious distinction between conventional tight and conventional reservoirs.

Therefore, conventional tight reservoirs are characterized by: distribution in structurally high areas; partial enrichment in zones of relatively high porosity and permeability; no direct connection between sources and reservoirs; overlaying cap rocks that play a significant role in sealing processes; a uniform gas–water interface; higher inner pressure than hydrostatic pressure under a stable state; continuously high pressure; and relatively small distribution areas and reserve scales. By trap type, conventional tight reservoirs can be subdivided into anticlinal tight gas reservoirs, fault block tight gas reservoirs, lithological tight gas reservoirs, and stratigraphic tight gas reservoirs (Pang et al. 2014).

4.3.3.2 Characteristics of "DA" tight gas reservoirs Unlike conventional and conventional tight reservoirs, deep tight gas reservoirs are characterized by the following features: close contact between source rocks and reservoirs; continuous widespread distribution; the presence of relatively deep depressions or sags and slope belts with large distribution areas (as buoyancy and cap rocks no longer determine natural gas accumulation processes); low well production but relatively large reserves; the absence of a uniform gas–water interface that can cause gas–water inversion; and pressure lower than stable-state hydrostatic pressure, showing stable low pressure anomalies (Jiang et al. 2006; Pang et al. 2014).

4.3.3.3 Characteristics of composite tight gas reservoirs Composite tight gas reservoirs are characterized by the following features: the accumulation of tight gas in structurally high and low areas; the accumulation of tight gas within sweet spots that are also gas-bearing in tight sandstone regions; gas-bearing properties within sand bodies that connect to the source regions; and high and low pressure anomalies within tight gas reservoir formation. Therefore, composite tight gas reservoirs are distributed widely, are not constrained by structural and trap conditions, and exhibit gas-bearing properties and enormous reserves but with complex gas–water distribution and no clear edge or bottom water areas. Hydrocarbon accumulation and preservation also significantly depend on subsequent tectonic movement processes (Pang et al. 2013, 2014).

5 Controlling factors and development models of tight sandstone gas reservoirs

5.1 Controlling factors and development models of "AD" tight sandstone gas reservoirs

Conventional tight gas reservoirs are formed from conventional gas reservoirs that experience subsequent densification due to compaction and diagenesis. Their formation mechanisms and distribution characteristics resemble those of conventional gas reservoirs. When gas accumulates, reservoirs have not yet been densified, i.e., gas charging and accumulation is completed in conventional gas reservoirs. Densification then occurs due to subsequent burial compaction and structural compression (Jiang et al. 2006). The development and distribution of "AD" tight sandstone gas reservoirs are mainly controlled by source kitchens (S), regional overlaying cap rocks (C), gas reservoirs (D) and low fluid potential areas (P), with the development model shown by "T-CDPS" (T means time). The spatiotemporal combination of these functional elements controls gas reservoir formation and distribution, and low fluid potential areas can be further divided into four sub-types: anticline (P_1), fault block (P_2), lithological (P_3), and stratigraphic (P_4) hydrocarbon reservoirs (Pang et al. 2012, 2014).

Conventional tight reservoirs develop prior to the reservoir densification period (RDP) and are characterized by two main stages (Fig. 8). Conventional gas reservoirs are formed during the first stage with buoyancy serving as the main accumulation force, while reservoir densification occurs during the second stage as a result of burial compaction, structural compression, and cementation. Groundwater circulation ceases after hydrocarbon charging, which results in the disturbance of chemical equilibrium and in the deceleration of cementation. However, natural gas charging has no effect on compaction, so compaction acts as the most important factor for reservoir densification. "Four high, two small, and one separation" principles are thus used as discriminant criteria of conventional tight gas reservoirs. "Four high" denotes hydrocarbon accumulation and sealing in structurally high positions, enrichment in zones of relatively high porosity and permeability, and high pressure reservoir formation. "Two small" denotes a small distribution area and small reserve scale. "One separation" denotes a general separation between sources and reservoirs, i.e., no direct connection between them (Pang et al. 2014).

5.2 Controlling factors and development models of "DA" tight sandstone gas reservoirs

Unlike conventional reservoirs, deep tight gas reservoirs are always found in structurally low areas (e.g., slopes of deep depressions or basins). "DA" deep tight gas forms after reservoir densification. Large quantities of gas expelled by effective source rocks migrates directly into adjacent tight reservoirs under the decisive force of hydrocarbon molecular expansive force (not buoyancy). Pore water is drained to extend the gas distribution area which is then aggregated to form large and continuously distributed tight deep gas reservoirs (Fig. 9). Deep tight gas reservoirs are mainly controlled by effective source rocks positioned adjacent to reservoirs (S) and characterized by continuous hydrocarbon expulsion, widely distributed reservoirs (D), stable tectonic settings (W), and universal reservoir densification (L), with the development model shown by "T-LWDS" (T means time). A stable tectonic setting is conducive to the preservation of deep tight gas reservoirs, and widely distributed and universally densified reservoirs promote the short-distance accumulation of natural gas and the displacement of pore water. Source rocks with continuous gas supplies also serve as a solid material base for the development and distribution of deep tight gas reservoirs. Spatiotemporal configuration of these four functional elements determines the timing and depth of deep tight gas reservoir development.

With reservoirs in deep depressions first reach the lower limit of buoyancy accumulation, deep tight gas reservoirs first form in deep depressions and gradually extend outward (Song et al. 2013). Deep tight gas reservoirs have

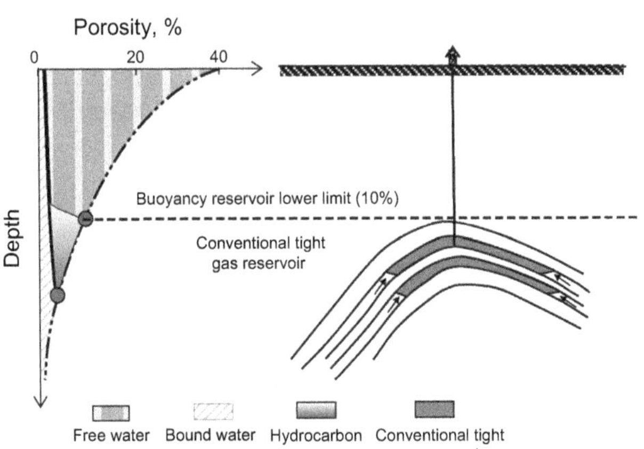

Fig. 8 Accumulation mechanisms and distribution features of conventional tight gas reservoirs (Pang et al. 2013)

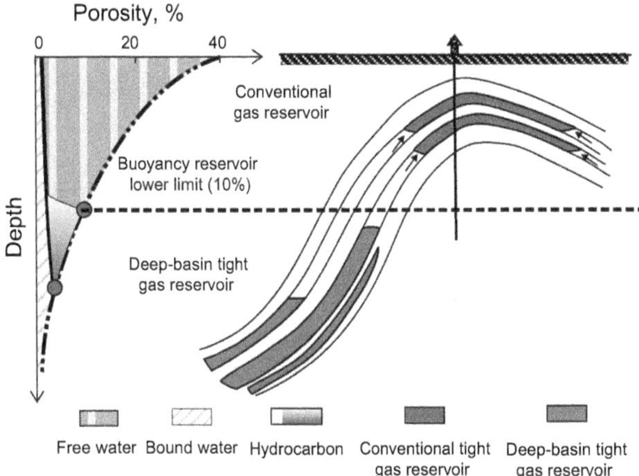

Fig. 9 Accumulation mechanisms and distribution features of deep tight gas reservoirs (Pang et al. 2013)

always expanded symmetrically from central areas of depressions (e.g., gas reservoirs in the Red Desert Basin of Wyoming and in the San Juan Basin of New Mexico, USA), from foreland lateral margin slopes (e.g., gas reservoirs in Elmworth of Alberta, Canada and in the Green River Basin of Wyoming, USA), and from structural slope belts (e.g., Milk River gas reservoirs in Alberta, Canada and Clinton sandstone gas reservoirs in the Appalachian Basin of eastern Ohio, USA). "Four low, two large, and one close contact" principles are thus used as discriminant criteria for deep tight gas reservoirs. "Four low" denotes hydrocarbon accumulation and inversion in structurally low positions, enrichment in zones of relatively low porosity and permeability, and stable reservoir formation under low pressure. "Two large" denotes large distribution area and large reserve scale. "One close contact" denotes close contact between source rocks and reservoirs (Pang et al. 2014).

5.3 Controlling factors and development models of composite tight gas reservoirs

Composite tight gas reservoirs that exhibit the composite features of conventional tight gas reservoirs and deep tight gas reservoirs are controlled by buoyancy and hydrocarbon molecular expansive force during the early and late accumulation stages, respectively. The main controlling factors of composite tight gas reservoir accumulation can thus be classified into two subsidiary sets. One set is associated with the accumulation of conventional tight gas reservoirs, mainly cap rocks (C), gas reservoirs (D), low fluid potential areas (P), and source kitchens (S). The other is associated with deep tight gas reservoirs, mainly universal reservoir densification (L), stable tectonic settings (W), widely distributed reservoirs (D), and source rocks in a close contact with reservoirs (S). Therefore, we infer that the development of composite tight gas reservoirs is driven by the following six main factors S, D, C, W, P, and L, with the accumulation process depending on the spatiotemporal combination of these functional elements, and with the development model shown by "T-CDPS + T-LWDS" (Pang et al. 2014). The genetic mechanisms of composite tight gas reservoirs are thus a recombination of conventional tight and deep tight gas reservoir accumulations, resulting in the complex gas–water relationships and full gas-bearing properties.

Composite tight gas reservoirs are distributed widely in anticline belts, deep depressions, and slope zones as a result of superimposition of conventional and deep tight gas reservoirs in deep basins, also including buoyancy-adjusted deep tight gas reservoirs reconstructed through subsequent fault or fracture development (e.g., subsequent inner-sag uplifting). Gas charging and accumulation processes are based on early buoyancy and then later on molecular expansive force. Composite tight gas reservoirs are formed through accumulation followed by densification and re-accumulation, and two genetic mechanisms can be identified. Driven by the first mechanism, conventional tight and deep tight gas reservoirs of the same depth form before and after reservoir densification, respectively. Given the order of gas accumulation processes, composite tight gas reservoirs are superimposed by different reservoir types following a period of multi-phase hydrocarbon expulsion. The other mechanism refers to partial uplifting in deep tight gas reservoirs induced by subsequent tectonic adjustments and improved reservoir physical properties close to those of conventional reservoirs (low porosity and high permeability) as a result of fault and fracture reconstruction. Buoyancy then occurs again, changing partial gas–water distribution and maintaining large gas-bearing areas in deep tight gas reservoirs, ultimately leading to the development of composite tight gas reservoirs of an adjusted genesis (Pang et al. 2013, 2014).

Three stages of composite tight gas reservoir formation are identified (Fig. 10). Conventional gas reservoirs always develop during the first stage (i.e., anticlinal, stratigraphic, and lithological gas reservoirs). Deep tight gas reservoirs form at the same time as or after the accumulation of conventional gas reservoirs during the second stage and are mainly distributed in depositional centers, slope belts, or marginal areas of deep basins. During the third stage, conventional gas reservoirs are transformed into conventional tight reservoirs with increasing the burial depth, and gas-bearing areas of deep tight gas reservoirs expand continuously, resulting in the final formation of composite tight gas reservoirs (Pang et al. 2014).

"Four high, four low, two large and one close contact" principles are used as discriminant criteria for composite tight gas reservoirs. "Four high and four low" denotes hydrocarbon accumulation in structurally high and low positions, enrichment in zones of both high and low porosity, high- and low-yield hydrocarbon bed development, and hydrocarbon beds found in high and low pressure settings. "Two large" denotes large distribution areas and reserve scale. "One close contact" denotes a close contact between source rocks and targeted reservoirs.

6 Resource potential of tight sandstone gas reservoirs

6.1 Resource potential of tight sandstone gas reservoirs worldwide

As a major exploration field of unconventional natural gas resources, tight sandstone gas shows great resource potential, and it is found around the world. According to

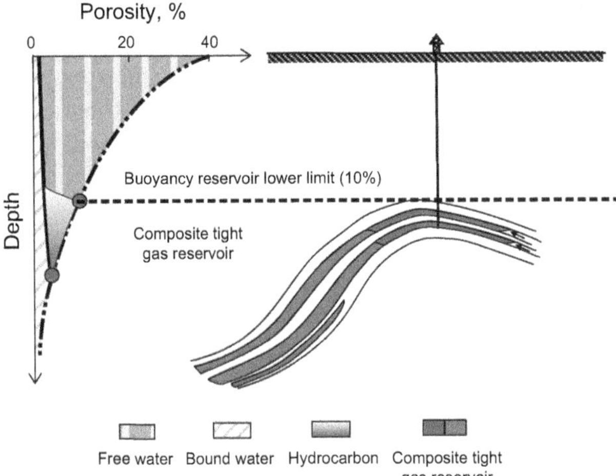

Fig. 10 Accumulation mechanisms and distribution features of composite tight gas reservoirs (Pang et al. 2013)

the USGS, roughly 70 tight sandstone gas reservoir basins have been identified around the world, and most are located in the Asia–Pacific region, North America, Latin America, the former Soviet Union, the Middle East, and North Africa, with geological reserves totaling 210×10^{12} m^3. Of this total, the Asia–Pacific region, North America, and Latin America account for 51.0×10^{12}, 38.8×10^{12}, and 36.6×10^{12} m^3, thus accounting for roughly 60 % of global tight gas reserves (IEA 2009).

Recently, numerous scholars have focused on recalculating global geological reserves of tight sandstone gas, which are significantly larger than the estimations presented above. Resource appraisals drawn from the Institut Français du Pétrole (IFP) show that tight gas reserves in the USA and Canada are $(402–442) \times 10^{12}$ m^3, and global total reserves amount to $(310–510) \times 10^{12}$ m^3 (Yang et al. 2012). Aguilera (2008) reported technical recoverable reserves of global tight sandstone gas of 428×10^{12} m^3, which roughly corresponds to the volume of conventional natural gas reserves (Qiu and Deng 2012). The resource potential of tight gas all over the world is thus evident.

6.2 Resource potential of tight sandstone gas reservoirs in China

6.2.1 Advantageous forming conditions of tight gas reservoirs in China

Based on the material basis, reservoir genesis, source–reservoir contact, and accumulation features of tight sandstone gas, the geological settings found in China favor the development of large-scale tight sandstone gas fields.

6.2.1.1 Major coal-measure source rocks Coal series are well developed in sedimentary basins of China, and three major coal-forming periods have been identified: the Late Paleozoic, Mesozoic, and Cenozoic (Li et al. 2012). The northern and southern China areas, respectively, include the Carboniferous-Permian series of the Ordos Basin and the Upper Permian Longtan and Changxing Formations, which developed during the Late Paleozoic, and the Lower-Middle Jurassic Yan'an Formation and Upper Triassic Xujiahe Formation in the Sichuan Basin that developed during the Mesozoic (the Lower Cretaceous Yingcheng Formation is referred to as coal-bearing series in fault basins of eastern China). Cenozoic coal-forming basins are mainly found along the west side of the Pacific and along the Neo-Tethys ocean shore.

Coal-measure source rocks are widely distributed in the Sulige gas field of the Ordos Basin, in the Xujiahe Formation of the central Sichuan Basin, and in the Kuqa gas-bearing areas of the Northern Tarim Basin. Coal-measure source rocks from the Taiyuan and Shanxi Formations of sea-land transitional facies in the Ordos Basin have a depositional area of approximately 13.8×10^4 km^2 that is characterized by coal layers and dark mudstones with a thickness of 10-14 m, TOC content of around 63 %, and hydrocarbon generation intensity of 15×10^8 m^3/km^2, serving as a favorable setting for the development of large-to medium-sized gas fields. Coal-measure series, as favorable source rocks, are mainly characterized by type III organic matter, high levels of abundance, gas generation capacities during highly thermal evolution phases, and sheet-like hydrocarbon generation and continuous charging. Carbon isotope diagrams presented by Dai et al. (2012) also show that all tight sandstone gas in China is coal-derived, supporting the fact of coal-measure source rocks as a favorable material basis.

6.2.1.2 Widely distributed tight reservoirs Tight reservoirs in China are mainly characterized by large distribution areas, deep burial depth, complex diagenetic evolutionary patterns, poor physical properties, significant heterogeneity, non-Darcy seepage migration patterns, and non-buoyancy accumulation. Continental hydrocarbon exploration efforts conducted in China show that widespread, multi-genetic sand bodies are well developed in the middle of large-scale lacustrine basins, thus acting as the most important prospecting targets for continental lithological hydrocarbon reservoirs (Li et al. 2012). Central sand bodies are considered to originate from shallow deltas and sandy debris flows, and their formation conditions, microfacies composition, distribution models, and genetic classification are still poorly understood. Shallow meandering channel deltas are well developed in modern lacustrine basins of China. The Ordos Basin underwent a denudation period that lasted over 100 Ma during the Caledonian, and a general flattening of topography in the

Late Paleozoic resulted in a small thickness difference of less than 30 m in the 8th member of the Shihezi Formation. This gentle delta with a slope obliquity of <1° favors the accumulation of slope-type tight sandstone gas reservoirs.

Tight reservoirs in large Chinese basins are mainly characterized by deep burial and intense compaction processes followed by densification as a result of structural compression and cementation. Natural gas develops as free gas within the micro-nano pore structures (Zhu et al. 2013), with diameters ranging from 0.03 to 2 μm.

In summary, tight sandstone gas reservoirs in China are characterized by sheet-like coal-measure source kitchens, widely distributed reservoirs, a close contact between source rocks and reservoirs, and 3D gas-bearing and partial accumulation. They are mainly found in the Ordos, Sichuan, Tarim, Bohai Gulf (deep zone), Qaidam, Songliao, southern Junggar, Chuxiong, and East China Sea Basins.

6.2.2 The promising potential of tight sandstone gas in China

Tight gas with considerable resource potential is widely distributed in petroliferous basins across China. Recent tight gas exploration efforts in China have developed rapidly, representing major fields of natural gas discovery and production. In addition to large tight gas fields found in the Upper Paleozoic of the Ordos Basin, in the Xujiahe Formation of the Sichuan Basin, and in the Kuqa deep depression of the Tarim Basin, a series of tight gas reservoirs has been recently found in the Turpan-Hami, Songliao, and Bohai Gulf Basins. Tight gas in China, which is ample and has excellent resource prospect, has been identified as the most feasible alternative unconventional natural gas resource.

The newly found proven reserves of tight gas are up to 3110×10^8 m^3 each year, accounting for roughly 52 % of total discovered natural gas reserves in the same period (Zou et al. 2014). By the end of 2011, the proven total reserves of tight gas in China reached 3.3×10^{12} m^3, roughly accounting for 39 % of total natural gas reserves, in which 96 % of the proven reserves of tight sandstone gas are from the Ordos and Sichuan Basins. The Sulige gas field, as the largest tight gas-bearing area, has proven geological reserves of 3.5×10^{12} m^3 and annual production of 169×10^8 m^3 (Zou et al. 2013). Tight gas resources yielded a total production of approximately 256×10^8 m^3 in 2011, accounting for nearly one-fourth of total natural gas production, with annual production in the largest Sulige gas field exceeding 137×10^8 m^3. Given the supplies and prospects of tight sandstone gas in China, we predict that the natural gas production will increase continuously and rapidly, which should reach 600×10^8 m^3 by 2020 (Li et al. 2012).

The resource evaluations of tight gas in China based on analog methods have shown widespread distribution of tight sandstone gas reservoirs. The favorable continental basin areas are roughly 32×10^4 km^2, geological reserves are roughly $(17.4\text{-}23.8) \times 10^{12}$ m^3, and recoverable resources are roughly $(8.8–12.1) \times 10^{12}$ m^3 (Jia et al. 2012). As feasible tight gas exploration areas, the Upper Paleozoic series of the Ordos Basin, the Upper Triassic Xujiahe Formation of the Sichuan Basin, and the Kuqa deep depression of the Tarim Basin were, respectively, found to have reserves of $(5.9–8.15) \times 10^{12}$, $(4.3–5.7) \times 10^{12}$, and $(2.7–3.4) \times 10^{12}$ m^3. We also identified four other potential target areas: the Lower Cretaceous Denglouku Formation of the Songliao Basin, the third and fourth members of the Paleogene Shahejie Formation of the Bohai Gulf Basin, the Jurassic series of the Turpan-Hami Basin, and the Permian and Jurassic series along the southern margin of the Junggar Basin.

Thus, tight sandstone gas reserves in China show considerable resource potential, which motivated natural gas exploration and production efforts in China, and tight sandstone gas plays a key role in the natural gas industry.

6.2.3 Tight sandstone gas distributed in basins of central and western China

Tight gas reservoirs are distributed widely across China and are mainly found in central and western regions of China (Wang 2002). The Ordos, Sichuan, and Tarim Basins are proven tight gas-bearing basins with considerable resource potential, with proven total reserves of roughly 3.6×10^{12} m^3, accounting for 40 % of natural gas proven reserves in China. The Junggar and Turpan-Hami Basins are also favorable areas.

Tight gas exploration in the Ordos Basin began in the late 1980s, and nine gas fields have been discovered. There are five gas-bearing reservoir-cap rock assemblages, and the average porosity and permeability are 4 %–8 % and $(0.5–1) \times 10^{-3}$ μm^2, respectively. The gas-bearing areas of tight sandstone gas cover approximately 10×10^4 km^2 and the geological reserves of 50×10^{12} m^3, accounting for over 90 % of the basin total reserves. The Upper Paleozoic Shihezi, Shanxi, and Taiyuan Formations are the major gas production series in the Sulige, Daniudi, Wushenqi, Shenmu, and Mizhi large-scale tight gas fields, with proven reserves exceeding 1×10^{11} m^3. The Sulige gas field has proven reserves of 2.85×10^{12} m^3 and has shown a rapid increase in production. In 2010, the gas production in the Sulige and Daniudi fields increased to 106×10^8 and 22.8×10^8 m^3, respectively, and those are expected to reach 230 × 10^8 and 35 × 10^8 m^3, respectively, by 2020.

Tight gas reservoirs in the Sichuan Basin are mainly found in the western depression and central zones, with the average porosity and permeability of 1.2 %–13.2 % and (0.01–0.82) × 10^{-3} μm^2, respectively, in the tight sandstone reservoir of the Upper Triassic Xujiahe Formation, and with total geological reserves estimated at 26 × 10^{12} m^3. The Guang'an, Hechuan, and Anyue gas fields with proven reserves exceeding 1 × 10^{11} m^3 were also discovered in the central zone and are characterized by poor physical properties of the Upper Triassic Xujiahe Formation evidenced by porosity and permeability of 6 %–10 % and (0.1–5) × 10^{-3} μm^2, respectively. Tight gas reservoirs in the central zone yield lower proven reserves than those of the Sulige gas field, which are 0.5 × 10^{12} m^3, and annual production of 12 × 10^8 and 15 × 10^8 m^3 was recorded for 2010 and 2011, respectively. Those are expected to reach 60 × 10^8 m^3 by 2020.

Tight sandstone gas in China presents great resource potential and is inferred to be the most feasible unconventional gas sources and an important component of the natural gas industry with the development of exploration theories and techniques. Tight gas exploration can now guide future unconventional natural gas exploration and production in China (Dai et al. 2012).

7 Conclusions

(1) Tight gas reservoirs, as a typical unconventional natural gas resource, are characterized by reservoir porosity of <10 %, in situ permeability of <0.1 × 10^{-3} μm^2 (air permeability <1.0 × 10^{-3} μm^2), superimposed and widespread distribution, sheet-like coal-measure source rocks, continuous hydrocarbon generation, adjacent source rocks and reservoirs, vertical migration across short distances, complex gas–water relationships that are not controlled by structural contours, an absence of uniform gas–water interface and pressure system, high or low pressure anomalies, large-scale distribution area in structurally high positions, deep depressions, core areas of synclines, and downdip regions in structural slope belts, high resource potential, and partial accumulation controlled by sweet spots and fractures.

(2) Water and gas inversion processes were detected by capillary glass tube and quartz sand modeling experiments, and critical geological conditions without buoyancy-driven mechanism can thus be proved. In light of the relationship between gas charging and reservoir densification periods during tight gas accumulation processes, we propose the following genetic classification: (a) "first accumulation then reservoir densification" conventional tight gas reservoirs, (b) "first reservoir densification then accumulation" deep tight gas reservoirs, and (c) composite tight gas reservoirs.

(3) "Accumulation–densification" gas charging occurs prior to reservoir densification, accumulating in structurally high positions under the action of buoyancy with the following main controlling factors, source kitchens (S), regional overlaying cap rocks (C), gas reservoirs (D), and low fluid potential areas (P). "Densification–accumulation" tight gas forms after reservoir densification under hydrocarbon generation expansive force (no buoyancy), and accumulates in depression and slope areas with the following controlling factors, effective source rocks (S), widely distributed reservoirs (D), stable tectonic settings (W), and universal reservoir densification (L). Composite tight gas reservoirs exhibiting features of both conventional tight gas and deep tight gas reservoirs are controlled by buoyancy and hydrocarbon molecular expansive force during the early and late accumulation phases, respectively, with the following main controlling factors, source kitchens (S), reservoirs (D), cap rocks (C), stable tectonic settings (W), low fluid potential areas (P), and universal reservoir densification (L), and are widely distributed in anticline belts, deep depressions, and slope areas.

(4) Tight sandstone gas, as an important unconventional natural gas resource, shows great resource potential and is widely distributed around the world with geological reserves reaching 210 × 10^{12} m^3. Based on the material basis, reservoir genesis, source–reservoir contact relations, and accumulation mechanisms of tight gas, we infer that tight gas in China exhibits favorable reservoir formation conditions and great resource potential. The tight gas resource is mainly found in the Ordos, Sichuan, Tarim, Junggar, and Turpan-Hami Basins of central-western China. This resource has promoted global unconventional natural gas exploration and production under existing technical conditions.

Acknowledgments This paper is supported by the National Natural Science Foundation of China (No. 41472112) and the National Major Projects (No. 2011ZX05018002).

References

Aguilera R. Role of natural fractures and slot porosity on tight gas sands. In: SPE Unconventional Reservoirs Conference. 10–12 Feb, Keystone, Colorado. 2008.

Chen G. Formation condition of synclinal hydrosealed gas pools in Qinshui Basin. Oil Gas Geol. 1998;19(4):302–6 **(in Chinese)**.

Chen JY, Tan DQ, Yang CP. Advances in the research and exploration of unconventional petroleum systems. Geol Sci Technol Inf. 2003;22(4):55–9 **(in Chinese)**.

Dai JX. Synclinal oil and gas pools. Acta Pet Sin. 1983;4(4):27–30 **(in Chinese)**.

Dai JX, Ni YY, Wu XQ. Tight gas in China and its significance in exploration and exploitation. Pet Explor Dev. 2012;39(3): 257–65 **(in Chinese)**.

Dai JX, Song Y, Zhang HF. The main controlling factors for the formation of large and middle sized gas reservoirs. Sci China Ser D. 1996;26(6):481–7 **(in Chinese)**.

Dong XX, Mei LF, Quan YW, et al. Types of tight sand gas accumulation and its exploration prospect. Nat Gas Geosci. 2007;18(3):351–5 **(in Chinese)**.

Gies RM. Case history for a major Alberta deep basin gas trap: the Cadomin Formation. In: Masters JA, editor. Case study of a deep basin gas field. AAPG Memoir 38. Tulsa: AAPG Bulletin; 1984. p. 115–40.

Guan DS, Niu JY. China unconventional oil and gas geology. Beijing: Petroleum Industry Press; 1995. p. 12–40 **(in Chinese)**.

Guo QL, Chen NS, Hu JW, et al. Geo-model of tight sandstone gas accumulation and quantitative simulation. Nat Gas Geosci. 2012;23(2):199–207 **(in Chinese)**.

Holditch SA. Tight gas sands. J Pet Technol. 2006;58(6):86–93.

IEA. World Energy Outlook. 2013.

IEA. World Energy Outlook. 2009.

Jia CZ, Zheng M, Zhang YF. Unconventional hydrocarbon resources in China and the prospect of exploration and development. Pet Explor Dev. 2012;39(2):129–36 **(in Chinese)**.

Jiang ZX, Lin SG, Pang XQ, et al. The comparison of two types of tight sand gas reservoir. Pet Geol Exp. 2006;28(3):210–4 **(in Chinese)**.

Jiang ZX, Pang XQ, Zhang JC, et al. Summarization of deep basin gas studies. Adv Earth Sci. 2000;15(3):289–92 **(in Chinese)**.

Jin ZZ, Zhang JC, Wang ZX. Some remarks on deep basin gas accumulation. Geol Rev. 2003;49(4):401–7 **(in Chinese)**.

Law BE. Basin-centered gas systems. AAPG Bull. 2002;86(11): 1891–919.

Law BE, Dickinson WW. Conceptual model for origin of abnormally pressured gas accumulations in low-permeability reservoirs. AAPG Bull. 1985;69(8):1295–304.

Lei Q, Wan YJ, Li XZ, et al. A study on the development of tight gas reservoirs in the USA. Nat Gas Ind. 2010;30(1):45–8 **(in Chinese)**.

Li JZ, Guo BC, Zheng M, et al. Main types, geological features and resource potential of tight sandstone gas in China. Nat Gas Geosci. 2012;23(4):607–15 **(in Chinese)**.

Liang DG, Chen JP, Zhang BM. Continental hydrocarbon generation of the Kuqa depression in the Tarim Basin. Beijing: Petroleum Industry Press; 2004 **(in Chinese)**.

Liu GY, Jin ZJ, Zhang LP. Simulation study on clastic rock diagenetic compaction. Acta Sedimentol Sin. 2006;24(3):407–13 **(in Chinese)**.

Ma ZZ. Dynamic mechanisms and distribution models of deep tight sandstone gas reservoirs. Ph.D. Thesis. China University of Petroleum, Beijing. 2008 **(in Chinese)**.

MacAvoy PW. Natural gas policy act of 1978. Washington: Federal Energy Regulatory Commission; 1978.

Masters JA. Deep basin gas trap, western Canada. AAPG Bull. 1979;63(2):152–81.

National Energy Administration. Geological evaluating methods for tight sandstone gas. SY/T6832-2011. Beijing: Geological Publishing House; 2011 **(in Chinese)**.

Nehring R. Growing and indispensable: the contribution of production from tight-gas sands to U.S. gas production. In: Cumella SP, Shanley KW, Camp WK, editors. Understanding, exploring, and developing tight-gas sands. 2005 Vail Hedberg Conference, vol. 3. Colorado: AAPG Hedberg Series; 2008. p. 5–12.

Pang XQ, Fang H, Tang LJ. Deep basin gas accumulation in the Ordos Basin. In: Fu CD, editor. Study of deep basin gas. Beijing: Petroleum Industry Press; 2001. p. 65–73 **(in Chinese)**.

Pang XQ, Jiang ZX, Huang HD, et al. Formation mechanisms, distribution models, and prediction of superimposed continuous hydrocarbon reservoirs. Acta Pet Sin. 2014;35(5):795–828 **(in Chinese)**.

Pang XQ, Jin ZJ, Jiang ZX, et al. Critical condition for gas accumulation in the deep basin trap and physical modeling. Nat Gas Geosci. 2003;14(3):207–14 **(in Chinese)**.

Pang XQ, Zhou XY, Dong YX, et al. Formation mechanism classification of tight sandstone hydrocarbon reservoirs in petroliferous basin and resources appraisal. J China Univ Pet. 2013;37(5):28–38 **(in Chinese)**.

Pang XQ, Zhou XY, Jiang ZX, et al. Hydrocarbon reservoirs formation, evolution, prediction and evaluation in the superimposed basins. Acta Geol Sin. 2012;86(1):53–7 **(in Chinese)**.

Qiu ZJ, Deng ST. Strategic position of unconventional natural gas resources in China. Nat Gas Ind. 2012;32(1):1–5 **(in Chinese)**.

Rose PR, Everett JR, Merin IS. Possible basin centered gas accumulation, Raton Basin, Southern Colorado. Oil Gas J. 1984;82(40):190–7.

Schmoker JW. National assessment report of USA oil and gas resources. Reston: USGS; 1995.

Schmoker JW. Resource-assessment perspectives for unconventional gas systems. AAPG Bull. 2002;86(11):1993.

Schmoker JW. US Geological Survey assessment concepts for continuous petroleum accumulations. US Geol Surv. 2005;1:1–9.

Shou JF, Zhu GH, Zhang HL. Lateral structure compression and its influence on sandstone diagenesis—a case study from the Tarim Basin. Acta Sedimentol Sin. 2003;21(1):90–5 **(in Chinese)**.

Song Y, Jiang L, Ma XZ. Formation and distribution characteristics of unconventional oil and gas reservoirs. J Palaeogeogr. 2013; 15(5):605–14 **(in Chinese)**.

Spencer CW. Geologic aspects of tight gas reservoirs in the Rocky Mountain region. J Pet Technol. 1985;37(8):1308–14.

Spencer CW. Review of characteristics of low-permeability gas reservoirs in western United States. AAPG Bull. 1989;73(5): 613–29.

Surdam RC. A new paradigm for gas exploration in anomalously pressured "tight gas sands" in the Rocky Mountain Laramide basins (in seals, traps, and the petroleum system). AAPG Memoir. 1997;67:283–98.

Walls JD. Tight gas sands—permeability, pore structure, and clay. J Pet Technol. 1982;34(11):2707–14.

Wang T. Deep Basin gas fields in China. Beijing: Petroleum Industry Press; 2002. p. 3–71 **(in Chinese)**.

Wang YC, Kong JX, Li HP, et al. Natural gas geology. Beijing: Petroleum Industry Press; 2004 **(in Chinese)**.

Wang ZM. Formation mechanism and enrichment regularities of Kelasu subsalt deep large gas field in Kuqa Depression, Tarim Basin. Nat Gas Geosci. 2014;25(2):153–66 **(in Chinese)**.

Wu HY, Liang XD, Xiang CF. Accumulation characteristics and mechanisms of synclines in the Songliao Basin. Sci China Ser D. 2007;37(2):185–91 **(in Chinese)**.

Wyman RE. Gas recovery from tight sands. SPE 13940. 1985.
Xiao ZH, Zhong NN, Huang ZL, et al. A study on hydrocarbon pooling conditions in tight sandstones through simulated experiments. Oil Gas Geol. 2008;29(6):721–5 **(in Chinese)**.
Xie GJ, Jin ZJ, Yang LN. A numerical simulation on accumulation mechanism of deep basin gas. J China Univ Pet. 2004;28:13–7 **(in Chinese)**.
Yang KM, Pang XQ. Developing mechanisms and detection methods of tight sandstone gas reservoirs: a case of the depression in the western Sichuan Basin. Beijing: Science Press; 2012 **(in Chinese)**.
Yang SY, Zhang JC, Huang WD, et al. "Sweet spot" types of reservoirs and genesis of tight sandstone gas in Kekeya area, Turpan-Hami Basin. Acta Pet Sin. 2013;34(2):272–82 **(in Chinese)**.
Yang T, Zhang GS, Liang K, et al. The exploration of global tight sandstone gas and forecast of the development tendency in China. Eng Sci. 2012;14(6):64–8 **(in Chinese)**.
Yang XN, Zhang HL, Zhu GH. Formation mechanism and geological implication of tight sandstones: a case of Well YN-2 in Tarim Basin. Mar Orig Pet Geol. 2005;10(1):31–6 **(in Chinese)**.
Yang XP, Zhao WZ, Zou CN, et al. Origin of low-permeability reservoir and distribution of favorable reservoir. Acta Pet Sin. 2007;28(4):57–61 **(in Chinese)**.
Yuan ZW, Xu HZ, Wang BS, et al. Deep tight gas research in the Alberta Basin. Beijing: Petroleum Industry Press; 1996 **(in Chinese)**.
Zeng JH. Experimental simulation of impacts of vertical heterogeneity on oil migration and accumulation in fining upwards sands. Pet Explor Dev. 2000;27(4):102–5 **(in Chinese)**.
Zhang JC. Source-contacting gas: derived from deep basin gas or basin-centered gas. Nat Gas Ind. 2006;26(2):46–8 **(in Chinese)**.
Zhang JC, Jin ZJ, Pang XQ. Formation conditions and internal features of deep basin gas accumulations. Explor Pet Geol. 2000;22(3):210–4 **(in Chinese)**.
Zhang RH, Yao GS, Shou JF, et al. An integration porosity forecast model of deposition, diagenesis and structure. Pet Explor Dev. 2011;38(2):145–51 **(in Chinese)**.
Zhao JZ, Fu JH, Yao JL, et al. Quasi-continuous accumulation model of large tight sandstone gas field in Ordos Basin. Acta Pet Sin. 2012;33(S1):37–52 **(in Chinese)**.
Zhao WZ, Zou CN, Wang ZC, et al. The intension and signification of "sag-wide oil-bearing theory" in the hydrocarbon-rich depression with terrestrial origin. Pet Explor Dev. 2004;31(2):5–13 **(in Chinese)**.
Zhou XY. Distribution regularities of natural gas in the Kuqa Foreland Basin. Beijing: Petroleum Industry Press; 2002 **(in Chinese)**.
Zhu RK, Bai B, Cui JW, et al. Research advances of microstructure in unconventional tight oil and gas reservoirs. J Palaeogeogr. 2013;15(5):615–23 **(in Chinese)**.
Zou CN, Tao SZ, Hou LH, et al. Unconventional petroleum geology. Beijing: Geological Publishing House; 2011a **(in Chinese)**.
Zou CN, Yang Z, Zhang GS, et al. Conventional and unconventional petroleum "orderly accumulation": concept and practical significance. Pet Explor Dev. 2014;41(1):14–30 **(in Chinese)**.
Zou CN, Zhang GS, Yang Z, et al. Geological concepts, characteristics, resource potential and key techniques of unconventional hydrocarbon: on unconventional petroleum geology. Pet Explor Dev. 2013;40(4):385–99 **(in Chinese)**.
Zou CN, Zhang GY, Tao SZ, et al. Geological features, major discoveries and unconventional petroleum geology in the global petroleum exploration. Pet Explor Dev. 2010;37(2):129–45 **(in Chinese)**.
Zou CN, Zhu RK, Bai B, et al. First discovery of nano-pore throat in oil and gas reservoir in China and its scientific value. Acta Pet Sin. 2011b;27(6):1857–64 **(in Chinese)**.
Zou CN, Zhu RK, Wu ST, et al. Types, characteristics, genesis and prospects of conventional and unconventional hydrocarbon accumulations: taking tight oil and tight gas in China as an instance. Acta Pet Sin. 2012;33(2):173–87 **(in Chinese)**.

9

Fractional thermoelasticity applications for porous asphaltic materials

Magdy Ezzat[1,2] · Shereen Ezzat[1]

Abstract A new mathematical model of poro-thermoelasticity has been constructed in the context of a new consideration of heat conduction with fractional order. One-dimensional application for a poroelastic half-space saturated with fluid is considered. The surface of the half-space is assumed to be traction-free, permeable, and subjected to heating. The Laplace transform technique is used to solve the problem. The inversion of the Laplace transform will be obtained numerically and the numerical values of the temperature, stresses, strains, and displacements will be illustrated graphically for the solid and the liquid.

Keywords Generalized thermo-poroelasticity · Asphaltic material · Thermal shock · Fractional calculus · Numerical results

List of symbols

c_s, c_f	Specific heat of the solid and the fluid phases
e_s	Strain component of the solid phase
e_f	Strain component of the fluid phase
F_i	Components of the body forces per unit mass
k_s	Thermal conductivity of the solid phase
k_f	Thermal conductivity of the fluid phase
q^*_{is}, q^*_{if}	Intensities of heat fluxes of the solid and fluid phases
q_{is}	$q_{is} = (1-\beta)q^*_{is}$, heat flux of the solid phase per unit area
q_{if}	$q_{if} = \beta q^*_{if}$, heat flux of the fluid phase per unit area
Q	Intensity of the heat sources per unit mass
$R_{11}, R_{12}, R_{21}, R_{22}$	Mixed and thermal coefficients
T_o	Reference temperature
u_{is}	ith component of the solid-phase displacement
u_{if}	ith component of the fluid-phase displacement
α_s, α_f	Coefficients of the thermal expansion of the phases
β	Porosity of the aggregate
σ_{ij}	Stress components in the solid
σ	Normal stress in the fluid
θ_s	$\theta_s = T_s - T_o$, temperature increment of the solid phase
θ_f	$\theta_f = T_f - T_o$, temperature increment of the fluid phase
ρ_s, ρ_f	Density of the solid and the liquid phases
ρ_1	$\rho_1 = (1-\beta)\rho_s$, density of the solid phase per unit volume of bulk
ρ_2	$\rho_2 = \beta\rho_f$, density of the fluid phase per unit volume of bulk
ρ_{11}	$\rho_{11} = \rho_1 - \rho_{12}$, mass coefficient of the solid phase

✉ Magdy Ezzat
maezzat2000@yahoo.com

1 Department of Mathematics, Faculty of Science and Letters in Al Bukayriyyah, Al-Qassim University, Al-Qassim, Saudi Arabia

2 Faculty of Education, Alexandria University, Alexandria, Egypt

Edited by Yan-Hua Sun

ρ_{22}	$\rho_{22} = \rho_2 - \rho_{12}$, mass coefficient of the fluid phase
ρ_{12}	Dynamics coupling coefficient
ρ	Displacements of the skeleton and fluid phases
ρ	$\rho = \rho_1 + \rho_2$
κ	Interface coefficients of the interphase heat conduction
$\tau_s,\ \tau_f$	Relaxation times of the solid and the fluid phases
$\lambda,\ \mu,\ R,\ G$	Poroelastic coefficients
η_s	$\eta_s = \frac{\rho_s c_s}{k_s}$, thermal viscosity of the solid
η_f	$\eta_f = \frac{\rho_f c_f}{k_f}$, thermal viscosity of the fluid
η	$\eta = \frac{\rho_{12} c_{sf}}{k}$, thermal viscosity couplings between the phases
P	$P = 3\lambda + 2\mu$
F_{11}	$F_{11} = \rho_s C_s$
F_{22}	$F_{22} = \rho_f C_f$
R_{11}	$R_{11} = \alpha_s P + \alpha_{fs} G$
R_{22}	$R_{22} = \alpha_f R + 3\alpha_{sf} G$
R_{12}	$R_{12} = \alpha_f G + \alpha_{sf} P$

1 Introduction

The concept of a poroelastic material was introduced by Biot (1955) in order to describe the mechanical behavior of water-saturated soil. Biot's material consists of a combination of a deformable solid and a fluid, the solid constituting a porous skeleton whose innumerable tiny cavities are interconnected and filled with the fluid. Apart from water-saturated soil, the class of poroelastic materials includes other materials such as polyurethane foam and sound-absorbing materials, osseous tissue, and so on.

Due to many applications in the fields of geophysics, plasma physics and related topics, increasing attention is being devoted to the interaction between fluid such as water and thermo elastic solids, which is the domain of the theory of poro-thermoelasticity. The field of poro-thermoelasticity has a wide range of applications especially in studying the effect of using waste materials on disintegration of asphalt concrete mixture (ACM).

Asphalt concrete pavements are made of asphalt concrete mixtures (ACMs) consisting of graded aggregates, asphalt binders, and air voids. Due to the complexity of internal composition, mechanical characteristics of ACM exhibit extremely non-linear constitutive behavior with respect to loading time, loading rate, and temperature. The experimental observations suggest that the total deformation behavior has recoverable and irrecoverable parts, which could occur simultaneously and are time-dependent functions of stress levels, strain rates, and temperature (Krishnan and Rajagopal 2003). Time-dependent properties of ACM are not only one of the main sources of pavement rutting (e.g., permanent deformation), but also plays a critical role in fatigue and low-temperature cracking, roughness, and corrugation.

Coupled thermal and poromechanical processes play an important role in a number of problems of interest in geomechanics such as stability of boreholes and hydraulic fracturing in geothermal reservoirs or high temperature petroleum-bearing formations. This is due to that fact that when rocks are heated/cooled, the bulk solid as well as the pore fluid tends to undergo expansion/contraction. A volumetric expansion can result in significant pressurization of the pore fluid depending on the degree of containment and the thermal and hydraulic properties of the fluid as well as the solid. The net effect is a coupling of thermal and poromechanical processes. A few analytical procedures have been developed and used to solve geomechanics problems of interest involving coupled thermal and poromechanical problems (Delaney 1982; Wang and Papamichos 1994; Li et al. 1998; Ghassemi and Diek 2002; Ghassemi and Zhang 2004). However, many problems formulated within the framework of poro-thermoelasticity are not amenable to analytical treatment and need to be solved numerically.

The coupled rock deformation and liquid flow and various factors that affect it need to be studied both theoretically and experimentally. Implementation of theoretical rheologies in numerical models enables scientific interpretation of field measurements and provides a powerful means for understanding and predicting phenomena related to the internal earth processes. This is of paramount significance to the development of renewable geothermal energy resources as well as the prediction of natural hazards. However, the traditional isotropic elastic, plastic, and viscous rheological models are inadequate to study Earth and its tectonics. This is especially true at transform plate boundaries, where deformation-enhanced weakening is thought to be responsible for localization of strain (Simakin and Ghassemi 2005). Therefore, there is strong scientific and industrial interest in developing new mechanical models of liquid saturated rocks such as poro-viscoelasticity with damage mechanics that can include the effects of fluid phase pressurization and transport.

Porous materials make their appearance in a wide variety of settings, natural, and artificial, and in diverse technological applications. As a consequence a number of problems arise dealing with, among other issues, statics and strength, fluid flow and heat conduction, and dynamics. In connection with the latter, we note that problems of this kind are encountered in the prediction of the behavior of sound-absorbing materials and in the area of exploration geophysics, the steadily growing literature bearing witness to the importance of the subject. The existence and uniqueness of the generalized solutions for the boundary value problems in elasticity of initially stressed bodies with voids (porous materials) are proved (Marin 2008).

The field of poro-thermoelasticity has a wide range of applications especially in studying the effect of using waste materials on disintegration of asphalt concrete mixture (Pecker and Deresiewiez 1973). The problem of a fluid-saturated porous material has been studied for many years. A short list of papers pertinent to the present study includes Biot (1956a, b), Biot and Willis (1957), Deresiewicz and Skalak (1963), Nur and Byerlee (1971), Sherief and Hussein (2012), Youssef (2007).

Mathematical modeling is the process of constructing mathematical objects whose behavior or properties correspond in some way to a particular real-world system. The term real-world system could refer to a physical system, a financial system, a social system, an ecological system, or essentially any other system whose behavior can be observed. In this description, a mathematical object could be a system of equations, a stochastic process, a geometric or algebraic structure, an algorithm or any other mathematical apparatus like a fractional derivative, integral or fractional system of equations. Fractional calculus and fractional differential equations serve as mathematical objects describing many real-world systems.

Fractional calculus has been used successfully to modify many existing models of physical processes. One can state that the whole theory of fractional derivatives and integrals was established in the second half of the 19th century. The first application of fractional derivatives was given by Abel, who applied fractional calculus in the solution of an integral equation that arises in the formulation of the tautochrone problem. The generalization of the concept of derivatives and integrals to a non-integer order has been subjected to several approaches and some various alternative definitions of fractional derivatives appeared elsewhere (Gorenflo and Mainardi 1997; Miller and Ross 1993; Samko et al. 1993; Oldham and Spanier 1974). In the last few years, fractional calculus has been applied successfully in various areas to modify many existing models of physical processes, e.g., chemistry, biology, modeling and identification, electronics, wave propagation, and viscoelasticity (Rossikhin and Shitikova 1967; Bagley and Torvik 1986). Fractional order models often work well, particularly for dielectrics and viscoelastic materials over extended ranges of time and frequency (Lakes 1999; Grimnes and Martinsen 2000). In heat transfer and electrochemistry, for example, the half-order fractional integral is the natural integral operator connecting the applied gradients (thermal or material) with the diffusion of ions with heat (Gorenflo et al. 2002). One can refer to Podlubny (1999) for a survey of applications of fractional calculus.

Ezzat (2010, 2011a, b, c) was the first writer who established a new formula of heat conduction law by using the new Taylor-Riemann series expansion of time-fractional order α developed by Jumarie (2010) as follows:

$$q(x,t) + \frac{\tau^{\upsilon}}{\upsilon!}\frac{\partial^{\upsilon} q}{\partial t^{\upsilon}} = -k\nabla T, \quad 0 < \upsilon \leq 1 \tag{1}$$

where q the heat flux vector and τ is the relaxation time.

Sherief et al. (2010) introduced a fractional formula of heat conduction and proved a uniqueness theorem and derived a reciprocity relation and a variational principle. El-Karamany and Ezzat (2011) introduced two general models of fractional heat conduction law for a non-homogeneous anisotropic elastic solid. Uniqueness and reciprocal theorems are proved and the convolutional variational principle is established and used to prove a uniqueness theorem with no restriction on the elasticity or thermal conductivity tensors except symmetry conditions. Ezzat and El-Karamany (2011a, b, c) introduced a new mathematical model for electro-thermoelasticity equations using the methodology of fractional calculus. The model is applied to one-dimensional problems to investigate the thermal behavior in thermoelectric solids. The verification process was done by comparison of model predictions with the previous work and shows good agreement is achieved. Abbas (2014, 2015) solved some problems on fractional order theory of thermoelasticity for a functional graded material. Some applications of fractional calculus to various problems in continuum mechanics are reviewed in the literature (Ezzat et al. 2012a, b; 2013a, b; 2014a, b, c, d; 2015).

In the current work, a modified law of heat conduction including fractional order of the time derivative is constructed and replaces the conventional Fourier's law in poro-thermoelasticity. The resulting non-dimensional coupled equations for poroelastic half-space saturated with fluid are considered. The general solution in the Laplace transform domain is obtained and applied in a certain asphalt material which is thermally shocked on its bounding plane. The inversion of the Laplace transform will be obtained numerically and the numerical values of the temperature, displacement, and stress will be illustrated graphically.

2 Derivation of fractional heat conduction equation in poro-thermoelasticity

Much effort has been devoted recently for determining conditions which guarantee that the assumption of local thermal equilibrium (LTE) is accurate when modeling heat transfer in porous media. When it is accurate, then the thermal field is well approximated by a single thermal energy equation. In other circumstances, the local thermal non-equilibrium (LTNE) prevails, and it is necessary to employ two energy equations, one for each phase (Nouri-Borujerdi et al. 2007). Fourie and Du Plessis (2003) showed that the intrinsic volume-averaged equilibrium temperature of the solid phase is equal to that of the fluid phase everywhere though that their values may differ locally.

The conventional poro-thermoelasticity is based on the principles of the classical theory of heat conductivity, specifically on classical Fourier's law, in which relates the heat flux vector q to the temperature gradient (Nowinski 1978)

$$q_{is}(x_i,t) = -k_s\theta_{s,i} \tag{2}$$

$$q_{if}(x_i,t) = -k_f\theta_{f,i} \tag{3}$$

The energy equation in terms of the heat conduction vectors q_{is} and q_{if} are

$$\begin{aligned}&\frac{\partial}{\partial t}(F_{11}\,\theta_s + T_oR_{11}\,e_s + T_oR_{21}\,e_f)\\&\quad= -\nabla q_{is} + \rho_1 Q_s - \kappa(\theta_s - \theta_f)\end{aligned} \tag{4}$$

$$\begin{aligned}&\frac{\partial}{\partial t}(F_{22}\,\theta_f + T_oR_{12}\,e_s + T_oR_{22}\,e_f)\\&\quad= -\nabla q_{if} + \rho_2 Q_f + \kappa(\theta_s - \theta_f)\end{aligned} \tag{5}$$

Although Fourier's law of heat conduction is well tested for most practical problems, it fails to describe the transient temperature field in situations involving short times, high frequencies, and small wavelengths (Lebon et al. 2008). To eliminate these anomalies, Cattaneo (1948) and Vernotte (1958) proposed a damped version of Fourier's law by introducing a heat flux relaxation term, by taking Taylor's series to expand $q_{is}(x_i,\,t+\tau_s)$, $q_{if}(x_i,\,t+\tau_f)$ and retaining terms up to the first order in τ_s and τ_f. The first well-known generalization of such a type (Sherief and Hussein 2012)

$$q_{is} + \tau_s\dot{q}_{is} = -k_s\theta_{s,i} \tag{6}$$

$$q_{if} + \tau_f\dot{q}_{if} = -k_f\theta_{f,i} \tag{7}$$

leads to the hyperbolic-type heat transport equation in the theory of poro-thermoelasticity

$$\begin{aligned}&k_s\theta_{s,ii} + \left(1+\tau_s\frac{\partial}{\partial t}\right)[\rho_1 Q_s - \kappa(\theta_s - \theta_f)]\\&\quad= \frac{\partial}{\partial t}\left(1+\tau_s\frac{\partial}{\partial t}\right)(F_{11}\,\theta_s + T_oR_{11}\,e_s + T_oR_{21}\,e_f)\end{aligned} \tag{8}$$

$$\begin{aligned}&k_f\theta_{f,ii} + \left(1+\tau_f\frac{\partial}{\partial t}\right)[\rho_2 Q_f + \kappa(\theta_s - \theta_f)]\\&\quad= \frac{\partial}{\partial t}\left(1+\tau_f\frac{\partial}{\partial t}\right)(F_{22}\,\theta_f + T_oR_{12}\,e_s + T_oR_{22}\,e_f).\end{aligned} \tag{9}$$

In the present work, the new fractional Taylor-Riemann series of time-fractional order υ is adopted to expand $q_{is}(x_i,\,t+\tau_s)$, $q_{if}(x_i,\,t+\tau_f)$ and retaining terms up to order υ in the thermal relaxation times τ_s and τ_f, we get

$$q_{is}(x_i,t+\tau_s) = q_{is}(x_i,t) + \frac{\tau_s^\upsilon}{\upsilon!}\frac{\partial^\upsilon q_{is}}{\partial t^\upsilon},\quad 0<\upsilon\leq 1 \tag{10}$$

$$q_{if}(x_i,t+\tau_f) = q_{if}(x_i,t) + \frac{\tau_f^\upsilon}{\upsilon!}\frac{\partial^\upsilon q_{if}}{\partial t^\upsilon},\quad 0<\upsilon\leq 1 \tag{11}$$

From a mathematical viewpoint, Fourier's laws (2) and (3) in poro-thermoelasticity theory of generalized fractional heat conduction are given by

$$q_{is}(x_i,t) + \frac{\tau_s^\upsilon}{\upsilon!}\frac{\partial^\upsilon q_{is}}{\partial t^\upsilon} = -k_s\theta_{s,i},\quad 0<\upsilon\leq 1 \tag{12}$$

$$q_{if}(x_i,t) + \frac{\tau_f^\upsilon}{\upsilon!}\frac{\partial^\upsilon q_{if}}{\partial t^\upsilon} = -k_f\theta_{f,i},\quad 0<\upsilon\leq 1. \tag{13}$$

Taking the partial time derivative of fraction order υ of Eqs. (4) and (5), we get (Tarasov 2008)

$$\begin{aligned}&\frac{\partial^{\upsilon+1}}{\partial t^{\upsilon+1}}(F_{11}\,\theta_s + T_oR_{11}\,e_s + T_oR_{21}\,e_f)\\&\quad= -\nabla\left(\frac{\partial^\upsilon q_{is}}{\partial t^\upsilon}\right) + \frac{\partial^\upsilon}{\partial t^\upsilon}[\rho_1 Q_s - \kappa(\theta_s - \theta_f)]\end{aligned} \tag{14}$$

$$\begin{aligned}&\frac{\partial^{\upsilon+1}}{\partial t^{\upsilon+1}}(F_{22}\,\theta_f + T_oR_{12}\,e_s + T_oR_{22}\,e_f)\\&\quad= -\nabla\left(\frac{\partial^\upsilon q_{if}}{\partial t^\upsilon}\right) + \frac{\partial^\upsilon}{\partial t^\upsilon}[\rho_2 Q_f + \kappa(\theta_s - \theta_f)]\end{aligned} \tag{15}$$

Multiplying Eqs. (14) and (15) by $\frac{\tau_s^\upsilon}{\upsilon!}$ and $\frac{\tau_f^\upsilon}{\upsilon!}$ and adding to Eqs. (4) and (5), respectively, we have

$$\begin{aligned}&k_s\theta_{s,ii} + \left(1+\frac{\tau_s^\upsilon}{\upsilon!}\frac{\partial^\upsilon}{\partial t^\upsilon}\right)[\rho_1 Q_s - \kappa(\theta_s - \theta_f)]\\&\quad= \frac{\partial}{\partial t}\left(1+\frac{\tau_s^\upsilon}{\upsilon!}\frac{\partial^\upsilon}{\partial t^\upsilon}\right)(F_{11}\,\theta_s + T_oR_{11}\,e_s + T_oR_{21}\,e_f)\end{aligned} \tag{16}$$

$$\begin{aligned}&k_f\theta_{f,ii} + \left(1+\frac{\tau_s^\upsilon}{\upsilon!}\frac{\partial^\upsilon}{\partial t^\upsilon}\right)[\rho_2 Q_f + \kappa(\theta_s - \theta_f)]\\&\quad= \frac{\partial}{\partial t}\left(1+\frac{\tau_s^\upsilon}{\upsilon!}\frac{\partial^\upsilon}{\partial t^\upsilon}\right)(F_{22}\,\theta_f + T_oR_{12}\,e_s + T_oR_{22}\,e_f)\end{aligned} \tag{17}$$

Equations (16) and (17) are the generalized energy equations of poro-thermoelasticity with fractional derivatives and taking into account the relaxation times τ_s and τ_f. Some theories of heat conduction law follow as limit cases for different values of the parameters τ_s, τ_f, and υ.

Taking into consideration

$$\frac{\partial^\upsilon}{\partial t^\upsilon}f(y,t) = \begin{cases} f(y,t) - f(y,0) & \upsilon\to 0\\ I^{\upsilon-1}\dfrac{\partial f(y,t)}{\partial t} & 0<\upsilon<1\\ \dfrac{\partial f(y,t)}{\partial t} & \upsilon = 1,\end{cases}$$

where the notion I^υ is the Riemann–Liouville fractional integral introduced as a natural generalization of the well-known n-fold repeated integral $I^n f(t)$ written in a convolution-type form (Mainardi and Gorenflo 2000)

$$\left.\begin{aligned} I^{\upsilon} f(t) &= \int_0^t \frac{(t-\varsigma)^{\upsilon-1}}{\Gamma(\upsilon-1)} f(\varsigma) d\varsigma \\ I^0 f(t) &= f(t) \end{aligned}\right\} \upsilon > 0$$

In the limit as υ tends to 1, Eqs. (16) and (17) reduce to the well-known Cattaneo-Vernotte (1948) law used by Lord and Shulman (1967) to derive the equation of the generalized theory of poro-thermoelasticity with one relaxation time. It is known that the classical entropy derived using this law instead of monotonically increasing behaves in an oscillatory way (Lebon et al. 2008; Jou et al. 1988). Strictly speaking, this result is not incompatible with Clausius' formulation of the second law, which states that the entropy of the final equilibrium state must be higher than the entropy of the initial equilibrium state. However, the non-monotonic behavior of the entropy is in contradiction with the local equilibrium formulation of the second law, which requires that the entropy production must be positive everywhere at any time (Lebon et al. 2008). During the last two decades, this became the subject of many research papers and resulted in the introduction of what is known now as extended irreversible thermodynamics. A review can be found in Jou et al. (1988).

2.1 Limiting cases

1. The heat Eqs. (16) and (17) for the two-phase system in the limiting case $\upsilon = 0$ transforms to the work of Biot (1984) in the context of coupled thermoelasticity (Biot 1956b).
2. The heat Eqs. (16) and (17) for the two-phase system in the limiting case $\upsilon = 1$ transforms to the work of Sherief and Hussein (2012) in the context generalized thermoelasticity with one relaxation time (Lord and Shulman 1967).

3 Mathematical models

The linear governing equations of isotropic, generalized poro-thermoelasticity in the absence of body forces and heat sources are as follows:

(i) Equations of motion

$$\mu u_{is,jj} + [(\lambda+\mu)\, e_s + Ge_f - R_{11}\theta_s - R_{12}\theta_f]_i = \rho_{11}\ddot{u}_{si} + \rho_{12}\ddot{u}_{fi} \tag{18}$$

$$[Ge_s + Re_f - R_{21}\theta_s - R_{22}\theta_f]_i = \rho_{12}\,\ddot{u}_{is} + \rho_{22}\,\ddot{u}_{if} \tag{19}$$

(ii) Fractional heat equation

$$k_s\theta_{s,ii} - \kappa\left(1+\frac{\tau_s^{\upsilon}}{\upsilon!}\frac{\partial^{\upsilon}}{\partial t^{\upsilon}}\right)(\theta_s - \theta_f) = \frac{\partial}{\partial t}\left(1+\frac{\tau_s^{\upsilon}}{\upsilon!}\frac{\partial^{\upsilon}}{\partial t^{\upsilon}}\right)(F_{11}\,\theta_s + T_o R_{11}\, e_s + T_o R_{21}\, e_f) \tag{20}$$

$$k_f\theta_{f,ii} + \kappa\left(1+\frac{\tau_s^{\upsilon}}{\upsilon!}\frac{\partial^{\upsilon}}{\partial t^{\upsilon}}\right)(\theta_s - \theta_f) = \frac{\partial}{\partial t}\left(1+\frac{\tau_s^{\upsilon}}{\upsilon!}\frac{\partial^{\upsilon}}{\partial t^{\upsilon}}\right)(F_{22}\,\theta_f + T_o R_{12}\, e_s + T_o R_{22}\, e_f) \tag{21}$$

(iii) Constitutive equations

$$\sigma_{ij} = 2\mu e_{ij} + \lambda e_s\delta_{ij} + (Ge_f - R_{11}\,\theta_s - R_{12}\,\theta_f)\,\delta_{ij} \tag{22}$$

$$\sigma = Re_f + Ge_s - R_{21}\theta_s - R_{22}\theta_f \tag{23}$$

$$e_{ij} = \frac{1}{2}\left(u_{is,j} + u_{js,i}\right), \quad e_s = e_{ii} = u_{is,i} \tag{24}$$

$$e_f = u_{if,i} \tag{25}$$

4 Formulation of the problem

We shall consider a homogeneous isotropic thermo-poroelastic medium occupying the region $x \geq 0$, where the x-axis is taken perpendicular to the bounding plane of half-space pointing inwards, subjected to traction-free time-dependent heating. The initial conditions of the problem are taken to be quiescent.

For the one-dimensional problems, all the considered functions will depend only on the space variables x and t. The displacement components take the form:

$$u_{xs} = u_s(x,t), \quad u_{ys} = u_{zs} = 0, \quad u_{xf} = u_f(x,t), \quad u_{yf} = u_{zf} = 0$$

Let us introduce the following non-dimensional variables

$$\begin{aligned} &(x', u_s', u_f') = c_1\eta(x, u_s, u_f), \\ &(t', \tau_s', \tau_s') = c_1^2\eta(t, \tau_s, \tau_s), \\ &c_1 = \sqrt{M/\rho_{11}}, \\ &(\sigma_{ij}', \sigma') = (\sigma_{ij}, \sigma)/M, \\ &(\theta_s', \theta_f') = R_{11}(\theta_s, \theta_f)/M, \\ &\eta = F_{11}/k_s, \\ &(q_s', q_f') = R_{11}M(k_s q_s', k_f q_f')/c_1\eta, \\ &M = \lambda + 2\mu \end{aligned}$$

In terms of these non-dimensional variables, Eqs. (18)–(25) take the following form (dropping primes for convenience):

$$\frac{\partial^2 u_s}{\partial x^2}+a_1\frac{\partial^2 u_f}{\partial x^2}-\frac{\partial\theta_s}{\partial x}-A_1\frac{\partial\theta_f}{\partial x}=\frac{\partial^2 u_s}{\partial t^2}+b_1\frac{\partial^2 u_f}{\partial t^2} \quad (26)$$

$$a_1\frac{\partial^2 u_s}{\partial x^2}+a_2\frac{\partial^2 u_f}{\partial x^2}-A_2\frac{\partial\theta_s}{\partial x}-A_3\frac{\partial\theta_f}{\partial x}=b_1\frac{\partial^2 u_s}{\partial t^2}+b_2\frac{\partial^2 u_f}{\partial t^2} \quad (27)$$

$$\frac{\partial^2\theta_s}{\partial x^2}=\left(\frac{\partial}{\partial t}+\frac{\tau_s^\upsilon}{\upsilon!}\frac{\partial^{\upsilon+1}}{\partial t^{\upsilon+1}}\right)\left(\theta_s+\varepsilon_2\frac{\partial}{\partial x}[u_s+A_2u_f]\right)+\varepsilon_1\left(1+\frac{\tau_s^\upsilon}{\upsilon!}\frac{\partial^\upsilon}{\partial t^\upsilon}\right)(\theta_s-\theta_f) \quad (28)$$

$$\frac{\partial^2\theta_f}{\partial x^2}=\left(\frac{\partial}{\partial t}+\frac{\tau_s^\upsilon}{\upsilon!}\frac{\partial^{\upsilon+1}}{\partial t^{\upsilon+1}}\right)\left(F\theta_f+\omega\varepsilon_2\frac{\partial}{\partial x}[A_1u_s+A_3u_f]\right)-\omega\varepsilon_1\left(1+\frac{\tau_s^\upsilon}{\upsilon!}\frac{\partial^\upsilon}{\partial t^\upsilon}\right)(\theta_s-\theta_f) \quad (29)$$

$$\sigma_{xx}=\frac{\partial u_s}{\partial x}+a_1\frac{\partial u_f}{\partial x}-\theta_s-A_1\theta_f \quad (30a)$$

$$\sigma_{yy}=\sigma_{zz}=a_3\frac{\partial u_s}{\partial x}+a_1\frac{\partial u_f}{\partial x}-\theta_s-A_1\theta_f \quad (30b)$$

$$\sigma_{xy}=\sigma_{xz}=\sigma_{zy}=0 \quad (30c)$$

$$\sigma=a_2\frac{\partial u_f}{\partial x}+a_1\frac{\partial u_s}{\partial x}-A_3\theta_f-A_2\theta_s \quad (30d)$$

$$q_s+\frac{\tau_s^\upsilon}{\upsilon!}\frac{\partial^\upsilon q_s}{\partial t^\upsilon}=-\frac{\partial\theta_s}{\partial x},\quad 0<\upsilon\le 1 \quad (31a)$$

$$q_f+\frac{\tau_f^\upsilon}{\upsilon!}\frac{\partial^\upsilon q_f}{\partial t^\upsilon}=-\frac{\partial\theta_f}{\partial x},\quad 0<\upsilon\le 1 \quad (31b)$$

The boundary conditions can be expressed as

$$\sigma_{xx}(0,t)=\sigma(0,t)=0 \quad (32a)$$

$$q_s(0,t)+B_s\theta_s(0,t)=f(t) \quad (32b)$$

$$q_f(0,t)+B_f\theta_f(0,t)=f(t), \quad (32c)$$

where B_s and B_f are the Biot's numbers and $f(t)$ represents the magnitude of surface heating.

5 The analytical solutions in the Laplace transform domain

Performing the Laplace transform defined by the relation

$$\bar{g}(s)=\int_0^\infty e^{-st}g(t)\,dt$$

of both sides Eqs. (26)–(32), with the homogeneous initial conditions

$$D^2\bar{u}_s+a_1D^2\bar{u}_f-D\bar{\theta}_s-A_1D\bar{\theta}_f=s^2(\bar{u}_s+b_1u_f) \quad (33)$$

$$a_1D^2\bar{u}_s+a_2D^2\bar{u}_f-A_2D\bar{\theta}_s-A_3D\bar{\theta}_f=s^2(b_1\bar{u}_s+b_2u_f) \quad (34)$$

$$D^2\bar{\theta}_s=s\left(1+\frac{\tau_s^\upsilon}{\upsilon!}s^\upsilon\right)\left(\bar{\theta}_s+\varepsilon_2D[\bar{u}_s+A_2\bar{u}_f]\right)+\varepsilon_1\left(1+\frac{\tau_s^\upsilon}{\upsilon!}s^\upsilon\right)\left(\bar{\theta}_s-\bar{\theta}_f\right) \quad (35)$$

$$D^2\bar{\theta}_f=s\left(1+\frac{\tau_s^\upsilon}{\upsilon!}s^\upsilon\right)\left(F\bar{\theta}_f+\omega\varepsilon_2D[A_1\bar{u}_s+A_3\bar{u}_f]\right)+\omega\varepsilon_1\left(1+\frac{\tau_s^\upsilon}{\upsilon!}s^\upsilon\right)\left(\bar{\theta}_s-\bar{\theta}_f\right) \quad (36)$$

$$\bar{\sigma}_{xx}=D\bar{u}_s+a_1D\bar{u}_f-\bar{\theta}_s-A_1\bar{\theta}_f \quad (37a)$$

$$\bar{\sigma}_{yy}=\bar{\sigma}_{zz}=a_3D\bar{u}_s+a_1D\bar{u}_f-\bar{\theta}_s-A_1\bar{\theta}_f \quad (37b)$$

$$\bar{\sigma}_{xy}=\bar{\sigma}_{xz}=\bar{\sigma}_{zy}=0 \quad (37c)$$

$$\bar{\sigma}=a_2D\bar{u}_f+a_1D\bar{u}_s-A_3\bar{\theta}_f-A_2\bar{\theta}_s \quad (37d)$$

$$\left(1+\frac{\tau_s^\upsilon}{\upsilon!}s^\upsilon\right)\bar{q}_s=-D\bar{\theta}_s,\quad 0<\upsilon\le 1 \quad (38a)$$

$$\left(1+\frac{\tau_s^\upsilon}{\upsilon!}s^\upsilon\right)\bar{q}_f=-D\bar{\theta}_f,\quad 0<\upsilon\le 1 \quad (38b)$$

$$\bar{\sigma}_{xx}(0,t)=\bar{\sigma}(0,t)=0 \quad (39a)$$

$$\left(1+\frac{\tau_s^\upsilon}{\upsilon!}s^\upsilon\right)[f(t)-B_s\theta_s(0,t)]=-D\bar{\theta}_s,\quad 0<\upsilon\le 1 \quad (39b)$$

$$\left(1+\frac{\tau_f^\upsilon}{\upsilon!}s^\upsilon\right)[f(t)-B_f\theta_f(0,t)]=-D\bar{\theta}_f,\quad 0<\upsilon\le 1 \quad (39c)$$

Eliminating $\bar{u}_s$, $\bar{u}_f$, and $\bar{\theta}_f$ between Eqs. (33) and (36), we obtain the following equation satisfied by $\bar{\theta}_s$:

$$\left(D^8+\ell_1D^6+\ell_2D^4+\ell_3D^2+\ell_4\right)\bar{\theta}_s=0. \quad (40a)$$

In a similar manner, we can get the following Eqs. for $\bar{u}_s$, $\bar{u}_f$, and $\bar{\theta}_f$ which satisfy Eq. (40) as follows:

$$\left(D^8+\ell_1D^6+\ell_2D^4+\ell_3D^2+\ell_4\right)\bar{u}_s=0 \quad (40b)$$

$$\left(D^8+\ell_1D^6+\ell_2D^4+\ell_3D^2+\ell_4\right)\bar{u}_f=0 \quad (40c)$$

$$\left(D^8+\ell_1D^6+\ell_2D^4+\ell_3D^2+\ell_4\right)\bar{\theta}_f=0, \quad (40d)$$

where $\ell_i,\ i=1,\ 2,\ 3,\ 4$ are the parameters depending on s.

Equation (40a) can be factorized as

$$\left(D^2-k_1^2\right)\left(D^2-k_2^2\right)\left(D^2-k_3^2\right)\left(D^2-k_4^2\right)\bar{\theta}_s=0, \quad (41)$$

where $k_m,\ m=1,\ 2,\ 3,\ 4$ are the roots of the characteristic equation of the system (40) which takes the form

$$k^8 + \ell_1 k^6 + \ell_2 k^4 + \ell_3 k^2 + \ell_4 = 0. \tag{42}$$

We can consider the general solutions of Eqs. (40), according to the bounded state of functions at infinity, in the following forms

$$(\bar{u}_s, \bar{u}_f, \bar{\theta}_s, \bar{\theta}_s) = \sum_{m=1}^{4} (L_m, M_m, N_m, H_m)\, e^{-k_m x}, \tag{43}$$

where L_m, M_m, N_m, and H_m are the parameters that depend on s.

Hence, the velocity for the fluid is given by

$$\bar{v}_f(x,s) = \sum_{m=1}^{4} s M_m(s)\, e^{-k_m x}. \tag{44}$$

Thus, the skin friction function τ_w (shear stress on the bounding plane surface of half-space) takes the expression (Ezzat 1994; Ezzat and Abd-Elaal 1997)

$$\tau_w(s) = -\left.\frac{\partial \bar{v}_f(x,s)}{\partial x}\right|_{x=0} = \sum_{m=1}^{4} s\, k_m(s)\, M_m(s). \tag{45}$$

Putting

$$M_m = g_{m1} L_m, N_m = g_{m2} L_m, \; H_m = g_{m3} L_m \tag{46}$$

Hence, from Eqs. (43) and (33)–(35), one can obtain the following relations:

$$g_{m1} = \left[k_m r_{m3}(A_1A_2 - A_3) - A_1 r_{m2} Z_3 - A_2 r_{m1} Z_4 + A_3 r_{m1} Z_3 + r_{m2} Z_4\right]/\Lambda,$$

$$g_{m2} = \left[k_m(A_1(r_{m2}Z_2 - r_{m3}Z_1) - A_3(r_{m1}Z_2 + r_{m2}r_{m3})) + Z_4(r_{m1}Z_1 + r_{m2}^2)\right]/k_m\Lambda,$$

$$g_{m3} = \left[k_m(A_2(r_{m1}Z_2 + r_{m2}r_{m3}) - r_{m2}Z_2 + r_{m3}Z_1) - Z_3(r_{m1}Z_1 + r_{m2}^2)\right]/k_m\Lambda.$$

where all constants are given in the "Appendix."

From conditions (37) and Eqs. (37a, 37d), (43) and (44), we obtain a linear system of equations whose solution (numerically) gives the parameters L_m, $m = 1, 2, 3, 4$.

This completes the solution in the Laplace transform domain.

6 Numerical results and discussion

The method, which is based on a Fourier series expansion proposed by Honig and Hirdes (1984) and is developed in detail in some literature (Ezzat et al. 1999; Sherief et al. 2010), is adopted to invert the Laplace transform in Eqs. (43). Numerical code has been prepared using the Fortran 77 programming language.

The asphaltic material saturated with water was chosen for the purpose of numerical evaluations. The basic data for this two-phase system are shown in Table 1 (Sherief and Hussein 2012).

The computations were carried out for the functions $f(t)$ where

$$f(t) = \begin{cases} \sin\left(\dfrac{\pi t}{a_o}\right) & 0 \le t \le a_o \\ 0 & \text{otherwise} \end{cases} \quad \text{or}$$

$$\bar{f}(s) = \frac{\pi a_o(s + e^{-a_o s})}{a_o^2 s^2 + \pi^2}$$

The temperature, stress, and displacement as well as the velocity of the fluid and the skin friction values for the two phases were calculated by using the numerical method of the inversion of the Laplace transform. The FORTRAN programming language was used on a personal computer. The precision maintained was five digits for the numerical program. The roots of the characteristic Eq. (42) were obtained analytically using the Mathematika 6.0 software. The system of linear equations resulting from applying the boundary conditions was solved numerically using the LU decomposition method. The results are displayed graphically at different positions of x as shown in Figs. 1, 2, 3, 4, 5, 6. In these figures, the solid line represents the solution obtained in the frame of the dynamic coupled theory ($\upsilon = 0$) (Biot 1955), and the dotted lines represent the solution obtained in the frame of the generalized thermoelasticity with thermal relaxation time ($\upsilon = 1.0$, $\tau_o = 0.02$) (Sherief and Hussein 2012), while the dashed lines represent the solution obtained in the frame of generalized thermoelasticity with fractional heat transfer ($0 < \upsilon < 1$, $\tau_o = 0.02$) (Ezzat 2012).

The important phenomenon observed in all figures is that the solution of any of the considered functions in the fractional theory is restricted in a bounded region. Beyond this region, the variations of these distributions do not take place. This means that the solutions according the generalized fractional thermoelasticity theory exhibit the behavior of finite speeds of wave propagation.

In the frame of the fractional theory, Figs. 1 and 2 indicate the variation in temperature in the elastic body and fluid for different values of fractional order $\upsilon = 0.0$, 1.0, 0.5. We notice that the temperature fields have been affected by the fractional order υ and the thermal waves cut the x-axis more rapidly when υ decreases. In the fractional theory of generalized poro-thermoelasticity, we observed that the thermal waves are continuous functions, smooth and reach steady state depending on the value of fractional υ, which means that the particles transport the heat to the other particles easily and this makes the decreasing rate of the temperature greater than the other ones.

Figures 3, 4, 5, 6 display the stress and displacement distributions in the elastic body and fluid, with distance

Table 1 Value of the constants

T_o, °C	27	ρ_s, mg cm^{-3}	2.35	τ_f, s	0.001
G, dyne cm^{-2}	0.4853×10^{11}	ρ_f, mg cm^{-3}	0.82	κ, W m^{-2} °C^{-1}	1.84×10^{3}
R, dyne cm^{-2}	0.0362×10^{11}	c_s, kJ kg^{-1} °C^{-1}	1.17152	ρ_{12}	-0.001ρ
α_s,° C^{-1}	0.0001	c_f, kJ mg^{-1} °C^{-1}	2.092	β	0.25
α_f, °C^{-1}	2.16×10^{-5}	k_s, W m^{-1} °C^{-1}	1.83	B_s	1
λ, dyne cm^{-2}	0.2160×10^{11}	k_f, W m^{-1} °C^{-1}	0.148	B_f	1
μ, dyne cm^{-2}	0.0926×10^{11}	τ_s, s	0.02		

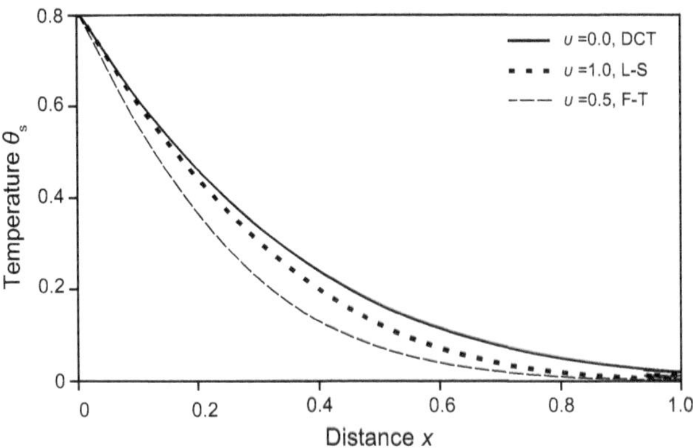

Fig. 1 Temperature distribution for the solid phase for different theories. (DCT: Dynamic coupled theory, L-S: Lord-Shulman theory, F-T: Fractional heat transfer model)

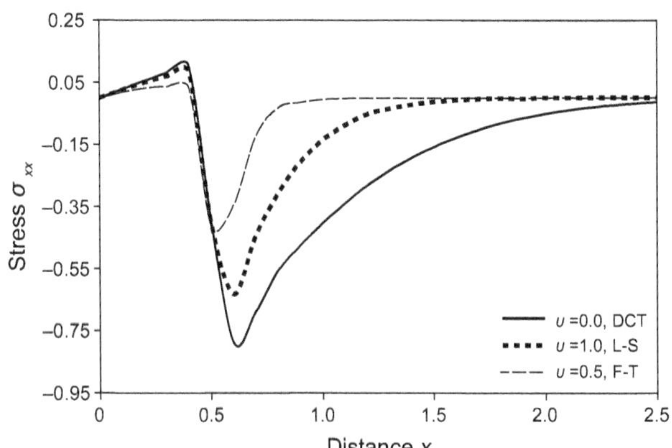

Fig. 3 Normal stress distribution for the solid phase for different theories

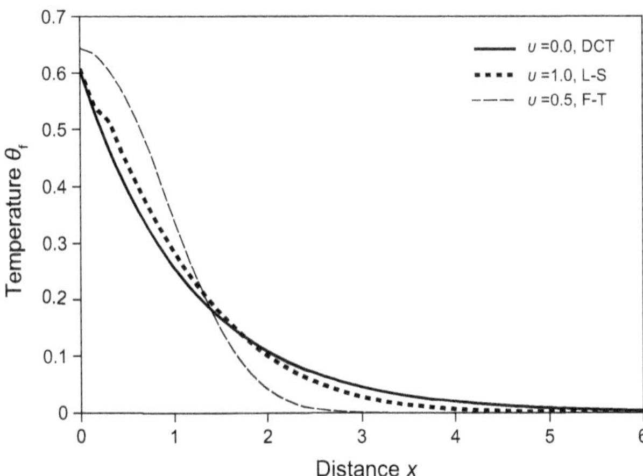

Fig. 2 Temperature distribution for the fluid phase for different theories

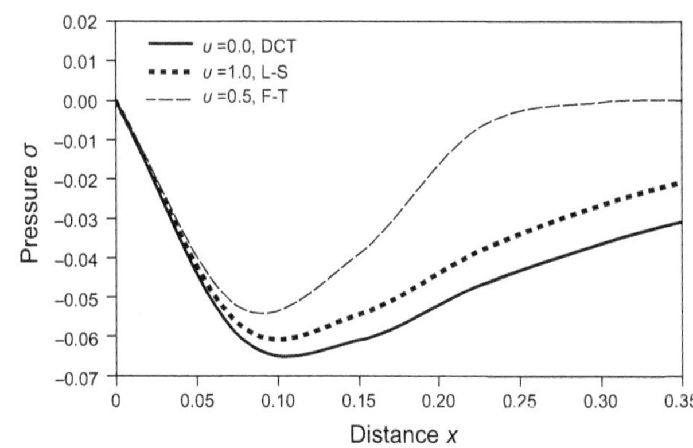

Fig. 4 Pressure for the fluid phase for different theories

x for different values of fractional order $\upsilon = 0.0,\ 1.0,\ 0.5$. We observe that the stress and displacement fields have the same behavior as the temperature and the absolute value of the maximum stress decreases. We note here that some of these figures show broken lines indicating discontinuities of the solutions. These discontinuities indicate the locations of the wave fronts of the shock waves resulting from the sudden heating of the surface.

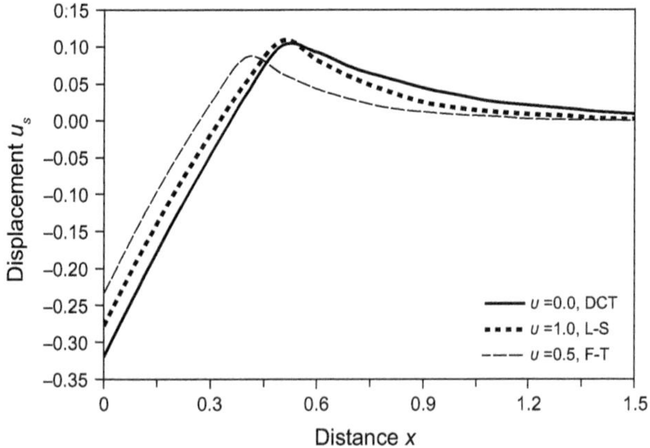

Fig. 5 Displacement distribution for the solid phase for different theories

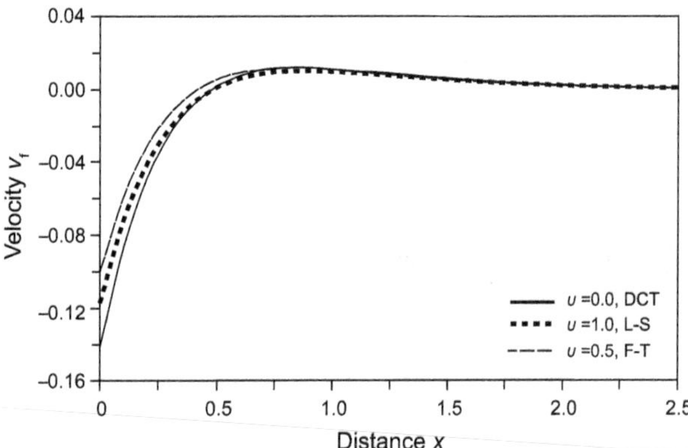

Fig. 6 Velocity distribution for the fluid phase

7 Conclusions

- The main goal of this work is to introduce a new mathematical model for Fourier law of heat conduction with time-derivative fractional order α (Fractional model). According to this new theory, we have to construct a new classification for materials according to their fractional parameter where this parameter becomes a new indicator of its ability to conduct heat.
- For the fractional model of thermoelasticity $0 < \upsilon < 1$, the solution seems to behave like the generalized theory of generalized thermoelasticity (Lord-Shulman theory). This result is very important that the new theory may preserve the advantage of the generalized theory that the velocity of waves is finite.
- As was mentioned in (Povstenko 2011) "From numerical calculations, it is difficult to say whether the solution for υ approaching 1 has a jump at the wave front or it is continuous with very fast changes. This aspect invites further investigation."
- The present model avoids the negative temperature defect of the Lord-Shulman model.
- The result provides a motivation to investigate thermoelastic materials as a new class of applicable materials.

Acknowledgments The first author is grateful for the support of Al-Qassim University (No. 2367), Al-Qassim, Saudi Arabia.

Appendix

$$F = F_{22}\omega / F_{11},$$

$$\omega = k_s / k_f,$$

$$b_1 = \rho_{12}/\rho_{11}, \quad b_2 = \rho_{22}/\rho_{11},$$

$$a_1 = G/M, \quad a_2 = R/M, \quad a_3 = \lambda/M,$$

$$A_1 = R_{12}/R_{11}, \quad A_2 = R_{21}/R_{11} A_3 = R_{22}/R_{11},$$

$$\varepsilon_1 = \kappa k_s \rho_{11}/M F_{11}^2, \quad \varepsilon_2 = T_0 R_{11}^2 / M F_{11},$$

$$\ell = a_1^2 - a_2,$$

$$\ell_1 = -\big(s^2(2b_1a_1 - b_2 - a_2) + s\varepsilon_2(2a_1(A_1A_3\omega\zeta_1 + A_2\zeta_0) \\ - a_2(A_1^2\omega\zeta_1 + \zeta_0) - A_2^2\zeta_0 - A_3^2\omega\zeta_1) + a(\zeta_2 + \zeta_3)\big)/\ell,$$

$$\ell_2 = \big(s^4(b_1^2 - b_2) + \varepsilon_2 s^3 \zeta_4 + s^2\zeta_5 + \varepsilon_2 s \zeta_7 \\ + \ell(\zeta_2\zeta_3 - \omega\zeta_0\zeta_1\varepsilon_1^2)\big)/a,$$

$$\ell_3 = -s^2\big(s^2(b_1^2 - b_2)(\zeta_2 + \zeta_3) + s\varepsilon_2\zeta_8 \\ + (2a_1b_1 - b_2 - a_2)(\zeta_2\zeta_3 - \omega\zeta_0\zeta_1\varepsilon_1^2)\big)/\ell,$$

$$\ell_4 = s^4(b_1^2 - b_2)(\zeta_2\xi_3 - \omega\zeta_0\zeta_1\varepsilon_1^2)/\ell,$$

$$\zeta_0 = 1 + \frac{\tau_s^\upsilon}{\upsilon!} s^\upsilon, \quad \zeta_1 = 1 + \frac{\tau_f^\upsilon}{\upsilon!} s^\upsilon,$$

$$\zeta_2 = \zeta_0(\varepsilon_1 + s), \quad \zeta_3 = \zeta_1(\varepsilon_1\omega + sA_3),$$

$$\zeta_4 = 2b_1(A_1A_3\omega\zeta_1 + A_2\zeta_0) - b_2(A_1^2\omega\zeta_1 + \zeta_0) - A_2^2\zeta_0 \\ - A_3^2\omega\zeta_1,$$

$$\zeta_5 = (\zeta_2 + \zeta_3)(2a_1b_1 - b_2 - a_2) - \omega\zeta_0\zeta_1\varepsilon_2^2(A_1A_2 - A_3)^2,$$

$$\zeta_6 = A_1\omega\zeta_1(A_2\zeta_0\varepsilon_1 + A_3\zeta_2) + \zeta_0(A_2\zeta_3 + A_3\omega\zeta_1\varepsilon_1),$$

$$\zeta_7 = 2a_1\zeta_6 - a_2\left(A_1^2\omega\zeta_1\zeta_2 + 2A_1\omega\zeta_0\zeta_1 + \zeta_0\zeta_3\right) - A_2^2\zeta_0\zeta_3 - A_3\omega\zeta_1(2A_2\zeta_0\varepsilon_1 + A_3\zeta_2),$$

$$\zeta_8 = 2b_1\zeta_6 - b_2(A_1^2\omega\zeta_1\zeta_2 + 2A_1\omega\zeta_0\zeta_1\varepsilon_1 + \zeta_0\zeta_3) - A_2^2\zeta_0\zeta_3 - A_3\omega\zeta_1(2A_2\zeta_0\varepsilon_1) + A_3\zeta_2,$$

$$Z_1 = a_2k_m^2 - b_2s^2, \quad Z_2 = s\zeta_0A_2k_m\varepsilon_2,$$

$$Z_3 = k_m^2 - \zeta_0(\varepsilon_1 + s), \quad Z_4 = \varepsilon_1\zeta_0,$$

$$r_{m1} = s^2 - k_m^2, \quad r_{m2} = b_1s^2 - a_1k_m^2, \quad r_{m3} = -sk_m\varepsilon_2\zeta_0,$$

$$\Lambda = k_mZ_2(A_1A_2 - A_3) - A_1Z_1Z_3 + A_2r_{m2}Z_4 - A_3r_{m2}Z_3 + Z_1Z_4.$$

References

Abbas IA. A problem on functional graded material under fractional order theory of thermoelasticity. Theor Appl Frac Mech. 2014;74(8):18–22. doi:10.1016/j.tafmec.2014.05.005.

Abbas IA. Generalized thermoelastic interaction in functional graded material with fractional order three-phase lag heat transfer. J Cen South Univ. 2015;22(5):1606–13. doi:10.1007/s11771-015-2677-5.

Bagley RL, Torvik PJ. On the fractional calculus model of viscoelastic behavior. J Rheol. 1986;30(1):133–55. doi:10.1122/1.549887.

Biot MA. Theory of elasticity and consolidation for a porous anisotropic solid. J Appl Phys. 1955;26(2):182–98. doi:10.1063/1.1721956.

Biot MA. Theory of propagation of elastic waves in a fluid-saturated porous solid. I: Low-frequency range. J Acoust Soc Am. 1956a;28(2):168–78. doi:10.1121/1.1908239.

Biot MA. Thermoelasticity and irreversible thermodynamics. J Appl Phys. 1956b;27(3):240–53. doi:10.1063/1.1722351.

Biot MA. New variational-Lagrangian irreversible thermodynamics with application to viscous-flow, reaction diffusion, and solid mechanics. Adv Appl Mech. 1984;24:1–91. doi:10.1016/S0065-2156(08)70042-5.

Biot MA, Willis DG. The elastic coefficients of the theory of consolidation. J Appl Mech. 1957;24(11):594–601.

Cattaneo C. Sulla conduzione del calore. Atti Sem Mat Fis Univ Modena. 1948;3(3):83–101.

Delaney PT. Rapid intrusion of magma into wet rock: groundwater flow due to pore pressure increases. J Geophys Res. 1982;87(B9):7739–56. doi:10.1002/(ISSN)2156-2202.

Deresiewicz H, Skalak R. On uniqueness in dynamic poroelasticity. Bull Seismol Soc Am. 1963;53(4):783–8.

El-Karamany AS, Ezzat MA. On fractional thermoelasticity. Math Mech Solids. 2011;16(3):334–46. doi:10.1177/1081286510397228.

Ezzat MA. State space approach to unsteady two-dimensional free convection flow through a porous medium. Can J Phys. 1994;72(5–6):311–7.

Ezzat MA. Thermoelectric MHD non-Newtonian fluid with fractional derivative heat transfer. Phys B. 2010;405(7):4188–94. doi:10.1016/j.physb.2010.07.009.

Ezzat MA. Thermoelectric MHD with modified Fourier's law. Int J Therm Sci. 2011a;50(4):449–55. doi:10.1016/j.ijthermalsci.2010.11.005.

Ezzat MA. Theory of fractional order in generalized thermoelectric MHD. Appl Math Model. 2011b;35(10):4965–78. doi:10.1016/j.apm.2011.04.004.

Ezzat MA. Magneto-thermoelasticity with thermoelectric properties and fractional derivative heat transfer. Phys B. 2011c;406(1):30–5. doi:10.1016/j.physb.2010.10.005.

Ezzat MA. State space approach to thermoelectric fluid with fractional order heat transfer. Heat Mass Trans. 2012;48(1):71–82. doi:10.1007/s00231-011-0830-8.

Ezzat MA, Abd-Elaal M. Free convection effects on a viscoelastic boundary layer flow with one relaxation time through a porous medium. J Frank Inst. 1997;334B(4):685–6.

Ezzat MA, El-Karamany AS. Fractional order theory of a perfect conducting thermoelastic medium. Can J Phys. 2011a;89(3):311–8. doi:10.1139/P11-022.

Ezzat MA, El-Karamany AS. Fractional order heat conduction law inmagneto-thermoelasticity involving two temperatures. ZAMP. 2011b;62(3):937–52. doi:10.1007/s00033-011-0126-3.

Ezzat MA, El-Karamany AS. Theory of fractional order in electro-thermoelasticity. Euro J Mech A/Solid. 2011c;30(4):491–500. doi:10.1016/j.euromechsol.2011.02.004.

Ezzat MA, El-Karamny AS, Ezzat SM, et al. Two-temperature theory in magneto-thermoelasticity with fractional order dual-phase-lag heat transfer. Nucl Eng Des. 2012a;252(11):267–77. doi:10.1016/j.nucengdes.2012.06.012.

Ezzat MA, El-Karamny AS, Fayik M, et al. Fractional ultrafast laser-induced thermo-elastic behavior in metal films. J Therm Stress. 2012b;35(7):637–51. doi:10.1080/01495739.2012.688662.

Ezzat MA, El-Bary AA, Fayik MA, et al. Fractional Fourier law with three-phase lag of thermoelasticity. Mech Adv Mater Struct. 2013a;20(8):593–602. doi:10.1080/15376494.2011.643280.

Ezzat MA, El-Karamny AS, El-Bary AA, Fayik M, et al. Fractional calculus in one-dimensional isotropic thermo-viscoelasticity. CR Mec. 2013b;341(7):553–66. doi:10.1016/j.crme.2013.04.001.

Ezzat MA, Abbas IM, El-Bary AA, Ezzat SM, et al. Numerical study of the Stokes' first problem for thermoelectric micropolar fluid with fractional derivative heat transfer. MHD. 2014a;50(3):263–77.

Ezzat MA, Alsowayan NS, Al-Mohiameed ZI, Ezzat SM, et al. Fractional modelling of Pennes' bioheat transfer equation. Heat Mass Trans. 2014b;50(7):907–14. doi:10.1007/s00231-014-1300-x.

Ezzat MA, El-Karamny AS, El-Bary AA, Fayik M, et al. Fractional ultrafast laser- induced magneto-thermoelastic behavior in perfect conducting metal films. J Electromagn Waves Appl. 2014c;28(1–2):64–82. doi:10.1080/09205071.2013.855616.

Ezzat MA, Sabbah AS, El-Bary AA, Ezzat SM, et al. Stokes' first problem for a thermoelectric fluid with fractional-order heat transfer. Rep Math Phys. 2014d;74(2):145–58. doi:10.1016/S0034-4877(15)60013-1.

Ezzat MA, El-Karamny AS, El-Bary AA, et al. On thermo-viscoelasticity with variable thermal conductivity and fractional-order heat transfer. Int J Thermophys. 2015;36(7):1684–97. doi:10.1007/s10765-015-1873-8.

Ezzat MA, Othman MI, Helmy K, et al. A problem of a micropolar magneto hydrodynamic boundary-layer flow Canad. J Phys. 1999;77(10):813–27.

Fourie J, Du Plessis J. A two-equation model for heat conduction in porous media. Transp Porous Med. 2003;53(2):145–61. doi:10.1023/A:1024071928123.

Ghassemi A, Diek A. Poro-thermoelasticity for swelling shales. J Pet Sci Eng. 2002;34(4–5):123–35. doi:10.1016/S0920-4105(02)00159-6.

Ghassemi A, Zhang Q. A transient fictitious stress boundary element method for porothermoelastic media. Eng Anal Boun Elem. 2004;28(11):1363–73. doi:10.1016/j.enganabound.2004.05.003.

Gorenflo R, Mainardi F. Fractional calculus: integral and differential equations of fractional orders, fractals and fractional calculus in continuum mechanics, vol. 378. Wien: Springer; 1997. p. 223–76.

Gorenflo R, Mainardi F, Moretti D, Paradisi P, et al. Time fractional diffusion: a discrete random walk approach. Nonlinear Dyn. 2002;29(1):129–43. doi:10.1023/A:1016547232119.

Grimnes S, Martinsen OG. Bioimpedance and bioelectricity basics. San Diego: Academic Press; 2000.

Honig G, Hirdes U. A method for the numerical inversion of the Laplace transform. J Comput Appl Math. 1984;10(1):113–32. doi:10.1016/0377-0427(84)90075-X.

Jou D, Casas-Vazquez J, Lebon G. Extended irreversible thermodynamics. Rep Prog Phys. 1988;51(9):1105–79. doi:10.1088/0034-4885/51/8/002.

Jumarie G. Derivation and solutions of some fractional Black-Scholes equations in coarse-grained space and time. Application to Merton's optimal portfolio. Comput Math Appl. 2010;59(3):1142–64. doi:10.1016/j.camwa.2009.05.015.

Krishnan JM, Rajagopal KR. Review of the uses and modeling of bitumen from ancient to modern times. Appl Mech Rev. 2003;56(2):149–214. doi:10.1115/1.1529658.

Lakes RS. Viscoelastic solids. Boca Raton: CRC Press; 1999.

Lebon G, Jou D, Casas-Vázquez J. Understanding non-equilibrium thermodynamics: foundations, applications, frontiers. Berlin: Springer-Verlag; 2008.

Li X, Cui L, Roegiers J-C. Thermoporoelastic modeling of wellbore stability in non-hydrostatic stress field. Int J Rock Mech Min Sci. 1998;35(4–5):584–8. doi:10.1016/s0148-9062(98)00079-5.

Lord H, Shulman Y. A generalized dynamical theory of thermoelasticity. J Mech Phys Solids. 1967;15(5):299–309.

Mainardi F, Gorenflo R. On Mittag-Leffler-type function in fractional evolution processes. J Comput Appl Math. 2000;118(1–2):283–99. doi:10.1016/S0377-0427(00)00294-6.

Marin M. Weak solutions in elasticity of dipolar porous materials. Math Prob Eng. 2008;. doi:10.1155/2008/158908.

Miller KS, Ross B. An introduction to the fractional integrals and derivatives—theory and applications. New York: John Wiley & Sons Inc; 1993.

Nouri-Borujerdi A, Noghrehabadi A, Rees A, et al. The effect of local thermal non-equilibrium on conduction in porous channels with a uniform heat source. Transp Porous Med. 2007;69(2):281–8. doi:10.1007/s11242-006-9064-5.

Nur A, Byerlee JD. An exact effective stress law for elastic deformation of rock with fluids. J Geophys Res. 1971;76(26):6414–9. doi:10.1029/JB076i026p06414.

Nowinski JL. Theory of thermoelasticity with applications. Alphen aan den Rijn: Sijthoff & Noordhoff International Publishers; 1978.

Oldham SG, Spanier J. The fractional calculus. New York: Academic Press; 1974.

Pecker C, Deresiewiez H. Thermal effects on wave in liquid-filled porous media. J Acta Mech. 1973;16(1):45–64. doi:10.1007/BF01177125.

Podlubny I. Fractional differential equations. New York: Academic Press; 1999.

Povstenko YZ. Fractional Cattaneo-type equations and generalized thermoelasticity. J Therm Stress. 2011;34(2):97–114. doi:10.1080/01495739.2010.511931.

Rossikhin YA, Shitikova MV. Applications of fractional calculus to dynamic problems of linear and nonlinear heredity mechanics of solids. Appl Mech Rev. 1967;50(1):15–67.

Samko SG, Kilbas AA, Marichev OI, et al. Fractional integrals and derivatives—theory and applications. Longhorne: Gordon & Breach; 1993.

Sherief HH, El-Said A, Abd El-Latief A, et al. Fractional order theory of thermoelasticity. Int J Solid Struct. 2010;47(2):269–75. doi:10.1016/j.ijsolstr.2009.09.034.

Sherief HH, Hussein EM. A mathematical model for short-time filtration in poroelastic media with thermal relaxation and two temperatures. Trans. Porous med. 2012;91(1):199–223. doi:10.1007/s11242-011-9840-8.

Simakin A, Ghassemi A. Modelling deformation of partially melted rock using a poroviscoelastic rheology with dynamic power law viscosity. Tectonophys. 2005;397(3–4):195–9. doi:10.1016/j.tecto.2004.12.004.

Tarasov VE. Fractional vector calculus and fractional Maxwell's equations. Ann Phys. 2008;323(11):2756–78. doi:10.1016/j.aop.2008.04.005.

Vernotte MP. Les paradoxes de la théorie continue de l'équation de la chaleur. CR Acad Sci. 1958;246(22):3154–5.

Wang Y, Papamichos E. Conductive heat flow and thermally induced fluid flow around a well bore in a poroelastic medium. Water Resour Res. 1994;30(12):3375–84. doi:10.1029/94WR01774.

Youssef HM. Theory of generalized porothermoelasticity. Int J Rock Mech Min Sci. 2007;44(2):222–7. doi:10.1016/j.ijrmms.2006.07.001.

10

Characteristics and control mechanisms of coalbed permeability change in various gas production stages

Da-Zhen Tang[1,2] · Chun-Miao Deng[1,2] · Yan-Jun Meng[1,2] · Zhi-Ping Li[1,2] · Hao Xu[1,2] · Shu Tao[1,2] · Song Li[1,2]

Abstract According to dimensionless analysis of the coalbed methane (CBM) production data of Fanzhuang block in southern Qinshui basin, the dimensionless gas production rate is calculated to quantitatively divide the CBM well production process into four stages, i.e., drainage stage, unstable gas production stage, stable gas production stage, and gas production decline stage. By the material balance method, the coal reservoir permeability change in different stages is quantitatively characterized. The characteristics and control mechanisms of change in coalbed permeability (CICP) during different production stages are concluded on five aspects, i.e., permeability trend variation, controlling mechanism, system energy, phase state compositions, and production performance. The study reveals that CICP is characterized by first decline, then recovery, and finally by increase and is controlled directly by effective stress and matrix shrinkage effects. Further, the duration and intensity of the matrix shrinkage effect are inherently controlled by adsorption and desorption features.

Keywords Production stage · Coalbed methane · Permeability · Dynamic change · Control mechanism

✉ Chun-Miao Deng
chunmiaocugb@gmail.com

1 School of Energy Resources, China University of Geosciences, Beijing 100083, China

2 Coal Reservoir Laboratory of National Engineering Research Center of Coalbed Methane Development & Utilization, Beijing 100083, China

Edited by Xiu-Qin Zhu

1 Introduction

Coalbed methane (CBM) is one of the most significant energy resources in the unconventional gas field. It has been developed successfully and commercially in the southern Qinshui basin and on the east edge of the Ordos basin in China (Tao et al. 2014; Meng et al. 2014b; Feng et al. 2014; Li et al. 2015). CBM development geology has become a research topic recently (Qin et al. 2014; Liu et al. 2010). As reported by many researchers, coal reservoir permeability is one of the key factors influencing the productivity of CBM wells (Song et al. 2012; Li et al. 2014; Fan et al. 2014). Thus, research on real-time quantitative prediction of characteristics of CICP during different CBM production stages has significant impacts upon the development and deployment, management of production, and productivity prediction of CBM fields in China. Many researchers have carried out studies on the theoretical model calculation and experimental simulation of CICP (Shi and Durucan 2005; Palmer 2009; Mazumder et al. 2012; Tao et al. 2012; Xu et al. 2014). However, so far the breakthrough of real-time monitoring and prediction of dynamic CICP has not been made, but it has great instructive meaning for CBM production. There is still an argument on the scheme of CBM well production stages division and their defined principles. In this paper, the production stages of CBM wells are quantitatively divided by the dimensionless analysis method, and the values of dynamic CICP in different CBM production stages are calculated by the material balance method based on production data. Finally, characteristics and control mechanisms of CICP during various stages are summarized on five aspects, including the permeability variation trend, controlling mechanism, system energy, phase state compositions, and the dynamic production.

2 CBM well production stages division

There are two main schemes for dividing the CBM well production stages as follows: (1) Four stages are divided based on the characteristics of the change in fluid phase, including saturated single-phase flow of water, unsaturated single-phase flow of water, two-phase flow of gas and water, and single-phase flow of gas stages (Ni et al. 2010). (2) Five stages, including dewatering, pressure hold, pressure control, high and stable yield, and depletion stages, are classified based on the in situ pressure change characteristics, such as bottom-hole flowing pressure, desorption pressure, and strata pressure (Qin et al. 2011; Li et al. 2011). The former scheme is more scientific and reasonable in theory than the latter one, but the unsaturated single-phase flow of water and single-phase flow of gas stages are difficult to distinguish in practical production. The latter scheme is more precise and practicable than the former one, but it has a relatively weak foundation in theory and the great influence of production control factors makes it difficult to be widely used.

Generally, CBM production curves are characterized by gas production changing over time, that is, the process from no gas production, start of gas production, rise of gas production, reaching a peak of gas production, decline of gas production, to gas depletion. If the CBM production stages are divided by gas production rate, then gas production data can be directly used for division of stages. This division method is feasible in practice and theory. Therefore, this paper proposes a new four-stage division scheme based on gas production status, i.e., drainage stage, unstable gas production stage, stable gas production stage, and gas production decline stage. How to define the dividing points becomes a key problem once the classification scheme is determined. Dimensionless treatment of production data is widely used to conventional oil/gas and CBM production analysis due to the elimination of the influence of dimension on data analysis (Clarkson et al. 2008; Aminian and Ameri 2009; Freeman et al. 2013). Hence, this paper uses the dimensionless method to process gas and water production data. And the dimensionless gas production rate is chosen as an appropriate parameter, which indicates the relative change in gas and water production rates.

2.1 Dimensionless gas and water production

The dimensionless gas production is defined as the ratio of gas production to the peak gas production, that is

$$N_{\mathrm{gD}} = \frac{q_{\mathrm{g}}}{q_{\mathrm{gmax}}}, \tag{1}$$

where N_{gD} is the dimensionless gas production; q_{g} is the CBM well gas production, m^3/days; and q_{gmax} is the peak gas production, m^3/days.

Similarly, dimensionless water production is defined as the ratio of water production to peak water production, that is

$$N_{\mathrm{wD}} = \frac{q_{\mathrm{w}}}{q_{\mathrm{wmax}}}, \tag{2}$$

where N_{wD} is the dimensionless water production; q_{w} is the CBM well water production, m^3/days; and q_{wmax} is the peak water production, m^3/days.

By dimensionless treatment, the value of gas or water production is limited in a certain range from 0 to 1 ($N_{\mathrm{gD}}, N_{\mathrm{wD}} \in [0, 1]$). For CBM wells with long production history, the peak gas or water production value can be obtained directly from their production data. While, for wells with a short history, these parameters can refer to analogous wells that have a long history and similar geological and engineering conditions.

2.2 Dimensionless gas production rate

The dimensionless gas production rate is defined as the ratio of dimensionless gas production to the sum of dimensionless gas production and dimensionless water production, that is

$$\eta_{\mathrm{gD}} = \frac{N_{\mathrm{gD}}}{N_{\mathrm{gD}} + N_{\mathrm{wD}}} = \frac{\frac{q_{\mathrm{g}}}{q_{\mathrm{gmax}}}}{\frac{q_{\mathrm{g}}}{q_{\mathrm{gmax}}} + \frac{q_{\mathrm{w}}}{q_{\mathrm{wmax}}}}, \tag{3}$$

where η_{gD} is the dimensionless gas production rate.

It can be seen from Eq. (3) that the value of the dimensionless gas rate is also between 0 and 1. According to the value of dimensionless gas rate, the CBM production process can be divided into four stages. In drainage stage, there is only water produced until the gas starts coming out. If we set η_1 as dimensionless gas rate of the time point at which gas begins to produce, then the dimensionless gas rate ranges from 0 to η_1 at this stage. It is worthy to note that each production well has different η_1 due to the various durations of the drainage stage. In the unstable gas production stage, the production of gas increases gradually from zero, and water production tends to decrease. The corresponding dimensionless gas rate also increases gradually during this stage. If we set η_2 as dimensionless gas rate at the time point when gas production begins to stabilize, then the dimensionless gas rate ranges from η_1 to η_2 in this stage. Generally, the dimensionless gas rate of this stage is larger than 0.5. In stable gas production stage, the production of gas remains at a high level with slight fluctuations. And the corresponding water production remains at a low level. If we set η_3 as dimensionless gas rate at the point when gas production begins to decline, then the dimensionless gas rate varies from η_2 to η_3 in this stage. In gas production decline stage, gas production tends to decline and water production decreases gradually from a very low level to 0. The dimensionless gas rate ranges from η_3 to 1 in this stage.

2.3 CBM well production stages division for the southern Qinshui basin

Fifteen CBM wells with stable and continuous production were selected from the Fanzhuang block of the southern Qinshui basin (China) for study of the CBM well production stages division. The basic production data of these 15 CBM wells are shown in Table 1. The production history of wells W1–W10 is relatively short, at about 330–770 days. W11–W15 have a longer production history than W1–W10, at about 2000 days. Besides, W4, W5, W6, W9, W14, and W15 are high-productivity wells.

By dimensionless treatment of the production data of 15 wells, the dimensionless gas and water production and the dimensionless gas production rate are calculated for each well. Production stages are divided for each well by the value of dimensionless gas production rate (Table 2). The original production curve and dimensionless production curve of W14 are shown in Figs. 1 and 2, respectively. All of the 15 wells have experienced drainage stage, unstable gas production stage, and stable gas production stage. Eight of the 15 wells have reached the gas production decline stage. In the drainage stage, the dimensionless gas production rate ranges from 0.107 to 0.426, which is related to the ratio of initial gas production to the highest gas production. Generally, the value of the dimensionless gas production rate of high-productivity wells is less than that of the low-productivity wells. The dimensionless gas production rate varies from 0.547 to 0.683 at the end of the unstable gas production stage, and from 0.713 to 0.909 at the end of the stable gas production stage.

3 Calculation of CICP at each stage

During the CBM production process, the coal reservoir permeability changes all the time. The CICP has an important influence on CBM production and on the adjustment of future CBM development plans. So, it is vital to understand the characteristics of CICP at each stage. Currently, there is no direct method to obtain coal reservoir permeability during the gas production process because almost no productivity test is applied in this process. However, the material balance method, one of the commonly used techniques in dynamic analysis in the reservoir engineering field, is applied very well in CBM reservoirs (King 1993; Lai et al. 2013). Based on the data of gas and water production, and relevant parameters of reservoir and fluid, the CICP during development process can be calculated by this method. Hence, in this paper we calculate the reservoir permeability at each production stage of CBM wells in the southern Qinshui basin (Table 3; Fig. 3). The calculations are based on the original production data, considering the effects of effective stress and matrix shrinkage.

Table 1 Basic production data of 15 CBM wells in the southern Qinshui basin (China)

Well name	Total pro-t, days	Drainage time, days	Gas pro-t, days	Total gas pro-, 10^3m^3	Total water pro-, m^3	Highest gas pro-, m^3/days	Highest water pro-, m^3/days	Average gas pro-, m^3/days	Average gas pro-, m^3/days
W1	407	82	325	18.036	617.1	1055	5.0	554.9	1.52
W2	336	129	207	28.324	1220.0	2123	6.3	1368.3	3.63
W3	336	128	208	20.678	1992.4	1679	10.5	1615.5	5.93
W4	770	128	642	190.159	1024.8	5806	7.1	2962.0	1.33
W5	770	52	718	175.009	455.1	4488	4.0	2437.4	0.59
W6	580	126	454	144.527	1455.9	5144	10.2	3183.4	2.51
W7	574	116	458	92.667	845.5	2714	4.2	2023.3	1.47
W8	771	74	697	124.676	705.8	4122	5.7	1788.8	0.92
W9	749	180	417	136.752	809.5	4235	11	2960.0	1.36
W10	409	186	223	26.135	451.5	2468	3.4	1171.9	1.10
W11	1908	186	1722	209.190	3281.8	2928	8.4	1214.8	1.72
W12	1910	185	1725	212.420	1780.6	1935	6.7	1231.4	0.93
W13	1890	224	1666	161.440	3624.7	1848	9.2	969.03	1.92
W14	1749	120	1629	595.010	2424.1	6344	8.3	3652.61	1.39
W15	1908	210	1698	342.870	10178.5	3272	9.5	2019.26	5.33

pro- production, *pro-t* production time

Table 2 Summary of the dimensionless gas production rate in each production stage of 15 CBM wells in the southern Qinshui basin

Well name	Drainage stage 0	Unstable gas pro-stage η_1	Stable gas pro-stage η_2	Gas pro-decline stage η_3
W1	0	0.174	0.607	0.904
W2	0	0.339	0.606	–
W3	0	0.321	0.547	–
W4	0	0.153	0.637	–
W5	0	0.127	0.646	–
W6	0	0.107	0.621	–
W7	0	0.353	0.659	–
W8	0	0.172	0.683	0.909
W9	0	0.124	0.627	0.866
W10	0	0.328	0.651	–
W11	0	0.393	0.635	0.713
W12	0	0.426	0.652	0.844
W13	0	0.406	0.607	0.867
W14	0	0.238	0.632	0.897
W15	0	0.279	0.621	0.891

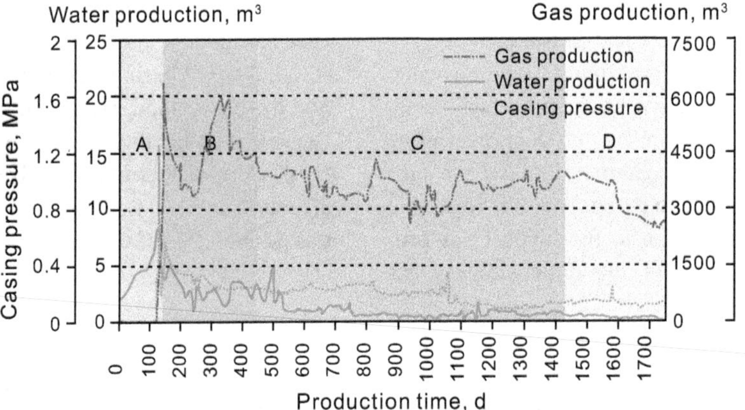

Fig. 1 Production curves of CBM well W14 at southern Qinshui basin, including gas and water production curves and reservoir pressure curve. *a* Drainage stage, *b* unstable gas production stage, *c* stable gas production stage, and *d* gas production decline stage

A number of important characteristics of CICP can be found from the calculated results and are summarized below: (1) In the drainage stage, the permeability of each well decreases, and CICP rate ranges from −0.92 % to −0.11 %, with an average value of −0.56 %, and CICP rate per unit pressure reduced (CICPR/PR) ranges from −8.68 % to −4.30 %/MPa, with an average value of −7.82 %/MPa. (2) In the unstable gas production stage, CICP rate ranges from −20.2 % to −0.41 %, with an average value of −8.53 %, and CICPR/PR ranges from −8.43 % to −4.00 %/MPa, with an average value of −6.94 %/MPa. During this stage, the reservoir permeability declines dramatically although some gas has been desorbed out. (3) In the stable gas production stage, the permeability change rate ranges from −5.53 to −0.15, with an average value of −2.73 %, and CICPR/PR ranges from −8.43 %/MPa to

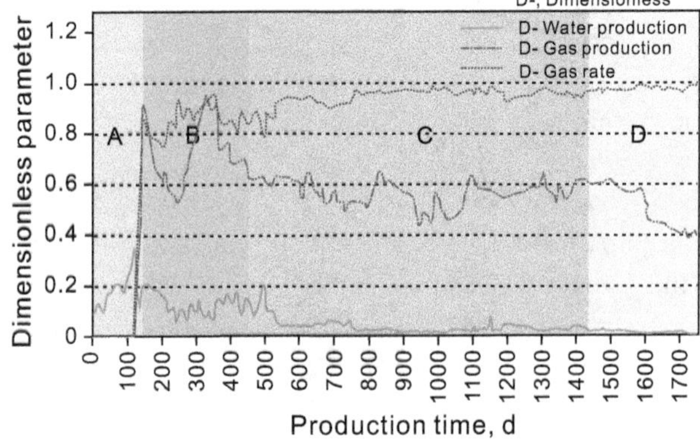

Fig. 2 Dimensionless production curves of CBM well W14 at southern Qinshui basin, including dimensionless gas production, water production, and gas production rate. *A* drainage stage, *B* unstable gas production stage, *C* stable gas production stage, and *D* gas production decline stage

−1.46 %/MPa, with an average value of −5.13 %/MPa. (4) In the gas production decline stage, five wells (from W11 to W15) show a relatively full decline stage as they have a long production history, but other wells show no decline or reaching an early decline stage. There are no obvious features of permeability change in three wells (W11, W12, and W13), due to applying production-enhanced measurement. Two high-productivity wells(W14 and W15) have a relatively long duration of their gas production decline stage, and their CICP does not decrease but increases in this stage, with CICP increasing by 0.78 % for W14 and 0.45 % for W15. A similar trend is found for the CICPR/PR of wells W14 and W15, with CICPR/PR increasing by 2.53 and 1.15 %/MPa, respectively.

By analyzing the calculated results of CICP of 15 wells, it is found that CICP decreases during the stages from drainage stage to stable gas production stage. Moreover, the CICP rate and the reduction value of permeability are the largest in unstable gas production stage. Similarly, the CICPR/PR and the value of permeability per unit pressure reduced decrease gradually during the stages from drainage stage to stable gas production stage, and the reduction of permeability is the largest in the drainage stage. With the progress of production, the permeability reduction phenomenon disappears gradually, and permeability can be recovered to the original level or even larger when production process reaches the gas production decline stage.

4 Characteristics and mechanisms of CICP in production stages

Some studies have demonstrated that effective stress and matrix shrinkage would cause the phenomenon of "decrease first and then increase" in coal reservoir permeability (Shi and Durucan 2005; Palmer 2009; Mazumder et al. 2012). However, this conclusion lacks verification with production data and systematic studies of control mechanisms of permeability. Therefore, this paper verifies the characteristics of dynamic change of permeability in high-rank coal reservoirs on the basis of production data. It also explains and concludes the characteristics and mechanisms of CICP during production stages from five aspects, i.e., the permeability variation trend, controlling mechanism, system energy, phase composition, and the production performance (Table 4; Fig. 4).

During the CBM production, phase compositions of the fluid change gradually from single-phase water to two-phase gas and water, and then to single-phase gas flow. Water production declines gradually over time, and gas production shows a trend of "start—rise—stable—decline" of gas production. With increasing cumulative gas production, the matrix shrinkage effect plays a more and more important role in the increase of the permeability. Thus, the largest reduction value of CICPR/PR occurs in the drainage stage. After this stage, the reduction of

Table 3 Summary of calculation results of permeability change in each stage of 15 CBM wells in southern Qinshui basin

Well name	Drainage stage			Unstable gas pro-stage			Stable gas pro-stage			Gas pro-decline stage		
	Δp, MPa	φk, %	ξk, %/MPa	Δp, MPa	φk, %	ξk, %/MPa	Δp, MPa	φk, %	ξk, %/MPa	Δp, MPa	φk, %	ξk, %/MPa
W1	0.069	−0.600	−8.662	0.275	−2.320	−8.430	0.275	−2.320	−8.430	–	–	–
W2	0.094	−0.814	−8.645	0.703	−5.747	−8.171	–	–	–	–	–	–
W3	0.106	−0.919	−8.638	0.643	−5.264	−8.192	–	–	–	–	–	–
W4	0.090	−0.778	−8.639	3.059	−20.18	−6.596	–	–	–	–	–	–
W5	0.040	−0.350	−8.677	2.893	−19.65	−6.793	–	–	–	–	–	–
W6	0.057	−0.494	−8.667	2.563	−17.96	−7.008	–	–	–	–	–	–
W7	0.058	−0.501	−8.666	1.804	−13.58	−7.524	–	–	–	–	–	–
W8	0.057	−0.491	−8.667	2.246	−16.24	−7.230	0.026	−0.145	−5.671	–	–	–
W9	0.097	−0.841	−8.643	1.477	−11.35	−7.686	0.907	−5.527	−6.095	–	–	–
W10	0.073	−0.633	−8.657	0.580	−4.800	−8.270	–	–	–	–	–	–
W11	0.144	−0.619	−4.297	0.103	−0.412	−3.997	0.21	−0.756	−3.601	0.088	−0.284	−3.233
W12	0.043	−0.318	−7.398	0.270	−1.907	−7.061	0.649	−4.196	−6.465	0.186	−1.114	−5.992
W13	0.081	−0.594	−7.339	0.165	−1.177	−7.132	0.438	−2.967	−6.774	0.535	−3.269	−6.110
W14	0.018	−0.105	−5.807	0.872	−4.107	−4.710	1.408	−2.055	−1.460	0.308	0.779	2.532
W15	0.050	−0.290	−5.805	0.619	−3.249	−5.249	1.493	−3.844	−2.574	0.395	0.454	1.149
Average	0.071	−0.556	−7.813	1.218	−8.530	−6.937	0.676	−2.726	−5.133	0.302	–	–

Δp cumulative pressure decline, φk CICP rate, ξk CICPR/PR, "−" minus sign indicates that permeability decreases comparing to the original value, "–" no data

CICPR/PR will decrease or become negative. The CICP rate changes with the production time, water production, and gas production in different production stages. Generally, during the unstable gas production stage, water production is relatively large and gas production relatively small. Besides, the duration of this stage is longer than that of drainage stage. Therefore, the largest value of reduction in permeability appears in this stage. Different types of CBM production wells have various characteristics of CICP. For wells with similar production time, the effect of matrix shrinkage is stronger in high-productivity wells than in low-productivity wells. Hence, the extent of recovery and improvement of permeability is better in high-productivity wells than in low-productivity wells.

In brief, the effective stress and the coal matrix shrinkage effects are the dominant mechanisms of CICP during CBM development process. Furthermore, the duration and degree of these two effects are the direct reasons for permeability change during different production stages. Besides, the duration and degree of matrix shrinkage, controlled by adsorption and desorption performance (CBM desorption efficiency), are direct controlling factors for the recovery of the coal reservoir permeability (Meng et al. 2014a). Phase composition provides a preliminary basis for coal reservoir permeability stage division. Dynamic production data provide a basis for the calculation of permeability of different stages.

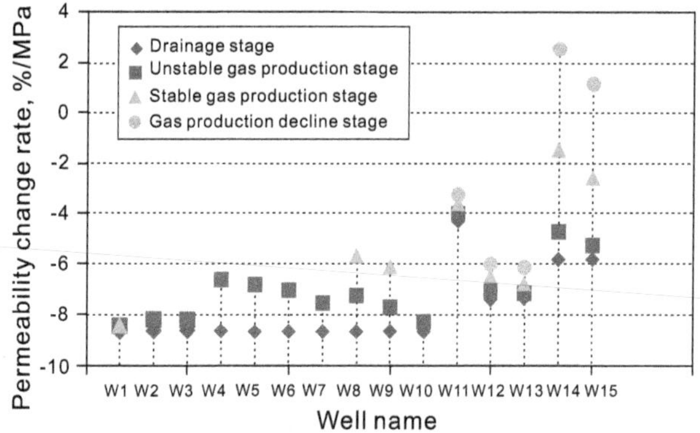

Fig. 3 CICP rate in each production stage of southern Qinshui basin

5 Conclusion

(1) Using dimensionless gas production rate, CBM production process is quantitatively divided into four stages, i.e., drainage stage, unstable gas production stage, stable gas production stage, and gas production decline stage. Calculation of dimensionless parameters for the Fanzhuang block indicates that dimensionless gas production rate ranges from 0.107 to 0.426 at the end of the drainage stage and the high-productivity wells generally have smaller dimensionless gas production rates than the low-productivity wells. The dimensionless gas production rate ranges from 0.547 to 0.683 and from 0.713 to 0.909, at the end of unstable gas production and stable gas production stages, respectively.

(2) The case study of the CICP during production stages of the Fanzhuang block shows that the permeability decreases in drainage stage, unstable gas production stage, and stable gas production stage. The highest permeability change rate and reduction of permeability are found in unstable gas production stage. The largest reduction of CICPR/PR occurs in drainage stage. With the progress of production, permeability reduction is gradually improved, and the permeability can be recovered to the original

Table 4 Summary of characteristics of CICP, control mechanisms, systematic energy, phase compositions, and production performance during production stages

Production stage	Permeability change trend	Controlling mechanism	System energy	Phase composition	Pro-performance
Drainage stage	Declines gradually	Effective stress effect	Inefficient and slow desorption	Single-phase water flow	No gas production
Unstable gas pro-stage	Declines in total, but recovers gradually	Effective stress effect and matrix shrinkage effect	Rapid and sensitive desorption	Gas–water two-phase flow	Gas production rise, unstable
Stable gas pro-stage	Recovers fast	Matrix shrinkage effect	Sensitive desorption	Gas phase flow	Gas production stable
Gas pro-decline stage	Recovers early, later remain unchanged	Early matrix shrinkage effect, later effective stress effect	Sensitive desorption	Single-phase gas flow	Gas production declines gradually

Notes System energy division is detailed in Meng et al. (2014a)

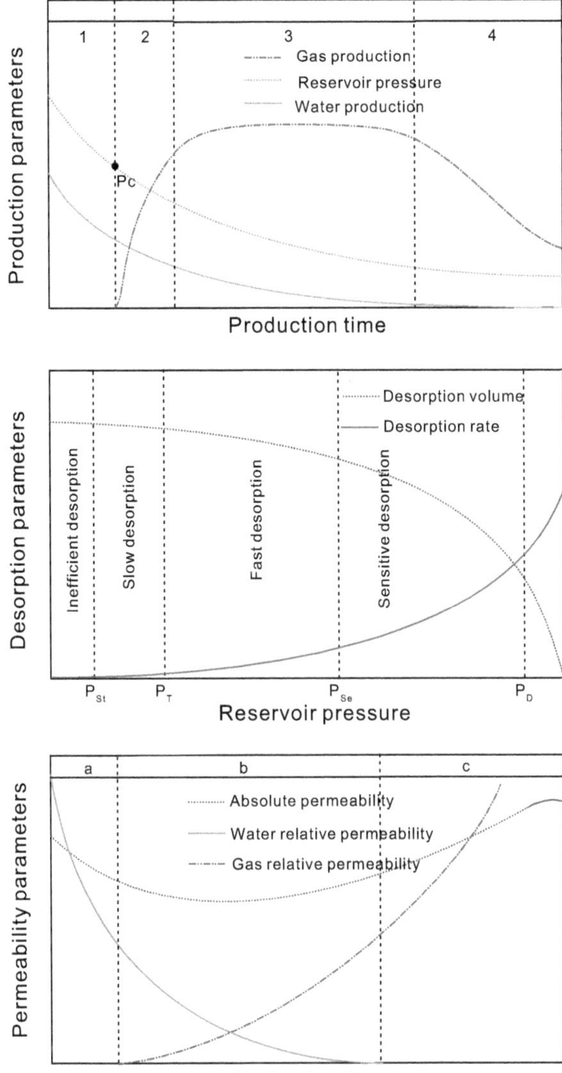

P_C, Critical desorption pressure; 1, drainage stage; 2, unstable gas production stage 3, stable gas production stage; 4, gas production decline stage

P_{St}, Starting pressure P_T, Turning pressure
P_{Se}, Sensitive pressure P_D, Decline pressure
a, effective stress effect control; b, matrix shrinkage effect control;
c, early matrix shrinkage effect and latter effective stress effect control

Fig. 4 Characteristics and mechanism of CBM permeability for different stages

level or even higher when production process reaches the gas production declinc stage. This study indicates that the characteristics of CICP during production stages are closely related to production of gas and water and the stability of production.

(3) Characteristics of CICP are directly controlled by the duration and degree of the effective stress effect and coal matrix shrinkage effect. From the point of system energy, the duration and degree of the matrix shrinkage effect are influenced by adsorption and desorption behavior. Phase composition provides a preliminary basis for coal reservoir permeability stage division. It also indicates the changes in effective gas permeability. Production performance reflects the characteristics of production in different stages. Dynamic production data provide a basis for the calculation of the permeability at different stages, making it useful for prediction of productivity.

(4) Study of characteristics of dynamic CICP on the basis of production data provides an effective solution for a difficult problem in real-time monitoring and forecasting CICP. In addition, the related calculation method and theoretical results can be used for the deployment of CBM well development, management of production, and prediction of productivity.

Acknowledgments The authors acknowledge financial support from the various funding agencies including the Major State Basic Research Development Program of China (973 Program, 2009CB219604), the National Natural Science Foundation of China (41272175), the Key Project of the National Science & Technology (2011ZX05034-001), and the China Scholarship Council.

References

Aminian K, Ameri S. Predicting production performance of CBM reservoirs. J Nat Gas Sci Eng. 2009;1(1):25–30.

Clarkson C, Jordan C, Gierhart R, et al. Production data analysis of coalbed-methane wells. SPE Reserv Eval Eng. 2008;11(2):311–25.

Fan T, Zhang G, Cui J. The impact of cleats on hydraulic fracture initiation and propagation in coal seams. Pet Sci. 2014;11(4): 532–9.

Feng D, Deng H, Zhou Z, et al. Paleotopographic controls on facies development in various types of braid-delta depositional systems in lacustrine basins in China[J]. Geosci Front. 2014. doi:10.1016/j.gsf.2014.03.007.

Freeman CM, Moridis G, Ilk D, et al. A numerical study of performance for tight gas and shale gas reservoir systems. J Petrol Sci Eng. 2013;108:22–39.

King GR. Material-balance techniques for coal-seam and devonian shale gas reservoirs with limited water influx. SPE Reserv Eng. 1993;8(1):67–72.

Lai F, Li Z, Fu Y, et al. A drainage data-based calculation method for coalbed permeability. J Geophys Eng. 2013;10(6):65–70.

Li M, Wang L, Cui X, et al. Output characteristics of vertical wells and dewatering control method used in Fanzhuang block of Qinshui CBM field. China Coalbed Methane. 2011;8(1):11–3 (in Chinese).

Li Y, Tang D, Xu H, et al. Geological and hydrological controls on water co-produced with coalbed methane in Liulin, eastern Ordos Basin, China. AAPG Bull. 2015;99(2):207–29.

Li Y, Tang D, Xu H, et al. In-situ stress distribution and its implication on coalbed methane development in the Liulin area, eastern Ordos basin, China. J Petrol Sci Eng. 2014;122:488–96.

Liu D, Yao Y, Cai Y, et al. Study progress on geological and dynamic evaluation of coal bed methane reservoir. Coal Sci Technol. 2010;38(11):10–6 (in Chinese).

Mazumder S, Scott M, Jiang J. Permeability increase in Bowen Basin coal as a result of matrix shrinkage during primary depletion. Int J Coal Geol. 2012;96(1):109–19.

Meng Y, Tang D, Xu H, et al. Division of coalbed methane desorption stages and its significance. Pet Explor Dev. 2014a;41(5):671–7.

Meng Y, Tang D, Xu H, et al. Geological controls and coalbed methane production potential evaluation: a case study in the Liulin area, eastern Ordos Basin, China[J]. J Nat Gas Sci Eng. 2014b;21:95–111.

Ni X, Su X, Zhang X. Coal bed methane development geology. Beijing: Chemical Industry Press; 2010 (in Chinese).

Palmer I. Permeability changes in coal: analytical modeling. Int J Coal Geol. 2009;77(1):119–26.

Qin Y, Li Y, Bai J, et al. Technologies in the CBM drainage and production of wells in the southern Qinshui basin with high-rank coal beds. Nat Gas Ind. 2011;31(11):22–5 (in Chinese).

Qin Y, Tang D, Liu D, et al. Geological evaluation theory and technology progress of coal reservoir dynamics during coalbed methane drainage. Coal Sci Technol. 2014;42(1):80–8 (in Chinese).

Shi JQ, Durucan S. A model for changes in coalbed permeability during primary and enhanced methane recovery. SPE Reserv Eval Eng. 2005;8(4):291–9.

Song Y, Liu S, Zhang Q, et al. Coalbed methane genesis, occurrence and accumulation in China[J]. Pet Sci. 2012;9(3):269–80.

Tao S, Tang D, Xu H, et al. Factors controlling high-yield coalbed methane vertical wells in the Fanzhuang Block, Southern Qinshui Basin[J]. Int J Coal Geol. 2014;134:38–45.

Tao S, Wang Y, Tang D, et al. Dynamic variation effects of coal permeability during the coalbed methane development process in the Qinshui Basin, China. Int J Coal Geol. 2012;93:16–22.

Xu H, Tang DZ, Tang SH, et al. A dynamic prediction model for gas–water effective permeability based on coalbed methane production data. Int J Coal Geol. 2014;121:44–52.

Contribution of moderate overall coal-bearing basin uplift to tight sand gas accumulation: case study of the Xujiahe Formation in the Sichuan Basin and the Upper Paleozoic in the Ordos Basin, China

Cong-Sheng Bian[1] · Wen-Zhi Zhao[1] · Hong-Jun Wang[1] · Zhi-Yong Chen[1] · Ze-Cheng Wang[1] · Guang-Di Liu[2] · Chang-Yi Zhao[1] · Yun-Peng Wang[3] · Zhao-Hui Xu[1] · Yong-Xin Li[1] · Lin Jiang[1]

Abstract Tight sand gas is an important unconventional gas resource occurring widely in different petroleum basins. In coal-bearing formations of the Upper Triassic in the Sichuan Basin and the Carboniferous and Permian in the Ordos Basin, coal measure strata and tight sandstone constitute widely distributed source–reservoir assemblages and form the basic conditions for the formation of large tight sand gas fields. Similar to most tight gas basins in North America, the Sichuan, and Ordos Basins, all experienced overall moderate uplift and denudation in Meso-Cenozoic after earlier deep burial. Coal seam adsorption principles and actual coal sample simulation experiment results show that in the course of strata uplift, pressure drops and desorption occurs in coal measure strata, resulting in the discharge of substantial free gas. This accounts for 28 %–42 % of total gas expulsion from source rocks. At the same time, the free gases formerly stored in the pores of coal measure source rocks were also discharged at a large scale due to volumetric expansion resulting from strata uplift and pressure drop. Based on experimental data, the gas totally discharged in the uplift period of Upper Paleozoic in the Ordos Basin, and Upper Triassic Xujiahe Formation in the Sichuan Basin is calculated as (3–6) $\times$ 10^8 m^3/km^2. Geological evidence for gas accumulation in the uplift period is found in the gas reservoir analysis of the above two basins. Firstly, natural gas discharged in the uplift period has a lighter carbon isotope ratio and lower maturity than that formed in the burial period, belonging to that generated at the early stage of source rock maturity, and is absorbed and stored in coal measure strata. Secondly, physical simulation experiment results at high-temperature and high-salinity inclusions, and almost actual geologic conditions confirm that substantial gas charging and accumulation occurred in the uplift period of the coal measure strata of the two basins. Diffusive flow is the main mode for gas accumulation in the uplift period, which probably reached 56 $\times$ 10^{12} m^3 in the uplift period of the Xujiahe Formation of the Sichuan Basin, compensating for the diffusive loss of gas in the gas reservoirs, and has an important contribution to the formation of large gas fields. The above insight has promoted the gas resource extent and potential of the coal measure tight sand uplift area; therefore, we need to reassess the areas formerly believed unfavorable where the uplift scale is large, so as to get better resource potential and exploration prospects.

Keywords Sichuan Basin · Ordos Basin · Tight sand gas · Stratigraphic uplift · Coal measure · Hydrocarbon accumulation mechanism · Diffusion

✉ Cong-Sheng Bian
bcs_1981@petrochina.com.cn

[1] Research Institute of Petroleum Exploration & Development, CNPC, Beijing 100083, China

[2] College of Geosciences, China University of Petroleum, Beijing 102249, China

[3] Guangzhou Institute of Geochemistry, Chinese Academy of Sciences, Guangzhou 510640, Guangdong, China

Edited by Jie Hao

1 Introduction

Tight sand gas resources are widely distributed in the world (Masters 1979; Law 2002; Holditch 2006). Statistics show that tight sand gas has been discovered or predicted in 70 basins in North America, Europe, and the Asian-Pacific region, with a resource extent of about 210 $\times$ 10^{12} m^3, showing huge potential in exploration and development (British Petroleum Company 2012). The United States has the highest annual gas yield from tight sandstone at present in the world, reaching 1.75×10^8 m^3 in 2010, about 29 % of total gas production of the country (British Petroleum Company 2012). In China, the Zhongba tight gas field was discovered in the western Sichuan Basin in 1971. With the progress in exploration technologies in recent years, new discoveries are continually being made. Two major tight gas provinces in the Sichuan and Ordos Basins and five breakthrough areas in the Kuche deep zone of the Tarim Basin have been found so far (Fig. 1), with technically recoverable gas resources of 8–11 TCM, and annual tight gas yield more than 25.6 BCM in 2011 (Li et al. 2012; Zou et al. 2012). Unconventional gas (i.e., tight gas and shale gas) is increasingly becoming one of the dominant resources in the global natural gas industry, and more efforts are being put into tight gas exploration in many countries (Kuuskraa and Bank 2003; Smith et al. 2009; Zeng 2010; Dai et al. 2012; He et al. 2013).

Statistics show that tight gas resources worldwide mostly originate from coal measure source rocks (Dai et al. 2012), e.g., the Piceance Basin (Johnson and Rice 1990; Zhang et al. 2008) and the San Juan Basin (Ayers 2002) in the United States, the Ordos and Sichuan Basins in China. As well, in the formation and development history of these petroleum basins, the strata were usually deposited and buried, followed by uplifting at a later stage. The study by Tissot and Welte (1984) shows that natural gas generated in gas source rocks due to high temperature in the process of burial was either discharged, migrated, and accumulated in the reservoirs to form gas pools or was retained in the gas source rocks to form shale gas and coalbed methane. There are limited publications at home and abroad regarding natural gas migration and accumulation under uplift tectonic settings (Tian et al. 2007). It is usually believed that strata denudation and faulting are likely to occur in the course of uplifting, resulting in damage to gas reservoirs and dissipation of the gas (Hao et al. 1995; Zhang et al. 1999). Based on studies of tight sand gas accumulation in coal-bearing strata in recent years, we have found that tight gas from coal measure strata not only can form gas reservoirs in the course of deposition and burial, but also can accumulate in the moderate uplift process if the regional seal has not been destroyed (Bian et al. 2009; Zhao et al. 2010). This paper mainly takes gas

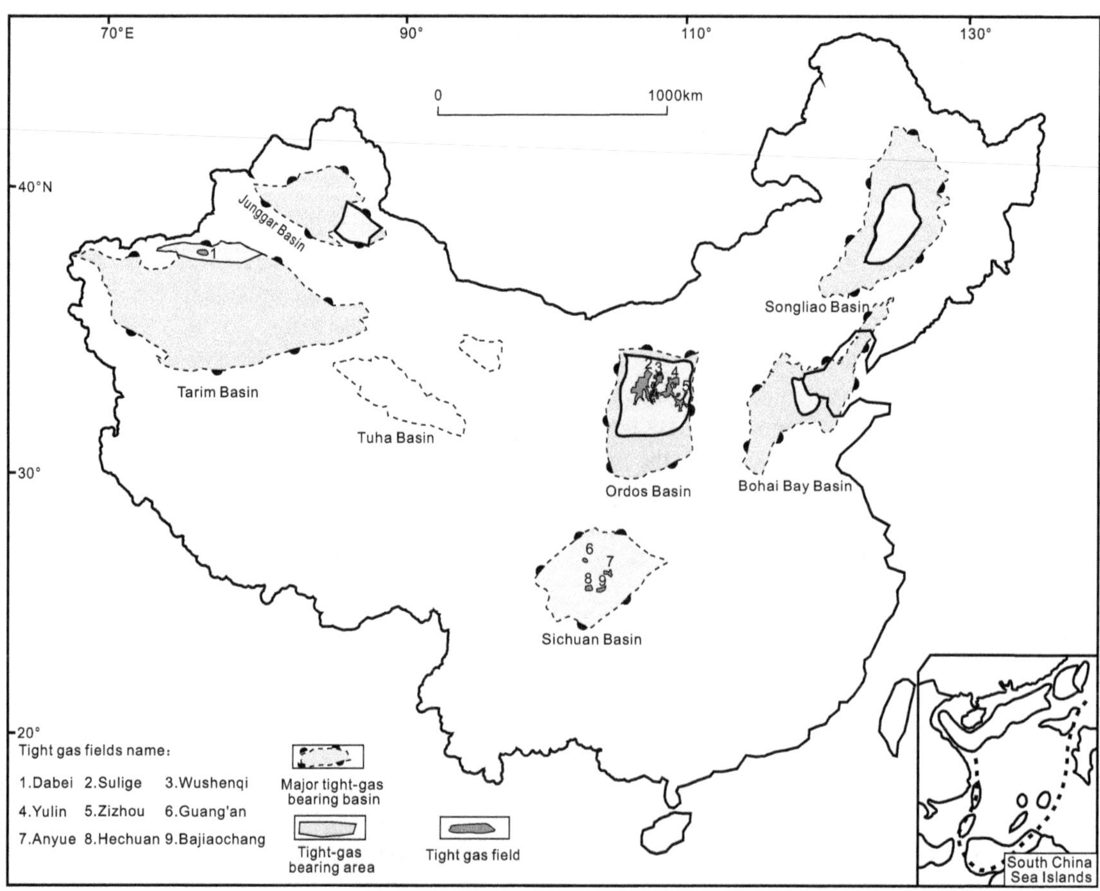

Fig. 1 Major tight sand gas basins in China

fields of the Upper Triassic Xujiahe Formation in the Sichuan Basin and the Upper Paleozoic in the Ordos Basin as examples to discuss this geological process.

2 Geological setting

Several large gas fields with reserves more than 100 BCM have been successively discovered in the Upper Paleozoic Permo-Carboniferous in the Ordos Basin and the Upper Triassic Xujiahe Formation in the central Sichuan Basin (Dai et al. 2012; Zhao et al. 2013) in recent years. By the end of 2011, large gas fields like Sulige, Wushenqi, and Yulin had been discovered in Upper Paleozoic of the Ordos Basin, with proved gas reserves (including basically proved reserves) approaching 3.0×10^{12} m^3 and gas-bearing area about 2.0×10^4 km^2. Large gas fields like Guang'an, Hechuan, and Anyue had been discovered in the Upper Triassic Xujiahe Formation of the Sichuan Basin, with proved gas reserves approaching 6000×10^8 m^3 and gas-bearing area of more than 3000 km^2. Tight sand gas resources of the two basins account for more than two-thirds of the present total tight gas resources in China (Dai et al. 2012).

A set of continental clastic coal-bearing formations covering the whole basin is developed in the Upper Paleozoic of the Ordos Basin and the Xujiahe Formation of the Sichuan Basin, respectively, and the source rocks mainly consist of coal seams and coaly mudstone with high gas generation potential. The Upper Paleozoic source rocks of the Ordos Basin are distributed in the Carboniferous Taiyuan Formation and Permian Shanxi Formation (Zhang et al. 2009b), with coal seam thicknesses of 10–25 m on average, 40 m locally, almost covering the whole basin stably. Coal measure mudstones have thicknesses of 60–130 m, almost the same distribution as the coal seams, and are usually 200-m thick in the west. The Upper Triassic Xujiahe Formation source rocks of the Sichuan Basin are distributed in the Xu1, Xu3, and Xu5 Members, with coal seam thicknesses of 5–15 m on average, distributed stably horizontally. Coal measure mudstones have thicknesses of 100–800 m, thinning out from the western to the central Sichuan Basin, basically covering the whole basin (Zhao et al. 2011). Thermal evolution history shows that substantial gas generation and expulsion occurred in the source rocks of the two basins in the geological history. The thermal evolution of source rocks has been at highly mature and overmature stages up to now, with cumulative gas generating strength of $(20–40) \times 10^8$ m^3/km^2 and $(40–100) \times 10^8$ m^3/km^2, respectively. This can provide abundant gas sources for large-scale gas accumulation in the tight sand reservoirs of the two large gas provinces. Their reservoirs are the Permian Shanxi and Shihezi Formations and the Upper Triassic Xu2, Xu4, and Xu6 Member tight sands, respectively (Zhang et al. 2009a), with porosities of 4 %–10 %, permeabilities of 0.01–1 mD and thicknesses of hundreds to a thousand meters. They closely and widely contact the source rocks, comprising a very favorable source–reservoir assemblage horizontally, which has laid the foundation for widespread gas accumulation in the reservoirs and then the discovery of large scale reserves (Figs. 2, 3).

Stratigraphic burial history shows that the tight sand gas zones in both the Ordos Basin and the Sichuan Basin all experienced deep burial before the Late Cretaceous (Zhao et al. 2005, 2010), with maximum burial depths up to 3500–4000 m and 4500–6000 m, respectively. From the Late Cretaceous to this day, overall tectonic uplift and strata denudation occurred in these two basins, and the hydrocarbon generation of the source rocks stopped. The denuded thickness is 800–1500 and 1500–2500 m, respectively, and the buried depth of tight gas zones of these two basins is 1800–4000 m at present. Fortunately, uplift and denudation have not resulted in the damage to the regional seal.

3 Theoretical model and experiment

Substantial gas infusion is a prerequisite for gas accumulation in the uplift period of tight sand formation. Coal petrography (Wang et al. 1995; Zhang et al. 2000) showed that microfractures and cleats are well developed in coals, especially in high-rank coals. Substantial free gas was stored in pores and microfractures at the time of substantial gas generating from coal seams in the burial period. This expanded volumetrically due to the pressure drop in the uplift period, and thus together with the desorbed and liberated coal seam gas, became the important gas source for gas accumulation in the uplift period (Cui et al. 2005; Zhao et al. 2010). Starting from the discussion of a theoretical model of coal seam adsorption–desorption, by means of thermal simulation and gas accumulation simulation experiments of actual coal samples, the geological process of gas accumulation in the uplift period of coal measure strata is demonstrated.

3.1 Theory of adsorption of gas in coal seams and the experimental model

Adsorption refers to the attachment of atoms or molecules of one substance on the surface of another substance (Busch et al. 2003; Bae and Bhatia 2006). The adsorption behavior of coal was observed in coal mining in the middle of the last century. After having successfully produced coalbed methane, the United States made an in-depth study

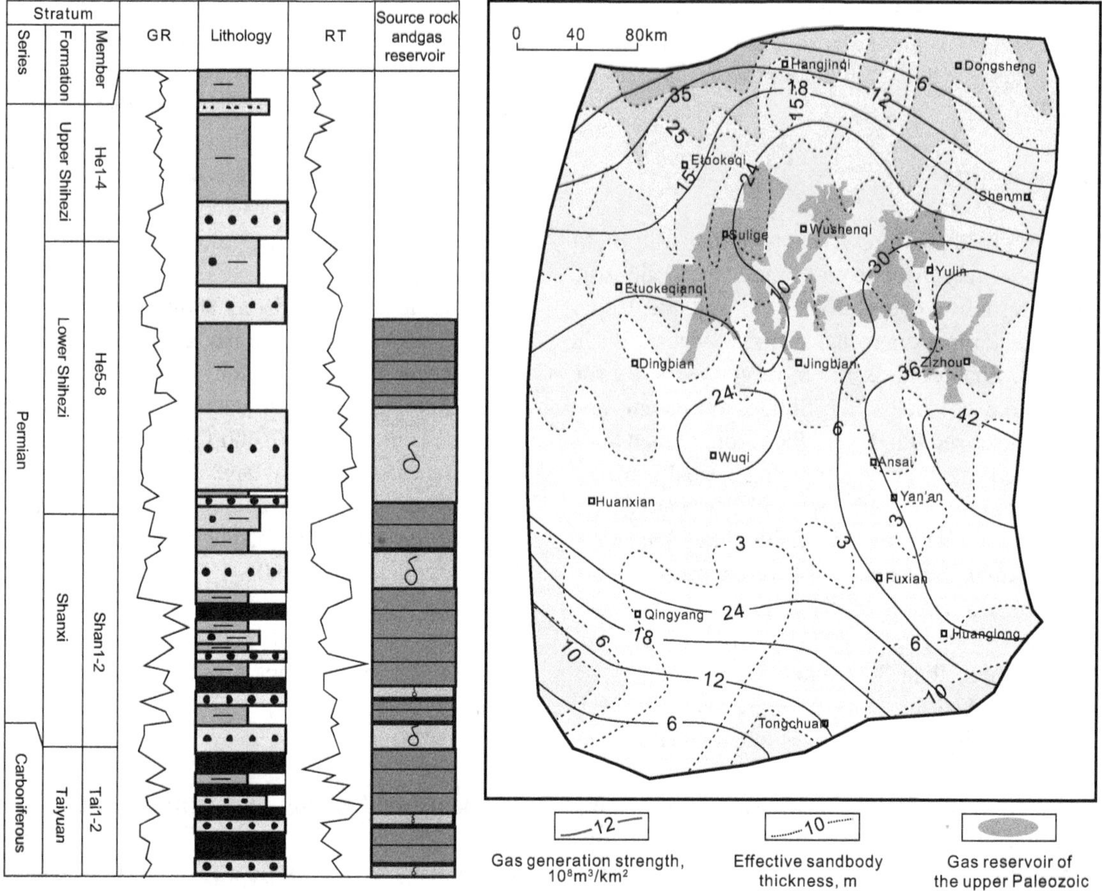

Fig. 2 Composite Upper Paleozoic stratigraphic section and coalbeds and He8 reservoirs, Ordos Basin

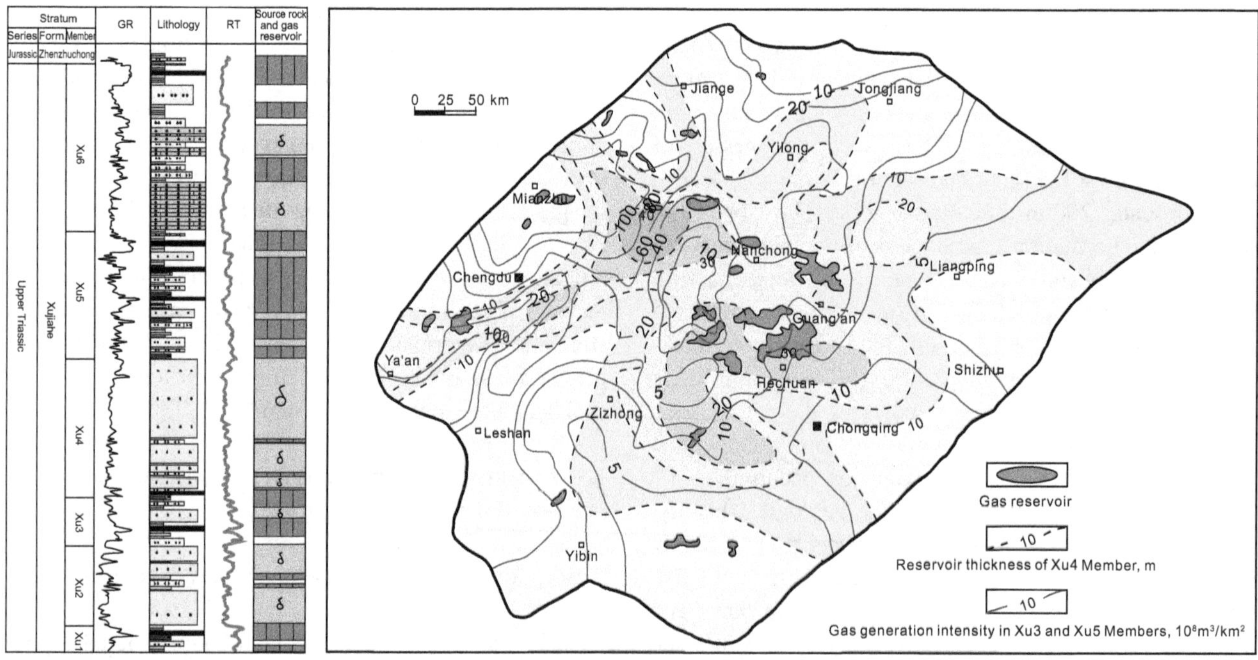

Fig. 3 Xujiahe stratigraphic section showing gas source rocks and reservoirs, Sichuan Basin

on the adsorptive action of coal seams (Radovic et al. 1997; Clarkson and Bustin 1999). Coal is universely believed at present to be a porous medium with a big surface area. The adsorption of gas by coal is a physical adsorption process; the adsorption energy (heat) is small, the adsorption rate is fast, and the adsorption and desorption procedure is reversible (Radovic et al. 1997; Clarkson and Bustin 1999). Therefore, the adsorptive force of coal to gas molecules is intermolecular force, i.e., there are both adsorption equilibrium and adsorption heat (energy). Previous studies (Mavor et al. 1990; Mukhopadhyay and Macdonald 1997; Crosdale et al. 1998) show that the adsorption by coal seams is affected by many factors like coal composition, temperature and pressure, gas properties, and coal rank. Under geological conditions, with the variation of buried depth of coal seams, the impact of temperature and pressure on it is maximum. Experimental results show that pressure is proportional to adsorptive capacity, whereas the influence of temperature is the reverse (Zhao et al. 2001).

Many researchers (Anderson et al. 1966; Ruppel et al. 1974; Yang and Saunders 1985) have conducted many theoretical and experimental studies on the adsorptive action of coal seams and established many theoretical models and mathematical expressions related to adsorptive capacity. These include the monomolecular layer adsorption model and Langmuir's equation, multimolecular layer adsorption model and BET's equation, Freundlieh's equation, Polomyi's adsorption potential theory, micropore filling theory, and Dubinin-Astak-hov's equation (Sang et al. 2005; Su et al. 2008). However, it is difficult to accurately describe the adsorptive properties of coal using a certain isothermal adsorption line or theoretical model due to the presence of micropores in coal. It is generally believed that gas is adsorbed on the surface of coal in a monomolecular state under geological conditions; therefore, a monomolecular layer adsorption model and Langmuir's equation are widely used to describe the adsorptive features of coal and have become a classical theory maturely used in the development of coalbed methane for the moment. The model is described as follows:

$$V = \frac{V_L \times P}{P_L + P} \quad (1)$$

where V represents the adsorptive capacity of coal, V_L is the Langmuir volume, P_L is Langmuir pressure, and P is pressure.

With increasing research, it is found that under subsurface high temperature and pressure conditions (pressure >15 MPa, temperature >80 °C) (Gregory and Karen 1986; Mavor et al. 1990), multimolecular layer adsorption of gas occurs in coal seams. As well, methane becomes a supercritical gas under high pressure, and its density increases significantly, resulting in a change of adsorptive capacity. This results in the phenomenon that the adsorptive capacity measured experimentally represents the maximum value and then declines, i.e., there is a large difference between the measured apparent adsorption capacity and the absolute adsorptive capacity. To help solve this problem, some correction models for calculating the adsorptive capacity are introduced (Haydel and Kobayashi 1967; Murata et al. 2001), and a relatively common method is as follows:

$$N_{ab} = N_{ap}(1 - \rho_{free}/\rho_{ad}) \quad (2)$$

where N_{ab} represents absolute adsorptive capacity, N_{ap} is excessive adsorptive capacity or apparent adsorptive capacity, ρ_{free} is free gas density under equilibrium conditions, and ρ_{ad} is the adsorbed gas density.

Because of equipment limitations, most coal adsorption experiments are conducted at temperatures less than 40 °C and pressures less than 15 MPa, and the relation between pressure and adsorptive capacity is basically measured at one temperature. There are few experimental studies with both temperature and pressure varying. Cui et al. (2005) derive a characteristic curve for coal-adsorbed methane based on adsorption potential theory and adsorption experiments under high temperature and high pressure and believe that this characteristic curve has uniqueness, i.e., there is only a unique peak value in adsorptive capacity, based on which a new adsorption model of coal is derived and can be used to estimate the adsorptive capacity at different temperature–pressure conditions.

$$\ln V = AT[2.7\ln T - \ln P - 12.6603] + B \quad (3)$$

where A and B stand for adsorption constant and can be obtained from experimental adsorptive data of coal at a given temperature. The above equation can be used to calculate the adsorptive capacity of coal at any temperature and pressure.

3.2 Simulation experiment of hydrocarbon expulsion due to coal seam uplift and pressure drop

For the sake of understanding gas desorption due to temperature and pressure drop under approximately real geological conditions, a specially designed autoclave was used to simulate the gas discharge process of coal seams at the time of dropping of temperature and pressure. Two and five sets of samples were selected from the Upper Triassic Xujiahe Formation of the Sichuan Basin and the Upper Paleozoic of the Ordos Basin, repectively, to conduct the experiments. The deep burial and gas generation process of coal seams was simulated by external heating and pressurizing; when gas generation and expulsion reached equilibrium in the autoclave, the temperature and pressure were reduced, and the gas discharge from coal was observed. The experimental results are listed in Table 1 and

Table 1 Gas expulsion of coals under different temperature and pressure conditions

Coal sample	Process	Temperature, °C	Pressure, MPa	Gas generation and expulsion, mL	Total gas generation and expulsion rate, mL/g	Staged gas generation and expulsion rate, mL/g	Staged gas generation and expulsion ratio, %
Xu3 member	Heating and pressurizing	420	105	75	48.8	32.7	67
	Cooling and depressurizing	320	50	37		16.1	33
Xu6 member	Heating and pressurizing	420	105	575	40	28.8	72
	Cooling and depressurizing	320	50	225		11.3	28
Liaohe lignite	Heating and pressurizing	450	104	475	19.4	11.9	61.3
	Cooling and depressurizing	350	60	300		7.5	38.7
Xianfeng lignite	Heating and pressurizing	450	6	850	73.3	42.5	58
	Cooling and depressurizing	350	2	615		30.8	42
Xianfeng lignite	Heating and pressurizing	450	8.2	950	79.5	47.5	59.7
	Cooling and depressurizing	350	4	640		32.0	40.3
Taiyuan Formation	Heating and pressurizing	450	4.1	826	60.8	41.3	67.9
	Cooling and depressurizing	350	2.4	390		19.5	32.1
Shanxi Formation	Heating and pressurizing	450	4.4	975	70.3	48.8	69.4
	Cooling and depressurizing	350	1	430		21.5	30.6

show that when the temperature and pressure of coals taken from the Xujiahe Formation are reduced from 420 °C and 105 MPa to 320 °C and 50 MPa, respectively (corresponding to uplifting of formation from 4000 to 2000 m), the desorbed and discharged gas is 11–16 mL/g, accounting for 28 %–33 % of total gas expulsion of coals. When the temperature and pressure of coals from the Upper Paleozoic of the Ordos Basin are reduced from 450 °C and 104–4 MPa to 350 °C and 60–1 MPa, respectively (corresponding to uplifting the formation from 4000–6000 m to less than 3000 m), the desorbed and discharged gas is 7.5–32 mL/g, accounting for 31 %–42 % of total gas expulsion, showing that quite a lot of gas is desorbed and discharged during the temperature and pressure decline process of coal seams in the uplift period.

3.3 Physical simulation of gas accumulation

Physical simulation is an important method to study and reproduce geological processes (Zeng and Jin 2002; Zhao et al. 2006). To better study the gas accumulation process of coal measure tight sand formations in the uplift period, an experiment was conducted to simulate gas migration and accumulation under strata uplift and pressure drop settings. To prove the existence of gas desorption and expulsion from coal measure formations in its uplift and pressure drop, a physical simulation experiment model was designed based on the configuration relation of source rocks and reservoirs of the Upper Triassic Xujiahe Formation in the Sichuan Basin. Coal samples were crushed to 150–200 mesh, sandstone was replaced by glass beads, and natural gas in the formation was approximated by pure methane. A 3D high-temperature and high-pressure physical simulation device independently designed by RIPED was used for experiment.

In the experiment (Table 2), the model was evacuated, then water was injected, and overburden pressure was applied to simulate the water discharge process of formations due to burial and compaction. When the overburden pressure was increased to 7 MPa, the fluid pressure reached 1.63 MPa at 60 °C, and the formation was saturated with water and was in a near-equilibrium state. Then, methane was injected into the bottom of the model to simulate the gas generation process. In the course of gas injection, water was continually discharged from the outlet. When the experiment had been conducted for about 20 h, gas started to appear at the outlet, showing that after the gas source rock had been saturated with adsorbed gas, free gas started to migrate. When the gas flowing out of the outlet reached 10 mL/min and the water yield decreased significantly, we stopped the gas injection. This simulates the uplift and pressure drop process, and the gas desorption and gas expulsion processes were observed. To better replicate the subsurface environment, we stopped any operation for 24 h after having ceased gas injection, to ensure the equilibration of gas filling and adsorption. This simulates the cessation of gas generation after the source rock reached its maximum buried depth. Subsequently, the overburden pressure, fluid pressure, and temperature of the model were reduced, and at 175–182 h later, gas flowed out of the outlet in an episodic type. This process corresponds to that after the formation had changed from burial to uplift and

Table 2 Experiment process and phenomena

Step	Process and phenomena	Step	Process and phenomena
1	Installed model under dry and water free conditions	7	Increased injection pressure, gas started to appear at outlet
2	Evacuated for 12 h	8	When the outflow rate of methane reached 10 mL/min, the injection pressure was 2 MPa, closed the gas inlet
3	Slowly injected water into model from the bottom until the internal pressure is at 1 MPa uniformly	9	The gas outflow rate gradually decreased below 0.1 mL/min, and the fluid pressure decreased gradually too
4	Allowed fluid pressure to be 1.63 MPa, increased overburden pressure to 7 MPa, and heated the model to 60 °C	10	When no further gas flowed out of the outlet, stood still for 24 h
5	Injected methane at pressure more than 1.63 MPa, water flow rate at outlet increased gradually	11	Gradually reduced the overburden pressure to 2 MPa, the fluid pressure did not change; reduced temperature to 30 °C, recorded the gas flow rate at outlet
6	Water flow rate decreased gradually after having been increased to 40 mL/h	12	Gradually reduced the fluid pressure of model to 1.63 MPa, the overburden pressure and temperature did not change, recorded the gas flow rate at outlet

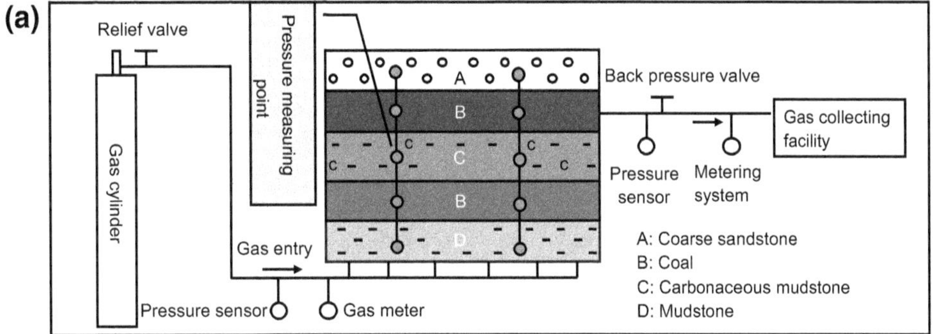

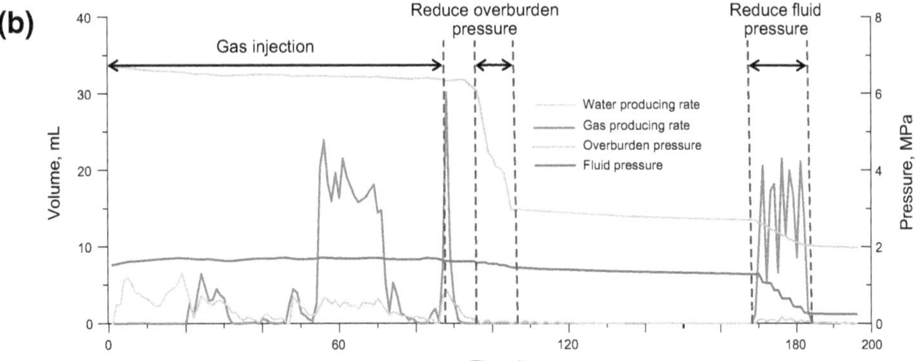

Fig. 4 Experimental device (**a**) and results (**b**) of simulating gas migration and accumulation in uplift and pressure drop environment

denudation when the temperature and pressure of the formation fluid had dropped. Desorption occurred in the coal measure formation, gas was liberated, and substantial gas migrated (Fig. 4a, b).

4 Experimental results and analysis

The above experiment shows that under uplift settings, although the gas generation process of gas source rocks had stopped, the substantial gas adsorbed on the particle surfaces inside the source rocks and free in the pores and highly compressed in deep strata suffered from desorption and bulk expansion due to pressure drop in the course of uplift. This generated power to force the gas to escape outward from the source rocks, and the gas can still largely migrate and accumulate. The existence of this process can enlarge the gas exploration realm to the "poor" gas accumulation area, i.e., gas reservoirs can still be found in the uplift area where gas reservoirs were formerly not believed to develop.

The gas discharged from different thickness coal seams in the two basins due to temperature and pressure drop in the course of moderate uplift at late stage can be

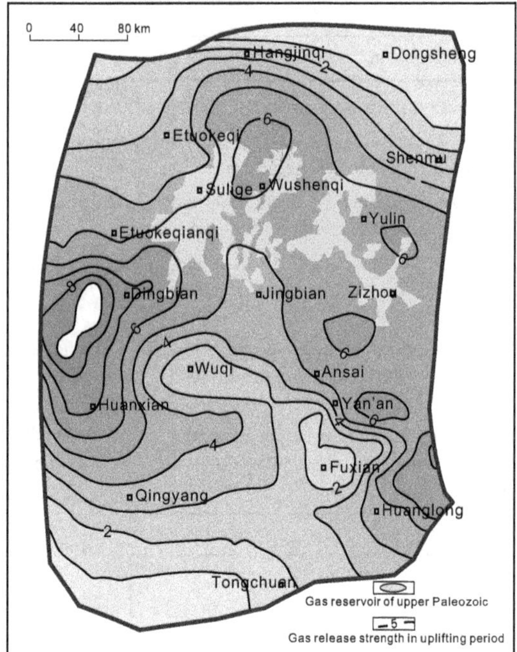

Fig. 5 Gas release strength in the uplift period of the Upper Paleozoic, Ordos Basin (unit: $\times 10^8$ m^3/km^2)

quantitatively estimated based on the simulation experiment results (Table 1) and the coal seam adsorption–desorption equations (Eqs. 2, 3). The strength of desorption and discharge of gas from coal seams in the uplift period is obtained based on calculation and statistics of data obtained from about 300 wells in the two basins, as shown in Figs. 5 and 6. Observed from these two figures, the gas expulsion area formed by pressure drop and desorption of coal seams of both the Ordos Basin and the Sichuan Basin in the uplift period can reach (15–18) $\times$ 10^4 km^2. The gas expulsion strength is basically (2–8) $\times$ 10^8 m^3/km^2, and the high value areas have corresponded well to the discovered gas fields, showing that the uplift period has provided important gas supply for large-scale gas accumulation in the two large gas provinces.

5 Geological evidence

5.1 Fluid inclusions

Fluid inclusions provide important means to study the gas accumulation process (Wang and Tian 2000; Rossi et al. 2002; Lu 2005). The study of inclusions from the Xujiahe Formation reservoir in the central Sichuan Basin (Fig. 7; Table 3) showed that two stages of hydrocarbon inclusions can be clearly identified based on their occurrence and fluorescent display characteristics. The first stage of hydrocarbon inclusions was developed in the early stage of quartz overgrowth, mainly occurring at the inner side of quartz overgrowth or along micro-fracture planes in the early diagenesis of quartz grains, orange-red or light brownish yellow in fluorescent light (upper part of Fig. 7),

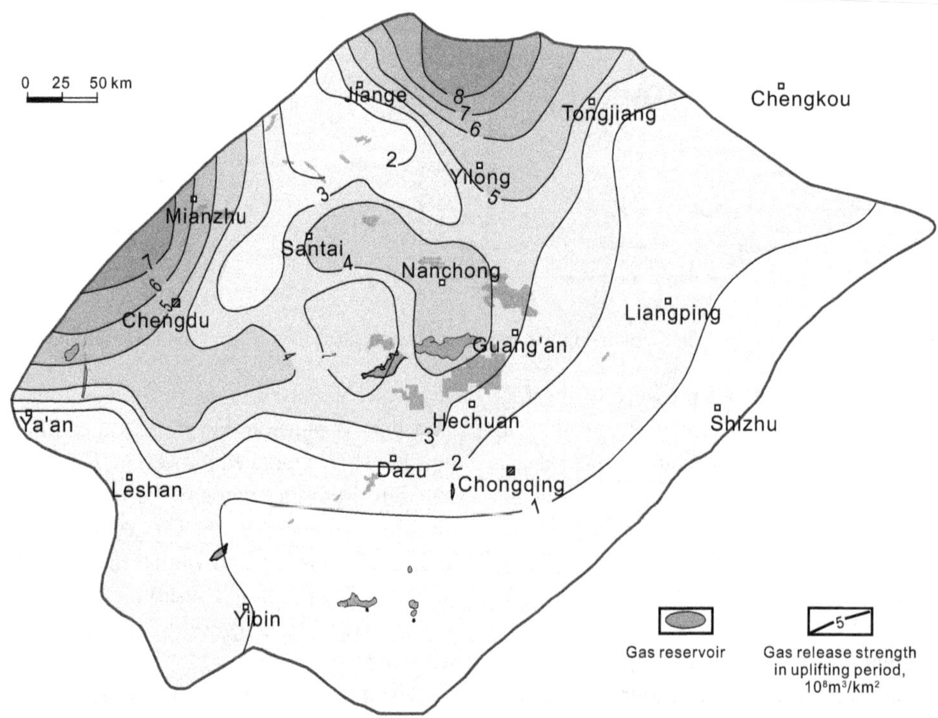

Fig. 6 Gas release strength in the uplift period of the Xujiahe coal measures and carbonaceous mudstones

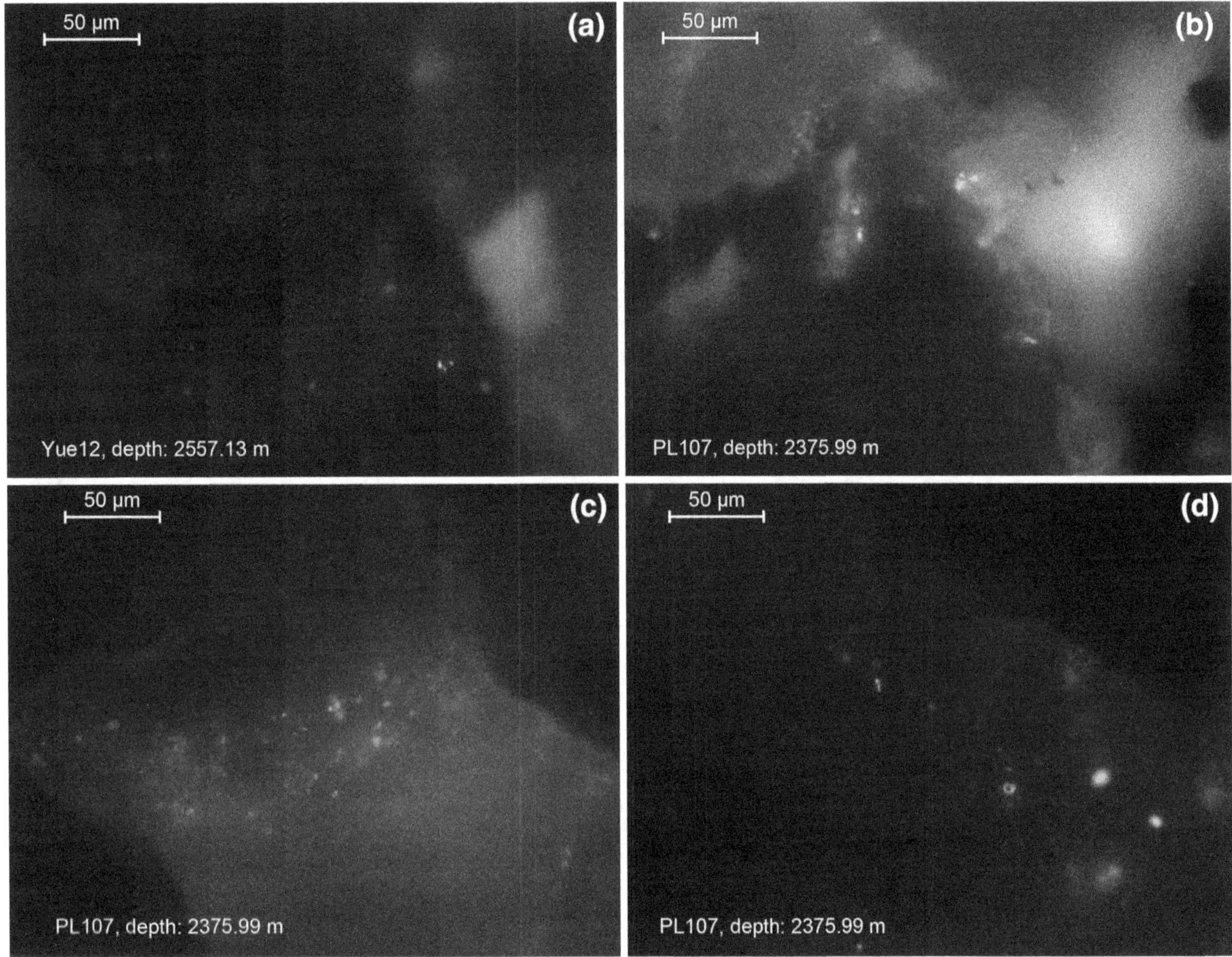

Fig. 7 Occurrence & fluorescent light features for fluid inclusions in Xu 2 Formation, Penglai area, Central Sichuan Basin. **a/b**-fluid inclusions in the inner side of quartz overgrowth or the micro-fissures, formed at early stage of diagenesis, *orange-red* or *light brown* in fluorescent light; **c/d**-fluid inclusions distributed in belts along micro-fissures cutting quartz grains, formed at the late stage of diagenesis, *light blue* or *light green* in fluorescent light

indicating that earlier heavy hydrocarbon exists in the inclusions, and the homogenization temperature peak of the associated brine inclusions is at 85–95 °C (Table 3). The second stage hydrocarbon inclusions were developed after the quartz overgrowth. They are mainly distributed in belts along post-diagenesis micro-fissures cutting quartz grains and are strong light blue and light bluish green in fluorescent light (lower part of Fig. 7). Laser Raman spectra showed that the main components of these inclusions are methane and higher hydrocarbons, and the homogenization temperature peak of the associated brine inclusions is between 110 and 130 °C (Table 3). This indicates that two stages of gas accumulation took place in the central Sichuan Basin. It is discovered by further analysis (Table 3) that the salinity of low temperature inclusions is lower (2 %–13 %), reflecting that the salinity of formation water was lower in this period, and they are the products of the early and middle stages of diagenesis when the source rocks started to become mature, generate substantial gas and enter the reservoir stage. The salinity of high-temperature inclusions is higher (16 %–22 %) and should be the record of the middle and late stages of diagenesis when the salinity increased with the substantial discharge of formation water, and the organic matter was discharged at mature and highly mature stages. The age of the former probably corresponds to the period before and after the end of the Jurassic and that of the latter corresponds to the uplift period occurring at the end of Cretaceous. It is discovered by studying the gas generation and expulsion history of Xujiahe Formation in central Sichuan Basin that there are a total of two stages of gas expulsion, migration, and accumulation: one stage occurred earlier, corresponding to the substantial gas generation period of gas source rocks, and the other stage occurred later, corresponding to tectonic uplift period. The two stages of gas accumulation are both characterized by large-scale

Table 3 Occurrence and test data of fluid inclusions from Xu2 Member, Penglai area

Occurrence in mineral deposit	Distribution pattern of inclusions	Type of fluid inclusions	Size, μm	Gas liquid ratio, %	Single phase	Homogenization temperature, °C	Salinity, wt%NaCl
At dust lane and inside of quartz overgrowth	Zonal	Hydrocarbon bearing brine inclusions	3 × 5	≤5	Liquid	89	2.90
	Zonal		4 × 7	≤5	Liquid	90	3.06
	Zonal		6 × 9	≤5	Liquid	92	3.06
	Zonal		3 × 3	≤5	Liquid	90	13.7
	Zonal		2 × 6	≤5	Liquid	89	13.6
	Zonal		3 × 4	≤5	Liquid	92	13.6
	Zonal		4 × 4	≤5	Liquid	90	13.7
	Zonal and lineal		10 × 12	≤5	Liquid	95	4.96
	Zonal and lineal		2 × 12	≤5	Liquid	94	5.71
	Zonal		26 × 15	≤5	Liquid	92	6.74
	Zonal		15 × 16	≤5	Liquid	92	6.88
	Zonal		6 × 8	≤5	Liquid	93	6.88
Along micro-fissures cutting quartz grains and overgrowth, formed at late stage of diagenesis	Zonal	Hydrocarbon bearing brine inclusions	2 × 6	≤5	Liquid	116	20.2
	Zonal		3 × 6	≤5	Liquid	118	20.2
	Zonal		1 × 7	≤5	Liquid	125	20.2
	Zonal		1 × 4	≤5	Liquid	125	20.2
	Zonal		3 × 7	≤5	Liquid	117	20.1
	Zonal and lineal		10 × 15	≤5	Liquid	128	16.9
	Zonal and lineal		2 × 3	≤5	Liquid	126	17.0
	Zonal and lineal		4 × 6	≤5	Liquid	130	17.0
	Zonal and lineal		3 × 4	≤5	Liquid	129	16.9
	Zonal		3 × 4	≤5	Liquid	115	22.4
	Zonal		4 × 5	≤5	Liquid	119	20.0
	Zonal		2 × 3	≤5	Liquid	116	20.0
	Zonal		2 × 10	≤5	Liquid	124	20.1
	Zonal		5 × 10	≤5	Liquid	129	20.1

accumulation, the former resulted from gentle structures in the central Sichuan Basin, whereas the latter resulted from overall tectonic uplift (Zhao et al. 2010) (Fig. 8).

5.2 Geochemical features

Experimental results show that with the decline of pressure, the carbon isotope ratios of hydrocarbon gas discharged from coal seams become lighter and lighter, and the $C_1/(C_1–C_5)$ coefficient becomes lower and lower. The gas generated at the late stage of formation burial is discharged first, having higher maturity, whereas the gas adsorbed in coal seams at early stage of formation burial is discharged last, having lower maturity (Table 4). The geochemical features of gas in the western Ordos Basin show that (Dai et al. 2005) the composition and carbon isotope ratios do not match the maturity of the coal measure source rocks. The lower carbon isotope ratio of methane and the lower $C_1/(C_1–C_5)$ coefficient constitute a contradiction with the higher maturity of source rocks. For instance, the maturity (R_o %) of Permo-Carboniferous coal seams in the Sulige region exceeds 2.0 %; however, the gas in some regions is wetter, with the $C_1/(C_1–C_5)$ coefficient being up to 86 %, and the carbon isotope ratio of methane is lighter, −29.96 to −36.45 ‰, showing that the gas accumulated in the Upper Paleozoic gas field mainly originates from adsorbed gas desorption and free gas expansion in the source kitchen in the uplift period, whereas the gas adsorbed inside source rocks mainly comes from the early and middle mature stage of source rocks, having lower maturity. Therefore, it is normal for it to be different from the current maturity of the source rocks. Moreover, such carbon isotope lightening gradually becomes apparent from the Sulige gas field in the west to the Yulin gas field in the east of the Ordos Basin, which is significantly related to the fact that the strata uplift gradually increased from 800 m in the west to 1400 m in the east at the late stage of the basin tectonics. This is because within a certain range, the larger the uplift, the more gas is desorbed from the coal measure, resulting in a large proportion of gas accumulated in the gas field in the uplift period (Fig. 9).

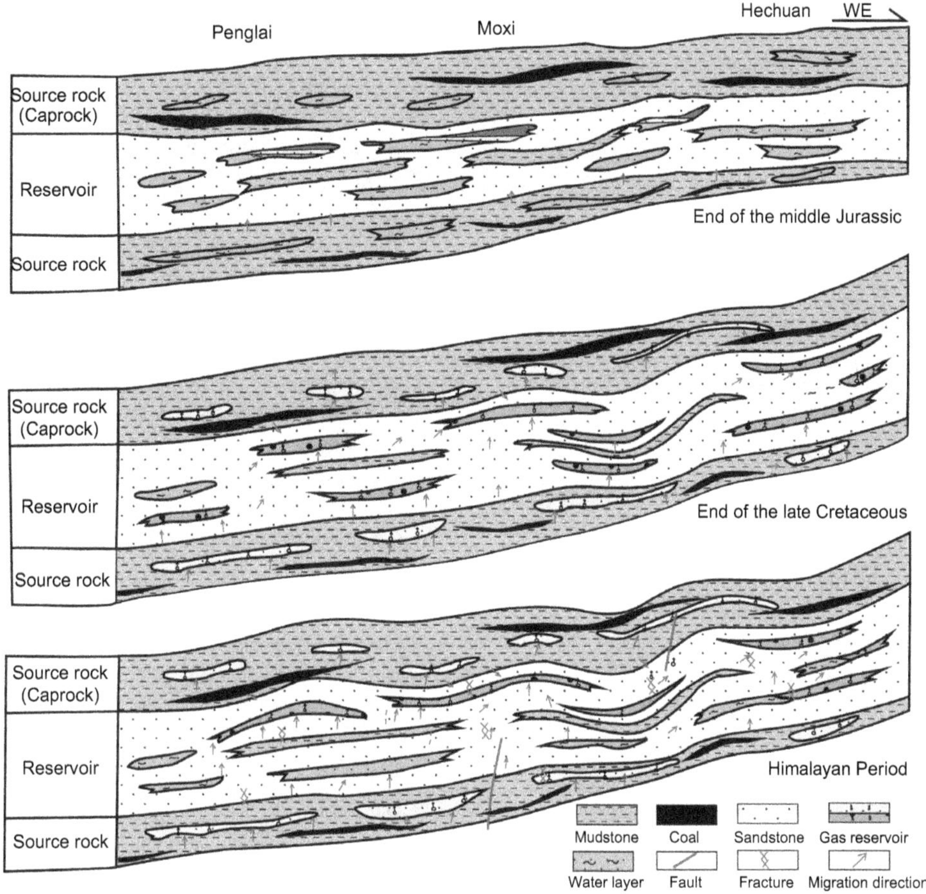

Fig. 8 Gas accumulation model of the Xujiahe Formation in the central Sichuan Basin

Table 4 Composition and carbon isotope of gas discharged by coal under different temperature and pressure conditions

Coal sample	Process	Temperature, °C	Gas yield, mL/g	Gas, %	C_1, %	C_{2+}, %	$C_1/(C_1–C_5)$ coefficient	$\delta^{13}C_1$, ‰	$\delta^{13}C_2$, ‰
Liaohe lignite	Heating and pressurizing	450	19.4	6.9	3.7	3.2	0.6	−34.4	−25.7
	Cooling and depressurizing	350		17.5	8.7	8.8	0.5	−40.6	−29.9
Xianfeng lignite	Heating and pressurizing	450	79.5	40.1	22.8	17.3	0.6	−31.8	−26.0
	Cooling and depressurizing	350		9.2	6.2	3	0.7	−34.5	−27.4
Taiyuan Formation	Heating and pressurizing	450	60.8	72.7	54.6	18.2	0.8	−30.1	−24.5
	Cooling and depressurizing	350		68.2	48.9	19.3	0.7	−34.6	−24.6
Shanxi Formation	Heating and pressurizing	450	70.25	76.3	61.6	14.7	0.8	−31.5	−23.0
	Cooling and depressurizing	350		71	59.4	11.6	0.8	−33.4	−23.8

This also suggests that the composition and carbon isotope ratios of the Xujiahe Formation gas in the central Sichuan Basin are inconsistent with the maturity of the source rocks. For instance, the carbon isotope ratio of the Xu2 Member methane in the Hechuan region is lighter (−39 to −42 ‰), but the maturity R_o of the lower coal measure source rocks is 1.1–1.3, belonging to the substantial gas generation stage of coal measure source rocks. Based on the statistical data from Dai (1992), the carbon isotope ratio of methane should be −32 to −36 ‰. In addition, the carbon isotopes of ethane and propane are also characterized by lightening, only the lightening amplitude decreases gradually. The experimental results show that such gas with low maturity was possibly formed and stored in the coal measure source rocks at the early stage, but it was discharged, became free gas, and accumulated due to tectonic uplift at the late stage (Fig. 9).

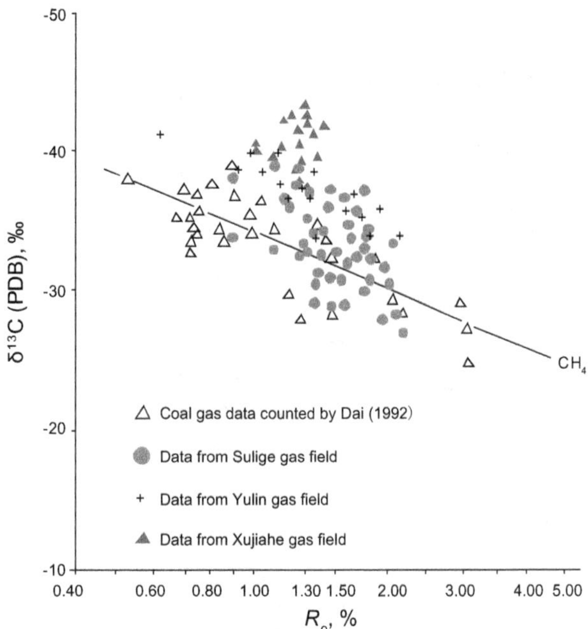

Fig. 9 Carbon isotope ratio versus R_O of Xujiahe Formation gas in the Sichuan Basin and upper Paleozoic gas in the Ordos Basin and typical coal-formed gas

6 Main gas accumulation pattern and geological significance of uplift period accumulation

It was generally believed that diffusion was one of the major factors damaging gas reservoirs, and studies show that diffusion is an important mode for gas to migrate with molecular motion (Nelson and Simmons 1992; Lu and Connell 2007; Lu et al. 2008; Korrani et al. 2012). With increasing study of low porosity and permeability reservoir gas, especially on tight gas, diffusion migration is believed as an important way for the gas to migrate and accumulate in tight reservoirs (Liu et al. 2012; Wang et al. 2014). (Liu et al. 2012) point out that the bulk flow under source–reservoir pressure differentials and the diffusion flow under hydrocarbon concentration differential are two important ways for gas migration and infusion. Experiments and geological analysis show that at the stage of strata uplift, gas generation in source rocks stopped, and the source–reservoir pressure differential dropped gradually. The combined action of desorption of coal seam adsorbed gas and expansion of free gas in the original pores of the coal seams significantly increased the gas concentration inside the source rocks and provided power for the gas to diffuse and migrate from source rocks to reservoirs.

On the basis of geological analysis of the Xujiahe Formation gas reservoirs in the Sichuan Basin, we calculated bulk flow charge, diffusion flow charge, and diffusion loss in tight sand gas reservoirs. The results show that in the strata uplift period, the average charge rate of diffusion flow reached 0.8×10^7 $m^3/(km^2 \cdot Ma)$, and the charge volume approached 56×10^{12} m^3, whereas the bulk flow charge of gas mainly occurred in the burial period of the basin, with a charge volume of 127×10^{12} m^3, and in the whole gas generation and accumulation history, the gas diffusion loss volume was estimated to be 135×10^{12} m^3. As a result, the gas diffusion loss cannot be compensated only by bulk flow charge, and it is hard to form the discovered Xujiahe Formation TCM scale gas field. Therefore, the diffusion charge in the uplift period of formation effectively compensates for the diffusion loss of gas and contributes more to the efficient gas accumulation and preservation of large tight sand gas fields.

The concept of gas accumulation in the uplift period has important theoretical and practical significances. Firstly, it breaks the conventional view that the uplift period is unfavorable for gas accumulation, and secondly it promotes the gas resource extent and potential of the tectonic uplift area. On this basis, many large uplift areas formerly believed to be unfavorable for exploration are reassessed, and the area of favorable gas exploration provinces has been increased.

7 Discussion

On the basis of analyzing the coal seam adsorbed and desorbed gas model, this paper uses a great deal of thermal simulation and physical simulation experiments to demonstrate the geological process of gas accumulation in the uplift period of coal-bearing formations. As well, a lot of fluid inclusions and geochemical evidence are available for the analysis of real gas fields, showing that recognition of gas accumulation due to pressure drop and desorption from coal-bearing formations in the uplift period has important theoretical and practical bases and is of great significance to promoting the resource potential of coal-bearing strata uplift areas. However, the high-pressure (>20 MPa) coal seam adsorbed and desorbed gas model needs further study, for the adsorption state of methane at high pressure has possibly changed. The ordinary coalbed methane adsorption and desorption simulation experiments are mainly undertaken at pressures of less than 20 MPa, and it is important to conduct higher pressure (20–40 MPa) adsorption–desorption gas tests in the future.

8 Conclusions

(1) The adsorption–desorption principle of coal seam gas emission and simulation experiment results confirms that pressure drop and desorption occur in the uplift process of coal measure formation, and the

substantial free gas discharged becomes the gas source in the uplift period. Gas supply strength is estimated as up to (3–6) $\times$ 10^8 m^3/km^2 in the Upper Paleozoic of the Ordos Basin and the Xujiahe Formation of the Sichuan Basin, which significantly increases the resource potential of tight sand gas. The diffusion charge of desorbed gas in the course of coal measure source rock uplift can compensate for the loss of gas due to diffusion after gas accumulation and contributes more to the formation of large gas fields.

(2) The carbon isotope ratios of gas discharged in the uplift period are lighter than those of gas formed in the burial period. High-temperature and high-salinity inclusions and the physical simulation experiment results obtained under nearly real geological conditions all confirm that large scale gas accumulation can occur in the uplift period of coal measure strata due to pressure drop, desorption, and gas expansion.

(3) This understanding enlarges the exploration potential of tight sand gas. Further geological study needs to be strengthened in regions where the stratigraphic uplift amplitude is large and that were formerly believed to be unfavorable gas accumulation conditions, so as to find more gas resources and reserves.

References

Anderson RB, Bayer J, Hofer LJE. Equilibrium sorption studies of methane on Pittsburgh Seam and Pocahontas No. 3 Seam Coal. Coal Sci. 1966;386–99.

Ayers WB. Coalbed gas systems, resources, and production and a review of contrasting cases from the San Juan and Powder River Basins. AAPG Bull. 2002;86:1853–90.

Bae JS, Bhatia SK. High-pressure adsorption of methane and carbon dioxide on coal. Energy Fuels. 2006;20(6):2599–607.

Bian CS, Wang HJ, Wang ZC, et al. Controlling factors for massive accumulation of natural gas in the Xujiahe Formation in central Sichuan Basin. Oil Gas Geol. 2009;30(5):548–55 (in Chinese).

British Petroleum Company. BP statistical review of world energy 2012. London: British Petroleum Company; 2012.

Busch A, Krooss BM, Gensterblum Y, et al. High-pressure adsorption of methane, carbon dioxide and their mixtures on coals with a special focus on the preferential sorption behavior. J Geochem Explor. 2003;78–79:671–4.

Clarkson CR, Bustin RM. The effect of pore structure and gas pressure upon the transport properties of coal: a laboratory and modeling study. 1. Isotherms and pore volume distributions. Fuel. 1999;78(11):1333–44.

Crosdale PJ, Beamish BB, Valix M. Coalbed methane sorption related to coal composition. Int J Coal Geol. 1998;35:147–58.

Cui YJ, Li YH, Zhang Q, et al. Characteristic curves of coal adsorbing methane and their application to coalbed methane gathering study. Chin Sci Bull. 2005;50(supplement I):76–81 (in Chinese).

Dai JX. Identification of various genetic natural gases. China Offshore Oil Gas (Geol). 1992;6(1):11–9 (in Chinese).

Dai JX, Li J, Luo X, et al. Alkane carbon isotopic composition and gas source in giant gas fields of Ordos Basin. Acta Petrolei Sin. 2005;26(1):18–26 (in Chinese).

Dai JX, Ni YY, Wu XQ. Tight gas in China and its significance in exploration and exploitation. Pet Explor Dev. 2012;39(3): 257–64 (in Chinese).

Gregory JB, Karen CR. Hysteresis of methane/coal sorption isotherms. In: SPE 15454, 1986.

Hao SS, Cheng ZM, Lv YF, et al. Natural gas reservoir formation and preservation. Beijing: Petroleum Industry Press; 1995 (in Chinese).

Haydel JH, Kobayashi R. Adsorption equilibria in the methane-propane-silica gel system at high pressures. Ind Eng Chem Fundam. 1967;6(4):546–54.

He DB, Jia AL, Ji G, et al. Well type and pattern optimization technology for large scale tight sand gas, Sulige gas field. Pet Explor Dev. 2013;40(1):257–64 (in Chinese).

Holditch SA. Tight gas sands. J Pet Technol. 2006;58:86–94.

Johnson RC, Rice DD. Occurrence and geochemistry of natural gases, Piceance Basin, Northwest Colorado. AAPG Bull. 1990;74(6): 805–29.

Korrani AKN, Gerami S, Ghotbi C, et al. Investigation on the importance of the diffusion process during lean gas injection into a simple synthetic depleted naturally fractured gas condensate reservoir. Pet Sci Technol. 2012;30(7):655–71.

Kuuskraa VA, Bank GC. Gas from tight sands, shales: a growing share of US supply. Oil Gas J. 2003;101(47):34–43.

Law BE. Basin-centered gas systems. AAPG Bull. 2002;86(11): 1891–919.

Li JZ, Guo BC, Zheng M, et al. Main types, geological features and resource potential of tight sandstone gas in China. Nat Gas Geosci. 2012;23(4):607–15 (in Chinese).

Liu GD, Zhao ZY, Sun ML, et al. New insights into natural gas diffusion coefficient in rocks. Pet Explor Dev. 2012;39(5): 559–66 (in Chinese).

Lu HZ. Fluid inclusions. Beijing: Science Press; 2005 (in Chinese).

Lu M, Connell LD. A model for the flow of gas mixtures in adsorption dominated dual porosity reservoirs incorporating multi-component matrix diffusion: Part I. Theoretical development. J Pet Sci Eng. 2007;59(1–2):17–26.

Lu M, Connell LC, Pan ZJ. A model for the flow of gas mixtures in adsorption dominated dual-porosity reservoirs incorporating multi-component matrix diffusion—Part II Numerical algorithm and application examples. J Petrol Sci Eng. 2008;62(3–4):93–101.

Masters JA. Deep basin gas trap, western Canada. AAPG Bull. 1979;63(2):152.

Mavor MJ, Owen LB, Pratt TJ. Measurement and evaluation of coal sorption isotherm data. In: SPE annual technical conference and exhibition, SPE-20728-MS, New Orleans, Louisiana, 23–26 September 1990.

Mukhopadhyay PK, Macdonald DJ. Relationship between methane/generation and adsorption potential, micropore system, and permeability with composition and maturity-examples from the Carboniferous coals of Nova Scotia, Eastern Canada. In: Proceeding of the 1997 coalbed methane symposium, Tuscaloosa, 1997, p. 183–93.

Murata K, El-Merraoui M, Kaneko K. A new determination method of absolute adsorption isotherm of supercritical gases under high pressure with a special relevance to density-functional theory study. J Chem Phy. 2001;114:4196–205.

Nelson JS, Simmons EC. The quantification of diffusive hydrocarbon losses through cap rocks of natural gas reservoirs—a reevaluation, discussion. AAPG Bull. 1992;76(11):1839–41.

Radovic LR, Menon VC, Leon LY, et al. On the porous structure of coals: evidence for an interconnected but constricted micropore system and implications for coalbed methane recovery. Adsorption. 1997;3(3):221–32.

Rossi C, Goldstein RH, Ceriani A, et al. Fluid inclusions record thermal and fluid evolution in reservoir sandstones, Khatatba Formation, Western Desert, Egypt: a case for fluid injection. AAPG Bull. 2002;86(10):1773–99.

Ruppel TC, Grein CT, Bienstock D. Adsorption of methane on dry coal at elevated pressure. Fuel. 1974;53:152.

Sang SX, Zhu YM, Zhang J. Solid-gas interaction mechanism of coal-absorbed gas (II)—Physical process and theoretical model of coal-adsorbed gas. Nat Gas Ind. 2005;25(1):16–8 (in Chinese).

Smith TM, Sayers CM, Sondergeld CH. Rock properties in low-porosity/low-permeability sandstones. Lead Edge. 2009;28(1): 48–59.

Su XB, Chen R, Lin XY, et al. Application of adsorption potential theory in the fractionation of coalbed gas during the process of adsorption/desorption. Acta Geol Sin. 2008;82(10):830–7 (in Chinese).

Tian FH, Jiang ZX, Zhang XB, et al. Preliminary study on contribution of rift-erosion to oil and gas accumulation. Acta Geol Sin. 2007;81(2):273–9 (in Chinese).

Tissot BP, Welte DH. Petroleum formation and occurrence. New York: Springer; 1984.

Wang HJ, Tian YC. Applications of general characteristics of fluid inclusions to the study of oil and gas pool formation. Pet Explor Dev. 2000;27(3):50–2 (in Chinese).

Wang HJ, Bian CS, Liu GD, et al. Two highly efficient accumulation models of large gas fields in China. Pet Sci. 2014; 11(1):21–31.

Wang SW, Chen ZH, Zhang M. Pore and microfracture of coal matrix block and their effects on the recovery of methane from coal. Earth Sci. 1995;20(5):557–61 (in Chinese).

Yang RT, Saunders JT. Adsorption of gases on coals and heat-treated coals at elevated temperature and pressure. Fuel. 1985;64: 616–25.

Zeng JH, Jin ZJ. The physical simulation of petroleum second migration and accumulation. Beijing: Petroleum Industry Press; 2002. p. 25–7.

Zeng LB. Microfracturing in the Upper Triassic Sichuan Basin tight-gas sandstones: tectonic, overpressure, and diagenetic origins. AAPG Bull. 2010;94(12):1811–25.

Zhang CL, Li TR, Xiong QH. Relationship between lithostructure and fracture distribution in coal. Coal Geol Explor. 2000;28(5): 26–30 (in Chinese).

Zhang E, Hill RJ, Katz BJ, et al. Modeling of gas generation from the Cameo coal zone in the Piceance Basin, Colorado. AAPG Bull. 2008;92(8):1077–106.

Zhang HF, Fang CL, Gao XZ, et al. Petroleum geology. Beijing: Petroleum Industry Press; 1999 (in Chinese).

Zhang LP, Bai GP, Luo XR, et al. Diagenetic history of tight sandstones and gas entrapment in the Yulin Gas Field in the central area of the Ordos Basin, China. Mar Pet Geol. 2009a;26(6):974–89.

Zhang SC, Mi JK, Liu LH, et al. Geological features and formation of coal-formed tight sandstone gas pools in China: cases from Upper Paleozoic gas pools, Ordos Basin and Xujiahe Formation gas pools, Sichuan Basin. Pet Explor Dev. 2009b;36(3):320–30 (in Chinese).

Zhao WZ, Bian CS, Xu CC, et al. Assessment on gas accumulation potential and favorable plays within the Xu-1, 3 and 5 Members of Xujiahe Formation in Sichuan Basin. Pet Explor Dev. 2011;38(4):1–12 (in Chinese).

Zhao WZ, Hu SY, Wang HJ, et al. Large-scale accumulation and distribution of medium-low abundance hydrocarbon resources in China. Pet Explor Dev. 2013;40(1):1–14 (in Chinese).

Zhao WZ, Wang HJ, Shan JZ, et al. Geological analysis and physical modeling of structural pumping in high effective formation of Kela 2 gas field. Sci China Ser D Earth Sci. 2006;49(10):1070–8.

Zhao WZ, Wang HJ, Xu CC, et al. Reservoir-forming mechanism and enrichment conditions of the extensive Xujiahe Formation gas reservoirs, central Sichuan Basin. Pet Explor Dev. 2010;37(2): 146–57 (in Chinese).

Zhao WZ, Wang ZC, Zhu YX, et al. Forming mechanism of low-efficiency gas reservoir in Sulige Gas Field of Ordos Basin. Acta Pet Sin. 2005;26(5):5–9 (in Chinese).

Zhao ZG, Tang XY, Zhang GM. Experiment and significance of isothermal adsorption of coal on methane under higher temperature. Coal Geol Explor. 2001;29(4):29–31 (in Chinese).

Zou CN, Zhu RK, Liu KY, et al. Tight gas sandstone reservoirs in China: characteristics and recognition criteria. J Pet Sci Eng. 2012;88–89:82–91.

12

Preparation and performance of fluorescent polyacrylamide microspheres as a profile control and tracer agent

Wan-Li Kang[1,2] · Lei-Lei Hu[1] · Xiang-Feng Zhang[1] · Run-Mei Yang[1] · Hai-Ming Fan[1] · Jie Geng[1]

Abstract Polyacrylamide microspheres have been successfully used to reduce water production in reservoirs, but it is impossible to distinguish polyacrylamide microspheres from polyacrylamide that is used to enhance oil recovery and is already present in production fluids. In order to detect polyacrylamide microspheres in the reservoir produced fluid, fluorescent polyacrylamide microspheres P(AM-BA-AMCO), which fluoresce under ultraviolet irradiation, were synthesized via an inverse suspension polymerization. In order to keep the particle size distribution in a narrow range, the synthesis conditions of the polymerization were studied, including the stirring speed and the concentrations of initiator, Na_2CO_3, and dispersant. The bonding characteristics of microspheres were determined by Fourier transform infrared spectroscopy. The surface morphology of these microspheres was observed under ultraviolet irradiation with an inverse fluorescence microscope. A laboratory evaluation test showed that the fluorescent polymer microspheres had good water swelling capability, thus they had the ability to plug and migrate in a sand pack. The plugging rate was 99.8 % and the residual resistance coefficient was 800 after microsphere treatment in the sand pack. Furthermore, the fluorescent microspheres and their fragments were accurately detected under ultraviolet irradiation in the produced fluid, even though they had experienced extrusion and deformation in the sand pack.

✉ Wan-Li Kang
kangwanli@126.com

[1] School of Petroleum Engineering, China University of Petroleum, Qingdao 266580, Shandong, China

[2] EOR Research Institute, China University of Petroleum (Beijing), Beijing 100249, China

Edited by Yan-Hua Sun

Keywords Inverse suspension polymerization · Fluorescence · Polyacrylamide microsphere · Narrow size distribution · Profile control performance

1 Introduction

Polyacrylamide microspheres have been successfully applied to reduce water production and enhance oil recovery in mature reservoirs with well-developed fractures or ultra-high permeability streaks/channels (Yao et al. 2012, 2013; Hua et al. 2013; Li et al. 2014). However, it is impossible to distinguish polyacrylamide microspheres from polyacrylamide that was used to enhance oil recovery and was already present in production fluids.

Many inter-well tracers have been widely used to monitor the displacement into reservoirs and to obtain information on the interaction between producer and injector. These include chemical tracers (Wang et al. 2004b), nuclear tracers (Bjørnstad et al. 1990; Lien et al. 1988) and isotope tracers (Selby et al. 2007; Osborn and McIntosh 2010). However, some chemical tracers are easily absorbed onto the surface of rocks in the stratum, and the instruments required to detect radioactive isotopes are extremely expensive.

In recent years, fluorescence-detection technology has attracted great interest in oil field operations because of its convenience, high accuracy, and low cost. Wang et al. (2004a) used fluorescein as a tracer to monitor inter-well connectivity. Because the cost of the potassium iodide is ¥200,000 per year, and the cost of the fluorescein is only ¥15,000 per year. Yu et al. (2009) used fluorescein and calcein as inter-well tracers. The fluorescent method was simple and convenient with high precision and accuracy. The detection ranges for fluorescein and calcein were 0–0.35 and 0–0.89 mg/L, respectively, with a measurement error within 2.5 %. Yan et al. (2012) synthesized a fluorescent tracer copolymer (AM-AMCO) by emulsion polymerization of acrylamide and coumarin in the presence of $(NH_4)_2S_2O_8$–$NaHSO_3$ as a redox initiator. Characterization of the copolymer (AM-AMCO) indicated that fluorescent polyacrylamide can be used as an oilfield tracer and has good heat and salt resistance.

Coumarin, a benzopyrone, which has a large Stokes shift, high fluorescence quantum yield, and excellent light stability, is an attractive choice as a fluorescent candidate. Its derivatives have been widely used as fluorescent probes (Kalyanaraman et al. 2012), fluorescent indicator (Suzuki et al. 2002), and as fluorescent dye (Koops et al. 2010). In this study, 7-(allyloxy)-4-methyl-2*H*-chromen-2-one (AMCO), a derivative of coumarin, was selected as the fluorescent monomer grafted onto a water-soluble polymer to form fluorescent polyacrylamide microspheres.

Various methods can be used to synthesize polymer microspheres, such as emulsion polymerization (Chern 2006; Thickett and Gilbert 2007), soap-free emulsion polymerization (Nagao et al. 2006; Nakabayashi et al. 2013), dispersion polymerization (Covolan et al. 2000), precipitation polymerization (Chen et al. 2012; Bai et al. 2004), inverse-microemulsion polymerization (Hemingway et al. 2010), seeded polymerization (Zhang et al. 2011; Li et al. 2013), suspension polymerization (Liu et al. 2011; Zhao and Qiu. 2011), seeded dispersion polymerization (Tokuda et al. 2013), and inverse suspension polymerization (Annaka et al. 2003).

The inverse suspension polymerization method has many advantages. These are the heat of reaction is easily excluded from the reaction system; a granular product can be obtained directly without a crushing process; the product is easy to dry and has good water absorption performance. Recently, inverse suspension polymerization has broad application prospects in China (Guo et al. 2001; Tian et al. 2012). In order to obtain micron-sized microspheres with a narrow size distribution, inverse suspension polymerization was selected.

However, preparation conditions of the inverse suspension polymerization can affect the average particle size, and the average particle size will affect the profile control behavior of microspheres. Thus, the effect of preparation conditions on the average particle diameter needs to be discussed, including the stirring speed, the concentrations of initiator and dispersant.

In addition, the polyacrylamide microspheres cut excess water production by profile modification and blocking of paths with high permeability to water and enhance oil recovery (EOR). So the swelling properties and the profile control performance of the fluorescent polyacrylamide microspheres were also investigated. As fluorescent polymer microspheres have the ability to fluoresce under ultraviolet irradiation; this can be widely used to detect the microspheres in the produced fluid.

2 Experimental

2.1 Materials

AMCO was synthesized by the method of Yan et al. (2012). Acrylamide (AM, a monomer), sorbitan monostearate (Span-60, an oil phase dispersant), anhydrous sodium carbonate (Na_2CO_3, an electrolyte), *N*, *N′*-methylenebisacrylamide (MBA, a cross-linking agent), ammonium persulfate (APS, water-soluble initiator), cyclohexane (a continuous phase), anhydrous ethanol, and other reagents used were all of analytical-reagent grade and commercially available. The total salinity of the brine used was 6726 mg/L, and its main components are listed in Table 1. The brine viscosity was 1 mPa s.

2.2 Methods

2.2.1 Synthesis and purification of fluorescent microspheres

Inverse suspension polymerization of fluorescent microspheres was carried out in a 250 mL four-neck flask. The flask was equipped with a mechanical stirrer, a reflux condenser, and a nitrogen inlet, and was incubated in a water bath. In inverse suspension polymerization, cyclohexane was used as a continuous phase and Span-60 as a nonionic polymeric surfactant. A synthesis route for fluorescent microspheres is shown in Scheme 1. Experimental procedures are as follows: (1) 0.07 g MBA, 14.06 g AM, and Na_2CO_3 were dissolved in 40 mL of distilled water. (2) This solution was then immediately poured into 60 mL of cyclohexane containing 0.014 g AMCO and Span-60, which was previously purged with dry nitrogen in the flask. (3) Air was flushed from the reactor by introducing nitrogen until the entire process was completed. (4) After aqueous droplets were observed in the continuous phase, 10 mL APS-water solution was added to the continuous phase to initiate polymerization. (5) Polymerization was allowed to proceed for 4 h at 50 °C. (6) Then stirring was stopped and the precipitate was washed with a large amount of methanol and filtered with qualitative filter paper (the pore size was about 1–3 μm). The product was dried in a vacuum oven at 45 °C for one day, and the obtained powder was fluorescent microspheres P(AM-BA-

Table 1 Brine components

Ion	$K^+ + Na^+$	Mg^{2+}	Ca^{2+}	Cl^-	SO_4^{2-}	HCO_3^-	CO_3^{2-}	Total salinity
Concentration, mg/L	2043	36	39	1337	10	3126	135	6726

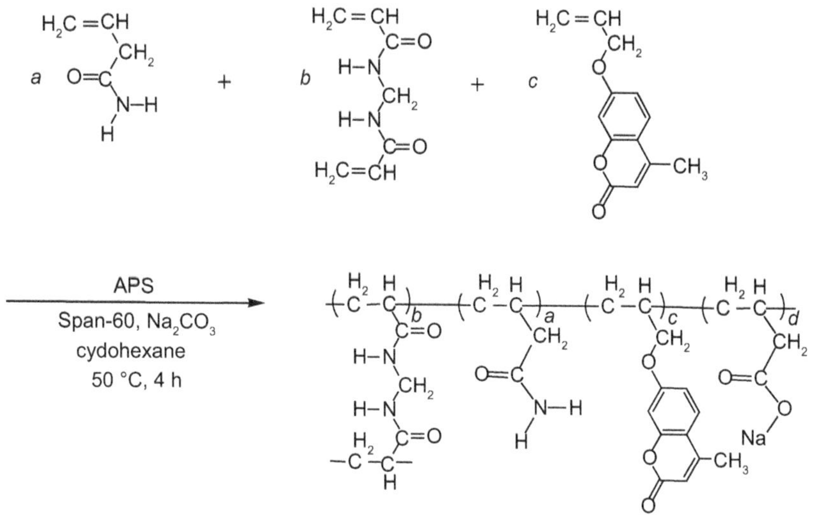

Scheme 1 Synthesis route of P(AM-BA-AMCO)

AMCO). The microspheres P(AM-BA) were prepared in the same way as above, without AMCO monomer in the reaction system.

2.2.2 Microsphere characterization

Fourier transform infrared spectroscopy (FTIR) spectra of P(AM-BA-AMCO), P(AM-BA), and acrylamide monomer were measured with a Nicolet model NEXUS670 spectrometer with samples prepared as KBr pellets. The surface morphology of the microspheres was observed in the DAPI fluorescence channel with an inverse fluorescence microscope (Leica DMI 3000B, Germany) at 25 °C.

The average particle size of microspheres, which is affected by different factors (i.e., stirring speed, concentrations of the initiator, dispersants, and Na_2CO_3) was determined at room temperature with a RISE-2006 laser particle size analyzer (Jinan Runzhi Technology Co., Ltd, China).

To measure the swelling ratio of microspheres, a tea bag (i.e., a 100 mesh nylon screen) containing pre-weighed dry samples was immersed entirely in the brine (with the composition given in Table 1). After being hydrated in brine for different times, the swollen microspheres were allowed to drain by removing the teabag from the brine and the excess surface water was removed with filter paper. The swelling ratio, S_w was calculated from the following equation:

$$S_w = (m_1 - m_0)/m_0, \quad (1)$$

where m_0 and m_1 are the weight of the dry and swollen microspheres, respectively.

2.2.3 Core flow experiment

Core flow experiments were designed to evaluate the effect of P(AM-BA-AMCO) on profile control. The core flow apparatus consisted of a constant-flux pump, a pressure sensor, and a sand pack ($d = 2.5$ cm, $L = 60$ cm, 20–50 mesh sand) (Fig. 1).

The core flow experiments were conducted conventionally at a flow rate of 5 mL/min at 45 °C. The sand pack was saturated with brine (1 PV) until the injection pressure stabilized. Then, 35 PV microsphere solution (2000 mg/L) was injected into the sand pack, followed by 25 PV brine. The differential pressures (ΔP, given in units of MPa) were measured during the whole injection process. The permeability of the sand pack in the displacement process (k, given in units of μm^2), and the plugging rate (η, expressed in units of %) were calculated using the following equations (Qin and Li 2006; Lei 2011):

$$k = \frac{Q\mu L}{10A\Delta P}, \quad (2)$$

$$\eta = \frac{k_0 - k'}{k_0} \times 100\%, \quad (3)$$

where Q is the injection rate, mL/s; μ the fluid viscosity, mPa s; L the length of the sand pack, cm; A is the sectional area of the sand pack, cm^2, $A = \pi d^2/4$; ΔP is the

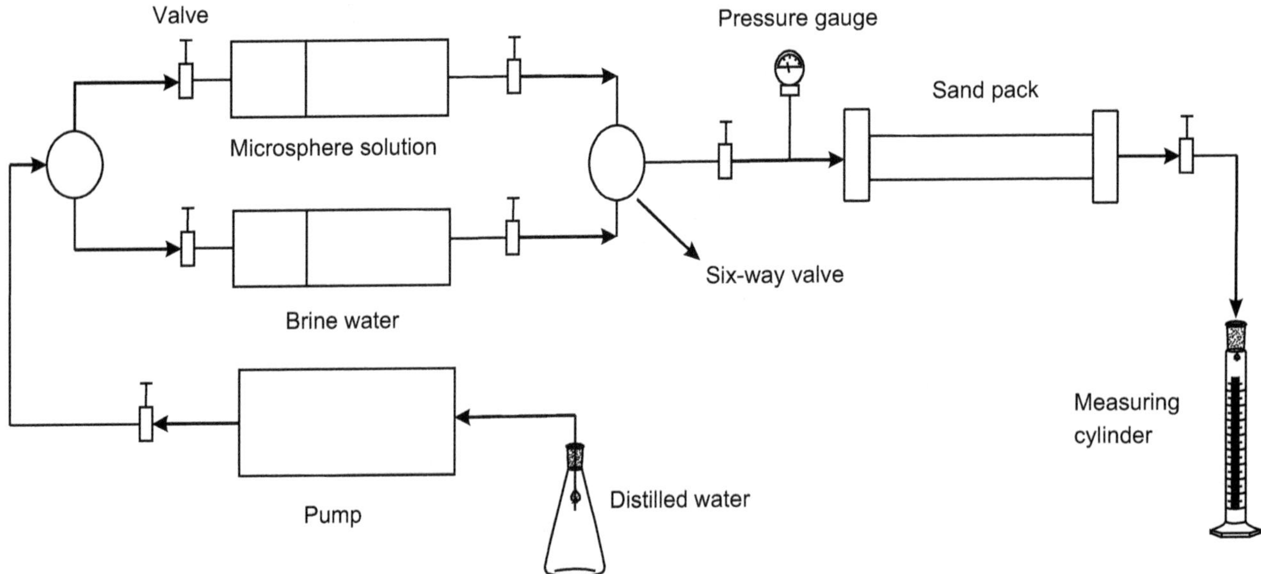

Fig. 1 Schematic illustration of the flow system

differential pressure, MPa; k_0 is the original permeability of the sand pack before microsphere treatment, μm^2; k' is the permeability of the sand pack after microsphere treatment, μm^2.

The residual resistance factor, F_{rr}, is a measure of permeability reduction and is determined from laboratory data using the following equation:

$$F_{rr} = \frac{(\Delta P/Q)_{\text{after microsphere flow}}}{(\Delta P/Q)_{\text{before microsphere flow}}}. \quad (4)$$

In this study the injection flow rate is kept constant before and after microsphere treatment, the Eq. (5) may be expressed as

$$F_{rr} = \frac{(\Delta P)_{\text{after microsphere flow}}}{(\Delta P)_{\text{before microsphere flow}}}. \quad (5)$$

3 Results and discussion

3.1 FTIR spectrum of polyacrylamide microspheres

Figure 2 shows FTIR spectra of polyacrylamide microspheres and AM. A double peak at 3100–3500 cm^{-1} in the spectrum of AM, which is attributed to N–H stretching vibrations, disappeared in the spectra of P(AM-BA) and P(AM-BA-AMCO); whereas a peak at 3400 cm^{-1}, which is due to N–H stretching vibrations, was observed in the spectra of P(AM-BA) and P(AM-BA-AMCO). This indicates that acrylamide units exist within the structure of P(AM-BA) and P(AM-BA-AMCO). Furthermore, the C=O bending vibration was shifted

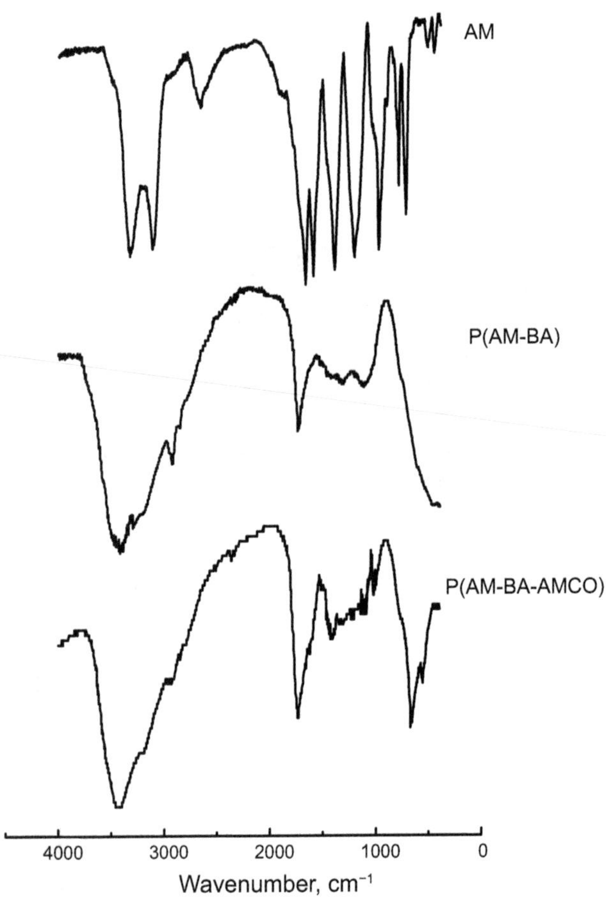

Fig. 2 A comparison of IR spectra of AM monomer, P(AM-BA), and P(AM-BA-AMCO)

from 1600 to 1740 cm^{-1}, which indicates that amide was partially hydrolyzed to carboxyl (Yang et al. 2007). A peak at 670 cm^{-1} is attributed to bending vibrations of aromatic hydrogen. The AMCO monomer which was connected to a side chain was at very low concentration. Therefore, other characteristic peaks of AMCO monomers are weak, such as from the benzene ring skeleton, methyl, and methylene.

3.2 Factors contributing to the average particle size of microspheres

We have found that with large amounts of monomers and high reaction temperature, bulk polymerization often took place in a very turbulent and even explosive form. It also has been found that a large amount of cross-linker could result in a reduction in swelling capability and mechanical toughness (Xia et al. 2003; Haraguchi et al. 2005). Therefore, the effect of reaction temperature and the concentrations of AM, MBA monomers were not investigated in this work.

3.2.1 Stirring speed

Figure 3a shows the effect of stirring speed on the average particle size of microspheres. The average particle size of microspheres decreased from 260 to 125 μm when the stirring speed increased from 150 to 300 r/min, but when the stirring speed increased from 300 to 500 r/min, the average particle size of microspheres increased greatly. At low stirring speeds (<300 r/min), big droplets are broken down and dispersed due to stirring, and the average particle size becomes smaller as a result. However, at high stirring speeds (300–500 r/min), the possibility of collisions between droplets is increased, which results in an increase in the average particle size. Besides, high stirring speeds would destroy the structure of microspheres, and would increase the particle size distribution of the microspheres. So, we set the mechanically stirring speed at 300 r/min in order to obtain an average particle size of 125 μm and to keep the particle size distribution in a narrow range.

3.2.2 Initiator (APS) concentration

The effect of APS concentration on the average particle size of microspheres is shown in Fig. 3b. The plot shows that with an increase in the APS concentration, the average particle size of microspheres increased significantly. This might be because, as the APS concentration increases, free radicals derived primarily from initiator (APS) increase within the time tested. Under the assumptions that each

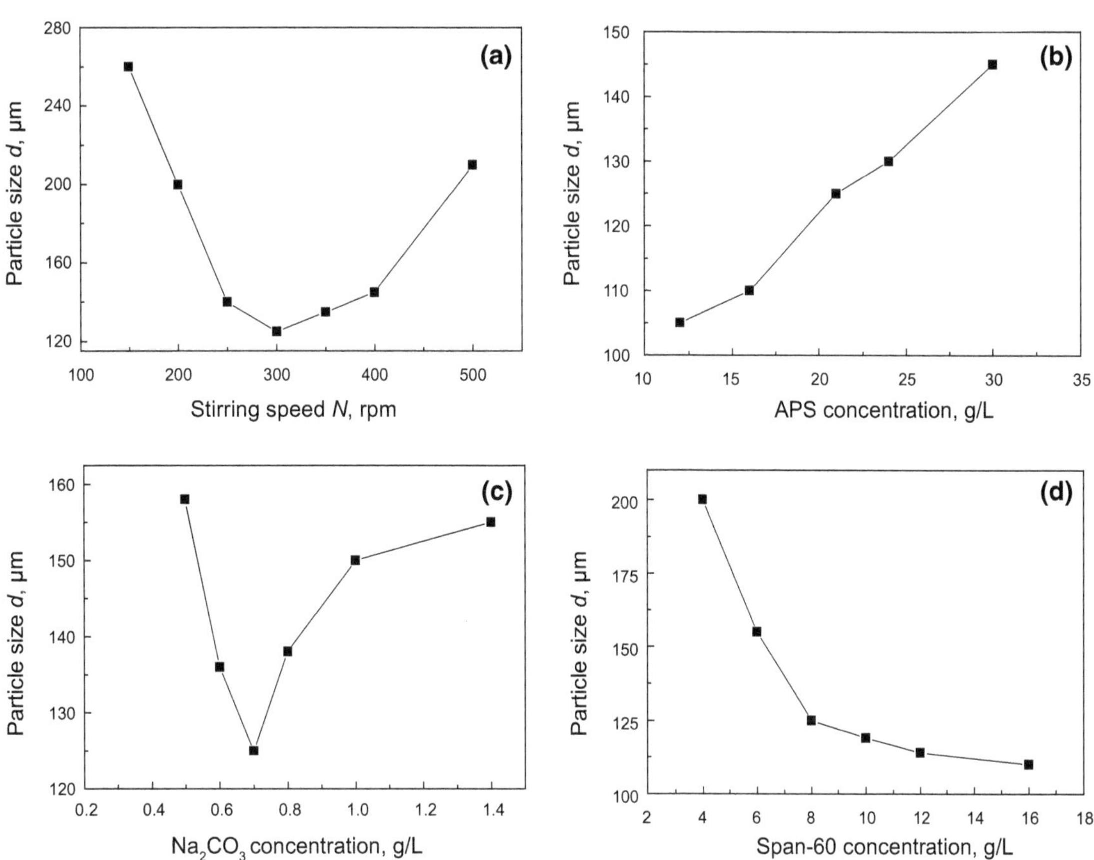

Fig. 3 Factors contributing to the average particle size. **a** Stirring speed; **b** APS concentration; **c** Na_2CO_3 concentration; **d** Span-60 concentration

polymer chain growth rate is same, this increase in free radicals results in a reduction in the relative molecular mass of the polymer, and a consequent increase in the polymer solubility in the aqueous phase. In order to generate microspheres of small average particle size, the APS concentration should be controlled between 20.0–24.0 g/L.

3.2.3 Na_2CO_3 concentration

Figure 3c shows that the average particle size of microspheres is reduced from 158 to 125 μm as the Na_2CO_3 concentration increased from 0.5 to 0.7 g/L. However, when the Na_2CO_3 concentration is increased from 0.7 to 1.4 g/L, the average particle size increased from 125 to 155 μm gradually.

First of all, APS as an initiator is decomposed into fragments and then adsorbs around the droplets to form electrical double layers, which must give rise to repulsion forces whenever two droplets approach each other to maintain the stability of droplets (this is the fundamental factor responsible for the stability of droplets). Secondly, the added Na_2CO_3 increases the ionic strength of the system, so the electric double layers formed at the droplet surfaces become thinner. Therefore, the electrostatic repulsion between droplets reduces, which increases the probability of initial intergranular coalescence, with the result that the average particle size of microspheres increases.

3.2.4 Dispersant (Span-60) concentration

Smaller microspheres may be produced by increasing the dispersant concentration. Figure 3d shows that with an increase in the Span-60 concentration from 4 to 16 g/L, the average particle size decreased gradually from 200 to 110 μm. This might be because the dispersants adsorbed onto the droplet surface obstruct the agglomeration between droplets.

In order to obtain relatively uniform fluorescent microspheres of an average particle size about 120 μm, the dispersant concentration (Span-60) should be kept at 8–10 g/L.

3.3 Size distribution of microspheres

When the weights of AM, MBA, AMCO were 14.06, 0.07, 0.014 g, the concentration of the APS, Na_2CO_3, Span-60 were 2, 0.7, 8 g/L, respectively, and the stirring speed was 300 r/min, the microspheres were obtained with a narrow size distribution. Figure 4 shows an optical micrograph and particle size distribution of microspheres in ethanol at 25 °C. It can be seen that the particle size of microspheres was 120–130 μm, with an average particle size of 125 μm.

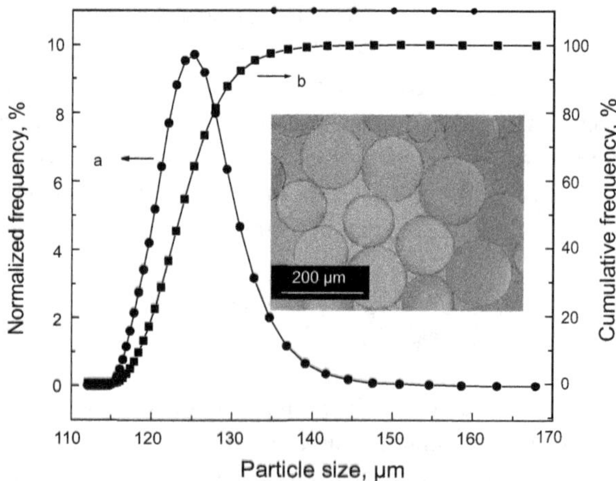

Fig. 4 Particle size distribution of elastic microspheres in ethanol at 25 °C

3.4 The swelling properties of microspheres

Figure 5 shows the swelling ratio of microspheres after being hydrated in brine. Within 24 h of hydration, the swelling ratio increased significantly to about 10 and gradually leveled off until 180 h, which indicates that brine can penetrate into the microsphere matrix. In fact, the microspheres are cross-linked particles with a three-dimensional network structure and many free hydrophilic groups (–$CONH_2$) inside. Polar water molecules can bond with these groups easily through hydrogen bonding and cause a significant increase in the size of the hydrated microspheres.

Furthermore, swelling behavior of P(AM-BA-AMCO) microspheres is similar to P(AM-BA) microspheres. This indicates that the swelling property of P(AM-BA-AMCO) has not been dramatically affected by the fluorescent group

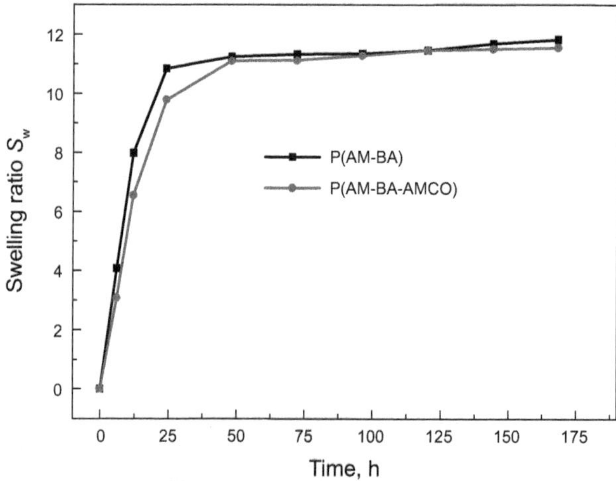

Fig. 5 Swelling properties of microspheres in brine at 45 °C

due to a low content of fluorescent group embedded in microspheres by grafting or coating.

3.5 Profile control performance of microspheres

The pressure drop during the core flow experiments was used to calculate residual resistance factor (F_{rr}) and the differential pressure was monitored, shown in Fig. 6. The differential pressure rose from around 1 kPa (stabilized pressure at the end of water flood) to around 1.8 MPa with the injection of the microsphere solution (0–35 PV). After 35 PV of microsphere injection, brine was injected into the sand pack. The differential pressure quickly dropped down to about 0.9 MPa (35–36 PV) thereafter gently drop at the range of 35–63 PV, and finally leveled off around 0.8 MPa.

The differential pressure fluctuates within a wide range at the second-stage (II), and this phenomenon was defined as "wave-type variation" (Yao et al. 2012). In the subsequent injection of brine, the pressure remains in this kind of variation in the range of 35–60 PV and then reaches a relatively high and constant level. In fact, when the brine-swelled microspheres were injected into the sand pack, they first adsorbed and accumulated near the inlet of the sand pack, causing the differential pressure to increase. Moreover, with the increased injection volume of microspheres, microspheres accumulated more and more, making the pressure rise faster. The differential pressure increased and fluctuated during the microsphere treatment stage (II) and the subsequent water flooding stage (III). This phenomenon indicates that the swelled microspheres have a certain degree of deformation ability and could deform under pressure to advance into the sand pack, to play the role of profile modification (Hua et al. 2013).

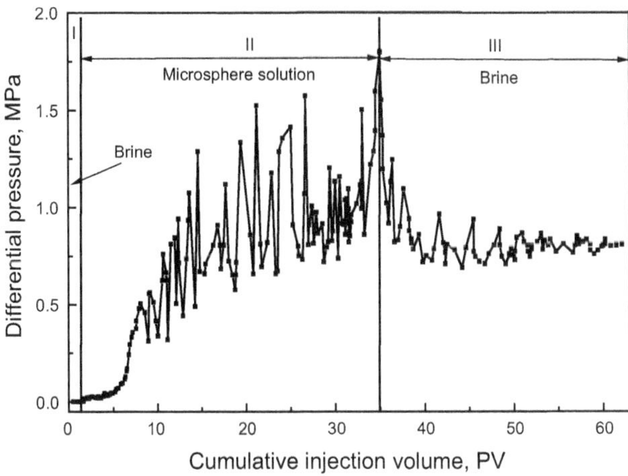

Fig. 6 Relationship between the differential pressure and injection volume

It can be inferred from Table 2 that at first brine could be injected into the sand pack at a relatively low pressure (1 kPa) due to high initial permeability of sand pack (102 μm^2). After injection of 1.0 PV brine and 35 PV microsphere solution, the sand pack permeability decreased to 0.06 μm^2, while the resultant differential pressure increased to 1.8 MPa. In the subsequent injection of brine, the sand pack permeability was restored to 0.13 μm^2 and the differential pressure was restored to 0.8 MPa. Meanwhile, the plugging rate was 99.8 % and the residual resistance factor was 800. This indicates that the microspheres injected into the sand pack would swell several times and deposit in pores and throats in the high-permeability unconsolidated sand packs, resulting in a reduction in permeability and an increase in residual resistance factor.

3.6 Surface morphology of P(AM-BA-AMCO) microspheres

The surface topography of P(AM-BA-AMCO) microspheres was observed with the inverse fluorescence microscope and shown in Fig. 7. The phase-contrast images of Fig. 7a, c was taken under natural light. Figure 7b, d was taken under ultraviolet irradiation. Fluorescent microspheres were spheroid-shaped, with a smooth surface and an average diameter of about 200 μm. Furthermore, fluorescence microscopy images demonstrated that the fluorescent monomers were uniformly distributed inside the microspheres and they had kept their fluorescent properties after swelling.

Figure 7c, d shows the images of P(AM-BA-AMCO) microspheres come out from the outlet of the unconsolidated sand pack after the subsequent water flooding experiment. It can be seen that the P(AM-BA-AMCO) microspheres and their fragments were accurately detected in the produced fluid. Such microspheres and fragments still released blue fluorescence under ultraviolet irradiation, even though they have experienced extrusion and deformation in the sand pack. These P(AM-BA-AMCO) microspheres may be potentially distinguished from the reservoir produced fluid.

4 Conclusions

(1) Coumarin fluorescent polyacrylamide microspheres [P(AM-BA-AMCO)], with an average particle size of 125 μm, were synthesized with the inverse suspension polymerization method.

(2) The microspheres had the ability to plug fractures and pores in the sand pack. The plugging rate was

Table 2 Changes in core permeability and injection pressure in plug process

Permeability, μm²			Differential pressure, MPa			Plugging rate, %	Residual resistance factor F_{rr}
Initial	At 35 PV	At 63 PV	Initial	At 35 PV	At 63 PV		
102	0.06[a]	0.13	0.001	1.8	0.8	99.8	800

[a] Permeability to water. The viscosity of the microsphere solution is 1 mPa s

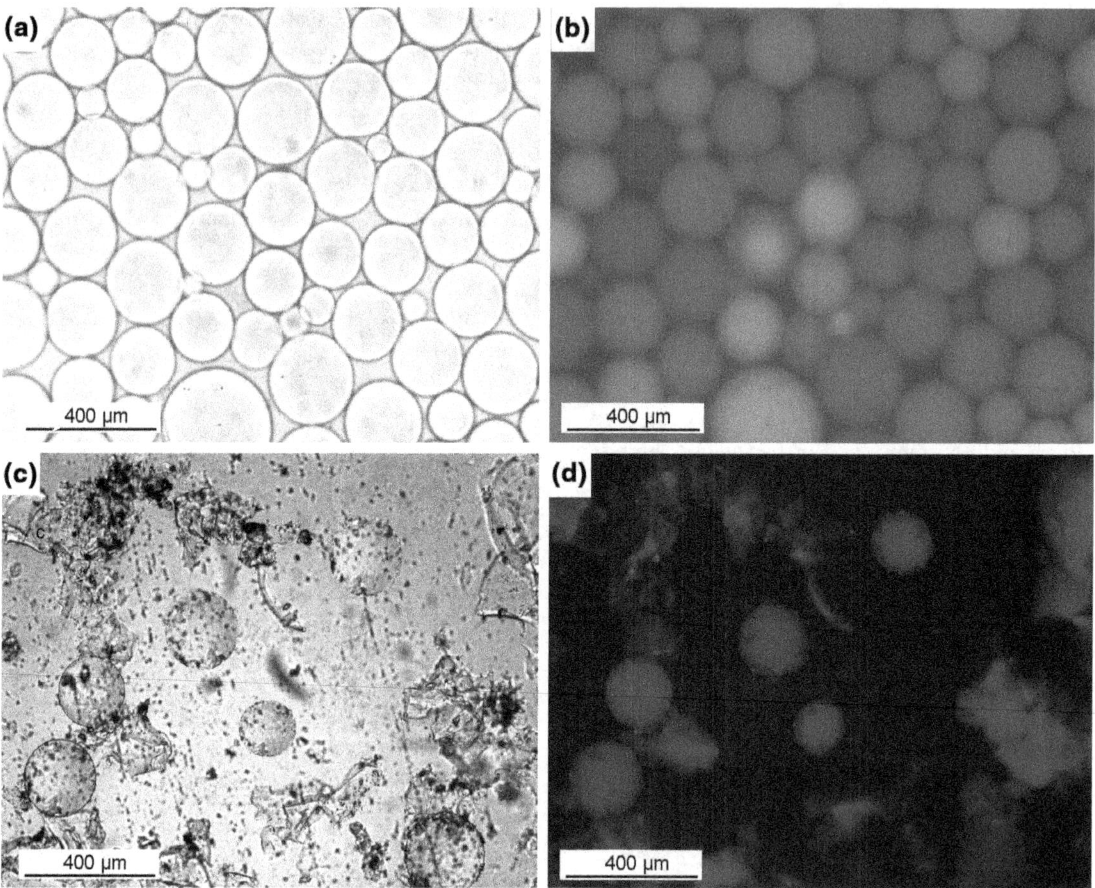

Fig. 7 Fluorescence microscope images of P(AM-BA-AMCO) microspheres. **a** Phase-contrast image of microspheres under the natural light after swelling for 24 h in brine; **b** Fluorescence micrographs for the (**a**) taken under ultraviolet irradiation; **c** Phase-contrast image of microspheres under the natural light which came out from the outlet of the unconsolidated sand pack after the subsequent water flooding experiment; **d** Fluorescence micrographs for the (**c**) taken under ultraviolet irradiation

99.8 % and the residual resistance factor was 800 after microsphere treatment.

(3) The microspheres and their fragments could be accurately detected in the produced fluid even though they experienced extrusion and deformation in the sand pack. This is a potential application value of fluorescent microspheres, and can be used to detect the polyacrylamide microspheres in produced liquid.

Acknowledgments This research was supported by the National Natural Science Foundation of China (No. 21273286), Doctoral Program Foundation of the Education Ministry (No. 20130133110005).

References

Annaka M, Matsuura T, Kasai M, et al. Preparation of comb-type *N*-isopropylacrylamide hydrogel beads and their application for size-selective separation media. Biomacromolecules. 2003;4(2):395–403.

Bai F, Yang XL, Huang WQ. Synthesis of narrow or monodisperse poly(divinylbenzene) microspheres by distillation-precipitation polymerization. Macromolecules. 2004;37(26):9746–52.

Bjørnstad T, Brendsdal E, Michelsen OB, et al. Analysis of radiolabelled thiocyanate tracer in oil field brines. Nucl Instrum Methods Phys Res, Sect A. 1990;299:629–33.

Chen Y, Tong LY, Zhang D, et al. Thermostable microspheres consisting of poly(*N*-phenylmaleimide-co-α-methyl styrene) prepared by precipitation polymerization. Ind Eng Chem Res. 2012;51:15610–7.

Chern CS. Emulsion polymerization mechanisms and kinetics. Prog Polym Sci. 2006;31(5):443–86.

Covolan VL, D'Antone S, Ruggeri G, et al. Preparation of aminated polystyrene latexes by dispersion polymerization. Macromolecules. 2000;33(18):6685–92.

Guo JW, Cui YD, Kang Z, et al. Synthesis of superabsorbent polymers by the inverse suspension method. Fine Chem. 2001;18(6):348–50 (in Chinese).

Haraguchi K, Li HJ, Matsuda K, et al. Mechanism of forming organic/inorganic network structures during in situ free-radical polymerization in PNIPA-clay nanocomposite hydrogels. Macromolecules. 2005;38(8):3482–90.

Hemingway MG, Gupta RB, Elton DJ. Hydrogel nanopowder production by inverse-miniemulsion polymerization and supercritical drying. Ind Eng Chem Res. 2010;49(20):10094–9.

Hua Z, Lin MQ, Guo JR, et al. Study on plugging performance of cross-linked polymer microspheres with reservoir pores. J Petrol Sci Eng. 2013;105:70–5.

Kalyanaraman B, Darley-Usmar V, Davies KJA, et al. Measuring reactive oxygen and nitrogen species with fluorescent probes: challenges and limitations. Free Radical Biol Med. 2012;52(1):1–6.

Koops SE, Barnes PRF, O'Regan BC, et al. Kinetic competition in a coumarin dye-sensitized solar cell: injection and recombination limitations upon device performance. J Phys Chem C. 2010;114(17):8054–61.

Lei GL. New deep profile control and flooding technology using pore-scale elastic microspheres. Dongying: China University of Petroleum Press; 2011 (in Chinese).

Li BX, Xu YF, Wang MZ, et al. Morphological control of multihollow polymer latex particles through a controlled phase separation in the seeded emulsion polymerization. Langmuir. 2013;29(48):14787–94.

Li JJ, Liu YZ, Na Z, et al. Investigation on the adaptability of the polymer microspheres for fluid flow diversion in porous media. J Dispers Sci Technol. 2014;35(1):120–9.

Lien JR, Graue A, Kolltveit K. A nuclear imaging technique for studying multiphase flow in a porous medium at oil reservoir conditions. Nucl Instrum Methods Phys Res, Sect A. 1988;3:693–700.

Liu ZD, Lu YC, Yang BD, et al. Controllable preparation of poly(butyl acrylate) by suspension polymerization in a coaxial capillary microreactor. Ind Eng Chem Res. 2011;50(21):11853–62.

Nagao D, Sakamoto T, Konno H, et al. Preparation of micrometer-sized polymer particles with control of initiator dissociation during soap-free emulsion polymerization. Langmuir. 2006;22(26):10958–62.

Nakabayashi K, Kojima M, Inagi S, et al. Size-controlled synthesis of polymer nanoparticles with tandem acoustic emulsification followed by soap-free emulsion polymerization. ACS Macro Lett. 2013;2:482–4.

Osborn SG, McIntosh JC. Chemical and isotopic tracers of the contribution of microbial gas in Devonian organic-rich shales and reservoir sandstones, northern Appalachian Basin. Appl Geochem. 2010;25(3):456–71.

Qin JS, Li AF. Reservoir physics. Dongying: China University of Petroleum Press; 2006 (in Chinese).

Selby D, Creaser RA, Fowler MG. Re–Os elemental and isotopic systematics in crude oils. Geochim Cosmochim Acta. 2007;71(2):378–86.

Suzuki Y, Komatsu H, Ikeda T, et al. Design and synthesis of Mg^{2+}-selective fluoroionophores based on a coumarin derivative and application for Mg^{2+} measurement in a living cell. Anal Chem. 2002;74(6):1423–8.

Thickett SC, Gilbert RG. Emulsion polymerization: state of the art in kinetics and mechanisms. Polymer. 2007;48(24):6965–91.

Tian JJ, Ma F, Yu XL, et al. Technology process investigation of superabsorbent resin by inverse suspension polymerization. J Wuhan Inst Tech. 2012;34(11):9–13 (in Chinese).

Tokuda M, Shindo T, Minami H. Preparation of polymer/poly(ionic liquid) composite particles by seeded dispersion polymerization. Langmuir. 2013;29(36):11284–9.

Wang BJ, Guo MR, Lv KF, et al. The application of fluorescein as an oil field tracer. Chinese patent: CN1415978A. 2004a (in Chinese).

Wang TG, He FJ, Li MJ, et al. Alkyldibenzothiophenes: molecular tracers for filling pathway in oil reservoirs. Chin Sci Bull. 2004b;49(22):2399–404.

Xia XH, Yih J, D'Souza NA, et al. Swelling and mechanical behavior of poly (N-isopropylacrylamide)/Na-montmorillonite layered silicates composite gels. Polymer. 2003;44(16):3389–93.

Yan LW, Peng CX, Ye ZB, et al. Synthesis and evaluation of a water-soluble copolymer (AM-AMCO) as a fluorescent tracer. Chem Res Appl. 2012;24(6):996–1001 (in Chinese).

Yang WB, Xia MF, Li AM, et al. Mechanism and behavior of surfactant adsorption onto resins with different matrices. React Funct Polym. 2007;67(7):609–16.

Yao CJ, Lei GL, Li L, et al. Selectivity of pore-scale elastic microspheres as a novel profile control and oil displacement agent. Energy Fuels. 2012;26(8):5092–101.

Yao CJ, Lei GL, Cheng MM. A novel enhanced oil recovery technology using pore-scale elastic microspheres after polymer flooding. Res J Appl Sci Eng Technol. 2013;6(19):3634–7.

Yu JT, Wang LY, Wang H, et al. Determination of interwell tracers by fluorospectrophotometry. Chem Ind Eng. 2009;26(3):256–61 (in Chinese).

Zhang MC, Lan Y, Wang D, et al. Synthesis of polymeric Yolk-Shell microspheres by seed emulsion polymerization. Macromolecules. 2011;44(4):842–7.

Zhao T, Qiu D. One-pot synthesis of highly folded microparticles by suspension polymerization. Langmuir. 2011;27(21):12771–4.

13

Main controlling factors and enrichment area evaluation of shale gas of the Lower Paleozoic marine strata in south China

Xian-Ming Xiao[1] · Qiang Wei[1,2] · Hai-Feng Gai[1] · Teng-Fei Li[1,2] · Mao-Lin Wang[1,2] · Lei Pan[1,2] · Ji Chen[1,2] · Hui Tian[1]

Abstract The Lower Paleozoic shale in south China has a very high maturity and experienced strong tectonic deformation. This character is quite different from the North America shale and has inhibited the shale gas evaluation and exploration in this area. The present paper reports a comprehensive investigation of maturity, reservoir properties, fluid pressure, gas content, preservation conditions, and other relevant aspects of the Lower Paleozoic shale from the Sichuan Basin and its surrounding areas. It is found that within the main maturity range (2.5 % < EqR_o < 3.5 %) of the shale, its porosity develops well, having a positive correlation with the TOC content, and its gas content is controlled mainly by the preservation conditions related to the tectonic deformation, but shale with a super high maturity (EqR_o > 3.5 %) is considered a high risk for shale gas exploration. Taking the southern area of the Sichuan Basin and the southeastern area of Chongqing as examples of uplifted/folded and faulted/folded areas, respectively, geological models of shale gas content and loss were proposed. For the uplifted/folded area with a simple tectonic deformation, the shale system (with a depth > 2000 m) has largely retained overpressure during uplifting without a great loss of gas, and an industrial shale gas potential is generally possible. However, for the faulted/folded area with a strong tectonic deformation, the sealing condition of the shale system was usually destroyed to a certain degree with a great loss of free gas, which decreased the pressure coefficient and resulted in a low production capacity. It is predicted that the deeply buried shale (>3000 m) has a greater gas potential and will become the focus for further exploration and development in most of the south China region (outside the Sichuan Basin).

Keywords Lower Paleozoic shale gas · Maturity · Main controlling factors · Tectonic deformation

1 Introduction

Within the Lower Paleozoic marine strata in south China, two sets of organic-rich Lower Silurian and Lower Cambrian shales are widely developed (Liang et al. 2008, 2009; Nie et al. 2009; Zou et al. 2010), covering areas of up to 42 × 10^4 km^2 and (30–50) × 10^4 km^2, respectively (Zou et al. 2010). The predicted geological resource potential of shale gas (<4500 m) is 17 × $10^{12}m^3$ and 35 × $10^{12}m^3$, respectively. Their sum accounts for about 39 % of the total onshore shale gas resource (Zhang et al. 2012), indicating a huge potential for shale gas exploration and development.

In recent years, shale gas exploration in the Lower Paleozoic marine strata in south China has been carried out actively. Great progress has been made in the Sichuan Basin where the tectonic deformation is relatively simple. Industrial shale gas production has been achieved from some areas, such as the Weiyuan, Changning, and Fuling blocks (Huang et al. 2012a, b; Guo and Liu 2013; Li et al. 2014; Wang et al. 2013c; Wang 2013). Especially in the Jiaoshiba area of the Fuling Block, initial gas production of (10–50) × 10^4 m^3/d has been reported for most tested

✉ Xian-Ming Xiao
xmxiao@gig.ac.cn

[1] State Key Laboratory of Organic Geochemistry, Guangzhou Institute of Geochemistry, Chinese Academy of Sciences, Guangzhou, Guangdong 510640, China

[2] University of Chinese Academy of Sciences, Beijing 100049, China

Edited by Jie Hao

wells, and the average yield of well JY1 in the first year was 6×10^4 m^3/d (Guo and Zhang 2014). The Jiaoshiba area is planned to the first commercial development demonstration base for shale gas in China. However, outside the Sichuan Basin, the gas content in shale is generally low, most of the tested wells do not reach an industrial capacity (Ma et al. 2014; Yi and Zhao 2014; Zhou et al. 2014a), and the exploration and development of shale gas still meets with challenges such as complicated structural conditions, poor resources, and variable geography (Wang 2013).

Based on the geological and geochemical characteristics of the Lower Paleozoic shale in south China, the present paper mainly focuses on the reservoir properties of the very high maturity shale, preservation conditions controlled by tectonic deformation, and their influence on shale gas content. The main controlling factors of shale gas content and the evaluation criteria for a favorable shale gas play are discussed. The purpose is to provide guidance for further exploration and development.

2 Geological characteristics of the Lower Paleozoic shale

There are a few similar geological and geochemical characteristics between the North America shale and the Lower Paleozoic shale in south China, such as the TOC (total organic carbon) content, organic type, shale thickness, mineral composition, brittle mineral content, Poisson's ratio, and Young's modulus (Li et al. 2013b; Wang et al. 2012a, b; Tu et al. 2014). This is the main basis for a huge shale gas potential of the Lower Paleozoic shale in south China predicted by scholars and organizations (Zhang et al. 2012; EIA 2011, 2013). However, the Lower Paleozoic shale is characterized by very high maturity and strong tectonic deformation (Nie et al. 2011; Wang et al. 2009b; Xiao et al. 2013), which shows an obvious difference from the North America shale (Table 1). Taking the southern area of the Sichuan Basin as an example (Fig. 1), the burial history and thermal maturation evolution of the lower Paleozoic strata can be largely divided into four stages:

Stage 1 (from Cambrian to Silurian): Marine strata with a thickness of 2500–3000 m were deposited. The Lower Cambrian and Lower Silurian shales underwent an early maturation at temperatures of 80–90 °C and 50–60 °C, respectively. Their EqR_o (equivalent vitrinite reflectance) values became 0.6 % and 0.4 %, respectively.

Stage 2 (from Devonian to Carboniferous): The strata underwent uplifting and erosion. The upper Silurian was generally eroded, with an eroded thickness of 500–1000 m.

Stage 3 (from Permian to Early Cretaceous): Deposition began from Permian, speeded up in Jurassic, and continued up to Early Cretaceous, with a total thickness of 5000–6000 m. During Early Cretaceous, the burial depth of the Lower Cambrian and Lower Silurian shales reached about 8000 and 7000 m, corresponding to a maximum temperature of 250–260 °C and 220–230 °C, respectively. The shales experienced strong thermal maturation and reached a very high maturity, with an EqR_o value of 2.5 % (the Lower Silurian shale) and 3.5 % (Lower Cambrian shale).

Stage 4 (from Late Cretaceous to present): Influenced by the tectonic deformation of the late Yanshanian–Himalayan movements, this area was uplifted and the strata were eroded strongly from the Late Cretaceous. The eroded thickness was up to 2500–4000 m generally.

For the Yangtze region outside the Sichuan Basin, the burial history and thermal maturation evolution of the Lower Paleozoic strata are basically consistent with the Sichuan Basin, but the uplifting occurring during the late Yanshanian–Himalayan period was earlier and the structural deformation was stronger, with an obvious regional differentiation (Mei et al. 2010). With a background of general strong uplifting, the intense tectonic deformation resulted in fold deformation and fault cutting, forming rather complex structural patterns (Hu et al. 2014; Tang and Cui 2011; Yan et al. 2000). For example, trough-like

Table 1 Main differences between the Lower Paleozoic shale in south China and the North America shale

Feature	Sichuan Basin	Outside the Sichuan Basin	North America
Age	Lower Silurian and Lower Cambrian	Lower Silurian and Lower Cambrian	Devonian, Carboniferous, and Cretaceous
Present burial depth, m	2000–6000	1000–6000	2000–4000
Burial depth with industrial shale gas production, m	2000–3500	No data	1500–3000
Maturity (R_o or EqR_o), %	2.5–3.5	2.5–3.5	1.0–2.5
Tectonic deformation	Strongly uplifted with an eroded thickness of 2000–4000 m	Strongly uplifted, folded, and faulted with an eroded thickness of 3000–5000 m	Simply uplifted with an eroded thickness of 1500–3000 m

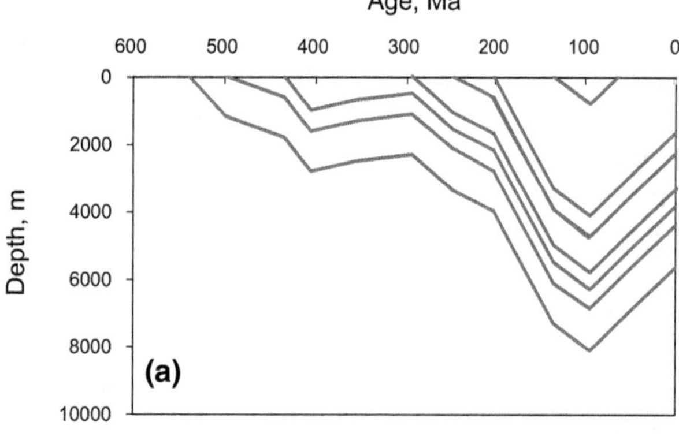

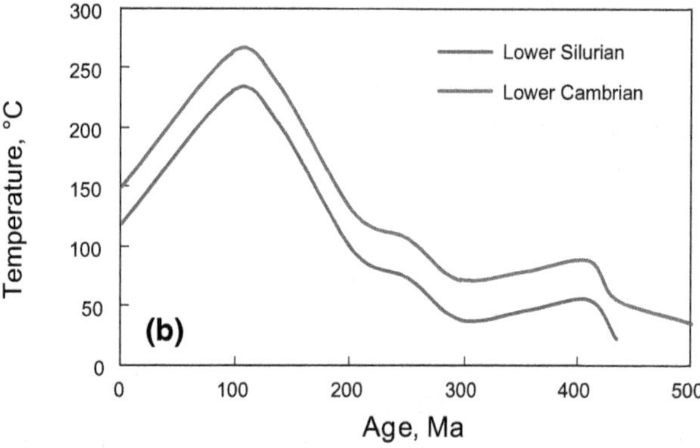

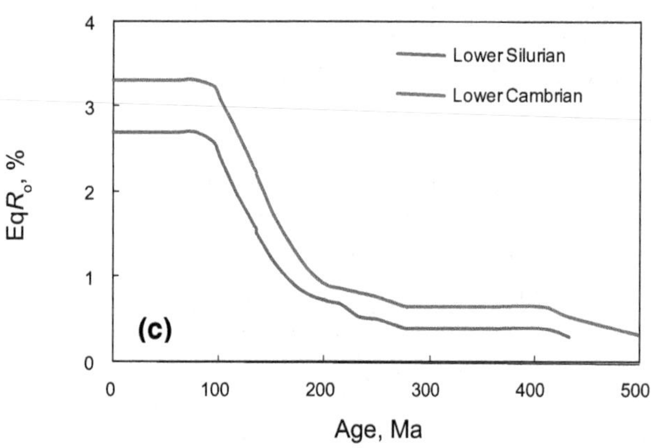

Fig. 1 Burial history of strata (**a**), paleogeotemperature of Lower Cambrian and Lower Silurian (**b**), and maturity evolution (**c**) from well Pan 1 in the southern area of the Sichuan Basin

folds (wide-gentle anticline and narrow-steep syncline) developed widely in the Zhaotong area to the south of the Sichuan Basin (Wang 2013). The structural styles are more complex in the southeast Chongqing area to the east of the Sichuan Basin. From the northwest to southeast, the structure changed gradually from comb-like folds (wide-gentle synclines and narrow-steep anticlines) to trough-like folds, and transitional folds composed of wide-gentle synclines and anticlines developed in the middle area (Li et al. 2012; Wei et al. 2014) (Fig. 2).

Maturity not only controls the gas generation of shale, but also affects the reservoir properties (Curtis et al. 2012; Fishman et al. 2012; Loucks et al. 2009; Mastalerz et al. 2013; Modica and Lapierre 2012; Valenza et al. 2013). The maturity of the Lower Paleozoic shale in south China is generally rather high, but it varies significantly in different regions (Nie et al. 2009). The Lower Cambrian and Lower Silurian shales have EqR_o values of, respectively, 2.7 %–6.2 % and 1.9 %–3.8 % in the whole Yangtze region (Cheng and Xiao 2013), and 2.5 %–3.5 % and 2.4 %–3.2 % in the Sichuan Basin (Wang et al. 2009a; Zou et al. 2014). Shale generally has a quite low porosity when it has evolved to the dry gas stage ($R_o > 2.0$ %) (Wang et al. 2013a), but its reservoir properties still change with further increasing maturity (Chen and Xiao 2014). Substantial uplifting and strata denudation reduce the temperature and pressure of a shale system, which will lead to an increase in its fluid pressure coefficient if the system is ideally confined (Zhou et al. 2014b). Faults and folds may damage the sealing condition of a shale system, and as a result a great loss of shale fluids occurs with a significant reduction of the fluid pressure coefficient (Hu et al. 2014; Guo 2014). These geological processes have a great impact on shale gas contents and yields, and become the primary concern for the shale gas exploration and development of the Lower Paleozoic strata in south China.

3 Maturity and gas content of shale

3.1 Influence of maturity on shale porosity

The relationship of maturity and gas contents in shale is mainly reflected by its impact on reservoir properties, which includes two diverging aspects: gas generation to form organic pores with an increase of total porosity, and compaction to reduce average pore size with a decrease of total porosity and permeability (Cander 2012; Curtis et al. 2012; Chalmers et al. 2012a; Mastalerz et al. 2013; Milliken et al. 2012, 2013). During burial history, continuous gas generation is of great significance to the enrichment of shale gas. On one hand, reservoir space will increase with a continuous formation of organic pores. On the other hand, continuously generated gas supplies a possible gas loss and balances a decrease of fluid pressure associated with the increase of free gas capacity caused by increasing burial depth, thus supporting the shale porosity (Milliken et al. 2013). When the burial depth is over the limit of gas generation, further burial will lead to a significant compaction, and a decrease of shale total porosity will occur due to the inevitable gas loss from the shale. Thus, the evolution of shale porosity has a close relationship with the gas generation.

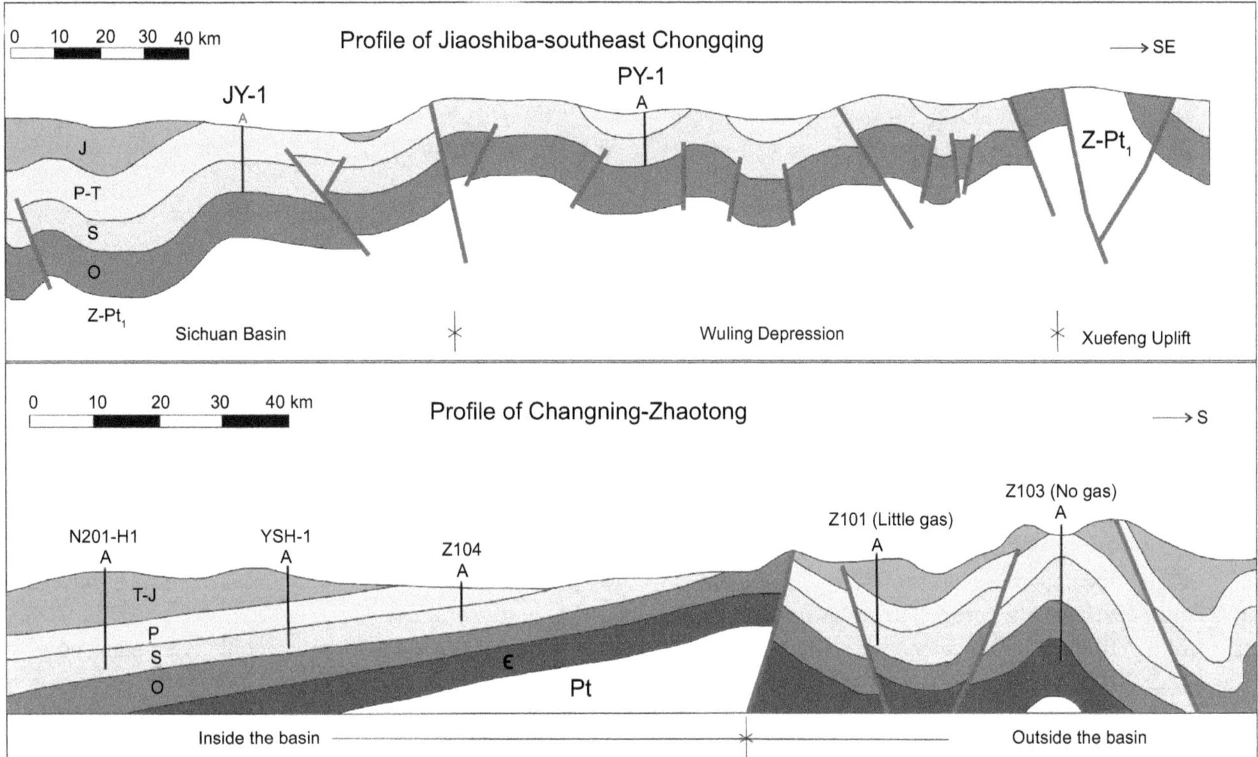

Fig. 2 Typical structural profiles across the Sichuan Basin and its surrounding areas (Modified after Guo 2014)

Using the hydrocarbon generation kinetics method, Cheng and Xiao (2013) simulated the hydrocarbon generation and evolution of the Lower Paleozoic shale in the Sichuan Basin, and suggested a general geological model for gas generation. According to their results, the main gas generation stage is within an EqR_o value range of 1.5 %–2.5 %, and the gas generation limit occurs at an EqR_o value of about 3.0 %. Gas generated after EqR_o > 2.5 % accounts for about 10 % of the total gas yield. This portion of gas would be important to the preservation of shale porosity in the very high maturity stage. Porosity data from the North America shale and the Lower Paleozoic shale of the southern area of the Sichuan Basin show an obvious decrease when R_o or EqR_o is over 3.0 % (Wang et al. 2013a), implying that this relationship indeed exists in natural shale systems. Figure 3 further shows that shale porosity varies widely from 3 % to 6 % when R_o or EqR_o value is between 1.0 % and 3.0 %, without a clear relationship with maturity, which reflects the combined effect of multiple factors (except for maturity), such as TOC content and mineral compositions on shale porosity (Curtis et al. 2010; Chalmers and Bustin 2008; Chalmers et al. 2012; Ross and Bustin 2009). When R_o or EqR_o is between 3 % and 5 %, shale porosity varies within a narrow range, mainly between 2 % and 4 %, having an obvious decreasing trend with increasing maturity.

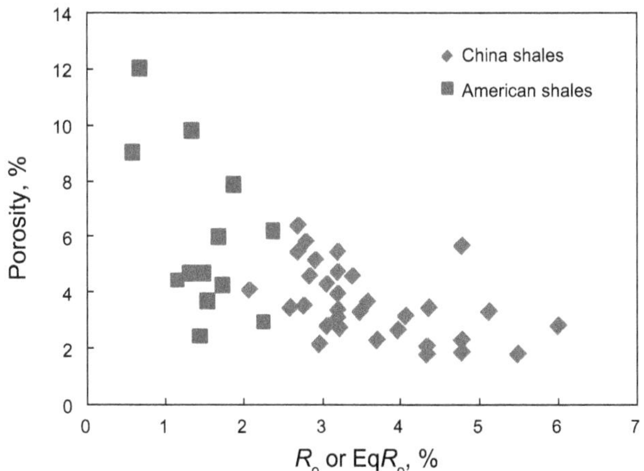

Fig. 3 Correlation between maturity and shale porosity (modified after Cheng and Xiao 2013, with some new data)

For shale with a very high maturity, inorganic matrix porosity is very low due to strong compaction. For instances, this value is approximately 1 % for the Marcellus Shale (R_o = 2 %) (Milliken et al. 2013) and 1.2 %–1.3 % for the Lower Paleozoic shale (EqR_o = 2.0 %–3.1 %) from the southern area of the Sichuan Basin (Wang et al. 2013a). Therefore, the development and evolution of organic pores becomes an important factor affecting shale porosity. According to the data from literature, shale porosity has a clear positive correlation with TOC content when TOC is within a definite range (Milliken et al. 2013; Pan et al. 2015; Tian et al. 2013; 2015; Wang et al. 2013a). However, this positive correlation will be changed or even reversed to be negative when TOC is very high, over this range (Milliken et al. 2013; Pan et al. 2015; Wang et al. 2013a). The TOC turning point occurs at about 5.6 % for the Marcellus Shale (Milliken et al. 2013), 5.0 % for the Lower Paleozoic shale from the southern area of the Sichuan Basin (Wang et al. 2013a), and 12 % for the Upper Permian shale from the Lower Yangtze region (Pan et al. 2015). This is attributed to an easier compaction of organic nano-pores for shale with a higher TOC content than shale with a lower TOC content (Milliken et al. 2013; Wang et al. 2013a; Pan et al. 2015). Compaction not only reduces total porosity, but also changes pore structure significantly. For example, the Lower Cambrian shale from well HY1 has a significantly lower porosity than the Lower Silurian shale from well PY1 (both shales have a similar maturity) (Tian et al. 2013, 2015), but the former has much higher level of nano-pores (<10 nm) as compared with the latter (Fig. 4), which indicates that the compaction effect is more evident for macropores and larger mesopores, resulting in a reduction of the average pore size and an increase of the micropore volume and specific surface area of the shale.

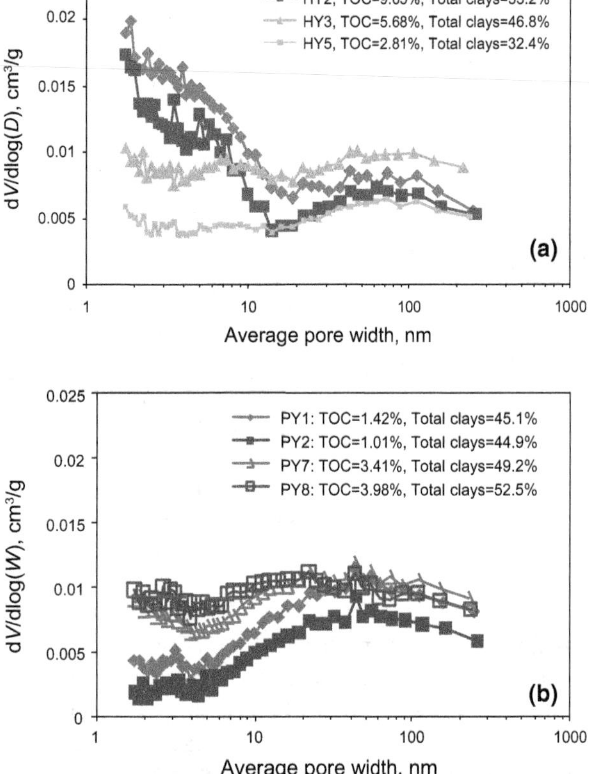

Fig. 4 Comparison of nano-pore structure between the Lower Cambrian shale from Well HY1 and the Lower Silurian shale from Well PY1 (simplified after Tian et al. 2015)

3.2 Porosity evolution model of shale

Based on the data from the Eagle Ford shale, Cander (2012) proposed an evolution model of inorganic matrix porosity and organic porosity in shale. According to his model, the inorganic matrix porosity significantly decreases with increasing burial depth up to 3000 m. When the burial depth is over 3000 m, the reduction of inorganic matrix porosity becomes less, but the organic pores start to develop and increase rapidly. The total porosity of shale does not change significantly within the range of 4000–5000 m. Combined with the gas generation evolution and porosity data of the Lower Paleozoic shale from the Southern China region, this model is modified and extended to a greater burial depth (9000 m) (Fig. 5). According to the present model, a turning point of organic porosity occurs at the EqR_o value of about 3.0 %, corresponding to a burial depth of about 7500 m during geological history. The organic porosity begins to diminish with further increasing maturity or burial depth, and this decrease becomes drastic when the EqR_o value is over 3.5 %, resulting in a very low total porosity for shale with an EqR_o value >4.0 %.

3.3 Gas content of shale with a very high maturity

The shale gas exploration and development carried out in the Lower Paleozoic shale of the Sichuan Basin over the past few years provide important data for estimating the gas content and yield of shale of a very high maturity (Table 2). The average gas content of the Lower Cambrian and Lower Silurian shales from the Weiyuan Block is 1.9 and 1.82 m^3/t, respectively (Huang et al. 2012a, b). For the Changning and Jiaoshiba blocks, the average gas content of the Lower Silurian shales is 1.93 and 2.96 m^3/t, respectively (Huang et al. 2012a; Guo and Zhang 2014). The gas content from the three blocks is similar to that of the Marcellus Shale and has a linear positive correlation with their TOC contents (Guo 2014; Wang et al. 2012a; Wang et al. 2013c). Wells with a high shale gas yield were reported from the three blocks. For example, the initial production of wells W204 in the Weiyuan Block, N201-H1 in the Changning Block, and JY1HF in the Jiaoshiba Block is 16.5×10^4 m^3/d, 15×10^4 m^3/d, and 20.3×10^4 m^3/d, respectively (Hu et al. 2014; Wang 2014), and the gas yield has a positive correlation with the fluid pressure coefficient of these shale systems (Hu et al. 2014). A comparison of their EqR_o values between the three blocks indicates that the sequence of maturity from high to low is Jiaoshiba,

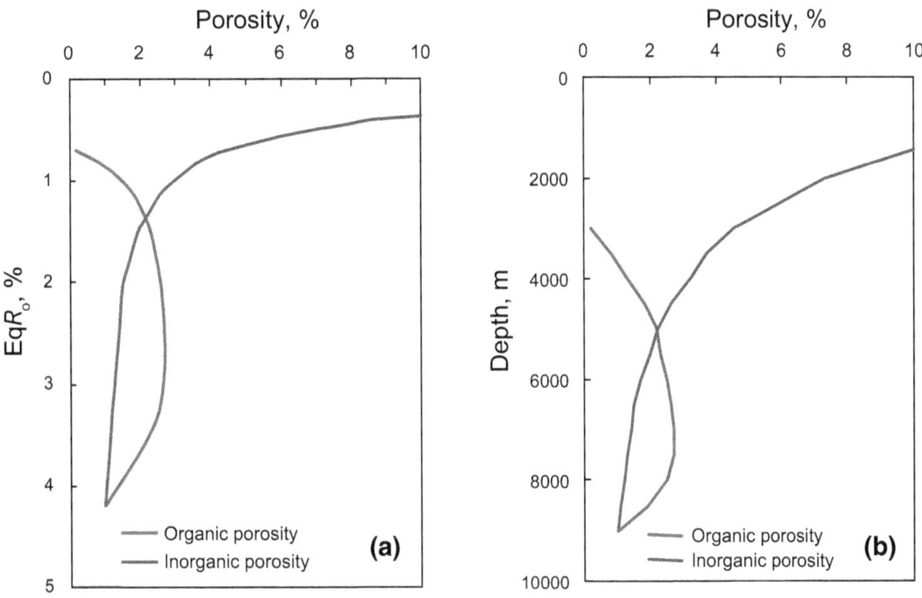

Fig. 5 Conceptual model of porosity evolution with maturity (**a**) and burial depth (**b**) for the Lower Paleozoic shale in south China

Table 2 Data of maturity and shale gas development of the Lower Paleozoic shale from the Sichuan Basin

Shale	Lower Silurian shale in the Weiyuan Block	Lower Cambrian shale in the Weiyuan Block	Lower Silurian shale in the Changning Block	Lower Silurian shale in the Jiaoshiba Block
EqR_o, %	1.8 (W106)[a] 2.45 (W201)[d]	2.68–2.80[c]; 2.85 (W117)[d]	3.21[e]	3.4[d]
Total porosity, %	(1.7–5.87)/4.2[a]	(0.82–4.86)/2.44[c]	(2.4–7.2)/5.5 (CX1)[f]	(1.2–8.0)/4.5 [h]
Gas content, m^3/t	(0.30–5.09)/1.82[a]	(0.27–6.02)/1.9[c]	(2.90–3.5)/1.93[a]	(0.44–5.19)/2.96[i]
Pressure coefficient of representative well	1.96 (well W204)[b]	0.92(W201)[b]	2.03 (N201)[g]	1.55 (JY1HF)[g]
Initial production of representative well, 10^4 m^3/d	16.5 (W204)[b]	2.83 (W201H3)[c]	15 (N201-H1)[g]	20.3 (JY1HF)[j]

[a] Huang et al. (2012a); [b] Yu (2014); [c] Huang et al. (2012b); [d] Data from the present study; [e] Wang et al. (2009b); [f] Wang et al. (2012a, b); [g] Hu et al. (2014); [h] Li et al. (2014); [i] Guo et al. (2014); [j] Wang (2014)

Changning, and Weiyuan. The EqR_o value of the Lower Silurian shales from the Jiaoshiba and Changning blocks is greater than 3.0 %, but their gas content and yield are still very high and are not affected by the maturity according to the available data (Table 2). The Marcellus Shale has two core gas areas: the southern west area and northern east area of the basin, with an R_o value of 1.5 %–2.5 % and 2.0 %–3.5 %, respectively. The initial production from some representative wells exceeds 20×10^4 m^3/d (Zagorski et al. 2011). These examples indicate that, within a definite range of maturity (R_o or EqR_o < 3.5 %), shale with a very high maturity still has a great shale gas potential.

Jarvie et al. (2007) proposed a positive correlation between maturity and gas production based on Barnett Shale data, but their model limited to an R_o value of up to about 2.5 % and lacked data from higher maturity shales. Nie et al. (2011), based on data of 41 shale samples from the Lower Paleozoic strata in the Sichuan Basin and its surrounding areas, found that maturity does not have a significant effect on the gas content of shale even if the EqR_o value is as high as 3.0 %. In the present study, a conceptual model of shale gas content versus maturity is preliminarily suggested, which is based on the available data from the Lower Paleozoic shale of the Sichuan Basin as well as the North American shale (Fig. 6). According to this model, within a maturity range of 2 %–3.5 % of R_o or EqR_o, shale gas content and yield and industrial shale gas potential would not be significantly reduced with increasing maturity, but exploration risk will increase greatly when the R_o or EqR_o value is over 3.5 % due to the significant decrease of total porosity. Based on the maturity

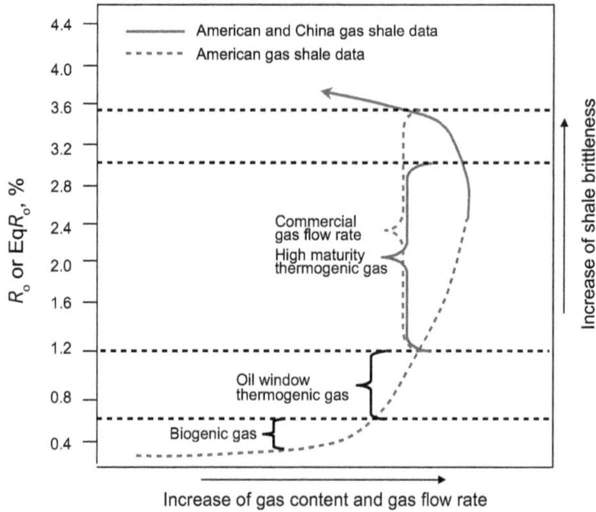

Fig. 6 A conceptual model of gas content versus maturity of shale (supplemented and modified after Jarvie et al. 2007)

distribution of the Lower Silurian and Lower Cambrian shales in south China, most of the Lower Silurian shale has an EqR_o value of 2.0 %–3.0 %, with few area >3.5 % (Nie et al. 2009; Wang et al. 2009a; 2013b), and the Lower Cambrian shale is mainly between 2.5 % and 3.5 % in the Sichuan Basin (Zou et al. 2014), while the Cambrian shale with an R_o value >3.5 %–4.0 % is basically located in the Lower Yangtze region (Nie et al. 2009), covering an area of less than 30 % of the whole Cambrian shale.

Therefore, in spite of the very high maturity of the Lower Paleozoic shale in south China, gas content and yield are not significantly affected within a definite maturity range (EqR_o < 3.5 %), and maturity is not the main controlling factor for gas content and yield in this area.

4 Tectonic deformation and gas content of shale

The Lower Paleozoic strata in south China experienced a few significant tectonic deformations related to the Caledonian, Indo-China, Yanshan, and Himalayan movements (Wang et al. 2012a; Li et al. 2013a), but the late Yanshan–Himalayan tectonic movement has had the greatest effect on shale gas exploration and development of the Lower Paleozoic shale (Li et al. 2013a). The tectonic deformation led to strong uplifting and denudation in the whole Yangtze region. Most of the area outside the Sichuan Basin underwent faulting and folding (Deng et al. 2009; Ma et al. 2006; Guo et al. 2014; Wang et al. 2013b; Wei et al. 2014). Similar to coal bed methane in China, the preservation of the Lower Paleozoic shale gas is mainly controlled by the intensity and pattern of the tectonic deformation (Guo and Zhang 2014; Guo 2014; Guo et al. 2014; Hu et al. 2014; Song et al. 2005, 2009). The tectonic deformation can be largely divided into two patterns. One is a simple uplifting, with or without folding (called an uplifting/folding pattern in the present paper), represented by the Sichuan Basin. Another is a combination of uplifting, folding, and faulting (called a faulting/folding pattern in the present paper), which may cover most of the Yangtze region outside the Sichuan Basin, represented by the shale gas blocks in southeast Chongqing. The present study will focus on the discussion of shale gas content and loss under the background of the two types of tectonic deformation.

4.1 Uplifting/folding and loss of shale gas

In the Sichuan Basin, the strong uplifting led to the erosion of the Lower Cretaceous–Jurassic strata after the late Cretaceous, with a total erosion thickness of 2000–5000 m (Hu et al. 2014; Wang and Xiao 2010; Zou et al. 2014), and, as a result, a portion of the Lower Paleozoic shale has a current burial depth suitable for shale gas exploitation (Li et al. 2013a). In the southeastern area of the Sichuan Basin, the shale system with a strong overpressure is far away from faults, eroded surfaces, and basin edges, with a pressure coefficient generally greater than 1.5 and up to 2.25 (e.g., well Y101) (Hu et al. 2014), which implies that simple uplifting/folding can basically retain the fluid pressure coefficient of a shale system. In the process of uplifting and erosion, the fluid pressure will increase with decreasing temperature and pressure (Zhou et al. 2014b), which may break its original dynamic balance with a significant loss of shale gas.

In the present study, the Lower Silurian shale from the Fuling Block was taken as an example to establish a geological model of shale gas storage capacity and loss (Fig. 7). For this model, the absorbed gas content and adsorbed phase volume were calculated using high-pressure adsorption experimental data at different temperatures (40–120 °C) of a Lower Silurian shale (total porosity = 3.25 %) from the southern area of the Sichuan Basin, the free gas content was simulated by PVTsim software (Zhou et al. 2014b), and the related geological parameters include a constant pressure coefficient of 1.5 during uplifting (Hu et al. 2014), a gas saturation of 80 % (Wei and Wei 2014), a geothermal gradient of 2.5 °C/100 m, and a surface temperature of 15 °C. The maximum burial depth of the Lower Silurian shale in the Fuling Block was estimated to be about 7000 m (Li et al. 2013a), with a total gas content of 3.79 m^3/t for this shale sample (calculated by using its gas-filling porosity of 2.6 % under the above conditions). During the uplifting, the loss of shale gas could be roughly divided into three stages. From 7000 to 4000 m, the lost gas is less than 5 %. From 4000 to 2000 m, the loss rate increases, with a cumulative amount of up to 20 %. With further uplifting, the gas loss is

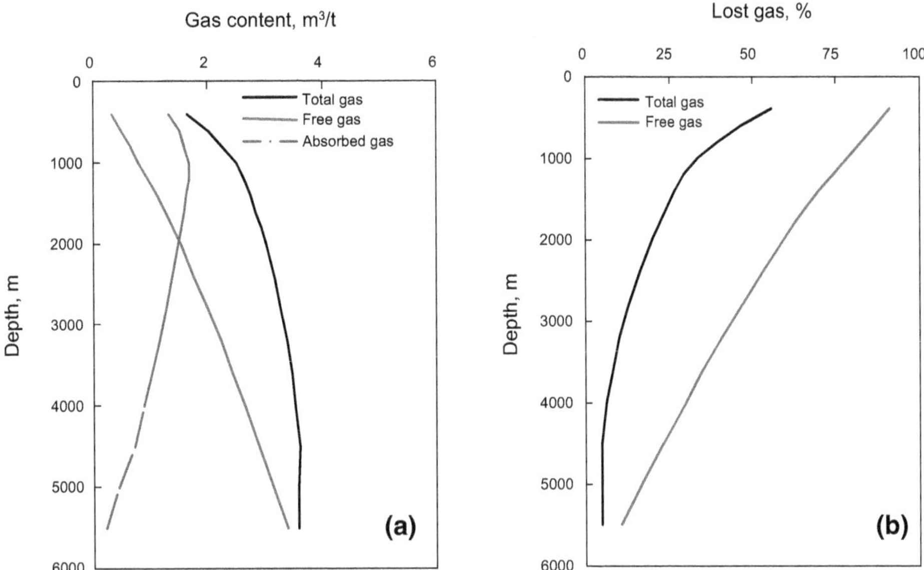

Fig. 7 A geological model of gas content (**a**) and loss (**b**) of the Lower Silurian shale in the Sichuan Basin (a geothermal gradient of 2.5 °C/100 m, a surface temperature of 15 °C, a fluid pressure coefficient of 1.5, and a gas-filling porosity of 2.6 %)

massive, with a cumulative loss of over 33 % as the depth decreases to 1000 m. According to the changes in both free gas and absorbed gas contents during the uplifting, it is obvious that a portion of the reduced free gas is transformed into absorbed gas, and the remaining portion is lost. When the burial depth is greater than 5000 m, the total gas is basically composed of free gas. When uplift reduces the burial depth to 4000, 2000, and 1000 m, the percentage of free gas becomes 75 %, 50 %, and 32 %, respectively. Therefore, the lost gas is mainly free gas during the uplifting, and a sufficient current burial depth (>2000–2500 m) is a key condition to obtain a high shale gas yield.

It should be pointed out that the illustration of Fig. 7 is only an example. It is limited by the geochemical and reservoir properties of the studied sample as well as the specific geological conditions. The geological model of shale gas content and loss would vary with different types of shale and their geological conditions, but the basic characteristics presented in Fig. 7 should be representative of the Lower Paleozoic shale in south China.

4.2 Faulting/folding and loss of shale gas

For the current exploration blocks outside the Sichuan Basin, the shale system of the Lower Paleozoic strata has a normal or negative pressure, with a pressure coefficient of 0.8–1.2 (Hu et al. 2014). Thus, the calculation of the lost gas in these areas should include two aspects: (1) gas loss related to the decrease of temperature and pressure caused by the uplifting as indicated above, and (2) gas loss related to the reduction of fluid pressure coefficient caused by the faulting and folding.

Taking the Lower Silurian shale in the southeast area of Chongqing as an example, it was assumed that the paleo-pressure coefficient of the shale system was 1.5 before the uplifting and the current pressure coefficient is 1.0 (Guo 2014). This decrease of the pressure coefficient is assumed to be attributed to the faulting and folding. In order to make a comparison with the gas model of the uplifting/folding (Fig. 7), a total porosity of 3.25 % and a gas saturation of 80 % (no available data in this area at present) were applied.

With the constraint of the above parameters, the high-pressure adsorption experimental data and free gas calculation method as described in the above section (Sect. 4.1) were applied to calculate the gas content at different burial depths for the faulting/folding block. Combined with the lost gas caused by the decrease of pressure coefficient in the shale system, the geological model of shale gas content and loss was established (Fig. 8).

Based on the maximum burial depth of 7000 m of the Lower Silurian shale in the southeast area of Chongqing during geological history, the total gas content was calculated to be 3.79 m^3/t before the uplifting. The current burial depth ranges mainly from 5000 to 500 m. At a depth of 5000–4000 m, the total gas content only slightly decreases and the reduced free gas is basically turned into absorbed gas. At a depth of 4000–2000 m, the total gas content decreases, most of the reduced free gas becomes absorbed gas, and the absorbed gas content exceeds the free gas content at a depth of about 2500 m. At a shallower depth

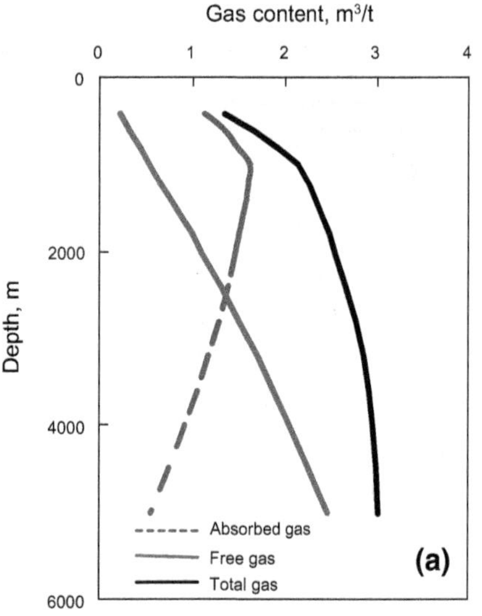

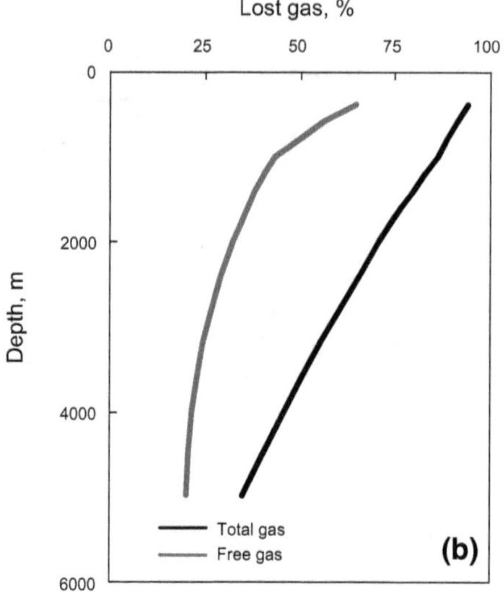

Fig. 8 A geological model of gas content (**a**) and loss (**b**) of the Lower Silurian shale in the southeast area of Chongqing (a geothermal gradient of 2.5 °C/100 m, a surface temperature of 15 °C, a decrease of fluid pressure coefficient from 1.5 to 1.0, a TOC content of 2.53 %, and a gas-filling porosity of 2.6 %)

(<2000 m), a large amount of gas is lost and the shale gas is dominated by the absorbed gas. Thus, the decrease of the fluid pressure coefficient in the shale system is accompanied by a great deal of free gas loss. At a depth of 4000, 3000, 2000, and 1000 m, the lost free gas amounts to 46 %, 58 %, 71 %, and 86 %, respectively.

By comparison between the two models (Figs. 7, 8), it can be seen that, for the same burial depth, the fluid pressure coefficient has no significant effect on the absorbed gas, but has a great influence on the free gas. For a shale system with a lower pressure coefficient, the free gas content is much less and the ratio of the free gas to absorbed gas is also obviously lower (Fig. 9). Therefore, free gas is controlled by the fluid pressure coefficient of the shale system. In a faulted–folded zone with a lower fluid pressure coefficient, the lost gas is basically free gas, which leads to a lower yield of shale gas. Deep burial shale with a relatively higher fluid pressure/a larger porosity, as a result of a higher content of free gas, would be a basic condition to achieve a high shale gas yield in the folded–faulted zone.

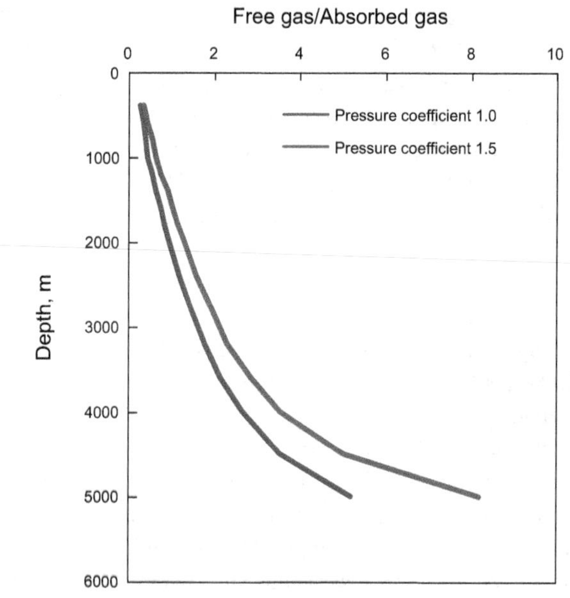

Fig. 9 Changes in the ratio of free gas to absorbed gas with burial depth for two shale systems with different pressure coefficients

5 Main controlling factors and enrichment area of shale gas

5.1 Main controlling factors of shale gas

Although a series of factors can affect shale gas production (Jarvie et al. 2007) and those factors may vary significantly for different basins or regions, a high content of gas-in-place (GIP) and effective fracturing properties of shale are mostly of importance. For a specific basin or block, there is a main controlling factor for shale gas production. For examples, maturity is the main controlling factor for the Barnett Shale and the high-yield wells are mostly located in areas with a higher maturity (R_o > 1.1 %–1.2 %) (Jarvie et al. 2007); the Marcellus Shale has two core areas and they have a very high TOC content (with an average of 4 %–10 %) although their maturity is quite different (Zagorski et al. 2011); the gas production of the Haynesville Shale is obviously restricted by its mineral compositions, and two core areas have a strong overpressure, located in quartz- and carbonate-rich zones, respectively (Buller and Dix, 2009; Hammes et al. 2010). The maturity of the lower Paleozoic shale in south China is very high, but the most is still within the maturity range (2.0 % < EqR_o < 3.5 %) with an industrial potential, and the preservation condition related to the tectonic deformation is the main factor to control gas content in this kind of area. Shale with a super high maturity (EqR_o > 3.5 %) is believed to be a high risk for industrial shale gas potential.

5.2 Prediction of enrichment area of shale gas

Referring to relevant literatures (Boyer et al. 2006; Burnaman et al. 2009; Curtis et al. 2009; Hill and Nelson 2000; Jarvie et al. 2007; Wang et al. 2012a, 2013c; Li et al. 2013b), the present study proposes a few parameters for predicting a shale gas enrichment area in the lower Paleozoic strata in south China (Table 3). As indicated in the front sections of the present paper, some basic geological and geochemical parameters of the lower Paleozoic shale in south China can be compared with those of the gas shale in North America, mainly including TOC content, brittle mineral content, clay mineral content, and effective shale thickness. These parameters can use the empirical data from the North America shale as presented by some Chinese authors (Wang et al. 2012a; Li et al. 2013b). Focusing on two characteristics (very high maturity and strong tectonic deformation) of the lower Paleozoic shale in south China, the present study modified the limiting value of the maturity and added the burial depth and pressure coefficient as evaluation parameters.

The lower limit of maturity for the North American shale with an industrial gas potential is usually considered to be an R_o value >1.2 % (Curtis et al. 2009; Jarvie et al. 2007), and the upper limit of maturity is not taken into account mainly because the maturity of the North American gas shale is usually lower, not exceeding the gas generation limit (Curtis 2002). As previously mentioned, the maturity of the lower Paleozoic shale in south China is very high, with some areas where the shale maturity is over the gas generation limit (Cheng and Xiao 2013). High gas production from shale with an R_o or EqR_o value of up to 3.0 %–3.5 % has been reported during recent few years (Zagorski et al. 2011; Guo et al. 2014), which has broken through the upper limit value of 3.0 % of R_o suggested by Burnaman et al. (2009) for a commercial shale gas reservoir. However, as shale maturity is significantly over its gas generation limit, a further burial may cause an obvious decrease of the porosity, which will reduce gas content and increase risk. When shale has matured to exceed the hydrocarbon gas preservation threshold (i.e., R_o > 4 %–5 %), there is no longer shale gas potential (Burnaman et al. 2009; Li et al. 2013a). Therefore, the EqR_o value of 3.5 % is proposed as the upper limit of the shale gas enrichment area.

Table 3 Predictive parameters of enrichment area of shale gas for the lower Paleozoic shale in south China

Parameters	Enrichment area	
	Uplifted/folded area (Sichuan Basin)	Folded–faulted area (outside the Sichuan Basin)
TOC, %	>2.0	>3.0
EqR_o, %	<3.5	<3.5
Total porosity, %	>4.0	>4.0
Brittle mineral, %	>40	>40
Thickness, m	>30	>30
Burial depth, m	>1500	>3000
Fluid pressure coefficient	>1.2	>1.0

Gas shale with commercial production in North America has a burial depth mainly in the range of 1500–3500 m, and the burial depth for economically recoverable shale gas is generally limited to 4000–5000 m (EIA 2013), which is mainly based on the consideration of the technology and cost related to shale gas exploitation. In fact, shale with a deeper burial has a greater gas content and free gas proportion than shale with a shallower burial (Figs. 7, 8). The evaluation parameters for the North America shale do not include the shallower limit of the burial depth, and successful commercial exploitation has been achieved in some shallow shales (<1000 m) (Curtis 2002), which is mainly attributed to their greater porosity with a high free gas content (Curtis 2002; Li et al. 2013b). Porosity of the Lower Paleozoic shale in south China is significantly lower than that of the North American shale (Wang et al. 2013a). To ensure a certain amount of gas storage capacity, especially free gas reserves, the Lower Paleozoic shale in south China needs a higher fluid pressure coefficient and/or a greater depth. Taking a gas content of 2.80 m^3/t as the lower limit of shale with a commercial shale gas potential suggested by Halliburton (Wang et al. 2012a), shale with a gas-filling porosity of 2.6 % should have a burial depth of about 1500 m to achieve this gas content under a pressure coefficient of 1.5 according to Fig. 7. When the pressure coefficient is 1.0, the corresponding depth is about 3000 m according to Fig. 8. Actually, except for the lower Silurian shale in the Sichuan Basin, the average gas-filling porosity of the other lower Paleozoic shales in the Upper Yangtze area rarely exceeds 2.6 % from the available data (Huang et al. 2012b; Huang and Shen 2015; Yi and Zhao 2014; Yu et al. 2014), which means that shale should have a greater burial depth to meet this gas content. Therefore, burial depth has an important significance for the Lower

Paleozoic shale gas exploration and development in south China. Within a definite maturity range (2.0 % < EqR_o < 3.5 %), the gas-in-place and free gas content increases with increasing burial depth. To exploit shale gas with a greater depth will become a trend. For example, the core shale gas area in north Canada has reached 5000 m (Ross and Bustin 2007; 2008). The present paper does not set the deeper limit of burial, but suggests a shallower burial limit of 1500 and 3000 m for the uplifted/folded and faulted/folded areas, respectively.

The evaluation of the North American shale does not include the pressure coefficient of a shale system as a parameter, which is also because of a greater porosity of the shale. Commercial shale gas production has been achieved in shale systems with a normal pressure, but an overpressure can still greatly increase the initial production, thus improving the commercial value (Curtis 2002; Curtis et al. 2008; Jarvie et al. 2007).

There exists a good positive correlation between the initial production of shale gas and the fluid pressure coefficient for the Lower Paleozoic shale in south China (Hu et al. 2014), and the high-yield wells in the Sichuan Basin are almost always associated with an overpressure. Wells outside the Sichuan Basin with an industrial shale gas production are reported at a normal pressure, such as well PY-1 with a yield of 2.5×10^4 m^3/d (a pressure coefficient of 0.90–1.0), but for a negative pressure shale system, there is no report of an industrial capacity, for example, well Z101 without any yield (a pressure coefficient of 0.80) (Guo 2014; Guo et al. 2014). To search for a block with better preservation conditions and a higher gas content from the faulted/folded area will be a real challenge for the exploration and development of the lower Paleozoic shale gas in south China. The pressure coefficient is an important parameter of shale gas preservation conditions, with a dominant control on shale gas content and production for the lower Paleozoic shale in south China. Pressure coefficients of 1.2 and 1.0 are suggested as the lower limit for shale gas enrichment in the uplifted/folded and faulted/-folded areas, respectively.

It should be particularly pointed out that shale gas exploration and development in China has only just started. There are only about 300–400 wells drilled specifically for shale gas. Among them, the number of development wells is less than 150. The targeted formation is mainly the lower Silurian shale, and few data are available for the lower Cambrian shale. Therefore, the proposed criteria are primarily intended for the evaluation of the lower Silurian shale, but whether it is suitable for the lower Cambrian shale requires the input of more data from further exploration and development.

6 Conclusions

Through a comprehensive investigation of the geochemical characteristics, reservoir properties, and gas content of the Lower Paleozoic marine shale from the Sichuan Basin and its surrounding areas in south China, we have arrived at the following conclusions:

(1) The shale has a very high maturity, with an EqR_o value of 1.9 %–6.2 %. Within the main maturity range (2.5 % < EqR_o < 3.5 %), the porosity develops well, mainly between 3 % and 5 %, and has a positive relationship with TOC content. Shale with a super high maturity (EqR_o > 3.5 %) has a quite low porosity and has a low potential for industrial shale gas.

(2) Shale gas content and yield are mainly constrained by the preservation conditions when the maturity is less than 3.5 % EqR_o. In the Sichuan Basin where the tectonic deformation is relatively gentle, the shale system is generally overpressure. The strong uplifting and erosion has not led to a big loss of shale gas (a burial depth > 2000 m), still with an industrial potential. In the faulted/folded zone outside the Sichuan Basin (a burial depth < 2000–3000 m), folding and faulting may lead to a big loss of shale gas, thus decreasing the production capacity.

(3) Burial depth, pressure coefficient, and maturity upper limit (EqR_o < 3.5 %) were suggested to be used as the evaluation parameters of shale gas enrichment areas. It was predicted that gentle synclines or anticlines where the shale has a burial depth >3000 m are the favorable areas for shale gas in the faulted–folded area outside the Sichuan Basin in south China, and these will become the focus for future exploration and development.

Acknowledgments This study was jointly supported by the National Key Basic Research Program of China (973 Program: 2012CB214700), the Strategic Priority Research Program of the Chinese Academy of Sciences (XDB10040300), and the National Natural Science Foundation of China (41321002).

appropriate credit to the original author(s) and the source, provide a link to the Creative Commons license, and indicate if changes were made.

References

Boyer C, Kieschnick J, Suarez-Rivera R, et al. Producing gas from its source. Oilfield Rev. 2006;18:36–49.

Burnaman M, Xia WW, Shelton J. Shale gas play screening and evaluation criteria. China Pet Explor. 2009;14:51–64.

Buller D, Dix MC. Petrophysical evaluation of the Haynesville shale in northwest Louisiana and northeast Texas. Gulf Coast Assoc Geol Soc Trans. 2009;59:127–43.

Cander H. Sweet spots in shale gas and liquids plays: prediction of fluid composition and reservoir pressure. AAPG Annual Convention and Exhibition, Long Beach, California, April 22–25, 2012.

Chalmers GR, Bustin RM. Lower Cretaceous gas shales in northeastern British Columbia, Part I: geological controls on methane sorption capacity. Bull Can Pet Geol. 2008;56(1):1–21.

Chalmers GR, Bustin RM, Power IM. Characterization of gas shale pore systems by porosimetry, pycnometry, surface area, and field emission scanning electron microscopy/transmission electron microscopy image analyses: examples from the Barnett, Woodford, Haynesville, Marcellus, and Doig units. AAPG Bull. 2012;96(6):1099–119.

Chen J, Xiao XM. Evolution of nanoporosity in organic-rich shales during thermal maturation. Fuel. 2014;129:173–81.

Cheng P, Xiao XM. Gas content of organic-rich shales with very high maturities. J China Coal Soc. 2013;8(5):737–41 **(in Chinese)**.

Curtis JB, Hill DG, Lillis PG. US Shale gas resources: classic and emerging plays, the resource pyramid and a perspective on future E&P. Prepared for presentation at AAPG Annual Convention. San Antonio, Texas. April 20–23, 2008.

Curtis JB, Jarvie DM, Ferworn KA. Applied geology and geochemistry of gas shales. AAPG Conference 2009, Denver Colorado, 2009.

Curtis JB. Fractured shale-gas systems. AAPG Bull. 2002;86: 1921–38.

Curtis ME, Ambrose RJ, Sondergeld CH, et al. Structural characterization of gas shales on the micro- and nano-scales. Canadian Unconventional Resources and International Petroleum Conference, 19–21 October, Calgary, Alberta, Canada. Society of Petroleum Engineers (SPE 137693), pp. 1–15, 2010.

Curtis ME, Cardott BJ, Sondergeld CH, et al. Development of organic porosity in the Woodford Shale with increasing thermal maturity. Int J Coal Geol. 2012;103:26–31.

Deng B, Liu SG, Liu S, et al. Restoration of exhumation thickness and its significance in Sichuan Basin, China. J Chengdu Univ Technol (Sci Technol Edn). 2009;36(6):675–86 **(in Chinese)**.

Fishman NS, Hackley PC, Lowers HA, et al. The nature of porosity in organic-rich mudstones of the Upper Jurassic Kimmeridge Clay Formation, North Sea, offshore United Kingdom. Int J Coal Geol. 2012;103:32–50.

Guo XS. Rules of two-factor enrichment for marine shale gas in southern China—Understanding from the Longmaxi Formation shale gas in Sichuan Basin and its surrounding area. Acta Geol Sin. 2014;88(7):1209–18 **(in Chinese)**.

Guo XS, Hu DF, Li YP, et al. Geological features and reservoiring mode of shale gas reservoirs in Longmaxi Formation of the Jiaoshiba Area. Acta Geol Sin. 2014;88(6):1811–21.

Guo TL, Liu RB. Implications from marine shale gas exploration breakthrough in complicated structural area at high thermal stage: taking Longmaxi Formation in well JY1 as an example. Nat Gas Geosci. 2013;24(4):643–51 **(in Chinese)**.

Guo TL, Zhang HR. Formation and enrichment mode of Jiaoshiba shale gas field, Sichuan Basin. Pet Explor Dev. 2014;41(1): 28–36 **(in Chinese)**.

Hill DG, Nelson CR. Gas productive fractured shales—an overview and update. Gas Tips. 2000;6(2):4–13.

Hu DF, Zhang HR, Ni K, et al. Main controlling factors for gas preservation conditions of marine shales in southeastern margins of the Sichuan Basin. Nat Gas Ind. 2014;34(6):17–23 **(in Chinese)**.

Huang JL, Zou CN, Li JZ, et al. Shale gas accumulation conditions and favorable zones of Silurian Longmaxi Formation in south Sichuan Basin, China. J China Coal Soc. 2012a;37(5):782–7 **(in Chinese)**.

Huang JL, Zou CN, Li JZ, et al. Shale gas generation and potential of the Lower Cambrian Qiongzhusi Formation in southern Sichuan Basin, China. Pet Explor Dev. 2012b;39(1):69–75 **(in Chinese)**.

Huang L, Shen W. Characteristics and controlling factors of the formation of pores of a shale gas reservoir: a case study from Longmaxi Formation of the Upper Yangtze region, China. Earth Sci Front. 2015;22(1):374–85 **(in Chinese)**.

Hammes U, Hamlin S, Eastwood R. Facies characteristics, depositional environments, and petrophysical characteristics of the Haynesville and Bossier shale-gas plays of east Texas and northwest Louisiana. Houston Geol Soc Bull. 2010;52(9):59–63.

Jarvie DM, Hill RJ, Ruble TE, et al. Unconventional shale-gas systems: the Mississippian Barnett Shale of north central Texas as one model for thermogenic shale gas assessment. AAPG Bull. 2007;91(4):475–99.

Li LJ, Yang F, Sun CM, et al. Tectonic movement and lower Silurian shale gas exploration potential in the southeast of Chongqing and west of Hunan and Hubei areas. J Oil Gas Technol. 2013a;35(4):68–71 **(in Chinese)**.

Li XT, Shi WR, Guo MY, et al. Characteristics of marine shale gas reservoirs in the Jiaoshiba area of Fuling shale gas field. J Oil Gas Technol. 2014;36(11):11–5 **(in Chinese)**.

Li YJ, Liu H, Zhang LH, et al. Lower limits of evaluation parameters for the lower Paleozoic Longmaxi shale gas in southern Sichuan Province. Sci China Earth Sci. 2013b;56(5):710–7.

Li YF, Fan TL, Gao ZQ, et al. Sequence stratigraphy of Silurian black shale and its distribution in the southeast area of Chongqing. Nat Gas Geosci. 2012;23(2):299–306 **(in Chinese)**.

Liang DG, Guo TL, Chen JP, et al. Some progresses on studies of hydrocarbon generation and accumulation in marine sedimentary regions, southern China (Part 2): geochemical characteristics of four suits of regional marine source rocks, south China. Mar Origin Pet Geol. 2009;14(1):1–15 **(in Chinese)**.

Liang DG, Guo TL, Chen JP, et al. Some progresses on studies of hydrocarbon generation and accumulation in marine sedimentary regions, southern China (Part 1): distribution of four suits of regional marine source rocks. Mar Origin Pet Geol. 2008;13(2):1–16 **(in Chinese)**.

Loucks RG, Reed RM, Ruppel SC, et al. Morphology, genesis, and distribution of nanometer-scale pores in siliceous mudstones of the Mississippian Barnett Shale. J Sediment Res. 2009;79(12): 848–61.

Ma YS, Lou ZH, Guo TL, et al. An exploration on a technological system of petroleum preservation evaluation for marine strata in south China. Acta Geol Sin. 2006;80(3):406–17 **(in Chinese)**.

Ma C, Ning N, Wang HY, et al. Exploration prospect of the Lower Cambrian Niutitang Formation shale gas in the west of Hunan and the east of Guizhou area. Spec Oil Gas Reserv. 2014;21(1):38–41 **(in Chinese)**.

Mastalerz M, Schimmelmann A, Drobniak A, et al. Porosity of Devonian and Mississippian New Albany Shale across a maturation gradient: insights from organic petrology, gas adsorption, and mercury intrusion. AAPG Bull. 2013;97:1621–43.

Mei LF, Liu ZQ, Tang JG, et al. Mesozoic intra-continental progressive deformation in western Hunan–Hubei–eastern Sichuan provinces of China: evidence from apatite fission track and balanced cross-section. Earth Sci. 2010;35(2):161–74 **(in Chinese)**.

Milliken KL, Esch WL, Reed RM, et al. Grain assemblages and strong diagenetic overprinting in siliceous mudrocks, Barnett Shale (Mississippian), Fort Worth Basin, Texas. AAPG Bull. 2012;96(9):1553–78.

Milliken KL, Rudnicki M, Awwiller DN, et al. Organic matter-hosted pore system, Marcellus Formation (Devonian), Pennsylvania. AAPG Bull. 2013;97(2):177–200.

Modica CJ, Lapierre SG. Estimation of kerogen porosity in source rocks as a function of thermal transformation: example from the Mowry Shale in the Powder River Basin of Wyoming. AAPG Bull. 2012;96(1):87–108.

Nie HK, Tang X, Bian RK. Controlling factors for shale gas accumulation and prediction of potential development area in shale gas reservoir of south China. Acta Pet Sin. 2009;30(4):484–91 **(in Chinese)**.

Nie HK, He FQ, Bao SJ. Peculiar geological characteristics of shale gas in China and its exploration countermeasures. Nat Gas Ind. 2011;31(11):111–6 **(in Chinese)**.

Pan L, Xiao XM, Tian H, et al. A preliminary study on the characterization and controlling factors of porosity and pore structure of the Permian shales in Lower Yangtze region, Eastern China. Int J Coal Geol. 2015;146:68–78.

Ross DJK, Bustin RM. Shale gas potential of the Lower Jurassic Gordondale Member, northeastern British Columbia, Canada. Bull Can Pet Geol. 2007;55(1):51–75.

Ross DJK, Bustin RM. Characterizing the shale gas resource potential of Devonian Mississippian strata in the Western Canada sedimentary basin: application of an integrated formation evaluation. AAPG Bull. 2008;92:87–125.

Ross DJK, Bustin RM. The importance of shale composition and pore structure upon gas storage potential of shale gas reservoirs. Mar Pet Geol. 2009;26(6):916–27.

Song Y, Zhao MJ, Liu SB, et al. The influence of tectonic evolution on the accumulation and enrichment of coalbed methane (CBM). Chin Sci Bull. 2005;50(supp.):1–6.

Song Y, Liu SB, Zhao MJ, et al. Coalbed gas reservoirs: boundary types, main controlling factors of gas pooling, and forecast of gas-rich areas. Nat Gas Ind. 2009;29(10):4–9 **(in Chinese)**.

Tang LJ, Cui M. Key tectonic changes, deformation styles and hydrocarbon preservation in middle-upper Yangtze region. Pet Geol Exp. 2011;33(1):12–6 **(in Chinese)**.

Tian H, Pan L, Xiao XM, et al. A preliminary study on the pore characterization of Lower Silurian black shales in the Chuandong Thrust Fold Belt, southwestern China using low pressure N_2 adsorption and FE-SEM methods. Mar Pet Geol. 2013;48:8–19.

Tian H, Pan L, Zhang TW, et al. Pore characterization of organic-rich Lower Cambrian shales in Qiannan Depression of Guizhou Province, Southwestern China. Mar Pet Geol. 2015;62:28–43.

Tu Y, Zou HY, Meng HP, et al. Evaluation criteria and classification of shale gas reservoirs. Oil Gas Geol. 2014;35(1):153–8 **(in Chinese)**.

U. S. Energy Information Administration (EIA). World shale gas resources: An initial assessment of 14 regions outside the United States [EB/OL]. 2011-04-05. http://www.eia.gov.

U. S. Energy Information Administration (EIA). Technically recoverable shale oil and shale gas resources: An assessment of 137 shale formations in 41 countries outside the United States [EB/OL]. 2013-06-13. http://www.eia.gov.

Valenza JJ, Drenzek N, Marques F, et al. Geochemical controls on shale microstructure. Geology. 2013;41(5):611–4.

Wang FY, Guan J, Feng WP, et al. Evolution of overmature marine shale porosity and implication to the free gas volume. Pet Explor Dev. 2013a;40(6):819–24.

Wang HY, Liu YZ, Dong DZ, et al. Scientific issues on effective development of marine shale gas in southern China. Pet Explor Dev. 2013b;40(5):574–9.

Wang YF, Xiao XM. An investigation of paleogeothermal gradients in the northeastern area of the Sichuan Basin. Mar Orig Petrol Geol. 2010;15(4):57-61 **(in Chinese)**.

Wang YM, Dong DZ, Li JZ, et al. Reservoir characteristics of shale gas in Longmaxi Formation of the Lower Silurian, southern Sichuan. Acta Pet Sin. 2012a;33(4):551–61 **(in Chinese)**.

Wang LS, Zou CY, Zheng P, et al. Geochemical evidence of shale gas existed in the Lower Paleozoic Sichuan Basin. Nat Gas Ind. 2009a;29(5):59–62 **(in Chinese)**.

Wang SJ, Yang T, Zhang GS, et al. Shale gas enrichment factors and the selection and evaluation of the core area. Chin Eng Sci. 2012b;14(6):94–100 **(in Chinese)**.

Wang SQ, Wang SY, Man L, et al. Appraisal method and key parameters for screening shale gas play. J Chengdu Univ Technol (Sci Technol Edn). 2013c;40(6):609–20 **(in Chinese)**.

Wang SQ. Shale gas exploration and appraisal in China: problems and discussion. Nat Gas Ind. 2013;33(2):13–29 **(in Chinese)**.

Wang SQ, Chen GS, Dong DZ, et al. Accumulation conditions and exploration prospects of shale gas in the Lower Paleozoic Sichuan Basin. Nat Gas Ind. 2009b;29(5):51–8 **(in Chinese)**.

Wang ZG. Practice and cognition of shale gas horizontal well fracturing stimulation in Jiaoshiba of Fuling area. Oil Gas Geol. 2014;35(3):425–30 **(in Chinese)**.

Wei XH, Wang T, Li C, et al. Exploration prospect of shale gas in complex geological areas, southeastern Chongqing. Nat Gas Technol Econ. 2014;8(1):24–32 **(in Chinese)**.

Wei ZH, Wei XF. Comparison of gas-bearing property between different pore types of shale: a case from the Upper Ordovician Wufeng and Longmaxi formations in the Jiaoshiba area, Sichuan Basin. Nat Gas Ind. 2014;34(6):37–41 **(in Chinese)**.

Xiao XM, Song ZG, Zhu YM, et al. Summary of shale gas research in North American and revelations to shale gas exploration of Lower Paleozoic strata in China south area. J China Coal Soc. 2013;38(5):721–7 **(in Chinese)**.

Yan DP, Wang XW, Liu YY. Analysis of fold style and its formation mechanism in the area of boundary among Sichuan, Hubei and Hunan. Geoscience. 2000;14(1):37–43 **(in Chinese)**.

Yi TS, Zhao X. Characteristics and distribution patterns of the Lower Cambrian Niutitang shale reservoirs in Guizhou, China. Nat Gas Ind. 2014;34(8):8–14 **(in Chinese)**.

Yu C, Nie HK, Zeng CL, et al. Shale reservoir space characteristics and the effect on gas content in Lower Palaeozoic Erathem of the eastern Sichuan Basin. Acta Geol Sin. 2014;88(7):1311–20 **(in Chinese)**.

Yu ZG. Introduction to shale gas exploration and development in the Sichuan Basin. 973 Program (2012CB214700) Academic Exchange Symposium. In August, 2014. Lanzhou, China.

Zagorski WA, Bowman DC, Emery M, et al. An overview of some key factors controlling well productivity in core areas of the Appalachian Basin Marcellus Shale play. AAPG Annual convention and exhibition, Houston, Texas, USA, oral presentation, 2011.

Zhang DW, Li YX, Zhang JC, et al. Nation-wide shale gas resource potential survey and assessment. Beijing: Geological Publishing House; 2012 **(in Chinese)**.

Zhou QH, Song N, Wang CZ, et al. Geological evaluation and exploration prospect of Huayuan shale gas block in Hunan Province. Nat Gas Geosci. 2014a;25(1):130–9 **(in Chinese)**.

Zhou Q, Xiao XM, Tian H, et al. Modeling free gas content of the Lower Paleozoic shales in the Weiyuan area of the Sichuan Basin, China. Mar Pet Geol. 2014b;56:87–96.

Zou CN, Dong DZ, Wang SJ, et al. Geological characteristics, formation mechanism and resource potential of shale gas in China. Pet Explor Dev. 2010;37(6):641–53 **(in Chinese)**.

Zou CN, Du JH, Xu CC, et al. Formation, distribution, resource potential and discovery of the Sinian—Cambrian giant gas field, Sichuan Basin, SW China. Pet Explor Dev. 2014;41(3):278–93 **(in Chinese)**.

14

Investigation of structural parameters and self-aggregation of Algerian asphaltenes in organic solvents

Asma Larbi[1] · Mortada Daaou[1] · Abassia Faraoun[1,2]

Abstract Elemental analysis, Fourier transform infrared (FTIR), ^{1}H-NMR, fluorescence spectroscopy, and surface tension methods have been used to characterize the molecular structure and the aggregation behaviors of two asphaltenic fractions derived, respectively, from an Algerian petroleum well and a corresponding storage tank deposit. Elemental analysis, FTIR, ^{1}H-NMR, and fluorescence spectroscopy were used to investigate the chemical composition and structural parameters of asphaltenes, while the surface tension method was used to measure the critical micelle concentration (CMC) in organic solvents with different solubility parameters and polarities in order to characterize the asphaltenes' aggregation behaviors. Results show that the unstable asphaltenes fraction extracted from the storage tank deposit possesses a higher polarity (higher heteroatoms content) and a lower aromaticity than stable asphaltenes from the petroleum well. The CMC results indicate that asphaltenes with high polarity and low aromaticity have a high solubility in polar solvents such as nitrobenzene, whereas asphaltenes with low polarity and high aromaticity are more soluble in solvents with weak polarity, like toluene. It is concluded that the difference of structure of asphaltene samples and polarity of solvents can lead to difference of aggregation behaviors.

Keywords Asphaltenes · Algerian petroleum · Structural parameters · CMC

✉ Mortada Daaou
daaou_mortada@yahoo.fr

1 Laboratoire de Synthèse organique, Physico-chimie, Biomolcules et Environnement (LSPBE), Department of Industrial Organic Chemistry, Faculty of Chemistry, University of Sciences and Technology of Oran, Po. Box 1505, 31000 El-M'naouer, Algeria

2 Laboratoire de Chimie Physique Macromoléculaire (LCPM), Department of Chemistry, Faculty of Sciences, University of Oran, Po. Box 1524, 31000 El-M'naouer, Algeria

Edited by Xiu-Qin Zhu

1 Introduction

Asphaltenes constitute the most complex fraction of crude oil. They are conventionally defined as a fraction of crude oil that is insoluble in *n*-alkanes but soluble in toluene. Because of their natural trend to form aggregates that may flocculate and precipitate, asphaltenes can cause severe problems in oil production, transportation, and refining. The propensity of asphaltenes to flocculate is a complex function of the crude oil composition and does not depend directly on the asphaltene content (Kokal and Sayegh 1995). Algerian crude oil is very unstable with respect to aggregation despite the low content of asphaltenes. It is believed that the structural features of asphaltene molecules leading to self-assembly should be taken into account in the study of the probable aggregation mechanism (Yang et al. 2004). According to chemical structure, asphaltenes are aromatic polycyclic molecules surrounded and linked by aliphatic chains and heteroatoms (sulfur, nitrogen, oxygen) as well as traces of metal elements like nickel, iron, and vanadium. The chemical composition and structure of single asphaltene molecules can be assessed using different analytical techniques (Calemma et al. 1995; Shirokoff et al. 1997; Pekerar et al. 1999; Bouhadda et al. 2000; Buenrostro-Gonzalez et al. 2002; Groenzin and Mullins 2000; Ancheyta et al. 2002; Ibrahim et al. 2003; Khan et al. 2003; Avid et al. 2004; Acevedo et al. 2005;

Siskin et al. 2006). Elemental analysis, mass spectroscopy, and ^{1}H NMR can be used for determination of the average structural parameters such as the aromatic carbon ratio, the average number of alkyl side chains, and the degree of peripheral aromatic carbon substitution (Dereppe et al. 1978; Dickinson 1980; Daaou et al. 2008, 2009).

The polarity and complex structure of asphaltenes can lead to their self-association, flocculation, and precipitation during the course of heavy oil processing. It has been realized that the first step in the formation of precipitated asphaltene particles is the self-aggregation of asphaltenes to form colloidal particles or pseudo-micelles in several solvents. A micelle is a reversible aggregate that is formed in a polar environment and which will remain constant in size and aggregation number for a given set of environmental constraints. Several studies have been carried out employing various methods to determine the "critical micelle concentration" (CMC) of asphaltenes, i.e., the concentration below which asphaltenes remain as monomer species (Taylor 1992). The experimental methods used, including calorimetry (Andersen and Bridi 1991; Andersen and Christensen 2000), surface tension (Sheu 1996; Carbognani et al. 1997; Mohamed et al. 1999; Rogel et al. 2000), viscosity (Storm et al. 1991), vapor pressure osmometry (Yarranton et al. 2000), NIR spectroscopy (Mullins 1990; Kyeongseok et al. 2004), and small-angle scattering measurements (Sheu and Storm 1995) show a variety of the CMC values and indicate that this parameter depends on the asphaltene origin and the solvent used. For instance, Andersen and Bridi (1991) have used microcalorimetric titration for measuring the CMCs of Venezuelan asphaltenes in different organic solvents. They have demonstrated that the breakpoint observed in the curve of the measured enthalpies of the asphaltenes concentrations represents the CMC that might vary from 0.5 to 15 g L^{-1}. Others investigators (Sheu 1996; Bouhadda et al. 2000; Mohamed et al. 1999; Rogel et al. 2000) have employed surface tension measurements to determine the CMC of asphaltenes from different origins in organic solvents. Sheu (1996) has determined the CMC of asphaltenes extracted from Ratawi vacuum residue in pyridine through surface tension measurements, and found a value around 0.4 g L^{-1}. Other surface tension studies have reported that "CMCs of asphaltene in toluene" are much higher, for example, 10 g L^{-1} (Brazilian asphaltenes) in one study (Da Silva Ramos et al. 2001) and 1.7 g L^{-1} (Algerian asphaltenes) in another study (Bouhadda et al. 2000). Rogel et al. (2000) have found that the CMC values of unstable asphaltenes in different solvents, obtained by surface tension measurements, vary in the range of 1–30 g L^{-1}. It is believed that the self-aggregation behavior of asphaltenes is related to the structure and chemical properties of asphaltenes, and by analogy, related to the behavior of surfactants in aqueous and non-aqueous solutions. On the other hand, some reports have indicated that structure and characteristics of asphaltenes are some of the main factors that determine their relative stability in crude oils (Carbognani et al. 1997). The intermolecular forces including electrostatic interactions, van der Waals forces, intermolecular charge transfer, exchange–repulsion interaction, induction effects, hydrogen bonding, and the so-called π–π interactions are found to be responsible for the asphaltenes aggregation (Murgich et al. 1996; Murgich 2002).

The purpose of this work is comparing structural properties and the aggregation behavior of two asphaltene fractions extracted from two different sources of Algerian petroleum. The first was obtained from the crude oil of the Hassi-Messaoud field. The second fraction was obtained from the oil storage tank deposit formed after the usual treatment (separation of water, salts, and mineral sediments) and pipeline transportation. The physical characteristics of the crude oil and the storage oils are summarized in Table 1.

The structural parameters of both asphaltene fractions were investigated using elemental analysis, Fourier transform infrared (FTIR), H-NMR, and fluorescence spectroscopy. The asphaltene aggregation was assessed determining the CMC in three solvents (toluene, pyridine, and nitrobenzene) with different solubility parameters and polarity using surface tension method.

2 Materials and methods

Two asphaltene samples from Hassi-Messaoud petroleum in Algeria were investigated. The first $(AS)_{DP}$ was extracted from the storage tank deposit, while the second $(AS)_{WL}$ was derived from a Hassi-Messaoud oil well. The toluene, pyridine, and nitrobenzene were from Fisher Chemicals with 99 % purity, while the *n*-heptane was from Prolabo Chemicals.

The asphaltenes were extracted by addition to petroleum or deposit an excess of *n*-heptane in 40:1 (volume: mass) ratio. The mixture was subjected to 24 h constant, gentle shaking to ensure complete dispersion of the material. The precipitated asphaltene was then removed by vacuum filtration through a 0.45 μm pore size membrane. The solid obtained was thoroughly washed with *n*-heptane to remove co-precipitated maltenes until the washing solvent was colorless.

Fourier transform infrared spectra were collected with a Fourier transform IR spectrometer (Spectrum one, Perkin-Elmer) coupled with attenuated total reflectance module (ATR) with a spectral resolution of 4 cm^{-1} in the 700–4000 cm^{-1} spectral range. The solid samples were

Table 1 General information for the Hassi-Messaoud well and storage tank oils

	Hassi-Messaoud well oil	Stock tank oil
Gravity, °API	45	42
Viscosity, cp 40°	2.23	1.5
Asphaltene, % w/w	0.70	0.15
Resin, % w/w	24.5	13.3
Total acidity, mg-KOH/g	0.96	–

introduced directly in ATR top plates which have composite zinc selenide (ZnSe) and diamond crystals.

^{1}H-NMR spectra were obtained from a 250 MHz spectrometer (Bruker 250) with pulse width of 3.5 μs (30° flip angle), recycle delay of 2 s at least 600 scans, tube diameter 5 mm, and spectral width of 18 ppm. Chemical shifts (δ) are reported relative to tetramethylsilane (TMS) used as an internal standard and the spectra were recorded in deuterated-dichloromethane (CD_2Cl_2).

The samples prepared by dissolution in toluene at a concentration of 5 mg L^{-1} were examined by emission fluorescence spectroscopy (Fluoromax-3) in the range from 350 to 700 nm with an excitation wavelength of 256 nm.

Carbon, hydrogen, nitrogen, and sulfur contents of asphaltene fractions were determined using a Thermo Finnigan EA 1112 elemental analyzer. Experimental accuracy of these measurements was 0.2 %.

With ^{1}H-NMR and elemental analysis measurements, the aromaticity factor (f_a), the average number of carbons per alkyl side chain (n), and shape factor of the aromatic sheet (Φ) of both asphaltene fractions were evaluated with the formulae reported in the literature (Dickinson 1980; Daaou et al. 2008, 2009; Matsumura and Sato 2009).

Surface tensions of asphaltenes solutions (0.05–10 g L^{-1}) in toluene, pyridine, or nitrobenzene were measured with a Cochon–Abrie tensiometer employing the Wilhelmy plate method. All the measurements were conducted in a temperature-controlled cell at 20 ± 0.1 °C, and each measure was repeated three times. The maximum error in these measurements was 0.1 mN m^{-1}.

3 Results and discussion

3.1 Structural characterization

The differences in the average chemical structures of the asphaltene molecules separated from the storage tank deposit, $(AS)_{DP}$ and the petroleum well, $(AS)_{WL}$ were studied using elemental analysis, FTIR, fluorescence spectroscopy, and proton nuclear magnetic resonance (^{1}H-NMR).

Results of C, H, N, S, and O (determined by difference) analysis of asphaltene fractions given in Table 2 showed a much higher content of heteroatoms, especially of oxygen and sulfur, in the asphaltenes extracted from the storage tank deposit $(AS)_{DP}$, suggesting that this fraction is more polar than $(AS)_{WL}$. In fact, the S/C and O/C ratio values are 6.0 and 1.7 times, respectively, higher in $(AS)_{DP}$. A small difference in N/C ratio between $(AS)_{DP}$ and $(AS)_{WL}$ was also observed. The greater value of H/C ratio observed for $(AS)_{DP}$ fraction indicates that this fraction is less aromatic than $(AS)_{WL}$.

Table 2 Elemental composition and corresponding atoms to carbon ratio of the asphaltenes samples

%wt	$(AS)_{DP}$	$(AS)_{WL}$
C	79.2	84.3
H	8.2	7.5
N	0.8	0.7
S	3.1	0.4
O[a]	8.7	7.1
E[b]	12.6	8.2
H/C	1.24	1.07
O/C	0.082	0.063
N/C	0.008	0.007
S/C	0.015	0.002
E/C	0.105	0.072

[a] Determined by difference

[b] Total heteroatoms (E = O + S + N)

The FTIR spectra of both the asphaltene fractions (Fig. 1) are very similar to the spectra of other asphaltenes published in the literature (Calemma et al. 1995; Ascanius et al. 2004; Daaou et al. 2008). The FTIR band assignment for asphaltene samples is summarized in Table 3. The spectra show peaks at 2935/1460 and 2850/1370 cm^{-1} which correspond to CH_3 and CH_2 aliphatic groups. The aromatic C=C and C=O (ketones, aldehydes, or carboxylic acids) bands are observed at around 1600 and at 1708 cm^{-1}, respectively. Particularly, the $(AS)_{DP}$ spectra display three peaks at 1770 and 1260/1150 cm^{-1} which are related to the presence of C=O and C–O of ester groups, respectively. A distinct band in the range between 3500 and 3100 cm^{-1} is also observed for $(AS)_{DP}$, indicating the presence of N–H and O–H groups.

In conclusion, these results are in agreement with those obtained by elemental analysis and justify the higher polarity of the $(AS)_{DP}$ fraction.

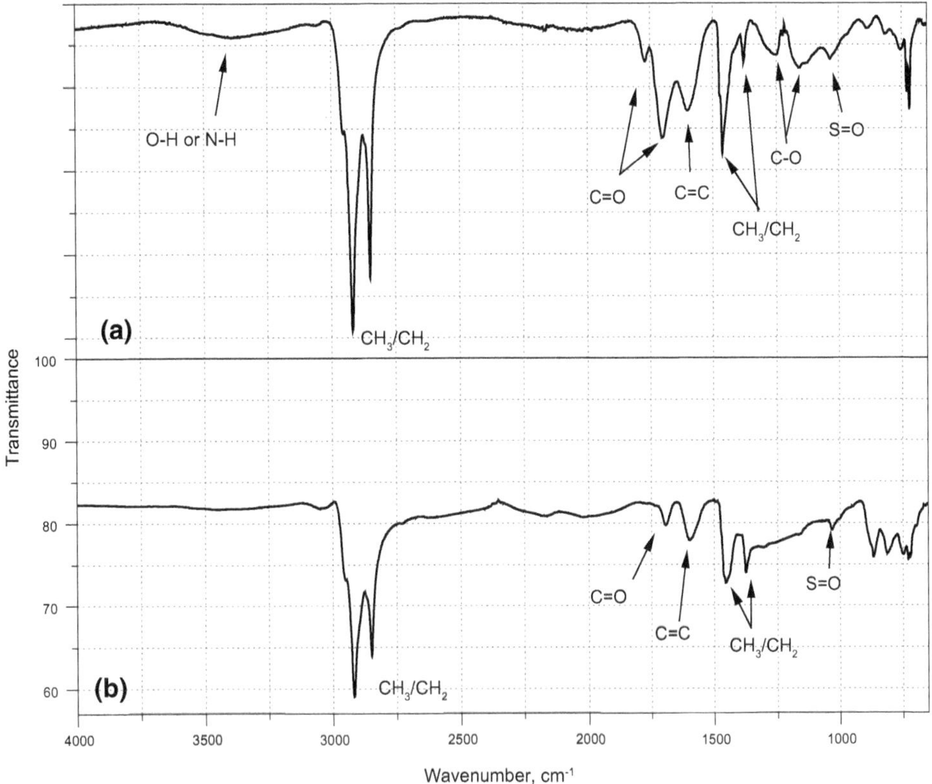

Fig. 1 FTIR spectrum of **a** AS_{DP} and **b** AS_{WL} asphaltenes

Table 3 Infrared spectral range assignments for asphaltene samples: strong (s), medium (m), weak (w)

Absorption frequency, cm^{-1}	Functional group	Asphaltenes characteristic bands	
		$(AS)_{DP}$	$(AS)_{WL}$
3100–3500	O–H or N–H (stretching)	3393 (w)	–
2800–3000	C–H (stretching)	2920 (s), 2850 (s)	2915 (s), 2845 (s)
1600–1800	C=O	1708 (m), 1770 (w)	1702 (w)
1590–1620	C=C (aromatic stretching)	1603 (m)	1595 (m)
1375–1450	(C–H deformation)	1460 (s), 1373 (s)	1450 (s), 1373 (s)
1300–1050	C–O of esters	1260/1150	–
~ 1000	S=O	1032 (w)	1030 (w)

For further interpretation of these results, the areas under the aliphatic stretching absorption in the range 3000–2770 cm^{-1} (A_1), the carbonyl stretching absorption in the range 1750–1650 cm^{-1} [for $(AS)_{WL}$] and in the range 1750–1650 and 1790–1750 cm^{-1} [for $(AS)_{DP}$] (A_2), and the aromatic C=C stretching vibrations in the range 1650–1520 cm^{-1} (A_3) were integrated. From these, two ratios A_3/A_1 and A_2/A_3 are evaluated and reported in Table 4. The first ratio is a measure indicating the degree of aromaticity in the molecular matrix and the second one representing the degree of carbonyl concentration per unit aromatic structure. From the results of these two ratios, it can be revealed that the $(AS)_{WL}$ asphaltene is more aromatic and less polar than $(AS)_{DP}$ sample.

Fluorescence emission spectra of asphaltenes samples presented in Fig. 2 are similar to those found in the literature (Ralston et al. 1996; Groenzin and Mullins 1999; Badre et al. 2006). The peak location exhibits a shift that can be explained in terms of the aromatic rings number and of the condensation rate of the polyaromatic structure. Indeed, the emission at shorter wavelength (λ_{em}) would, in principle, imply the presence of smaller aromatic rings (Ralston et al. 1996), while the displacement of the bands to longer λ_{em} (red shift) (Pesce et al. 1971) can correspond to larger aromatic rings. The comparison between the asphaltene samples' emission spectra shows that the $(AS)_{WL}$ spectrum is shifted to greater value of λ_{em} value (490 nm) with respect to $(AS)_{DP}$ (455 nm). This indicates

Table 4 Areas under aliphatic (A_1), carbonyl (A_2), and aromatic (A_3) bands determined from FTIR spectra and corresponding aromatic (A_3/A_1) and carbonyl (A_2/A_3) indexes for the both asphaltenes samples

Area	A_1 3000–2770, cm^{-1}	A_2 1790–1750, cm^{-1} and/or 1750–1650, cm^{-1}	A_3 1650–1520, cm^{-1}	A_3/A_1	A_2/A_3
$(AS)_{DP}$	275	188	136	0.49	1.39
$(AS)_{WL}$	272	124	160	0.59	0.77

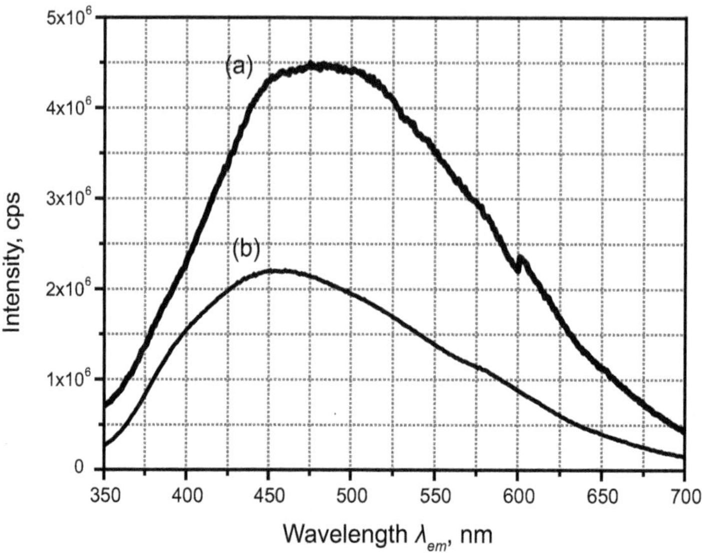

Fig. 2 Fluorescence spectrum in emission mode of **a** $(AS)_{DP}$, and **b** $(AS)_{WL}$ asphaltenes at $\lambda_{ex} = 265$ nm

that the two investigated asphaltene fractions were different in aromatic hydrocarbon ring number and in the aromatic ring arrangement. The analysis of the present results in terms of the literature data of the shortest emission wavelength of a series of polyaromatic compounds (Badre et al. 2006) suggests that the $(AS)_{DP}$ and $(AS)_{WL}$ asphaltene samples contain 3–4 and 5–6 fused aromatic rings, respectively.

^{1}H-NMR spectra presented in Fig. 3 show four signals observed at 0–1.0, 1.0–2.0, 2.0–4.0, and 6.5–9.0 ppm and corresponding, respectively, to aliphatic protons in γ (H_γ), β (H_β), and α (H_α) positions to aromatic ring and aromatic protons (H_{ar}). The atomic abundances of each proton type obtained from their corresponding integrated band areas are reported in Table 5. From these results together with the elemental analysis data (Table 2), the main molecular parameters of asphaltene fractions, such as aromaticity factor (f_a), alkyl side chain (n), and shape factor of the aromatic sheet, (Φ) are calculated according to the formulas proposed by Speight (Speight 1970).

$$f_a = \frac{c_A}{c_T} = 1 - \left(\frac{H}{c}\right)\left(\frac{H_\alpha + H_\beta}{2} + \frac{H_\gamma}{3}\right) \quad (1)$$

$$\phi = \frac{c_A}{c_T} = \frac{\left(H_{ar} + \frac{H_\alpha}{2}\right)}{\left(\frac{H}{c}\right)\left(\frac{H_\alpha + H_\beta}{2} + \frac{H_\gamma}{3}\right)} \quad (2)$$

$$n = \frac{H_\alpha + H_\beta + H_\gamma}{H} \quad (3)$$

where C_A, C_T, and C_p are the aromatic, the total, and the peripheral condensed aromatic carbons numbers, respectively. H_α, H_β, and H_γ are the percentages of the aliphatic protons in α, β, and γ positions obtained as the corresponding ratio of the ^{1}H-NMR spectrum integration area to the total integration area of the ^{1}H-NMR spectrum. H_{ar} is the ratio of the ^{1}H NMR spectrum integration area of the aromatic proton to the total integration area of the ^{1}H NMR spectrum.

The H_α, H_β, H_γ, and H_{ar} percentages and the average structural parameter results of both asphaltene fractions are given in Table 5. The results show a notable difference in both aromatic sheet shape and aromaticity factors for $(AS)_{DP}$ and $(AS)_{WL}$. The latter sample has the higher aromaticity ($f_a = 0.53$) and the smaller aromatic sheet shape factors ($\Phi = 0.35$) in comparison with $(AS)_{DP}$ ($f_a = 0.43$ and $\Phi = 0.55$). The smaller aromatic sheet shape factor value can indicate that the aromatic rings are very condensed in $(AS)_{WL}$ molecules with respect to $(AS)_{DP}$. These results agree with the conclusions drawn from elemental analysis (H/C), FTIR (A_3/A_1), and fluorescence (λ_{em}) measurements. The alkyl side chain is relatively long for $(AS)_{DP}$ asphaltenes.

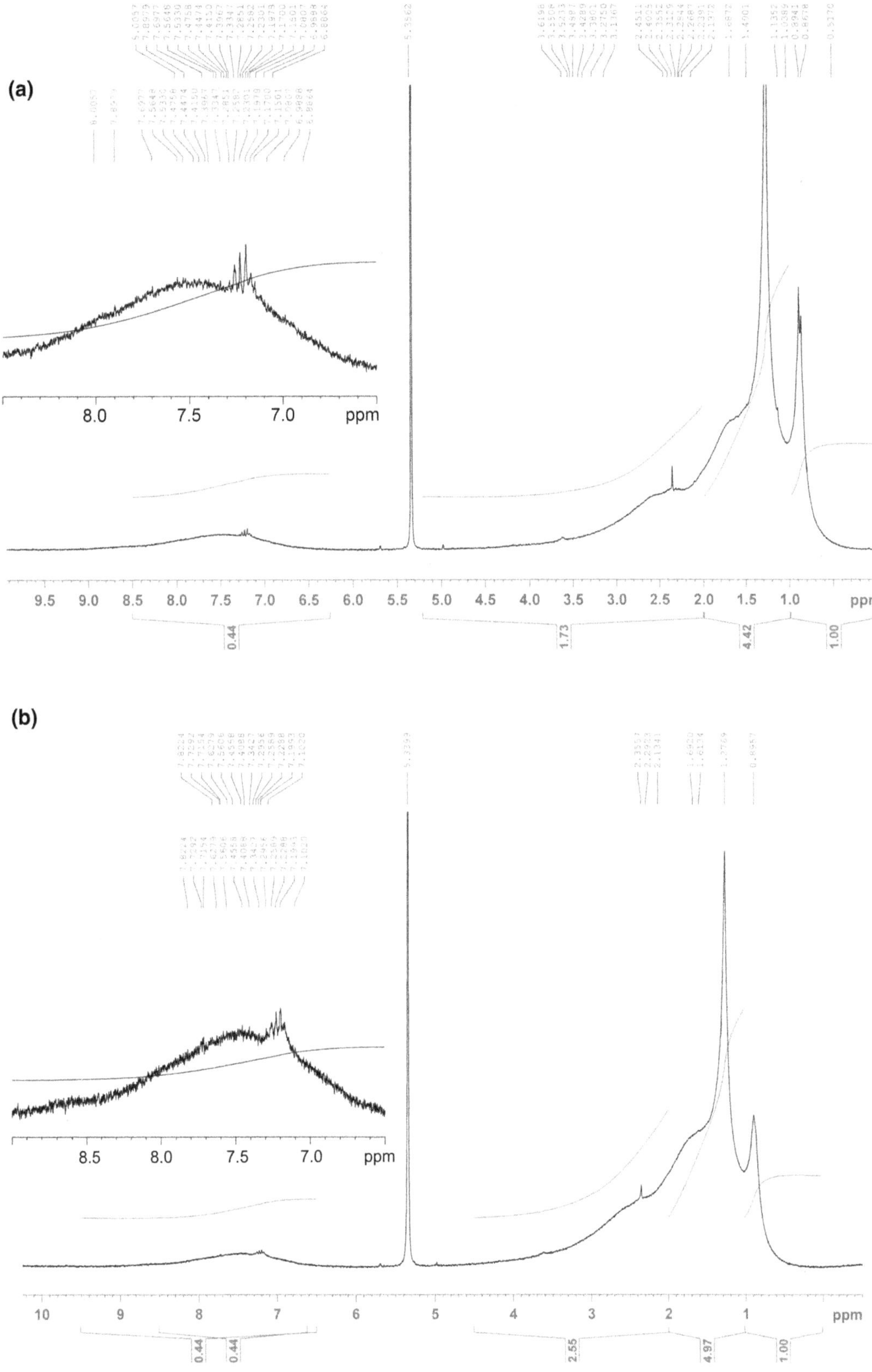

Fig. 3 Liquid state ^{1}H NMR spectrum of **a** AS_{DP} and **b** AS_{WL} asphaltenes

Table 5 Assignments of proton chemical shift NMR spectra and average molecular parameters of $(AS)_{DP}$ and $(AS)_{WL}$ asphaltenes molecules calculated from the correlation Eqs. (1)–(3)

%wt	$(AS)_{DP}$	$(AS)_{WL}$
H_{α}	28.46	22.79
H_{β}	55.47	58.23
H_{γ}	11.16	13.20
H_{ar}	04.91	05.60
H_{al}	95.09	94.40
f_a	0.43	0.53
Φ	0.55	0.35
n	3.34	4.10

Speight (1970)

3.2 Surface tension

Results of the surface tension measurements of $(AS)_{DP}$ and $(AS)_{WL}$ solutions in toluene, pyridine, and nitrobenzene are plotted in Fig. 4a, b, respectively. The variation of surface tension is small as already observed by other authors for various asphaltenes (Bardon et al. 1996; Sheu 1996). Upon increase of asphaltene concentration, surface tension curves showed distinct discontinuities similar to those observed in surfactant solutions, and with this in mind, these discontinuities were considered as an indication of the occurrence of an aggregation phenomenon, in accordance with the Gibbs equation and defined a CMC for asphaltenes in each solvent (Loh et al. 1999; Mohamed et al. 1999). It was established that this parameter (CMC) is an indicator of asphaltene molecule association. Indeed, below the CMC value, the asphaltenes are in a molecular state, while above CMC, associated asphaltenes are formed (Priyanto et al. 2001). The CMC values are given in Table 6 and indicate that for each asphaltene sample, the CMC value increases with increasing solvents' solubility parameter and polarity. Indeed, in a more polar solvent, nitrobenzene (21.8 $MPa^{1/2}$), both $(AS)_{DP}$ and $(AS)_{WL}$ can form micelles in higher concentration than in a less polar solvent, toluene (18.5 $MPa^{1/2}$). These results are in good agreement with those found by Rogel et al. (2000) and Kyeongseok et al. (2004). Comparison of the studied asphaltenic fractions shows that $(AS)_{WL}$ is more soluble in toluene than $(AS)_{DP}$, but less soluble in nitrobenzene. This can be explained in terms of

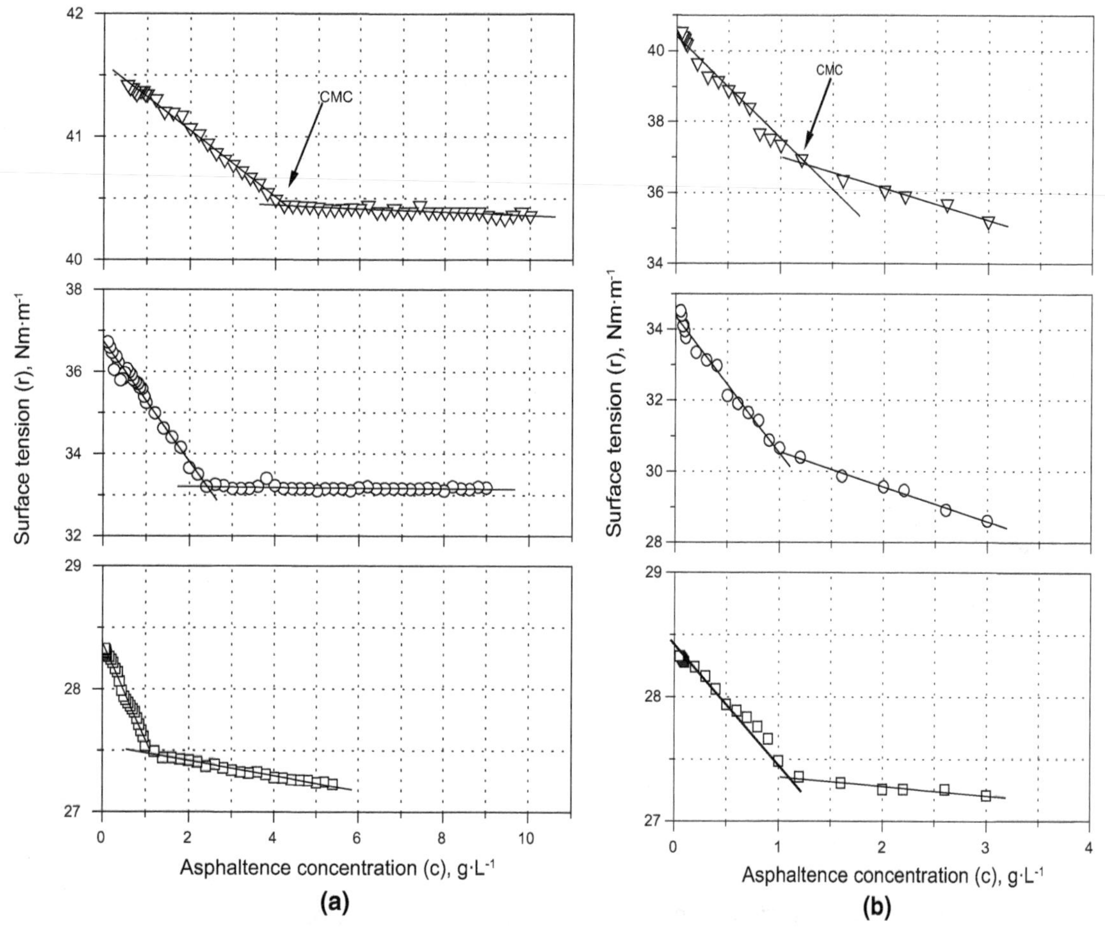

Fig. 4 Surface tension as a function of concentration in g L^{-1} for **a** AS_{DP} and **b** AS_{WL} asphaltenes in nitrobenzene (*inverted triangle*), pyridine (*circle*), and toluene (*square*). Full lines correspond to the linear fit

Table 6 CMC (g L^{-1}) and molecular areas A_a (Å^2) values of asphaltenes samples in different solvents

Asphaltenes	Toluene		Nitrobenzene		Pyridine	
	CMC, g L^{-1}	A_a, Å^2	CMC, g L^{-1}	A_a, Å^2	CMC, g L^{-1}	A_a, Å^2
$(AS)_{WL}$	1.10	774	1.31	175	1.06	166
$(AS)_{DP}$	0.98	434	4.17	489	2.36	177

structure parameters determined above. In fact, the more polar asphaltene with a lower aromaticity, $(AS)_{DP}$, corresponds to the higher solubility in a polar solvent (nitrobenzene). The less polar and higher aromaticity one, $(AS)_{WL}$, is more soluble in solvent with low polarity (toluene). These results indicate that the association of more polar and less aromatic asphaltene $(AS)_{DP}$ is probably dominated by strong specific polar and hydrogen-bonding forces (Porte et al. 2003) rather than other interactions like as π–π bonding of aromatic rings and Van der Waals interactions of long-chain hydrocarbons (Spiecker et al. 2003). These findings confirm suggestions of Rogel et al. (2000) and indicate that the self-aggregation behavior is related to the structural and compositional characteristics. Moreover, the same properties determine the solubility parameters of asphaltenes, and as a consequence, their solubility behavior. The content and the nature of heteroatoms influence the polarity of asphaltenes. Higher polarity favors association and self-assembly. Therefore, asphaltenes with a higher content of heteroatoms such as O, N, and S display a higher degree of self-assembling depending on the polarity of the surrounding medium.The molecular area of the asphaltene (A_a) adsorbed to the liquid surface may be calculated using the Gibbs Eq. (4):

$$A_a = \frac{1}{N_A \Gamma} \tag{4}$$

where N_A is Avogadro's number and Γ is the surface concentration of the adsorbed species and can be obtained from the curve by evaluating the slope just below the CMC, using the Gibbs equation (Sheu 1996).

Results presented in Table 6 show the molecular area, A_a, of both asphaltenes $(AS)_{DP}$ and $(AS)_{WL}$ in different solvents. The values obtained vary in the range 165–774 Å^2. It may be noticed that A_a of $(AS)_{DP}$ is higher than that obtained for $(AS)_{WL}$ in polar solvents such as pyridine and nitrobenzene. In a non-polar solvent-like toluene, the situation is reversed. Thus, the polarities of the solvent and of asphaltenes determine the arrangement of asphaltene molecules at the surface.

4 Conclusion

Two asphaltenes samples derived from different sources of Algerian petroleum have been characterized to compare their structure and aggregation behaviors in the presence of three solvents (toluene, pyridine, and nitrobenzene) with different solubility parameters and polarities.

Elemental analysis, FTIR, ^{1}H-NMR, and fluorescence spectroscopy methods indicate that there is a notable difference between their structures. This difference appears principally in both aromaticity and polarity (heteroatoms content) properties. The unstable asphaltenes $(AS)_{DP}$, which were extracted from the stage tank deposit, have a higher heteroatoms content ($E = 12.6$ wt%), lower aromaticity factor ($f_a = 0.43$), and higher aromatic sheet shape factor ($\Phi = 0.55$) in comparison with the stable sample derived directly from the petroleum well $(AS)_{WL}$ ($E = 8.2$ wt%, $f_a = 0.53$, and $\Phi = 0.35$). The alkyl side chain is relatively long for $(AS)_{DP}$ asphaltenes ($n = 4$). These results confirm, as already suggested, that asphaltene instability is principally governed by both aromaticity and polarity properties.

The surface tension method allowed investigation of the asphaltenes aggregation. It was found that the CMC and corresponding molecular areas values ranged from 0.98 to 4.17 g L^{-1} and 166 to 775 Å^2, respectively. It may be concluded that the aggregation behavior is governed by asphaltene structure and polarity as well as by the polarity of the solvent. Indeed, the more polar the solvent, the higher is the CMC with both $(AS)_{DP}$ and $(AS)_{WL}$ fractions. Consequently, it was found that the more polar asphaltene with a lower aromaticity, $(AS)_{DP}$ displays the higher solubility and the higher specific area at the surface in a polar solvent (nitrobenzene). $(AS)_{WL}$ asphaltenes that are less polar and display a higher aromaticity can form aggregates at lower concentrations (low CMC) in a non-polar solvent (toluene) and display a lower specific area at the surface.

These results confirm the recommendations of Rogel et al. (2000) and indicate that the self-aggregation behavior is related to the structural and compositional characteristics (polarity and aromaticity) which determine their solubility parameters, and as a consequence, their solubility behavior.

Acknowledgments The authors thank Complexe de Sonatrach Hassi-Messaoud for supplying the general information for the Hassi-Messaoud well and storage tank oils.

References

Acevedo S, Gutierrez LB, Negrin G, Pereira JC, Delolme BMF, Dessalces G, Broseta D. Molecular weight of petroleum asphaltenes: a comparison between mass spectrometry and vapor pressure osmometry. Energy Fuels. 2005;19:1548–60.

Ancheyta J, Centeno G, Trejo F, et al. Extraction and characterization of asphaltenes from different crude oils and solvents. Energy Fuels. 2002;16:1121–7.

Andersen SI, Christensen SD. The critical micelle concentration of asphaltenes as measured by calorimetry. Energy Fuels. 2000;14: 38–42.

Andersen SI, Bridi KS. Aggregation of asphaltenes as determined by calorimetry. J Colloid Interface Sci. 1991;142:497–502.

Ascanius BE, Garcia DM, Andersen SI. Analysis of asphaltenes subfractionated by *n*-methyl-2-pyrrolidone. Energy Fuels. 2004;18:1827–31.

Avid B, Sato S, Takanohashi T, et al. Characterization of asphaltenes from Brazilian vacuum residue using heptane-toluene mixtures. Energy Fuels. 2004;18:1792–7.

Badre S, Goncalves CC, Norinaga K, et al. Molecular size and weight of asphaltene and asphaltene solubility fractions from coals crude oils and bitumen. Fuel. 2006;85:1–11.

Bardon C, Barré L, Espinat D, et al. The colloidal structure of crude oils and suspensions of asphaltenes and resins. Fuel Sci Technol Int. 1996;14:203–42.

Bouhadda Y, Bendedouch D, Sheu E, et al. Some preliminary results on a physico-chemical characterization of a Hassi Messaoud petroleum asphaltene. Energy Fuels. 2000;14:845–53.

Buenrostro-Gonzalez E, Andersen SI, Garcia-Martinez JA, et al. Solubility/molecular structure relationships of asphaltenes in polar and nonpolar media. Energy Fuels. 2002;16:732–41.

Calemma V, Iwanski P, Nali M, et al. Structural characterization of asphaltenes of different origins. Energy Fuels. 1995;9:225–30.

Carbognani L, Espidel J, Izquierdo A. In: Yen TF, Chilingarian GV, editors. Asphaltenes and asphalts: developments in petroleum science. Amsterdam: Elsevier; 1997 (**chapter 13**).

Daaou M, Bendedouch D, Bouhadda Y, et al. Explaining the flocculation of Hassi Messaoud asphaltenes in terms of structural characteristics of monomers and aggregates. Energy Fuels. 2009;23:5556–63.

Daaou M, Modarressi A, Bendedouch D, et al. Characterization of the non-stable fraction of Hassi-Messaoud asphaltenes. Energy Fuels. 2008;22:3134–41.

Da Silva Ramos AC, Haraguchi L, Nostripe FR, et al. Interfacial and colloidal behavior of asphaltenes obtained from Brazilian crude oils. J Pet Sci Eng. 2001;32:201–16.

Dereppe JM, Moreaux C, Castex H. Analysis of asphaltenes by carbon and proton nuclear magnetic resonance spectroscopy. Fuel. 1978;57:435–41.

Dickinson EM. Structural comparison of petroleum fractions using proton and ^{13}C-NMR spectroscopy. Fuel. 1980;59:290–4.

Groenzin H, Mullins OC. Petroleum asphaltene molecular size and structure. J Phys Chem A. 1999;103:11237–45.

Groenzin H, Mullins OC. Molecular size and structure of asphaltenes from various sources. Energy Fuels. 2000;14:677–84.

Ibrahim YA, Abdelhameed MA, Al-sahhaf TA, et al. Structural characterization of different asphaltenes of Kuwaiti origin. Pet Sci Technol. 2003;21:825–37.

Khan MA, Ahmed I, Ishaq M, et al. Spectral characterization of liquefied products of Pakistani coal. Fuel Process Technol. 2003;85:63–74.

Kokal SL, Sayegh SG. Asphaltenes: the cholesterol of petroleum. March, 1995, Presented at the SPE Middle East Oil Show; Society of Petroleum Engineers: Bahrain; SPE 29787.

Kyeongseok O, Terry AR, Milind DD. Asphaltene aggregation in organic solvents. J Colloid Interface Sci. 2004;271:212–9.

Loh W, Mohamed RS, Ramos AC. Aggregation of asphaltenes obtained from a Brazilian crude oil in aromatic solvents. Pet Sci Technol. 1999;17:147–63.

Matsumura A, Sato S. Estimation of carbon aromaticity for asphaltenes by elemental analysis and proton NMR. Journal of the Japan Institute of Energy. 2009;88:586–91.

Mohamed RS, Ramos ACS, Loh W. Aggregation Behavior of two asphaltenic fractions in aromatic solvents. Energy Fuels. 1999;13:323–7.

Mullins OC. Asphaltenes in crude oil: absorbers and/or scatterers in the near-infrared region? Anal Chem. 1990;62(5):508–14.

Murgich J, Rodriguez J, Aray Y. Molecular recognition and molecular mechanics of micelles of some model asphaltenes and resins. Energy Fuels. 1996;10:68–76.

Murgich J. Intermolecular forces in aggregates of asphaltenes and resins. Pet Sci Technol. 2002;20:983–97.

Pekerar S, Lehmann T, Mendez B, et al. Mobility of Asphaltene samples studied by ^{13}C NMR spectroscopy. Energy Fuels. 1999;13:305–8.

Pesce AJ, Rosen CG, Pasby TL. Fluorescence spectroscopy. New York: Marcel Dekker; 1971. p. 54.

Porte G, Zhou H, Lazzeri V. Reversible description of asphaltene colloidal association and precipitation. Langmuir. 2003;19:40–7.

Priyanto S, Ali Mansoori G, Suwono A. Measurement of property relationships of nano-structure micelles and coacervates of asphaltene in a pure solvent. Chem Eng Sci. 2001;56:6933–9.

Ralston CY, Kirtley SM, Mullins OC. Small population of one to three fused-aromatic ring moieties in asphaltenes. Energy Fuels. 1996;10:623–30.

Rogel E, Leon O, Torres G, Espidel J. Aggregation of asphaltenes in organic solvents using surface tension measurements. Fuel. 2000;79:1389–94.

Sheu EY, Storm DA. Colloidal properties of asphaltenes in organic systems. In: Sheu EY, Mullins OC, editors. Asphaltenes: fundamentals and applications. New York: Plenum Press; 1995. p. 1.

Sheu EY. Physics of asphaltene micelles and microemulsions—theory and experiment. J Phys: Condens Matter. 1996;8:A125.

Shirokoff JW, Siddiqui MN, Ali MF. Characterization of the structure of Saudi Crude asphaltenes by X-ray diffraction. Energy Fuels. 1997;11:561–5.

Siskin M, Kelemen SR, Eppig CP, et al. Asphaltene molecular structure and chemical influences on the morphology of coke produced in delayed coking. Energy Fuels. 2006;20:1227–34.

Speight JG. A structural investigation of the constituents of Athabasca bitumen by proton magnetic resonance spectroscopy. Fuel. 1970;49:76–90.

Spiecker PM, Gawrys KL, Kilpatrick PK. Aggregation and solubility behavior of asphaltenes and their subfractions. J Colloid Interface Sci. 2003;267:178–93.

Storm DA, Barresi RJ, De Canio SJ. Colloidal nature of vacuum residue. Fuel. 1991;70:779–82.

Taylor SE. Use of surface tension measurements to evaluate aggregation of asphaltenes in organic solvents. Fuel. 1992;71:1338–9.

Yang X, Hamza H, Czarnecki J. Investigation of subfractions of Athabasca asphaltenes and their role in emulsion stability. Energy Fuels. 2004;18:770–7.

Yarranton HW, Alboudwarej H, Jakher R. Investigation of asphaltene association with vapor pressure osmometry and interfacial tension measurements. Ind Eng Chem Res. 2000;39:2916–24.

15

CO_2-triggered gelation for mobility control and channeling blocking during CO_2 flooding processes

De-Xiang Li[1] · Liang Zhang[1] · Yan-Min Liu[1] · Wan-Li Kang[1] · Shao-Ran Ren[1]

Abstract CO_2 flooding is regarded as an important method for enhanced oil recovery (EOR) and greenhouse gas control. However, the heterogeneity prevalently distributed in reservoirs inhibits the performance of this technology. The sweep efficiency can be significantly reduced especially in the presence of "thief zones". Hence, gas channeling blocking and mobility control are important technical issues for the success of CO_2 injection. Normally, crosslinked gels have the potential to block gas channels, but the gelation time control poses challenges to this method. In this study, a new method for selectively blocking CO_2 channeling is proposed, which is based on a type of CO_2-sensitive gel system (modified polyacrylamide-methenamine-resorcinol gel system) to form gel in situ. A CO_2-sensitive gel system is when gelation or solidification will be triggered by CO_2 in the reservoir to block gas channels. The CO_2-sensitivity of the gel system was demonstrated in parallel bottle tests of gel in N_2 and CO_2 atmospheres. Sand pack flow experiments were conducted to investigate the shutoff capacity of the gel system under different conditions. The injectivity of the gel system was studied via viscosity measurements. The results indicate that this gel system was sensitive to CO_2 and had good performance of channeling blocking in porous media. Advantageous viscosity-temperature characteristics were achieved in this work. The effectiveness for EOR in heterogeneous formations based on this gel system was demonstrated using displacement tests conducted in double sand packs. The experimental results can provide guidelines for the deployment of the CO_2-sensitive gel system for field applications.

✉ Shao-Ran Ren
rensr@upc.edu.cn

[1] School of Petroleum Engineering, China University of Petroleum, Qingdao 266580, Shandong, China

Edited by Yan-Hua Sun

Keywords CO_2 flooding · Gas channeling · CO_2 sensitivity · Sweep efficiency · Enhanced oil recovery · Mobility control

1 Introduction

In recent years, sequestration and utilization of CO_2 is becoming an important research topic (Ren et al. 2010; Zhang et al. 2010). With the promotion of CO_2 capture and storage technology, the problem of CO_2 gas source can be solved and an increasing number of oil reservoirs will be candidates for CO_2 EOR projects (Ren et al. 2014). Injection of CO_2 into oil and gas reservoirs may have both economic and environmental benefits (Elsharkawy et al. 1996; Sweatman et al. 2009; Talebian et al. 2014). Targets for reducing greenhouse gas emission were also emphasized during the 21st session of the conference of the parties (COP21) which was held in 2015. Carbon dioxide capture and storage (CCS) is a promising way of making low-carbon energy solutions sustainable. However, CCS has made limited progress. An important route to making CCS a more sustainable option is via CO_2 EOR as part of carbon capture, utilization and storage (CCUS) systems (http://www.bbc.co.uk/news/science-environment-35086346). In the southern states of the United States, CO_2 flooding has been one of the main techniques for EOR due to their abundant natural CO_2 sources (Zhu et al. 2011). It has been demonstrated that CO_2 flooding can be one of the most effective EOR technologies proven in field tests and indoor experiments. There is a large-scale CO_2 EOR project in the Weyburn Oilfield of Canada to deal with the industrial carbon emissions. Since 2000, about 1.8 million tons of CO_2 per year captured from a coal gasification plant in the North Dakota have been transported to the Weyburn Oilfield through pipelines for injection to meet the requirements of storage and enhancing oil recovery. The potential of CO_2 storage in the field is over 50 million tons and the incremental oil recovery can be achieved is approximately 9.8 % OOIP (Preston et al. 2005). Also a large-scale CO_2 EOR project has been under way in China since 2006 in the Jinlin Oilfield (Northeast China) (Li and Fang 2007), in which CO_2 produced and separated from natural gas reservoirs nearby is utilized, and about 0.3–0.5 million tons of CO_2 per year have been injected into oil reservoirs since then. Meanwhile, in the Shengli Oilfield CO_2 is captured from a coal-fired electricity plant for EOR and the capacity of CO_2 injection will be over 0.5 million tons per year.

For the oil and gas industry, CO_2 EOR has been demonstrated as a feasible technology to improve oil recovery. The EOR mechanisms include interfacial tension (IFT) reduction, CO_2 dissolution into oil to reduce its viscosity and swell oil, permeability improvements as the reaction of carbonic acid (due to CO_2 dissolution in water) with the limestone/dolomite, and wettability alteration due to asphaltene precipitation. It is notable that CO_2 and oil may become miscible when the reservoir pressure exceeds the minimum miscible pressure (MMP) which can greatly reduce the residual oil saturation and improve oil recovery (Li et al. 2015). However, viscous fingering (due to the viscosity contrast of CO_2 and oil) and gravity segregation (due to the lower density of CO_2) can develop quickly and lead to lower macroscopic sweep efficiency during the process of CO_2 injection. Reservoir heterogeneity leads to more severe early breakthrough of CO_2 in production wells (Hamouda et al. 2009; Hou and Yue 2010; Enick et al. 2012). Early breakthrough of CO_2 at production wells can reduce the sweep efficiency and undermine the oil recovery factor (Chakravarthy et al. 2006; Nezhad et al. 2006). A key issue of CO_2 flooding is to mitigate the problem of gas channeling, especially for the reservoirs associated with high-permeability zones (thief zones) (Xing et al. 2010). Different techniques have been developed for blocking channels with high permeability and controlling the mobility of CO_2, such as the water-alternating-gas (WAG) process (Christensen et al. 2001; Luo et al. 2013; Majidaie et al. 2015), injection of CO_2 foam (Khalil and Asghari 2006; Yu et al. 2012; Zhang et al. 2013a), and use of CO_2 thickener (Enick et al. 2012; Zhang et al. 2011; Mclendon et al. 2012). However, the technologies mentioned above have their own limitations, for example technical feasibility, economy, environmental friendliness, and safety. Gel polymer systems have been applied successfully in oil and gas wells to control unwanted water production in recent years. A gel system consists of polymer, crosslinker, water as the solvent, and some additive materials. The polymers used in gel systems mainly include polyacrylamides (PAM), partially hydrolyzed polyacrylamides (PHPA), xanthan gum, carboxymethylcellulose, furfural alcohol, acrylic/epoxy resins, silicate based gels, and block copolymers. Meanwhile, the crosslinker can be classified into two groups: organic (such as phenol, formaldehyde, and polyethyleneimine (PEI)) and inorganic (such as Cr^{3+}, Al^{3+}, Zr^{3+}) (Dalrymple et al. 1994; Prada 1998; Niu et al. 2013). For inorganically crosslinked gels, nanotechnology was used to extend the gelation time of PHPA/Cr^{3+} systems (Cordova et al. 2008). The particles based on PEI and dextran sulfate (DS) can sequester Cr^{3+}. In addition, PHPA/Cr^{3+} system assisted by foam was evaluated to block deep wormholes (Asghari et al. 2005). Copolymers of PAM were also studied for crosslinking with Cr^{3+} (Prada et al. 2000). Colloidal dispersion gels (CDGs) were investigated for blocking of "thief zones" in heterogeneous formations (Chang et al. 2006). With respect to the organically crosslinked gels, a system was designed by the PDVSA Research and Development Center that can exhibit blocking performance at temperature up to 160 °C. The PAtBA (PAM *tert*-butyl acrylate)/PEI system was developed for near-wellbore treatments, and it also can tolerate temperature up to 160 °C in fields. In order to enhance the strength of gels, materials were also added into the PAtBA/PEI system like cement, silica flour, and rigid-setting materials (RSMs). For the deep modification of injection profiles, two gel systems were identified: microspheres using PAM monomers crosslinked with *N,N′*-methylenebisacrylamide and microspheres produced by crosslinking 2-acrylamido-2-methylpropane sulfonic acid (AMPS) with diacrylamides and methacrylamides of diamines (El-Karsani et al. 2014). Polymer microsphere emulsion attracts considerable attention for water shutoff material with relatively low viscosity that can be injected continuously and prepared using formation water produced from oilfields (Guo et al. 2014). Preformed particle gels (PPGs) were also used to overcome different drawbacks, which include the uncertainty of gelling due to shear degradation and gelant composition changes caused by dilution by formation water and chromatographic fractionation, inevitable in situ gelation systems (Elsharafi and Bai 2016). Although generally gel treatment is performed with waterflooding for conformance control, several laboratory experiments and field applications have been conducted to divert CO_2 (Martin and Kovarik 1987; Martin et al. 1988; Seright 1995; Hughes et al. 1999; Karaoguz et al. 2007; Hou and Yue 2010; Pipes and Schoeling 2014). However, gelation time limits the application of this technology mentioned above. In matrix treatments, it must be long enough so that the gel can penetrate into the reservoir to ensure sufficiently deep placement. It is difficult to control the gelation time with various factors affecting it, such as temperature, salinity of mixing water, and concentration of polymer. Hence, new types of gel systems need to be developed for blocking the gas channeling during the CO_2 injection process with more reliable operation.

In this study, a new method for selectively controlling CO_2 channeling and reducing its mobility is proposed, which is based on a CO_2 sensitive gel system (modified polyacrylamide-methenamine-resorcinol) to form gel in situ. The CO_2 sensitive gel system is gelation that will be triggered by CO_2 in the reservoir, and the chemical can be dissolved in water and injected separately via a simple and economic slug injection technique. This method based on CO_2 sensitivity makes the gelation process more reliable because the gelation needs not only the gelation time but also the presence of CO_2. In this paper, the mechanisms of the gelation and their rheological behavior under different conditions were described. The CO_2-sensitivity of the gel system was demonstrated using transparent tubes and the gel strength and gelation time were evaluated based on a gel strength code method. By means of viscosity measurements, the injectivity of the gel system was investigated. The effectiveness of the gel system for blocking gas channels in porous media was studied in sand pack flow experiments. The performance of EOR in simulated heterogeneous formations assisted by this gel system was also exhibited through displacement tests conducted in double sand packs.

2 Gelation mechanisms

A modified polyacrylamide-methenamine-resorcinol gel system investigated in this study has been used successfully as a water shutoff gel system (Zhang and Yang 1988). Its mechanism of gelation can be described as below under reservoir conditions in the presence of CO_2.

In an acidic environment and at a high temperature, methenamine can release methanal (formaldehyde), and the released methanal can react with polyacrylamide (PAM) and resorcinol to generate phenolic resin via a polycondensation process. Phenolic resin can react further with PAM to produce linear polymers that can block channels (Noller 1965; Xing et al. 2005). Under normal reservoir conditions, an acid (such as HCl solution) is injected to trigger the chain reactions. While in the process of CO_2 flooding, CO_2 dissolves in formation water and reduces its pH to 2–4 (an acidic environment) (Raje et al. 1999; Hild and Wackowski 1999; Cai 2010; Zhang et al. 2013b, c), then the gel system can become CO_2-sensitive that only works in high-permeability zones where CO_2 can readily breakthrough or channel out. Therefore, the gel system can selectively block the high-permeability zones or gas channels. The reaction mechanisms involved are as follows: (1) CO_2 dissolves in formation water and generates carbonic acid in situ. Carbonic acid creates an acidic environment that is conductive to the formation of gels. (2) Methenamine releases methanal in the acidic environment at relatively high temperatures. (3) Multi-hydroxymethyl resorcinol may be formed through the reaction between methanal and resorcinol, and via polycondensation of multi-hydroxymenthyl resorcinol, phenolic resin is formed. (4) Large linear polymers may be formed through further polycondensation between phenolic resin and polyacrylamide that can make the gel more stable and strong.

3 Experimental

3.1 Materials

Modified polyacrylamide (PAM) was supplied by the Sinopharm Chemical Reagent Co., Ltd, with an average molecular weight (MW) over 3×10^6 and a solid content of 85.0 % above. Methenamine (colorless or white crystals, MW 140.9, purity $\geq$99.0 %), resorcinol (white needle-like crystals, MW 110.1, purity $\geq$99.5 %), and $CaCl_2$ (purity $\geq$96.0 %) were also supplied by the Sinopharm Chemical Reagent Co., Ltd. CO_2 (purity $\geq$99.8 %), N_2 (purity $\geq$99.999 %), NaCl (purity $\geq$99.5 %), and deionized water were used in experiments. In the experiments, the simulated formation water was prepared using $NaCl_2$ and $CaCl_2$, with total salinity of 20,000 ppm or 200,000 ppm that included 1000 ppm of $CaCl_2$, representative of the typical formation (Zhang et al. 2013a). In order to provide enough crosslinker for the gelation in the presence of CO_2, relatively high concentrations of resorcinol (0.1 wt%) and methenamine (0.4 wt%) were applied in these experiments. The gel solution (modified PAM-methenamine-resorcinol solution) was prepared with the simulated formation water. The oil used was taken from TP block in the Tahe Oilfield, China, with a viscosity of 72.3 mPa s at 50 °C. Quartz sand with grain sizes in the range of 60–100 mesh was used to make sand packs in order to investigate the effect of the gel system on permeability reduction and EOR performance in porous media.

3.2 Methods

3.2.1 Improved bottle test

Sydansk's gel strength code (GSC), which is an intuitive, rapid, and semi-quantitative method to evaluate the gelation rate and gel strength through visual observations in the gel formation process (Sydansk and Argabright 1987), was used in this paper. In this study, the transparent tubes for bottle tests were improved, with a pressure tolerance up to 1.5 MPa. The correspondence of the gel state and strength and the strength codes were described in Table 1.

The experimental apparatus for measuring gelation rate and gel strength at reservoir conditions was shown in Fig. 1. The key devices were the transparent tubes (inner diameter 2.5 cm, length 20.5 cm, effective volume 100 mL). In order to observe the gelation process more easily and directly, the transparent tubes were made from polymethyl methacrylate.

Experimental procedures are as follows. (1) The gel solution (modified PAM with different concentrations, methenamine of 0.4 wt%, and resorcinol of 0.1 wt%) of 25 mL was pumped into the transparent tubes. The concentration of the modified PAM solution (prepared with the formation water) were 0.5, 1.0, 1.5, and 2.0 wt%, respectively. The salinity of the formation water used in bottle tests was 20,000 ppm (with 1000 ppm Ca^{2+}). (2) CO_2 and N_2 were separately injected into the two transparent tubes at a pressure of 1 MPa. (3) The tubes were then placed in the air bath at a given temperature and a pressure of 1 MPa for gelation. Please note, in the case of CO_2 experiments, the pressure of 1 MPa is high enough for CO_2 to dissolve in water and create an acid environment (pH < 4) in terms of safety considerations. (4) At regular intervals, the gelation processes in the transparent tubes were monitored, and the gelation time and gel strength were recorded. As time passed, the state and strength of the gel system changed until it stabilized. (5) The experiments were repeated at different temperatures and PAM concentrations.

3.2.2 Sand pack experiments

An experimental set-up was built to evaluate the capability of the CO_2-sensitive gel system for water or gas shutoff and

Table 1 Sydansk's gel strength codes (after Sydansk and Argabright 1987)

Strength codes	Name of the gel
A	No detectable gel formed
B	Highly flowing gel
C	Flowing gel
D	Moderately flowing gel
E	Barely flowing gel
F	Highly deformable nonflowing gel
G	Moderately deformable nonflowing gel
H	Slightly deformable nonflowing gel
I	Rigid gel
J	Ringing rigid gel

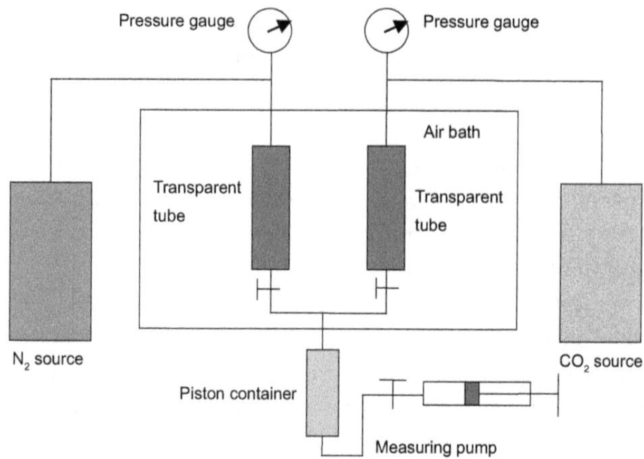

Fig. 1 Schematic diagram of the apparatus for improved bottle testing

EOR performance in simulated heterogeneous reservoirs, as shown in Fig. 2. This set-up consisted of two sand packs, a fluid injection system, an air bath, and pressure and temperature control systems. Before each experiment, the sand packs filled with quartz sand were connected into the experimental system and high-pressure CO_2 was injected into the system to carry out a leak test and ensure the gas tightness of the system.

3.2.2.1 Blocking performance experiments Experimental procedures are as follows. (1) In these tests, only one sand pack (No. 1, as shown in Fig. 2) was used and saturated with the simulated formation water of different salinities (20,000 and 200,000 ppm) at a backpressure of 10.28 MPa under different temperatures. The pressure difference across the sand pack, ΔP_1, was measured for calculating the initial permeability of the sand pack. The water injection rate was maintained at 1 mL/min. (2) CO_2 was injected into the sand pack until a gas breakthrough occurred at the outlet of the sand pack. (3) A slug of the gel solution (modified PAM-methenamine-resorcinol gel system) (normally 0.3 pore volume (PV)) was injected into the sand pack. The concentration of the modified PAM used in brine was 1.0 wt%. (4) A slug of CO_2 (0.3 PV) was then injected into the sand pack, and all the valves were turned off for gel reactions. (5) After 8 h, the simulated formation water was injected into the sand pack at an injection rate of 1 mL/min, and the pressure difference (ΔP_2) was measured at different injection volumes. The backpressure regulator may control the water flow more steadily than a CO_2 flow in sand packs and the pressure difference can be measured with relatively higher accuracy using water flow. Hence, the performance of the gel system for water shutoff was employed to reflect the blocking capacity to CO_2 based on this gel system indirectly.

Normally, the capability of the gel system for water or gas shutoff can be evaluated by the reduction in

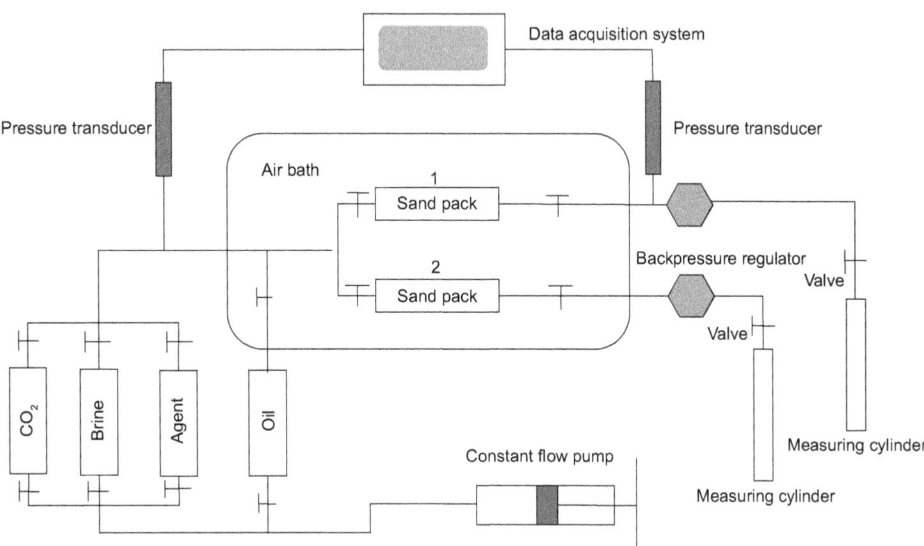

Fig. 2 Schematic diagram of the apparatus for flow tests

permeability of the sand pack after the gel treatment, which is defined as

$$K_R = \frac{k_1 - k_2}{k_1} \times 100\,\%, \tag{1}$$

where k_1 is the initial permeability to water before the gel treatment; k_2 is the permeability to water after the gel treatment. In this study, the injection rate was kept constant before and after the gel treatment, Eq. (1) may be expressed as follows:

$$K_R = \frac{\Delta P_2 - \Delta P_1}{\Delta P_2} \times 100\,\%, \tag{2}$$

where K_R is a measure of the flow resistance after the formation of gels in the sand pack. The permeability reduction can directly reflect the changes of permeability in porous media. It is feasible and effective for evaluating the gel performance for water shutoff treatments. However, the backpressure and temperature should be set to the initial condition in order to achieve the parameter of k_2. Therefore, the parameter K_R achieved in Eq. (2) was convenient and used in these experiments for exhibiting the gel performance in porous media dynamically.

3.2.2.2 Displacement tests in simulated heterogeneous formations Double sand packs were used to simulate heterogeneous formations. In the displacement tests, two sand packs (Nos. 1 and 2 sand packs as they were exhibited in Fig. 2) with different permeability were used to evaluate the EOR performance assisted by the gel system in the simulated heterogeneous reservoir. The permeability values of these two sand packs were 1993.2 and 150.6 mD, respectively. Therefore, the permeability ratio of these two sand packs was greater than 13 with different compaction degrees. The pore volume of the high-permeability sand pack was 130 mL and that of the low permeability one was 80 mL. The experimental procedures are as follows. (1) The temperature of the air bath was set to 90 °C and the backpressure was 10.28 MPa to simulate reservoir conditions. (2) The formation water (with a total salinity of 20,000 ppm that included 1000 ppm Ca^{2+}) was injected into each sand pack separately until a steady state was reached. The water injection rate was maintained at 1 mL/min. (3) Oil was injected into each sand pack until no more water was produced at the end of the sand packs to establish initial oil saturation. (4) CO_2 was injected into the double sand packs simultaneously when no more oil was produced at the end of each sand pack. (5) The modified PAM-methenamine-resorcinol gel solution (0.3 PV) was then injected into the double sand packs and then all the valves were turned off for gel reactions. The PAM concentration in this solution was 1.0 wt%. (6) After 8 h, CO_2 was injected into the sand pack at an injection rate of 1 mL/min, and the EOR performance assisted by this technology was achieved until no more oil was produced.

4 Experimental results and discussion

4.1 CO_2 solubility in water

CO_2 solubility in formation water is the key factor for the trigger of CO_2-sensitive chemicals. It directly influences the pH value of the formation water with CO_2 dissolved. The CO_2 solubility in the formation water has significant effect on the performance of CO_2-sensitive chemicals. To measure and predict the CO_2 solubility in water, a large number of experiments have been done and many types of prediction models built (Duan and Sun 2003; Duan et al. 2006; Bastami et al. 2014). The accuracy of the Duan and Sun model is relatively high with an average relative error of 2.86 % (Duan and Sun 2003; Duan et al. 2006). So the Duan and Sun model was used to evaluate the factors influencing CO_2 solubility in formation water. The prediction results are shown in Fig. 3. CO_2 solubility in water is a function of temperature, pressure, and salinity. CO_2 solubility decreases with an increase in temperature and salinity. An increase in pressure increased the solubility of CO_2 in water. Although the high salinity of formation water may inhibit the dissolution of CO_2, the effective CO_2 solubility (0.0634 mol CO_2/kg H_2O) in the formation water of salinity of 200,000 ppm can be achieved at a relatively high pressure of 1 MPa and at 90 °C.

4.2 CO_2-sensitivity

In the improved bottle tests, the gelation time and strength of gels formed were investigated at 1 MPa N_2 or CO_2 pressure and at different temperatures (50–80 °C). The gel solution contained 1.0 wt% PAM, 0.4 wt% methenamine, and 0.1 wt% resorcinol. The results in Table 2 and Fig. 4 show that gels were formed in a CO_2 atmosphere, with a high gel strength; while no gel was formed or very weak gels were formed in an N_2 atmosphere. The phenomena of gelation processes with the presence of CO_2 were exhibited in the right tube and the phenomena in a N_2 atmosphere were achieved in left tube in Fig. 4. This indicates that the gel solution used in these experiments was sensitive to CO_2 and the injection of CO_2 can induce crosslinking reaction. Under experimental conditions, gaseous CO_2 can dissolve in water and distribute uniformly, which makes the gels generated with a good strength (strength code of H). Under reservoir conditions, CO_2 is in a supercritical state and its mass transfer can be greatly improved (Li et al. 2015). It is expected that the gelation condition can be improved.

4.3 Influence of temperature

As seen in Table 2, the gelation time of the gel system was shortened with an increase in temperature (50–80 °C). The gel formed in the CO_2 atmosphere was strong, strength code H, and the gelation rate improved with an increase in temperature. However, little or no gel was formed in the N_2 atmosphere.

In general, the temperature had a positive effect on the gelation process at experimental temperatures. The solubility of CO_2 in water decreases with increasing temperature, which may affect the pH value of the solution. However, the molecular motion of CO_2 may be accelerated with increasing temperature, a low pH environment may appear quickly with the improved mass transfer of CO_2, which reduces the gelation time. Meanwhile, the release rate of formaldehyde from methenamine becomes higher under relatively higher temperatures, which may promote the formation of gel with a more timely supply of crosslinkers. So, the positive effects of temperature dominated the gelation process in the experimental range.

4.4 Influence of the PAM concentration

Table 3 lists the effect of the PAM concentration on gelation time and gel strength in CO_2 and N_2 atmospheres at 80 °C. The PAM concentrations in the modified PAM-methenamine-resorcinol gel solution were 0.5, 1.0, 1.5, and 2.0 wt%, respectively. In the N_2 atmosphere, no gel was observed using the modified PAM-methenamine-resorcinol gel solution with different PAM concentrations during the 24-h experimental period. The gelation time was postponed with increasing PAM concentration in the CO_2 atmosphere. Specifically, the strength of the formed gel can be high with the code of H once the PAM concentration was equal to or higher than 1.0 wt%. A highly deformable nonflowing gel (strength code F) was observed for the gel system with

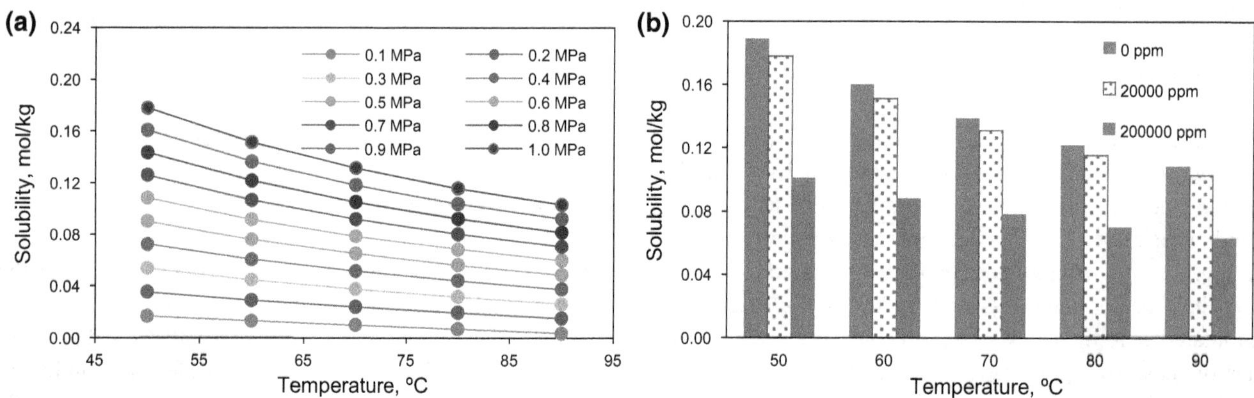

Fig. 3 CO_2 solubility in water at different conditions. **a** Solubility at the salinity of 20,000 ppm. **b** Solubility at the pressure of 1 MPa

Table 2 Gel performance at different temperatures (experimental pressure, 1 MPa)

Temperature, °C	In a N_2 atmosphere		In a CO_2 atmosphere	
	Gelation time, h	Strength code	Gelation time, h	Strength code
50	No gel formed over 24 h	A	13.93	H
60	No gel formed over 24 h	A	7.83	H
70	No gel formed over 24 h	A	4.78	H
80	No gel formed over 24 h	A	2.08	H

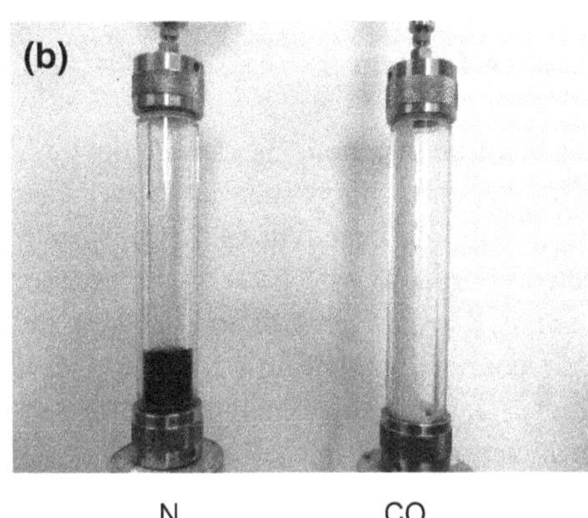

Fig. 4 Observed gelling occurrences in the presence of CO_2 and N_2 for the gel solution with 1.0 wt% PAM at 70 °C. **a** Before gas injected. **b** After gas injected and the gel formed

Table 3 Gel performance at different PAM concentrations (experimental temperature 80 °C)

PAM concentration, wt%	In a N_2 atmosphere		In a CO_2 atmosphere	
	Gelation time, h	Strength code	Gelation time, h	Strength code
0.5	No gel formed over 24 h	A	1.78 (gel disappeared after 6 h)	F-B
1.0	No gel formed over 24 h	A	2.08	H
1.5	No gel formed over 24 h	A	2.27	H
2.0	No gel formed over 24 h	A	2.53	H

0.5 wt% PAM concentration after 1.78 h and the gel structure disappeared after 6 h, with the generation of dispersed brown floccules. Under the same temperature condition, the viscosity of the gel system would be thickened with an increase in the PAM concentration and the increase in viscosity may slow the molecular motion (including CO_2, formaldehyde, and hydroxyl resorcinol, etc.) and diffusion rate down in the whole gel system. So the gel formation rate was reduced with increasing PAM concentration. Although a part of the gel solution can form gel structure with the formation of a gelation environment, the gel structure was weak and unstable when the PAM concentration was low (such as the PAM concentration of 0.5 wt%). When the PAM concentration exceeded 1.0 wt%, the gel strength did not improve with an increase of the PAM concentration.

4.5 Influence of pressure

In the evaluation of the influence of pressure on gel performance, the PAM concentration in the modified PAM-methenamine-resorcinol gel solution was 1.0 wt%. As exhibited in Table 4, the gelation time of the gel system became shorter with an increase in pressure (0.2–1 MPa). The gel formed in the CO_2 atmosphere had high levels of strength, code H or F, when the pressure was higher than 0.2 MPa, and the gelation rate improved with increasing pressure. However, little or no gel formed in the N_2 atmosphere.

Generally speaking, the increase in pressure had a positive effect on the gelation process at experimental pressures. The solubility of CO_2 in water increases with increasing pressure as exhibited in Fig. 3, which decreases the pH of the solution. At the same time, the concentration of CO_2 may be increased with increasing pressure, a low pH environment may appear quickly with the higher concentration of CO_2, which reduces the gelation time. Meanwhile, the release rate of formaldehyde from methenamine becomes higher under an enhanced environment of low pH with relatively higher CO_2 pressure, which may also promote the formation of gel with a more timely supply of crosslinkers. So increased pressure has positive effects on the gel performance in the experimental range.

4.6 Injectivity of the gel system

During CO_2 and chemical injection, the formation temperature in the vicinity of the wellbore is usually lower than that in the deep formation. The injectivity of the gel system should be considered in order to reduce energy dissipation and pumping pressure. The viscosity of the gel systems with different PAM concentrations were evaluated with a Brookfield viscometer and shown in Fig. 5. The viscosity of the gel system increased with an increase in the PAM concentration, but decreased with an increase in temperature. These characteristics are beneficial to migration of the gel system to the deep formation. Generally speaking, the near-wellbore temperature is relatively lower, but the fluid absorption ability is comparatively higher after some kind of stimulation and EOR works. The relatively higher viscosity of the gel system in the near-wellbore area is good for the control of the gel system injection and preventing injected fluid channeling. With the gel system flowing into the formation, the viscous effect accumulates and the resistance to fluid flow increases which makes the injection process more difficult. At the same time, the temperature is relatively higher in extended horizontal direction and this can decrease the viscosity of the gel system. So the decreased viscosity of the gel system is expected to improve the injectivity, which can contribute to the conformance improvement of deep reservoirs.

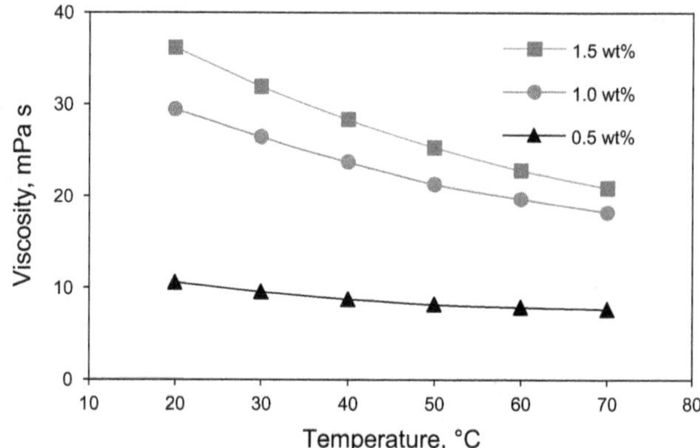

Fig. 5 Injectivity of the gel solutions with different PAM concentrations

4.7 Shutoff capacity of the gel system in porous media

The shutoff capacity of the gel system was evaluated via flow tests in the sand packs. As shown in Table 3, the gelation time and gel strength obtained from gel systems with a polymer concentration of 1.0, 1.5, and 2.0 wt% were similar, so the PAM concentration of 1.0 wt% was recommended for further examinations via sand pack experiments in terms of better injectivity of the gel system. The sand packs used were 60 cm in length and 2.5 cm in diameter. The experimental conditions for shutoff capacity in porous media and experimental results are shown in Table 5 and Fig. 6.

The data of ΔP_2 were collected after a constant flow of water was obtained at the outlet of the sand pack (approximately 0.5 PV water injected). Table 5 and Fig. 6 show that the permeability of the sand pack reduced by 94.5 % at 80 °C after treatment with the PAM-methenamine-resorcinol gel solution. This indicates that the PAM-methenamine-resorcinol gel solution, a CO_2-sensitive gel, can effectively block the CO_2 channels in the sand pack. With more formation water injected into the sand pack, the sand pack permeability changed slightly and the final permeability reduction remained at 93.8 %.

It was worth noting that this type of CO_2-sensitive gel also had remarkable resistance to scouring compared to other CO_2-sensitive shutoff systems such as sodium aluminate (with 2.0 PV water injected, less than 83 % permeability reduction) and resol phenol–formaldehyde resin (with 2.0 PV water injected, less than 31 % permeability reduction because the curing resin is brittle).

Table 4 Gel performance at 80 °C and different pressures

Pressure, MPa	In a N_2 atmosphere		In a CO_2 atmosphere	
	Gelation time, h	Strength code	Gelation time, h	Strength code
0.2	No gel formed over 24 h	A	No gel formed over 24 h	A
0.4	No gel formed over 24 h	A	10.36	F
0.6	No gel formed over 24 h	A	4.23	H
0.8	No gel formed over 24 h	A	2.24	H
1.0	No gel formed over 24 h	A	2.08	H

Table 5 Shutoff capacity of the gel system with 1.0 wt% PAM concentration

System	Temperature, °C	Salinity of the formation water, ppm	Sand pack		Water injection volume, PV	Maximum permeability reduction, %	Final permeability reduction, %
			Pore volume, mL	Initial permeability, mD			
PAM-methenamine-resorcinol gel solution	90	200,000	110	1698.5	3	92.7	90.0
	80	20,000	100	19.4	2	94.5	93.8
	70	20,000	100	59.6	2	98.9	98.2
	70	20,000	109	120.2	2	97.8	97.3
Sodium aluminate	80	20,000	98	31.6	2	89.3	82.9
Resol phenol–formaldehyde resin	80	20,000	105	26.1	2	87.3	30.5

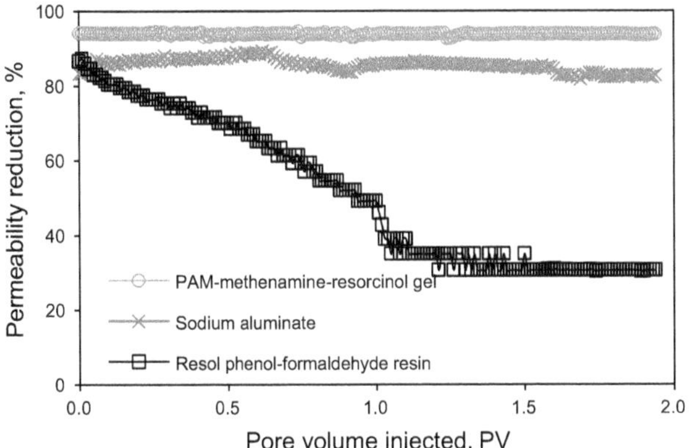

Fig. 6 Changes of shutoff capacity of different CO_2-sensitive systems during water flooding at 80 °C

Table 5 also shows that after the low-permeability sand packs were treated with 0.3 PV PAM-methenamine-resorcinol gel solution the permeability reduced by 97–99 % at 70 °C. For the high-permeability sand pack treated by PAM-methenamine-resorcinol gel solution, the final permeability reduced by above 90 % at 90 °C even after 3 PV formation water of high salinity (200,000 ppm with 1000 ppm Ca^{2+}) was injected into the sand pack. This demonstrates that the PAM-methenamine-resorcinol gel had good stability at high salinity (200,000 ppm) and at high temperatures. In the second round of CO_2 injection for enhancing the gelation environment, the permeability reduction rate increased with the injection of CO_2 and can reach to more than 80 %, which was observed during the experimental processes. This performance indicates that gel with a certain extent of strength can be achieved under the condition of fluid flow.

4.8 EOR performance in simulated heterogeneous formation

In the displacement tests conducted in double sand packs, the EOR performance was evaluated after the simulated heterogeneous reservoir was treated with the gel system (modified PAM-methenamine-resorcinol gel system). The experimental results are shown in Fig. 7. Before the gel treatment, 1 PV CO_2 was injected into the sand packs, the oil recovery from the high-permeability sand pack (1993.2 mD) was higher than that from the low-permeability one (150.6 mD). The final oil recovery by CO_2 flooding was 71 % in the high-permeability sand pack, but only 15.5 % oil was recovered in the low-permeability sand pack after injection of approximately 1 PV of CO_2. In this stage (0–1 PV), the oil recovery in the higher permeability sand pack was the main contributor to the comprehensive oil recovery. The so-called comprehensive oil recovery represents total oil recovery in the simulated heterogeneous formation. In order to further displace or mobilize the oil in the sand packs, 0.3 PV of the gel system was injected into these heterogeneous sand packs (simulated heterogeneous formations) to block the channeling. The oil recovery increased slightly both in the high- and low-permeability sand packs during the injection of the gel system. After gelation for 8 h, CO_2 was then injected into the sand packs again at an injection rate of 1 mL/min, and the EOR performance was evaluated until no more oil was produced at the end of the sand packs. After the gel treatment, the oil recovery increased slightly in the high-permeability sand pack. Meanwhile, the oil recovery increased sharply in the low-permeability sand pack and finally reached 62 %. The oil recovery increase in lower permeability sand pack dominated the increase in the comprehensive oil recovery in this stage (1.3–2.5 PV). This indicates that gel was formed in the high-permeability sand pack due to the high CO_2 saturation and blocked the channels and pores. Therefore, CO_2 subsequently injected was diverted into the low-permeability sand pack from the high-permeability sand pack to increase the oil recovery of the low-permeability sand pack.

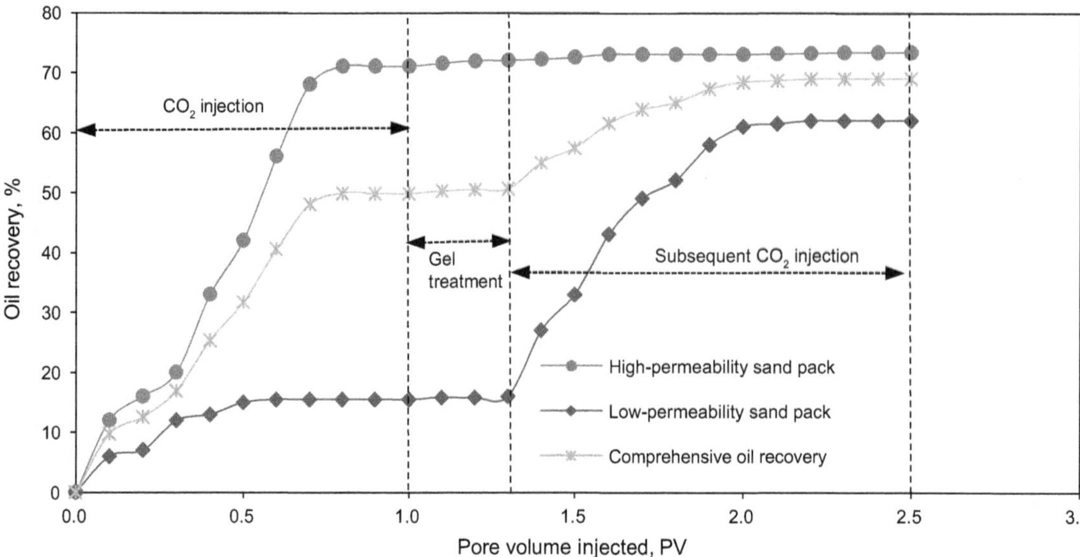

Fig. 7 Cumulative oil recovery via CO_2 injection in the simulated heterogeneous formation

5 Conclusions

(1) A modified polyacrylamide-methenamine-resorcinol system was investigated as CO_2-sensitive chemicals to form gels in situ in the presence of CO_2 under reservoir conditions. Bottle tests demonstrate that this gel system had good CO_2-sensitivity, and strong gel was formed with a strength code of H at temperatures of up to 80 °C.

(2) The improved bottle tests indicate that the gelation time was reduced with an increase in temperature, which was attributed to the promoting effects of high temperature on the molecular motion of CO_2 and the release of the crosslinkers. The concentration of polymer (PAM) also influenced the gelation process and the strength of gels formed. The higher the PAM concentration, the stronger the gel formed. However, a high polymer concentration can reduce the injectivity of the gel system. A PAM concentration of 1.0 wt% in the solution was recommended. The increase in pressure had a positive effect on the gelation process at experimental pressures.

(3) The analysis of injectivity shows good viscosity-temperature characteristics of the CO_2-sensitive gel system. With the temperature increasing, the viscosity of this gel system decreased which can contribute to deep conformance improvement. An effective viscosity can be maintained under relatively higher temperatures which may inhibit the channeling of the injection system in deep formations.

(4) The water shutoff capacity of the gel system was determined through sand pack flow tests, the test results show that the 90 % final permeability reduction was achieved at high salinity (200,000 ppm with 1000 ppm Ca^{2+}) and at temperature up to 90 °C in the high-permeability sand pack (1698.5 mD).

(5) The modified PAM-methenamine-resorcinol solution can flow into and form strong gel in the high-permeability zone in a heterogeneous formation in a CO_2 atmosphere. This performance would increase the sweep efficiency of injected CO_2, and improve the final oil recovery in heterogeneous formations.

Acknowledgments The authors would like to acknowledge financial support from the National Basic Research Program of China (2015CB251201), the Fundamental Research Funds for the Central Universities (15CX06024A), and the Program for Changjiang Scholars and Innovative Research Team in University (IRT1294 and IRT1086).

References

Asghari K, Taabbodi L, Dong M. A new gel-foam system for water shut-off purposes in wormhole reservoirs. In: SPE international thermal operations and heavy Oil symposium, 1–3 November, Calgary, Alberta, Canada; 2005. doi:10.2118/97765-MS.

Bastami A, Allahgholi M, Pourafshary P. Experimental and modelling study of the solubility of CO_2 in various $CaCl_2$ solutions at different temperatures and pressures. Pet Sci. 2014;11(4):569–77. doi:10.1007/s12182-014-0373-1.

Cai SZ. Study of CO_2 mobility control using cross-linked gel conformance control and CO_2 viscosifiers in heterogeneous media. Master Thesis. Texas A & M University; 2010.

Chakravarthy D, Muralidaharan V, Putra E, et al. Mitigating oil bypassed in fractured cores during CO_2 flooding using WAG and polymer gel injections. In: SPE/DOE symposium on improved oil recovery, 22–26 April, Tulsa, Oklahoma, USA; 2006. doi:10.2118/97228-MS.

Chang HL, Sui X, Xiao L, et al. Successful field pilot of in-depth colloidal dispersion gel (CDG) technology in Daqing oil field. SPE Reserv Eval Eng. 2006;9(6):664–73. doi:10.2118/89460-PA.

Christensen JR, Stenby EH, Skauge A. Review of WAG field experience. SPE Reserv Eval Eng. 2001;4(2):97–106. doi:10.2118/71203-PA.

Cordova M, Cheng M, Trejo J, et al. Delayed HPAM gelation via transient sequestration of chromium in polyelectrolyte complex nanoparticles. Macromolecules. 2008;41(12):4398–404. doi:10.1021/ma800211d.

Dalrymple D, Tarkington JT, Hallock J. A gelation system for conformance technology. In: SPE annual technical conference and exhibition, 25–28 September, New Orleans, Louisiana, USA; 1994. doi:10.2118/28503-MS.

Duan ZH, Sun R. An improved model calculating CO_2 solubility in pure water and aqueous NaCl solutions from 273 to 533 K and from 0 to 2000 bar. Chem Geol. 2003;193(3–4):257–71. doi:10.1016/S0009-2541(02)00263-2.

Duan ZH, Sun R, Zhu C, et al. An improved model for the calculation of CO_2 solubility in aqueous solutions containing Na^+, K^+, Ca^{2+}, Mg^{2+}, Cl^-, and SO_4^{2-}. Mar Chem. 2006;98(2–4):131–9. doi:10.1016/j.marchem.2005.09.001.

El-Karsani KSM, Al-Muntasheri GA, Hussein IA. Polymer systems for water shutoff and profile modification: a review over the last decade. SPE J. 2014;19(1):135–49. doi:10.2118/163100-PA.

Elsharafi MO, Bai B. Influence of strong preformed particle gels on low permeable formations in mature reservoirs. Pet Sci. 2016;13(1):77–90. doi:10.1007/s12182-015-0072-3.

Elsharkawy AM, Poettmann FH, Christiansen RL. Measuring CO_2 minimum miscibility pressures: slim-tube or rising bubble method? Energy Fuels. 1996;10(2):443–9. doi:10.1021/ef940212f.

Enick RM, Beckman EJ, Johnson JK. Synthesis and evaluation of CO_2 thickeners designed with molecular modeling. Technical report prepared for National Energy Technology Laboratory (U.S DOE, under Contact No. DE-FG26-04NT-15533), University of Pittsburgh, Pittsburgh PA; 2010.

Enick RM, Olsen DK, Ammer JR, et al. Mobility and conformance control for CO_2 EOR via thickeners, foams, and gels-a literature review of 40 years of research and pilot tests. In: SPE improved oil recovery symposium, 14–18 April, Tulsa, Oklahoma, USA; 2012. doi:10.2118/154122-MS.

Guo A, Geng Y, Zhao L, et al. Preparation of cationic polyacrylamide microsphere emulsion and its performance for permeability reduction. Pet Sci. 2014;11(3):408–16. doi:10.1007/s12182-014-0355-0.

Hamouda AA, Chukwudeme EA, Mirza D. Investigating the effect of CO_2 flooding on asphaltenic oil recovery and reservoir wettability. Energy Fuels. 2009;23(2):1118–27. doi:10.1021/ef800894m.

Hild GP, Wackowski RK. Reservoir polymer gel treatments to improve miscible CO_2 flood. SPE Reserv Eval Eng. 1999;2(2):196–204. doi:10.2118/56008-PA.

Hou Y, Yue X. Research on a novel composite gel system for CO_2 breakthrough. Pet Sci. 2010;7(2):245–50. doi:10.1007/s12182-010-0028-6.

Hughes TL, Friedmann F, Johnson D, et al. Large-volume foam-gel treatments to improve conformance of the Rangely CO_2 flood. SPE Reserv Eval Eng. 1999;2(1):14–24. doi:10.2118/54772-PA.

Karaoguz OK, Topguder NN, Lane RH, et al. Improved sweep in Bati Roman heavy-oil CO_2 flood: bullhead flowing gel treatments plug natural fractures. SPE Reserv Eval Eng. 2007;10(2):164–75. doi:10.2118/89400-PA.

Khalil F, Asghari K. Application of CO_2-foam as a means of reducing carbon dioxide mobility. J Can Pet Technol. 2006;45(5):37–42. doi:10.2118/06-05-02.

Li DX, Ren B, Zhang L, et al. CO_2-sensitive foams for mobility control and channeling blocking in enhanced WAG process. Chem Eng Res Des. 2015;102:234–43. doi:10.1016/j.cherd.2015.06.026.

Li XC, Fang ZM. Status quo of connection technologies of CO_2 geological storage in China. Rock Soil Mech. 2007;28(10):2229–33, 2239 (in Chinese).

Luo P, Zhang YP, Huang S. A promising chemical-augmented WAG process for enhanced heavy oil recovery. Fuel. 2013;104:333–41. doi:10.1016/j.fuel.2012.09.070.

Majidaie S, Onur M, Tan IM. An experimental and numerical study of chemically enhanced water alternating gas injection. Pet Sci. 2015;12(3):470–80. doi:10.1007/s12182-015-0033-x.

Martin FD, Kovarik FS. Chemical gels or diverting CO_2: baseline experiments. In: SPE annual technical conference and exhibition, 27–30 September, Dallas, Texas, USA; 1987. doi:10.2118/16728-MS.

Martin FD, Kovarik FS, Chang PW, et al. Gels for CO_2 profile modification. In: SPE enhanced oil recovery symposium, 16–21 April, Tulsa, Oklahoma, USA; 1988. doi:10.2118/17330-MS.

McLendon WJ, Koronaios P, McNulty S, et al. Assessment of CO_2-soluble surfactants for mobility reduction using mobility measurements and CT imaging. In: SPE improved oil recovery symposium, 14–18 April, Tulsa, Oklahoma, USA; 2012. doi:10.2118/154205-MS. .

Nezhad SAT, Mojarad MRR, Paitakhti SJ, et al. Experimental study of applicability of water-alternating-CO_2 injection in the secondary and tertiary recovery. In: International oil conference and exhibition in Mexico, August 31–September 2, Cancun, Mexico; 2006. doi:10.2118/103988-MS.

Niu LW, Lu XG, Xiong CM, et al. Experimental study of gelling property and plugging effect of inorganic gel system (OMGL). Pet Explor Dev. 2013;40(6):780–4. doi:10.1016/S1876-3804(13)60104-4.

Noller CR. Chemistry of organic compounds. Philadelphia: Saunders; 1965. p. 221.

Pipes JW, Schoeling LG. Performance review of gel polymer treatments in a miscible CO_2 enhanced recovery project, SACROC unit Kelly-Snyder Field. In: SPE improved oil recovery symposium, 12–16 April, Tulsa, Oklahoma, USA; 2014. doi:10.2118/169176-MS.

Prara A. Experimental study of the effectiveness of gels for profile modification. Master Thesis. The University of Oklahoma, Norman, Oklahoma; 1998.

Prada A, Civan F, Dalrymple ED. Evaluation of gelation systems for conformance control. In: SPE/DOE improved oil recovery symposium, 3–5 April, Tulsa, Oklahoma, USA; 2000. doi:10.2118/59322-MS.

Preston C, Monea M, Jazrawi W, et al. IEA GHG Weyburn CO_2 monitoring and storage project. Fuel Process Technol. 2005;86(14–15):1547–68. doi:10.1016/j.fuproc.2005.01.019.

Raje M, Asghari K, Vossoughi S, et al. Gel systems for controlling CO_2 mobility in carbon dioxide miscible flooding. SPE Reserv Eval Eng. 1999;2(2):205–10. doi:10.2118/55965-PA.

Ren SR, Zhang L, Zhang L. Geological storage of CO_2: overseas demonstration projects and their implications for China. J China Univ Pet (Edn Nat Sci). 2010;34(1):93–8 (in Chinese).

Ren SR, Li DX, Zhang L, et al. Leakage pathways and risk analysis of carbon dioxide in geological storage. Acta Pet Sin. 2014;35(3):591–601. doi:10.7623/syxb201403024 (in Chinese).

Seright RS. Reduction of gas and water permeabilities using gels. SPE Prod Facil. 1995;10(2):103–8. doi:10.2118/25855-PA.

Sweatman RE, Parker ME, Crookshank SL. Industry experience with CO_2-enhanced oil recovery technology. In: SPE international conference on CO_2 capture, storage, and utilization, 2–4 November, San Diego, California, USA; 2009. doi:10.2118/126446-MS.

Sydansk RD, Argabright PA. Conformance improvement in a subterranean hydrocarbon-bearing formation using a polymer gel. U.S. Patent No. 4683949; 1987.

Talebian SH, Masoudi R, Tan IM, et al. Foam assisted CO_2-EOR: a review of concept, challenges and future prospects. J Pet Sci Eng. 2014;120:202–15. doi:10.1016/j.petrol.2014.05.013.

Xing DZ, Wei B, Trickett K, et al. CO_2-soluble surfactants for improved mobility control. In: SPE improved oil recovery symposium, 24–28 April, Tulsa, Oklahoma, USA; 2010. doi:10.2118/129907-MS.

Xing QY, Pei WW, Xu RQ, et al. Fundamental organic chemistry. Beijing: Higher Education Press; 2005. p. 359 (in Chinese).

Yu JJ, An C, Mo D, et al. Foam mobility control for nanoparticle stabilized CO_2 foam. In: SPE improved oil recovery symposium, 14–18 April, Tulsa, Oklahoma, USA; 2012. doi:10.2118/153336-MS.

Zhang L, Ren SR, Wang RH, et al. Feasibility study of associated CO_2 geological storage in a saline aquifer for development of Dongfang 1-1 gas field. J China Univ Pet (Edn Nat Sci). 2010;34(3):89–93. doi:10.3969/j.issn.1673-5005.2010.03.019 (in Chinese).

Zhang MS, Yang ZR. Water shut-off technology using polyacrylamide-methenamine-resorcinol gel system. Oil Drill Prod Technol. 1988;7:161–70 (in Chinese).

Zhang SY, She YH, Gu YA. Evaluation of polymers as direct thickeners for CO_2 enhanced oil recovery. J Chem Eng Data. 2011;56(4):1069–79. doi:10.1021/je1010449.

Zhang Y, Song H, Li DX, et al. Experiment on high pressure CO_2 foam stability of nonionic surfactants. J China Univ Pet (Edn Nat Sci). 2013a;37(4):119–23, 128 (in Chinese).

Zhang YM, Chu ZL, Dreiss CA, et al. Smart wormlike micelles switched by CO_2 and air. Soft Matter. 2013b;9:6217–21. doi:10.1039/C3SM50913C.

Zhang YM, Feng YJ, Wang YJ, et al. CO_2-switchable viscoelastic fluids based on a pseudogemini surfactant. Langmuir. 2013c;29(13):4187–92. doi:10.1021/la400051a.

Zhu YZ, Liao CH, Wang CQ. Mitigation and recovery of carbon dioxide. Beijing: Chemical Industry Press; 2011. p. 110–6 (in Chinese).

16

An amplitude-preserved adaptive focused beam seismic migration method

Ji-Dong Yang[1] · Jian-Ping Huang[1] · Xin Wang[1] · Zhen-Chun Li[1]

Abstract Gaussian beam migration (GBM) is an effective and robust depth seismic imaging method, which overcomes the disadvantage of Kirchhoff migration in imaging multiple arrivals and has no steep-dip limitation of one-way wave equation migration. However, its imaging quality depends on the initial beam parameters, which can make the beam width increase and wave-front spread with the propagation of the central ray, resulting in poor migration accuracy at depth, especially for exploration areas with complex geological structures. To address this problem, we present an adaptive focused beam method for shot-domain prestack depth migration. Using the information of the input smooth velocity field, we first derive an adaptive focused parameter, which makes a seismic beam focused along the whole central ray to enhance the wavefield construction accuracy in both the shallow and deep regions. Then we introduce this parameter into the GBM, which not only improves imaging quality of deep reflectors but also makes the shallow small-scale geological structures well-defined. As well, using the amplitude-preserved extrapolation operator and deconvolution imaging condition, the concept of amplitude-preserved imaging has been included in our method. Typical numerical examples and the field data processing results demonstrate the validity and adaptability of our method.

✉ Ji-Dong Yang
yangjidong_china@163.com

[1] School of Geoscience, China University of Petroleum, Qingdao 266580, Shandong, China

Edited by Jie Hao

Keywords Gaussian beam · Adaptive focused beam · Amplitude-preserved migration · Depth imaging

1 Introduction

With the development of petroleum exploration, seismic surveys have gradually extended into areas with complex geological conditions, such as regions with continental faulted basins and offshore salt deposits. This situation presents new challenges for seismic imaging, which compel us to explore a more efficient, accurate, and robust migration method than the existing ones.

Over the past decades, dramatical progress has been achieved in the field of prestack depth imaging. Ray-based Kirchhoff migration has been developed from 2D to 3D (Hubral et al. 1996; Epili and McMechan 1996; Sun et al. 2000), from single-arrival migration (including first arrival, most energy arrival etc.) to multi-arrival migration (Brandsberg-Dahl et al. 2001; Xu et al. 2001) and from kinematic migration to true-amplitude migration (Schleicher et al. 1993; Albertin et al. 1999; Xu and Lambaré 2006). Due to its high efficiency and flexibility, Kirchhoff migration is still the workhorse in practical applications, especially for land seismic data. On the other hand, one-way wave equation migration accuracy has also been improved significantly using different numerical algorithms, such as split-step Fourier, Fourier finite difference, and generalized screen propagator (Stoffa et al. 1990; Ristow and Rühl 1994; Chen and Ma 2006; Li et al. 2008; Kaplan et al. 2010; Huang and Fehler 2000; Liu and Yin 2007; Zhu et al. 2009; De Hoop et al. 2000; Wu et al. 2001; Le Rousseau and De Hoop 2003; Liu et al. 2012). Reverse time migration based on two-way wave equations has become more and more practical in actual projects as computer technology develops (Baysal et al. 1983), and many geophysicists have provided lots of constructive suggestions about the problems of time-consuming defects and low frequency imaging noise (Fletcher et al. 2006; Symes 2007; Chattopadhyay and McMechan 2008; Abdelkhalek et al. 2009; Li et al. 2010).

Gaussian beam migration (GBM) is an elegant and effective depth imaging method, which not only retains the advantage of ray-based migration, such as flexibility and efficiency, but also has an imaging accuracy comparable with wave equation migration. Ever since the basic framework of GBM was presented by Hill (1990, 2001), it has been extended to irregular topographic conditions (Gray 2005; Yue et al. 2012; Yang et al. 2014) and true-amplitude migration (Gray and Bleistein 2009). In addition, many new seismic beam imaging methods have been developed, such as fast beam migration (Gao et al. 2006, 2007), focused beam migration (Nowack 2008), and laser beam migration (Xiao et al. 2014), which expand the members of the beam migration family. Most of them, however, are based on the GBM framework and use a constant initial beam parameter, which makes a seismic beam focused either at the initial position or at a certain depth. Thus, the imaging accuracy varies along the central ray, highest at the focus point and decreasing as moving away from the focus point.

Aiming at this problem, Nowack (2009) has provided a dynamically focused beam method, which improves the deep imaging quality to some extent. Because this method is achieved using a unified beam width for all the subsurface imaging points and performing many local slant stacks and quadratic phase corrections for each beam, it does not consider the effects of velocity variation on the beam width and is very time-consuming. Hu and Stoffa (2009) have implemented a modified GBM for low-fold seismic data acquired sparsely, utilizing the instantaneous slowness of the local plane wave. Derived from the Maslov wave equation solution, Zhu (2009) also proposed a complex-ray beam method for exploration areas with complex topography. All methods mentioned above use the Fresnel zone information to limit the seismic beam energy and improve the imaging quality, which provides new options for beam migration.

In this paper, we have proposed another way to optimize the beam propagation shape and implemented an adaptive focused beam migration method for common-shot data, which could improve the shallow and deep imaging quality simultaneously. Using the information of the input smoothed velocity field, we derived an adaptive focused parameter that makes a seismic beam focused at the whole central ray, and then use it to construct a Green function in the acoustic medium and to solve the seismic migration problem. Unlike the dynamically focused beam method proposed by Nowack, our method uses the single input-trace imaging approach of classic Kirchhoff migration to avoid repeating local slant stacks for a beam at different imaging points, which is helpful to speed up the migration process. However, compared with Hill's GBM, there is a tradeoff in computational efficiency, as now the number of emergent beams increases. In addition, using the amplitude-preserved extrapolation formula and deconvolution imaging condition, we have included the concept of amplitude-preserved imaging in our method. Typical numerical examples and the field data processing results demonstrate the feasibility and validity of the proposed method.

2 Theory

2.1 Adaptive focused beam

Considering a Gaussian beam from $Q(s_0, 0)$ to $P(s, 0)$ in a 2D acoustic medium, its ray propagation matrix can be written as

$$\pi(s; s_0) = \begin{bmatrix} q_1(s) & q_2(s) \\ p_1(s) & p_2(s) \end{bmatrix}, \tag{1}$$

where (s, n) are ray-centered coordinates as shown in Fig. 1, $(p_1(s), q_1(s))$ and $(p_2(s), q_2(s))$ are two fundamental solutions of dynamic ray tracing equation system (Červený et al. 1982).

If the beam focused at P and its beam width equals $l(s)$, then the complex dynamic parameter at the initial position can be calculated by

$$\begin{bmatrix} q(s_0) \\ p(s_0) \end{bmatrix} = \pi(s; s_0)^{-1} \begin{bmatrix} -i\omega_{\text{ref}} l^2(s) \\ 1 \end{bmatrix}, \tag{2}$$

here ω_{ref} denotes the referenced frequency, $i = \sqrt{-1}$ and $l(s)$ takes the following form

$$l(s) = 2\pi v(s)/\omega_{\text{ref}}, \tag{3}$$

where $v(s)$ is the velocity of central ray.

Further, using the relation

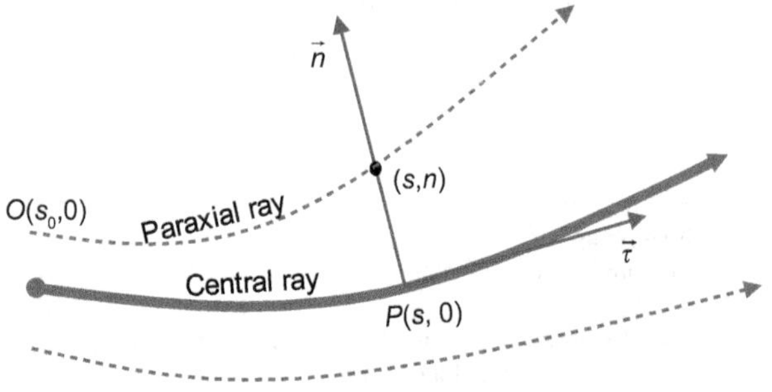

Fig. 1 Ray-centered coordinate system

$$\varepsilon = q(s_0)/p(s_0), \tag{4}$$

We obtain a new initial beam parameter

$$\varepsilon(s) = \frac{-q_2(s) - i\omega_{\text{ref}} l^2(s) p_2(s)}{q_1(s) + i\omega_{\text{ref}} l^2(s) p_1(s)}. \tag{5}$$

Unlike the Gaussian beam and focused beam, the initial parameter in Eq. (5) is no longer a constant, but a function of ray arc length. This choice makes the main energy of the beams focus in the range of a wavelength along the whole ray. For the convenience of discussion, we define the beam determined by $\varepsilon(s)$ in Eq. (5) as the adaptive focused beam.

Now we analyze the propagation property of the adaptive focused beam briefly. Considering an inhomogeneous medium as shown in Fig. 2a, which includes a high-speed layer and a low-speed layer in a constant-gradient velocity background, the adaptive focused beam keeps a narrow beam width and plane wave-front along the whole central ray (see Fig. 2b). Besides, it has a small beam width at the low-speed layer and large beam width at the high-speed layer marked by the blue ellipses in Fig. 2b, which is helpful to resolve the steep-dip reflectors and velocity abnormal bodies in seismic imaging. For a Gaussian beam, however, the beam width increases quickly and the wave-front diffuses rapidly (see Fig. 2c), which result in inaccurate travel-time and amplitude extrapolated from central ray at deep parts, especially in a medium with strong lateral velocity variation.

2.2 Green function represented with the adaptive focused beam integral

With the initial parameter $\varepsilon(s)$ in Eq. (5), we can write the expression of the adaptive focused beam in the frequency domain as

$$U(s,n,\omega) = \sqrt{\frac{\varepsilon(s_0)v(s)}{[\varepsilon(s)q_1(s) + q_2(s)]v(s_0)}} \times \exp\left[i\omega\left(\tau(s) + \frac{1}{2}\frac{\varepsilon(s)p_1(s) + p_2(s)}{\varepsilon(s)q_1(s) + q_2(s)}n^2\right)\right]. \tag{6}$$

where ω denotes the circular frequency, τ is the travel-time of central ray.

According to Müller (1984), Green's function at point M shown in Fig. 3 can be approximately represented with an integral over all the rays of beams, i.e.,

$$G(M,\omega) = \int_{\phi_0-\pi/2}^{\phi_0+\pi/2} \Phi(\phi,s) U_\phi(s,n,\omega)\,\mathrm{d}\phi, \tag{7}$$

where $\Phi(\phi,s)$ is an integral weight coefficient for the adaptive focused beam with emergence angle ϕ.

If the weight coefficient $\Phi(\phi,s)$ was known, Eq. (7) could be used to calculate the seismic wave-field at any point of the medium. Now we determine the function $\Phi(\phi,s)$ in a homogenous medium with constant velocity v_0. Denoting

$$\begin{aligned} F(\phi) &= \Phi(\phi,s)\sqrt{\frac{\varepsilon(\phi,s_0)v(s)}{[\varepsilon(\phi,s)q_1(s) + q_2(s)]v(s_0)}} \\ f(\phi) &= -i\left[\tau(s) + \frac{1}{2}\frac{\varepsilon(\phi,s)p_1(s) + p_2(s)}{\varepsilon(\phi,s)q_1(s) + q_2(s)}n^2\right]. \end{aligned} \tag{8}$$

Equation (7) can be rewritten as

$$G(M,\omega) = \int_{\phi_0-\pi/2}^{\phi_0+\pi/2} F(\phi)\exp[-\omega f(\phi)]\,\mathrm{d}\phi. \tag{9}$$

In a homogeneous medium, we have (see Fig. 3)

$$\begin{aligned} &\tau(s) = r\cos(\phi-\phi_0)/v_0,\ n = r\sin(\phi-\phi_0) \\ &q_1 = p_2 = 1,\ p_1 = 0,\ q_2 = v_0 r\cos(\phi-\phi_0). \end{aligned} \tag{10}$$

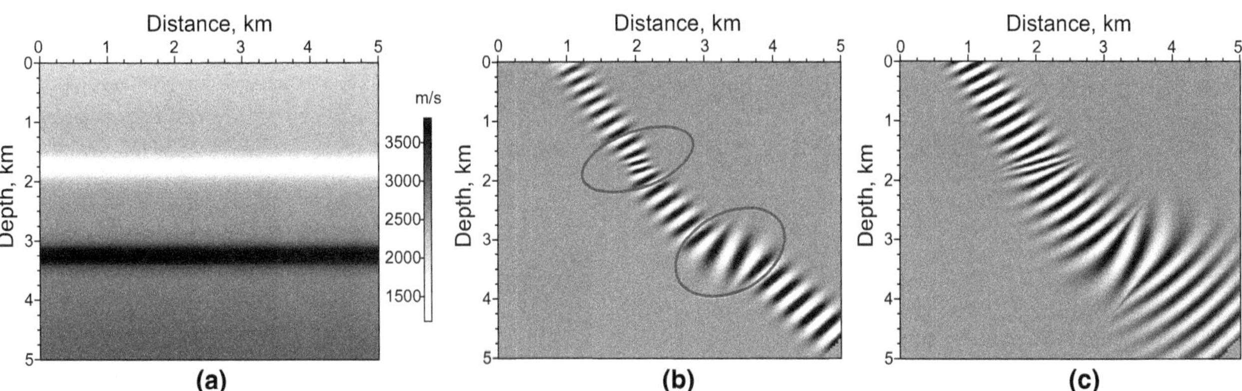

Fig. 2 The propagation of different seismic beams. **a** Velocity model, **b** adaptive focused beam, **c** Gaussian beam

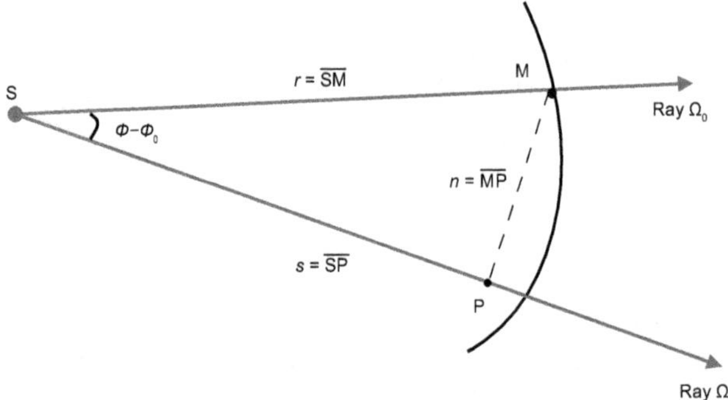

Fig. 3 Green's function constructed with adaptive focused beam summation (referenced to Červený et al. 1982)

Inserting the latter expression in the Eq. (8), we obtain

$$F(\phi)=\Phi(\phi,s)\sqrt{\frac{\varepsilon(\phi,s_0)}{\varepsilon(\phi,s)+v_0 r\cos(\phi-\phi_0)}}$$
$$f(\phi)=-i\left[r\cos(\phi-\phi_0)/v_0+\frac{1}{2}\frac{r^2\sin^2(\phi-\phi_0)}{\varepsilon(\phi,s)+v_0 r\cos(\phi-\phi_0)}\right]. \tag{11}$$

In a smoothed medium, $\varepsilon_\phi(\phi,s)$ can be expanded with Taylor's formula as follows

$$\varepsilon(\phi,s)=\varepsilon(\phi_0,s)+\varepsilon_\phi(\phi_0,s)(\phi-\phi_0), \tag{12}$$

where $\varepsilon_\phi(\phi_0,s)$ denotes the partial derivative of $\varepsilon(\phi_0,s)$ with respect to ϕ.

It is easy to see from Eq. (11) that the saddle point of Eq. (9), defined by $f_\phi(\phi)=0$, is $\phi=\phi_0$. So we assume that in the vicinity of ϕ_0, which contributes most to Eq. (9), $\Phi(\phi,s)$ can be replaced by $\Phi(\phi_0,s)$. Then, the corresponding approximation of $F(\phi)$ and $f(\phi)$ takes the form

$$F(\phi)=\Phi(\phi_0,s)\sqrt{\frac{\varepsilon(\phi_0,s_0)+\varepsilon_\phi(\phi_0,s_0)(\phi-\phi_0)}{\varepsilon(\phi_0,s)+\varepsilon_\phi(\phi_0,s)(\phi-\phi_0)+v_0 r}}$$
$$f(\phi)=\frac{-ir}{v_0}+\frac{ir}{2v_0}\left\{\frac{\varepsilon(\phi_0,s)(\phi-\phi_0)^2}{\varepsilon(\phi_0,s)+v_0 r}+\frac{v_0 r\varepsilon_\phi(\phi_0,s)(\phi-\phi_0)^3}{[\varepsilon(\phi_0,s)+v_0 r]^2}\right\}. \tag{13}$$

Under the far-field condition, the latter expressions can be further reduced to

$$F(\phi)=\Phi(\phi_0,s)\sqrt{\frac{\varepsilon(\phi_0,s_0)+\varepsilon_\phi(\phi_0,s_0)(\phi-\phi_0)}{v_0 r}}$$
$$f(\phi)=\frac{-ir}{v_0}+\frac{i}{2v_0^2}\left[\varepsilon(\phi_0,s)(\phi-\phi_0)^2+\varepsilon_\phi(\phi_0,s)(\phi-\phi_0)^3\right]. \tag{14}$$

Therefore, we obtain the approximate expression of Green's function represented by the adaptive focused beam integral as

$$G(r,\phi_0)\approx\Phi(\phi_0,s)\frac{\exp(i\omega r/v_0)}{\sqrt{v_0 r}}\Theta, \tag{15}$$

where

$$\Theta=\int_{\phi_0-\pi/2}^{\phi_0+\pi/2}\mathrm{d}\phi\left[\varepsilon(\phi_0,s)+\varepsilon_\phi(\phi_0,s)(\phi-\phi_0)\right]^{1/2}$$
$$\exp\left\{\frac{i\omega}{2v_0^2}\left[\varepsilon(\phi_0,s)(\phi-\phi_0)^2+\varepsilon_\phi(\phi_0,s)(\phi-\phi_0)^3\right]\right\}. \tag{16}$$

Comparing Eq. (15) with the leading term in the expansion of Green's function obtained by asymptotic ray theory (ART):

$$G\approx\frac{\exp[i\omega r/v_0+i\,\mathrm{sgn}(\omega)\pi/4]}{2\sqrt{2\pi|\omega|r/v}}, \tag{17}$$

we obtain

$$\Phi(\phi_0,s)=\frac{v_0\exp(i\,\mathrm{sgn}(\omega)\pi/4)}{2\sqrt{2\pi|\omega|}\Theta}. \tag{18}$$

In order to simplify the integral of Eq. (16), here we consider two extreme cases:

Case 1 $\varepsilon_\phi(\phi_0,s)=0,\ \varepsilon(\phi_0,s)\neq 0$

$$\Theta=v_0\sqrt{\frac{2\pi}{i\omega}},\ \Phi(\phi_0,s)=\frac{i}{4\pi} \tag{19}$$

Case 2 $\varepsilon_\phi(\phi_0,s)\neq 0,\ \varepsilon(\phi_0,s)=0,$

$$\Theta=\frac{2}{3}v_0\sqrt{\frac{2\pi}{i\omega}},\ \Phi(\phi_0,s)=\frac{i}{6\pi} \tag{20}$$

Case (1) means there is no variation of $\varepsilon(\phi,s)$ in the neighborhood of ϕ_0, and the corresponding integral coefficient is consistent with that proposed by Červený et al. (1982). Case (2) means there is strong variation of $\varepsilon(\phi,s)$ around ϕ_0, but the integral coefficient in this case is only the 2/3 times of that of case (1), which means $\Phi(\phi_0,s)$ has only a relatively mild dependence on the beam parameter $\varepsilon(\phi,s)$. Thus, it appears reasonable to use the results of case (1), and we obtain the expression of Green's function represented by the adaptive focused beam integral

$$G(M,\omega)=\frac{i}{4\pi}\int_{\phi_0-\pi/2}^{\phi_0+\pi/2}U_\phi(s,n,\omega)\mathrm{d}\phi. \tag{21}$$

In order to demonstrate the validity of Eq. (21), we compare it with the analytic Green's function calculated with ART in a homogeneous medium, where the velocity is

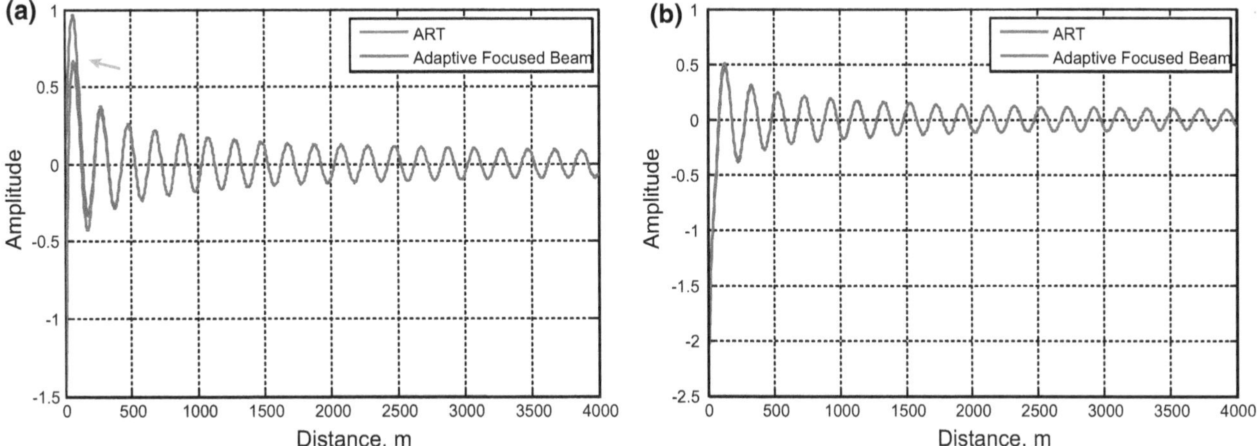

Fig. 4 Real (**a**) and imaginary (**b**) parts of Green's function calculated with ART and adaptive focused beam methods

2000 m/s and the reference frequency and the data frequency are all 10 Hz. The results are shown in Fig. 4. It is easy to see that both the real and imaginary parts of the Green's function calculated with the two methods are similar for most distances, except the real part in the shallow zone marked by the green arrow. One possible explanation is that an adaptive focused beam appears a plane wave at the initial point and its beam width is about a wavelength, which is not consistent with the point source wave-field calculated with Eq. (17). But the near-source error can be neglected in practical seismic modeling and imaging applications.

2.3 Adaptive focused beam migration formula

Nowack (2009) has used many local slant stacks and quadratic phase corrections for every beam in the dynamically focused beam migration, which is time-consuming and complex in designing the implementation algorithm. Here we adopt the single input-trace imaging approach of classical Kirchhoff migration to implement the adaptive focused migration. As a result, our method is more computer intensive than Hill's method, but it is more efficient than Nowack's method and can be accepted given the current level of computer power for an accurate depth image of subsurface geological structures.

The main point of our method is replacing the Green's function by its approximate form in terms of the adaptive focused beam integral. From the latter section, we know the Green's function from $\mathbf{x}'$ to $\mathbf{x}$ can be written as

$$G(\mathbf{x};\mathbf{x}';\omega) = \frac{i}{4\pi}\int \mathrm{d}\phi A(\mathbf{x};\mathbf{x}')\exp[i\omega T(\mathbf{x};\mathbf{x}')]. \tag{22}$$

Here $A(\mathbf{x};\mathbf{x}')$ and $T(\mathbf{x};\mathbf{x}')$ are the complex amplitude and travel-time of the adaptive focused beam respectively, and they have the following form

$$\begin{aligned} A(\mathbf{x};\mathbf{x}') &= \sqrt{\frac{\varepsilon(\mathbf{x}')v(\mathbf{x})}{[\varepsilon(\mathbf{x})q_1(\mathbf{x})+q_2(\mathbf{x})]v(\mathbf{x}')}} \\ T(\mathbf{x};\mathbf{x}') &= \tau(\mathbf{x};\mathbf{x}')+\frac{1}{2}\frac{\varepsilon(\mathbf{x})p_1(\mathbf{x})+p_2(\mathbf{x})}{\varepsilon(\mathbf{x})q_1(\mathbf{x})+q_2(\mathbf{x})}n_{\mathbf{x}}^2, \end{aligned} \tag{23}$$

where $\varepsilon(\mathbf{x})$ is the adaptive focused beam parameter defined in Eq. (5).

According to the true-amplitude GBM formula presented by Gray and Bleistein (2009), the up-going and down-going wave-fields can be written as

$$\begin{aligned} P_U(\mathbf{x}^{\mathrm{P}};\mathbf{x}^{\mathrm{S}};\omega) &= -2i\omega\int \mathrm{d}r\frac{\cos\phi^{\mathrm{R}}}{v^{\mathrm{R}}}G^*(\mathbf{x}^{\mathrm{P}};\mathbf{x}^{\mathrm{R}};\omega)P_U(\mathbf{x}^{\mathrm{R}};\mathbf{x}^{\mathrm{S}};\omega) \\ P_D(\mathbf{x}^{\mathrm{P}};\mathbf{x}^{\mathrm{S}};\omega) &= -2i\omega\frac{\cos\phi^{\mathrm{S}}}{v^{\mathrm{S}}}G(\mathbf{x}^{\mathrm{P}};\mathbf{x}^{\mathrm{S}};\omega), \end{aligned} \tag{24}$$

where $\mathbf{x}^{\mathrm{S}}$, $\mathbf{x}^{\mathrm{R}}$, and $\mathbf{x}^{\mathrm{P}}$ are the Cartesian coordinates of source, receiver, and imaging point, respectively, ϕ^{S} and ϕ^{R} are the ray emergence angles from surface at source and receiver (see Fig. 5), $P_U(\mathbf{x}^{\mathrm{R}};\mathbf{x}^{\mathrm{S}};\omega)$ is the recorded wave-field, v^{S} and v^{R} are the velocity at source and receiver separately, "*" denotes the complex conjugate.

Inserting Eq. (22) into Eq. (24) and using the deconvolution imaging condition

$$R(\mathbf{x}^P,\mathbf{x}^{\mathrm{S}}) = \frac{1}{2\pi}\frac{\cos\phi^{\mathrm{S}}}{v^{\mathrm{S}}}\int \mathrm{d}\omega i\omega\frac{P_U(\mathbf{x}^P;\mathbf{x}^{\mathrm{S}};\omega)P_D^*(\mathbf{x}^P;\mathbf{x}^{\mathrm{S}};\omega)}{P_D(\mathbf{x}^P;\mathbf{x}^{\mathrm{S}};\omega)P_D^*(\mathbf{x}^P;\mathbf{x}^{\mathrm{S}};\omega)}. \tag{25}$$

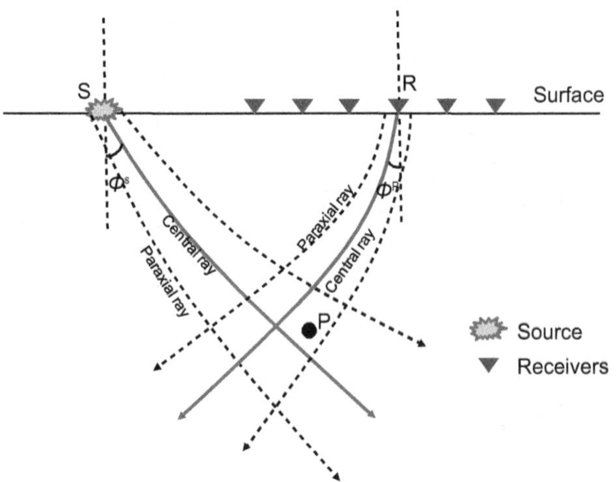

Fig. 5 Scheme of adaptive focused beam migration

We obtain the shot-domain adaptive focused beam migration formula

$$R(\mathbf{x}^P, \mathbf{x}^S) = \frac{-1}{8\pi^3}\frac{\cos\phi^S}{v^S}\int d\omega \int dr \frac{i\omega^3}{P_D P_D^*}\iint d\phi^S d\phi^R \times \frac{\cos\phi^S}{v^S}\frac{\cos\phi^R}{v^R}A_S^* A_R^* \exp(-i\omega T^*)P_U(\mathbf{x}^R, \mathbf{x}^S, \omega). \quad (26)$$

To reduce the computational cost, Gray and Bleistein (2009) used the stationary-phase approximation to simplify the double integral about the emergence angles of source and receiver. Here we adopt this accelerating strategy. Denoting

$$\phi^m = \phi^S + \phi^R$$
$$\phi^h = \phi^R - \phi^S, \quad (27)$$

Equation (26) can be rewritten as

$$R(\mathbf{x}^P, \mathbf{x}^S) = \frac{-1}{16\pi^3}\frac{\cos\phi^S}{v^S}\int d\omega \int dr \frac{i\omega^3}{P_D P_D^*}\iint d\phi^m d\phi^h \times \frac{\cos\phi^S}{v^S}\frac{\cos\phi^R}{v^R}A_S^* A_R^* \exp(-i\omega T^*)P_U(\mathbf{x}^R, \mathbf{x}^S, \omega). \quad (28)$$

Further, using the stationary-phase approximation to calculate the inner integral with respect to ϕ^h and the term of source illumination $P_D P_D^*$, we obtain the final adaptive focused beam migration formula

$$R(\mathbf{x}^P, \mathbf{x}^S) = \frac{-1}{4\pi\sqrt{2\pi}}\int d\omega \int dr\sqrt{i\omega}\omega \int d\phi^m \exp(-i\omega T^*) \times \left(\frac{\cos\phi^S}{v^S}\right)^2 \frac{\cos\phi^R}{v^R}\frac{A_S^* A_R^*}{|A_S|^2}\frac{|T_S''(\phi_0^S)|}{\sqrt{T^{*\prime\prime}(\phi_0^h)}}P_U(\mathbf{x}^R, \mathbf{x}^S, \omega), \quad (29)$$

where

$$T = T(\mathbf{x}^P; \mathbf{x}^S) + T(\mathbf{x}^P; \mathbf{x}^R)$$
$$T_S''(\phi_0^S) = \left[\frac{q(\mathbf{x}^S)q_2(\mathbf{x}^P)}{q(\mathbf{x}^P)v^2(\mathbf{x}^S)}\right] \quad (30)$$
$$T^{*\prime\prime}(\phi_0^h) = \left[\frac{q(\mathbf{x}^S)q_2(\mathbf{x}^P)}{q(\mathbf{x}^P)v^2(\mathbf{x}^S)} + \frac{q(\mathbf{x}^R)q_2(\mathbf{x}^P)}{q(\mathbf{x}^P)v^2(\mathbf{x}^R)}\right]^*,$$

ϕ_0^h is the stationary value, i.e., when ϕ^m is fixed, ϕ_0^h minimizes the imaginary part of the total time T. ϕ_0^S and ϕ_0^R are the corresponding emergence angles of source and receiver determined by Eq. (27).

3 Numerical examples

We present a hierarchy of numerical examples in this section. First, we use a simple horizontal layered model to test the amplitude-preserved property of the proposed method. Then, we carry out two applications of this method to the Marmousi dataset and the field data of a survey in East China, respectively.

To show how the adaptive focused beam migration works, we first apply it to a layered model in a medium with constant velocity of 2000 m/s, which simulates reflections from density contrasts as shown in Fig. 6a. Three horizontal reflectors with identical reflection coefficients are placed at depths of 2000, 3000, and 4000 m. A common-shot record shown in Fig. 6b, with a recording aperture of 3750 m on either side of the source, is migrated with our method, and the result is displayed in Fig. 6c. The normalized migration amplitude of reflectors and distances is shown in Fig. 7. Half-opening angles were limited to 50° in the migration.

It is easy to see that the proposed method has eliminated the influences of offset and images the reflectors accurately (see Fig. 6c). On the other hand, the amplitude-preserved extrapolation formula and deconvolution imaging conditions have eliminated the reflection amplitude difference caused by the different incident angles (see Fig. 7a) and compensated the deep energy loss caused by geometric spreading (see Fig. 7b), leading to the migration amplitude being proportional to the vertical reflection coefficients in effective aperture. Migration aperture truncation artifacts marked by red arrows in Fig. 7a begin to interfere with the amplitudes at the distance corresponding to half-opening angles approaching 50°.

The second example is a synthetic dataset from the 2D Marmousi model as shown in Fig. 8a. The model is about 9.2 km long and 3 km deep, and is characterized by strong lateral velocity variations that cause complicated multi-pathing of the seismic energy. Because the smoothed velocity is required in ray-based migration for numerical

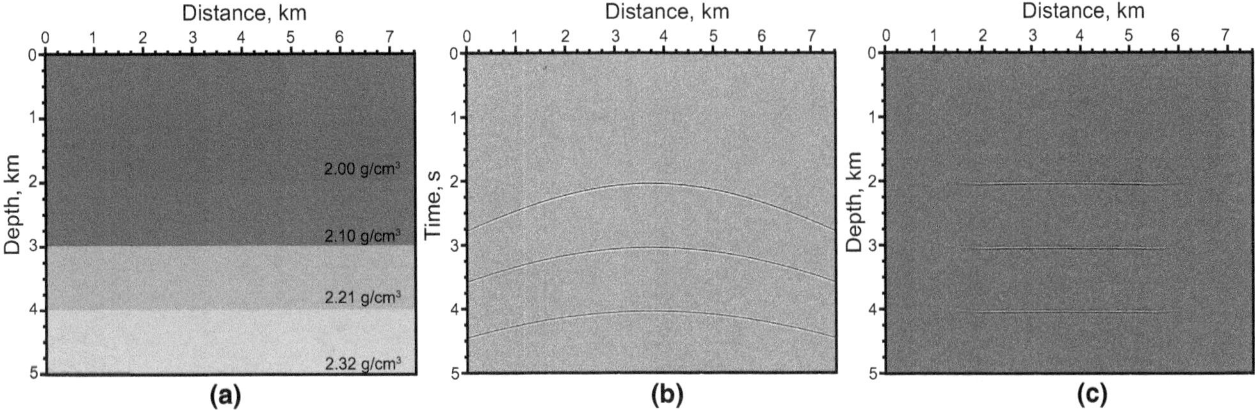

Fig. 6 Layered model and its migration result. **a** Density model, **b** single-shot record, **c** depth image migrated with the proposed method

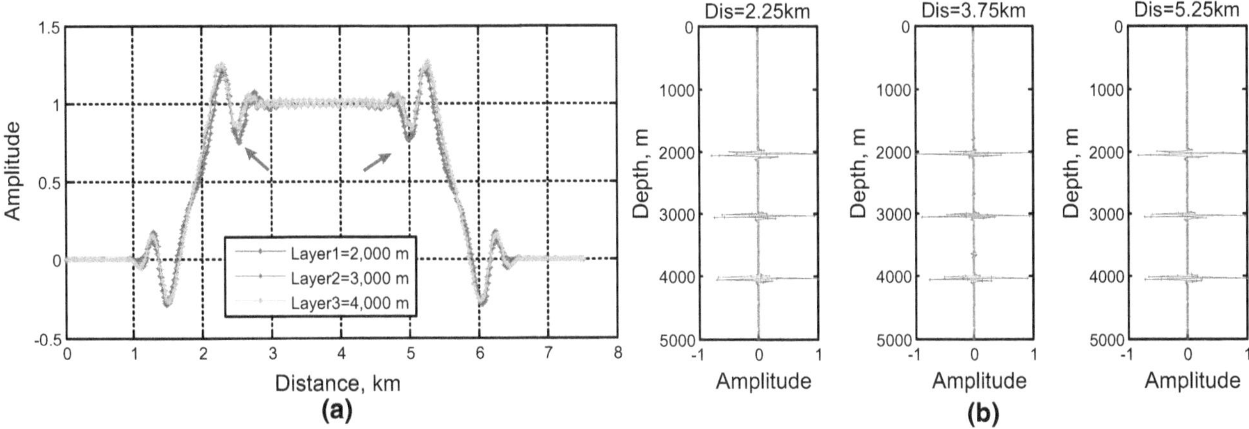

Fig. 7 Normalized amplitude along reflectors (**a**) and at different distances (**b**) in Fig. 6c

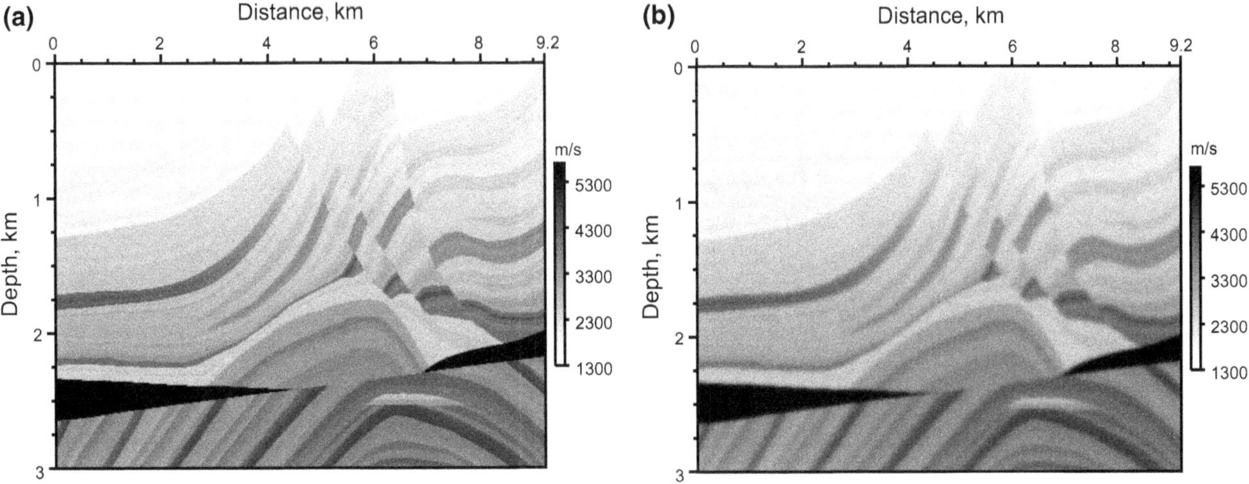

Fig. 8 Marmousi model: **a** the velocity used for simulation, **b** the smoothed velocity used for migration

stability in ray tracing, here we use a damped least squares algorithm to smooth the velocity (see Fig. 8b), which permits us to specify the degree of smoothing of the first- and second-order derivatives of the velocity (Popov et al. 2010). The depth images migrated with the Kirchhoff method and GBM with different initial widths l_0, one-way

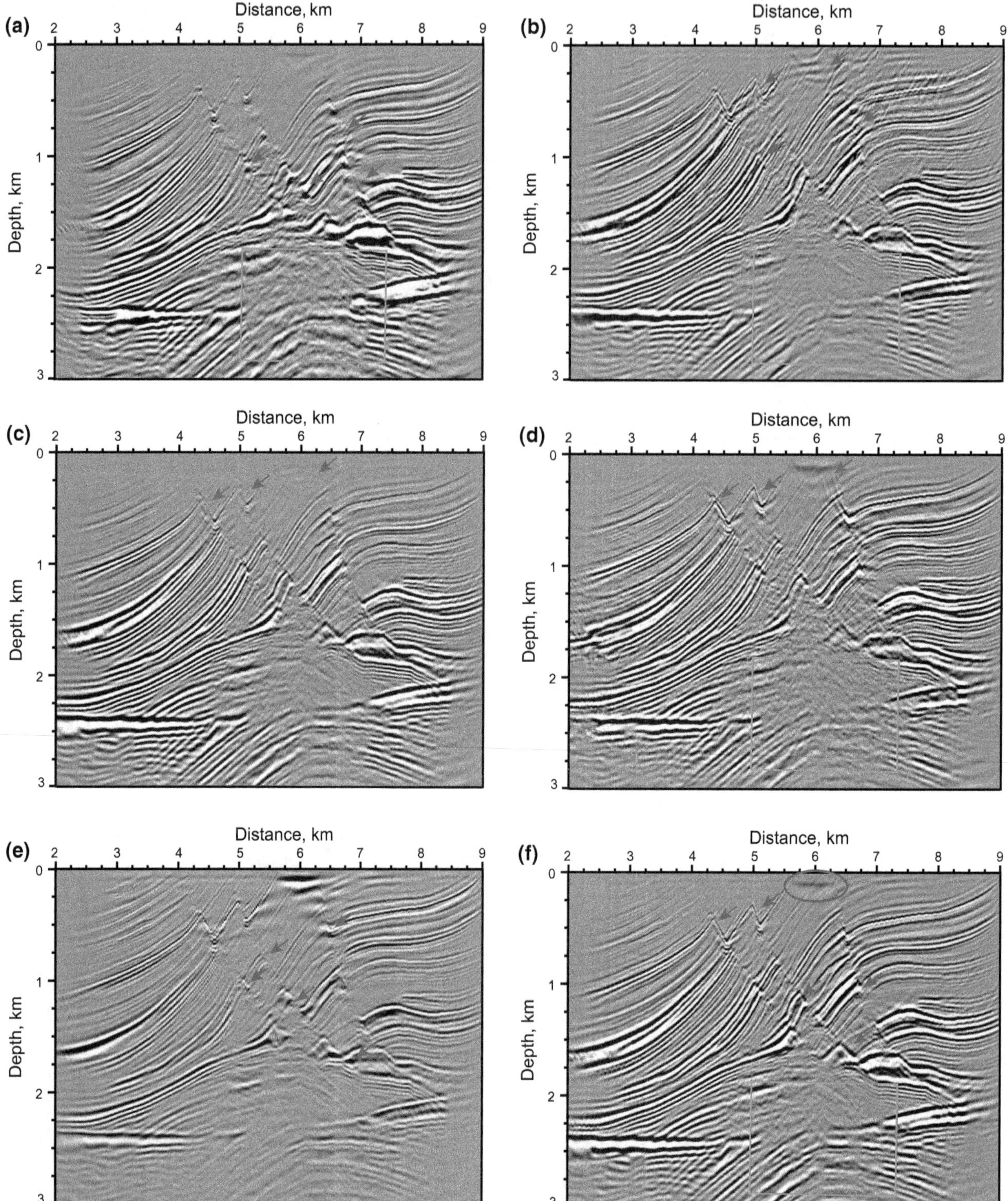

Fig. 9 Depth images migrated from the Marmousi dataset: **a** Kirchhoff migration, **b** GBM with $l_0 = 0.25\lambda_{avg}$, **c** GBM with $l_0 = \lambda_{avg}$, **d** GBM with $l_0 = 2\lambda_{avg}$, **e** one-way wave equation migration, **f** adaptive focused beam migration. λ_{avg} denotes the average wavelength of the input velocity field and the reference frequency for GBM and our method is 10 Hz

wave equation migration and the proposed method are shown in Fig. 9, where the migration aperture is about 5 km and frequency range is from 5 to 60 Hz.

In general, all four methods have imaged the three main faults, pinch-outs, and the anticline at the bottom of the model. The Kirchhoff result shows lots of swing noise in

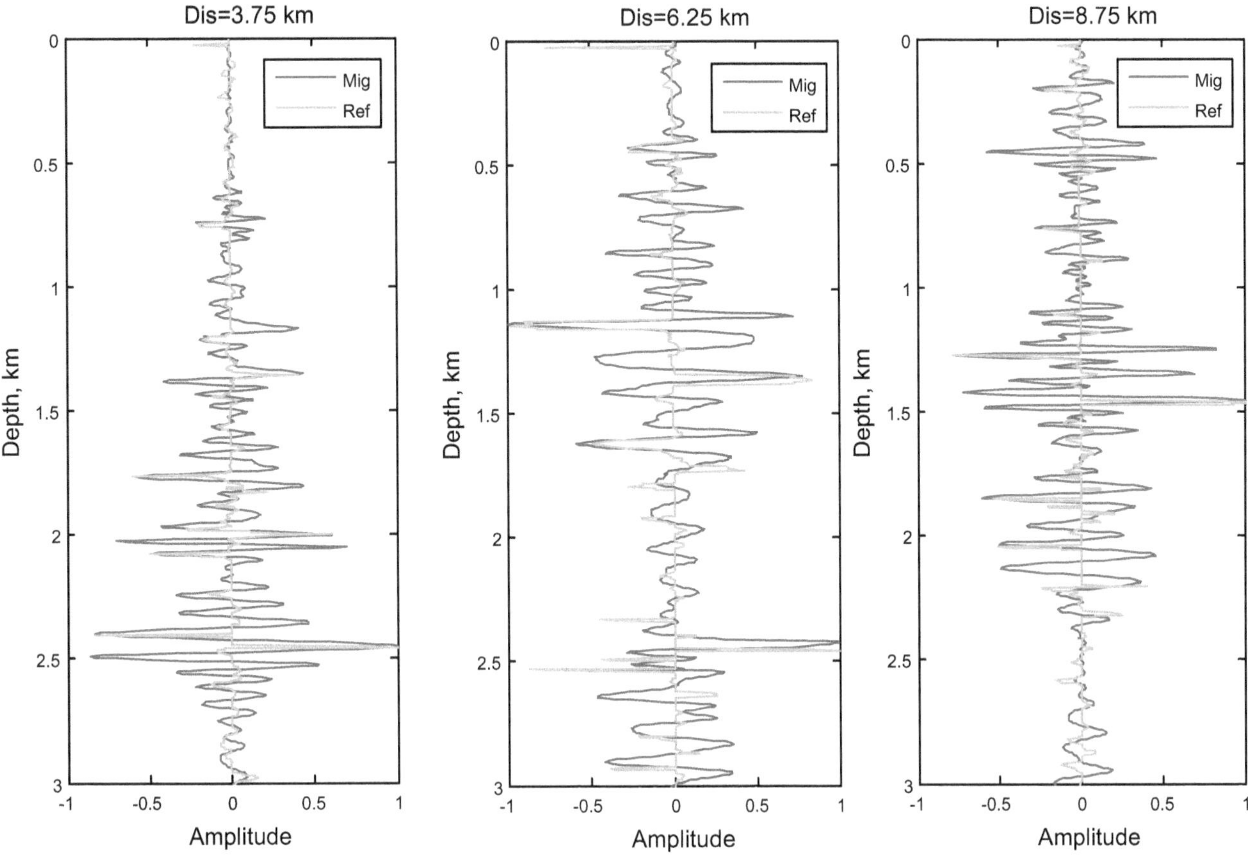

Fig. 10 Comparison of the vertical reflection coefficient (the *green line*) with migration result produced with our method (the *blue line*) for the Marmousi model

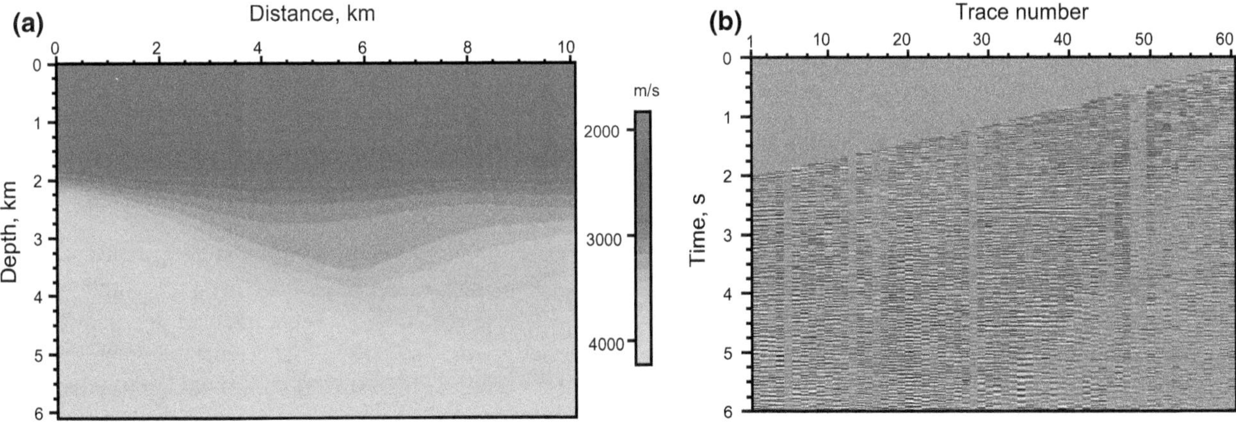

Fig. 11 A seismic survey in east China: **a** velocity model, **b** a single-shot record

the central deep parts with large lateral velocity variation, which blurs the steep-dip fault boundaries and the anticline (marked by blue arrows and green rectangle in Fig. 9a). Figure 9b–d shows the effects of the initial width on the imaging quality of GBM. It is noticed that both the shallow and the deep structures are blurred and destroyed in the image with initial width $l_0 = 2\lambda_{\text{avg}}$ (see the blue arrows and green rectangle in Fig. 9b). The reason is that the large initial width of the Gaussian beam makes its travel-time and amplitude extrapolation inaccurate in the vicinity of the whole ray. The migrated result with initial width $l_0 = \lambda_{\text{avg}}$ (the parameter proposed by Hill) appears relatively cleaner at depth, but small-scale geological structures in the shallow part are not defined well (see the blue arrows in

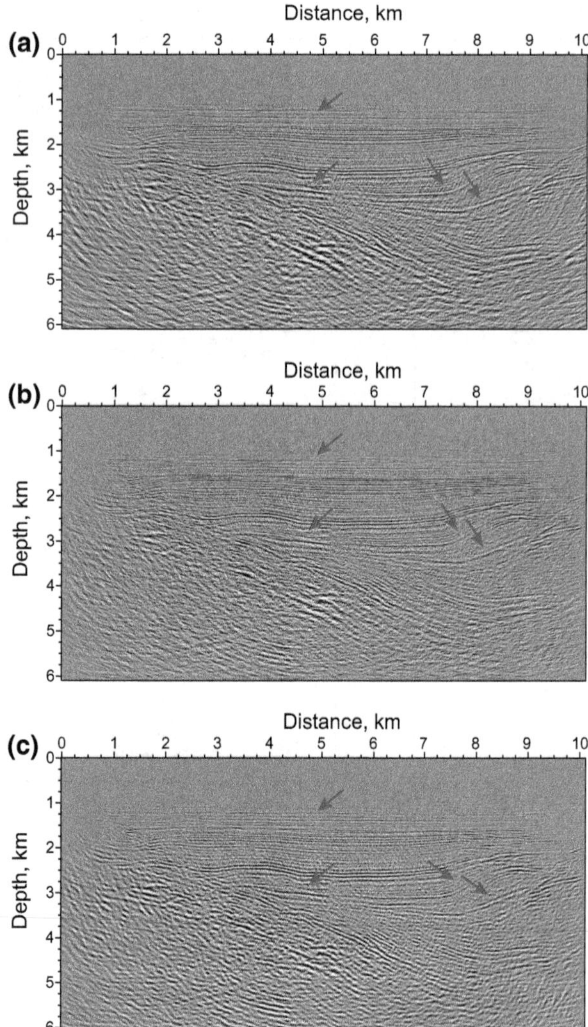

Fig. 12 Depth images of a 2D seismic survey in east China: **a** Kirchhoff method, **b** Gaussian beam method, **c** adaptive focused beam method. The images' frequency is from 5 to 50 Hz, the migration aperture is about 5500 m, the initial width of GBM equals the average wavelength of the velocity model and the reference frequency of GBM and our method is also 10 Hz

Fig. 9c). That is because the Gaussian window width used in plane wave decomposition, which is calculated by the initial beam width, is still larger than the scale of shallow structures. Figure 9d shows good resolution in the shallow zone with initial width $l_0 = 0.25\lambda_{avg}$ (marked by the blue arrows in Fig. 9d), but low imaging accuracy at depth (see the green rectangle in Fig. 9d), which is caused by the beam width increasing rapidly when using a small initial width and becoming excessively large at deeper parts of the beam. The wave equation migration resolves the simple structures well, but it is unable to image the steep-dip faults accurately due to its one-way approximation (see the blue arrows in Fig. 9e). Adaptive focused beam migration, on the other hand, produces little migration noise (see the green rectangle in Fig. 9c), and defines the steep-dip structures clearly (see blue arrows in Fig. 9c). The reason is that adaptive focused beam is narrow and stable in propagation along the whole central ray, which is critical for imaging the structures with strong lateral velocity variations. It is worthy to note that our method also resolves the shallow part small-scale structures accurately (see the red ellipse in Fig. 9f). One possible explanation is that without slant stack, it reduces the beam spacing to trace spacing, which has the potential to improve the shallow resolution. As shown in Fig. 10, the proposed method has compensated the deep energy loss and makes the peak amplitude of the events basically consistent with the main reflection coefficient.

The computer run time of our method is about four times of that of GBM, which is caused by the increased beam number and the removal of local slant stack. Under the current computer power levels, however, it is acceptable for an accurate depth image.

The final example is from a 2D seismic survey in east China, which covers a basin edge with complex geological structures. As shown in Fig. 11a, the grid size of the velocity model is 809 × 1525 with a CDP spacing of 12.5 m and a depth sample of 4 m. A single-shot record of the field dataset processed with muting, traces-killing, and band-pass filtering is shown in Fig. 11b. The data were migrated using Kirchhoff migration, GBM, and our method, and the migrated results are shown in Fig. 12. Compared with Kirchhoff and GBM sections, the image migrated with our method is in general cleaner and the events are more continuous and better defined as indicated by the red arrows both at the shallow and the deep parts, which makes it more convenient for subsequent interpretation.

4 Conclusions

Using the information of the input velocity field to control the beams shape, we have developed an amplitude-preserved adaptive focused beam method for shot-domain prestack depth migration. This method represents an improvement over GBM while still retaining its advantages over Kirchhoff and wave equation migration, and is more robust in imaging geologically complex structures. The example with the constant-velocity model shows that our method can eliminate the effects of geometric spreading and incident angles on the migrated amplitude, producing a depth section with correct kinematic and dynamic information. The second example demonstrates that the adaptive focused beam method is superior to Kirchhoff and Gaussian beam methods in imaging steep-dip structures and producing less swing noise and fewer migration artifacts. More importantly, it shows that our method has high shallow resolution, which is helpful to define the shallow small-scale geological structures. The example of field data processing further confirms the conclusions obtained from the previous two examples and shows its potential in practical applications. Thus, adaptive focused beam migration provides a new robust and flexible tool for seismic depth imaging in geologically complex areas.

References

Abdelkhalek R, Calandra H, Coulaud O, et al. Fast seismic modeling and reverse time migration on a GPU cluster. In: High performance computing and simulation, international conference on IEEE. 21–24 June 2009. p. 36–43.

Albertin U, Jaramillo H, Yingst D, et al. Aspects of true amplitude migration. In: 69th annual international meeting, SEG technical program expanded abstracts. 1999. p. 1358–61.

Baysal E, Kosloff DD, Sherwood JWC. Reverse time migration. Geophysics. 1983;48(11):1514–24.

Brandsberg-Dahl S, De Hoop MV, Ursin B. Imaging-inversion with focusing in dip. In: 63rd EAGE conference and exhibition. 2001.

Červený V, Popov MM, Pšenčík I. Computation of wave fields in inhomogeneous media—Gaussian beam approach. Geophys J Int. 1982;70(1):109–28.

Chattopadhyay S, McMechan GA. Imaging conditions for prestack reverse-time migration. Geophysics. 2008;73(3):S81–9.

Chen SC, Ma ZT. The local split-step Fourier propagation operator in wave equation migration. Comput Phys. 2006;23(5):604–8 **(In Chinese)**.

De Hoop MV, Le Rousseau JH, Wu RS. Generalization of the phase-screen approximation for the scattering of acoustic waves. Wave Motion. 2000;31(1):43–70.

Epili D, McMechan GA. Implementation of 3-D prestack Kirchhoff migration, with application to data from the Ouachita frontal thrust zone. Geophysics. 1996;61(5):1400–11.

Fletcher RP, Fowler PJ, Kitchenside P, et al. Suppressing unwanted internal reflections in prestack reverse-time migration. Geophysics. 2006;71(6):E79–82.

Gao F, Zhang P, Wang B, et al. Fast beam migration-a step toward interactive imaging. In: 2006 SEG Annual Meeting. 2006.

Gao F, Zhang P, Wang B, et al. Interactive seismic imaging by fast beam migration. In: 69th EAGE conference and exhibition. 2007.

Gray SH, Bleistein N. True-amplitude Gaussian-beam migration. Geophysics. 2009;74(2):S11–23.

Gray SH. Gaussian beam migration of common-shot records. Geophysics. 2005;70(4):S71–7.

Hill NR. Gaussian beam migration. Geophysics. 1990;55(11):1416–28.

Hill NR. Prestack Gaussian-beam depth migration. Geophysics. 2001;66(4):1240–50.

Hu C, Stoffa PL. Slowness-driven Gaussian-beam prestack depth migration for low-fold seismic data. Geophysics. 2009;74(6): WCA35–45.

Huang LJ and Fehler MC. Globally optimized Fourier finite-difference migration method. In: SEG 70th annual meeting. 2000. p. 802–5.

Hubral P, Schleicher J, Tygel M. A unified approach to 3-D seismic reflection imaging, part I: basic concepts. Geophysics. 1996;61(3):742–58.

Kaplan ST, Routh PS, Sacchi MD. Derivation of forward and adjoint operators for least-squares shot-profile split-step migration. Geophysics. 2010;75(6):S225–35.

Le Rousseau JH, De Hoop MV. Generalized-screen approximation and algorithm for the scattering of elastic waves. Q J Mech Applied Math. 2003;56(1):1–33.

Li B, Liu HW, Liu GF, et al. Computational strategy of seismic pre-stack reverse time migration on CPU/GPU. Chin J Geophys. 2010;53(12):2938–43 **(In Chinese)**.

Li F, Lü B, Wang YC, et al. True-amplitude split-step Fourier pre-stack depth migration. Oil Geophys Prospect. 2008;43(4): 387–90 **(In Chinese)**.

Liu DJ, Yin XY. Amplitude-preserved Fourier finite difference pre-stack depth migration method. Chin J Geophys. 2007;50(1): 268–76 **(In Chinese)**.

Liu DJ, Yang RJ, Luo SY, et al. Stable amplitude-preserved high-order general screen seismic migration method. Chin J Geophys. 2012;55(7):2402–11 **(In Chinese)**.

Müller G. Efficient calculation of Gaussian-beam seismograms for two-dimensional inhomogeneous media. Geophys J Int. 1984;79(1):153–66.

Nowack RL. Dynamically focused Gaussian beams for seismic imaging. In: Proceedings of the project review, geo-mathematical imaging group. West Lafayette IN: Purdue University; 2009. vol. 1. p. 59–70.

Nowack RL. Focused gaussian beams for seismic imaging. In: 2008 SEG annual meeting. 2008.

Popov MM, Semtchenok NM, Popov PM, et al. Depth migration by the Gaussian beam summation method. Geophysics. 2010;75(2):S81–93.

Ristow D, Rühl T. Fourier finite-difference migration. Geophysics. 1994;59(12):1882–93.

Schleicher J, Tygel M, Hubral P. 3-D true-amplitude finite-offset migration. Geophysics. 1993;58(8):1112–26.

Stoffa PL, Fokkema JT, De Luna Freire RM, et al. Split-step Fourier migration. Geophysics. 1990;55(4):410–21.

Sun Y, Qin F, Checkles S, et al. 3-D prestack Kirchhoff beam migration for depth imaging. Geophysics. 2000;65(5):1592–603.

Symes WW. Reverse time migration with optimal check pointing. Geophysics. 2007;72(5):SM213–21.

Wu RS, Jin SW, Xie XB. Generalized screen propagator and its application in seismic wave migration imaging. Oil Geophys Prospect. 2001;36(6):655–64 **(In Chinese)**.

Xiao X, Hao F, Egger C, et al. Final laser-beam Q-migration. In: 2014 SEG annual meeting. 2014.

Xu S, Lambaré G. True amplitude Kirchhoff prestack depth migration in complex media. Chin J Geophys. 2006;49(5):1431–44 **(In Chinese)**.

Xu S, Chauris H, Lambaré G, et al. Common-angle migration: a strategy for imaging complex media. Geophysics. 2001;66(6):1877–94.

Yang JD, Huang JP, Wang X, et al. Amplitude-preserved Gaussian beam migration based on wave field approximation in effective vicinity under rugged topography condition. In: 2014 SEG annual meeting. 2014.

Yue YB, Li ZC, Qian ZP, et al. Amplitude-preserved Gaussian beam migration under complex topographic conditions. Chin J Geophys. 2012;55(4):1376–83 **(In Chinese)**.

Zhu SW, Zhang JH, Yao ZX. High-order optimized Fourier finite difference migration. Oil Geophys Prospect. 2009;44(6):680–4 **(In Chinese)**.

Zhu T. A complex-ray Maslov formulation for beam migration. In: 2009 SEG annual meeting. 2009.

17

Geological characteristics and "sweet area" evaluation for tight oil

Cai-Neng Zou[1] · Zhi Yang[1] · Lian-Hua Hou[1] · Ru-Kai Zhu[1] · Jing-Wei Cui[1] · Song-Tao Wu[1] · Sen-Hu Lin[1] · Qiu-Lin Guo[1] · She-Jiao Wang[1] · Deng-Hua Li[1]

Abstract Tight oil has become the focus in exploration and development of unconventional oil in the world, especially in North America and China. In North America, there has been intensive exploration for tight oil in marine. In China, commercial exploration for tight oil in continental sediments is now steadily underway. With the discovery of China's first tight oil field—Xin'anbian Oilfield in the Ordos Basin, tight oil has been integrated officially into the category for reserves evaluation. Geologically, tight oil is characterized by distribution in depressions and slopes of basins, extensive, mature, and high-quality source rocks, large-scale reservoir space with micro- and nanopore throat systems, source rocks and reservoirs in close contact and with continuous distribution, and local "sweet area." The evaluation of the distribution of tight oil "sweet area" should focus on relationships between "six features." These are source properties, lithology, physical properties, brittleness, hydrocarbon potential, and stress anisotropy. In North America, tight oil prospects are distributed in lamellar shale or marl, where natural fractures are frequently present, with TOC > 4 %, porosity > 7 %, brittle mineral content > 50 %, oil saturation of 50 %–80 %, API > 35°, and pressure coefficient > 1.30. In China, tight oil prospects are distributed in lamellar shale, tight sandstone, or tight carbonate rocks, with TOC > 2 %, porosity > 8 %, brittle mineral content > 40 %, oil saturation of 60 %–90 %, low crude oil viscosity, or high formation pressure. Continental tight oil is pervasive in China and its preliminary estimated technically recoverable resources are about (20–25) $\times$ 10^8 t.

Keywords Tight oil · Geological features · "Sweet area" evaluation · Tight reservoirs · Unconventional oil and gas · Shale oil

1 Introduction

Tight oil refers to the oil preserved in tight sandstone or tight carbonate rocks with overburden pressure matrix permeability less than or equal to 0.1 $\times$ 10^{-3} μm^2 (air permeability less than 1 $\times$ 10^{-3} μm^2). Individual wells generally have no natural productivity or their natural productivity is lower than the lower limit of industrial oil flow, but industrial oil production can be obtained under certain economic conditions and technical measures (Jia et al. 2012a, b; Zou et al. 2012; Hao et al. 2014). Such measures include acid fracturing, multi-stage fracturing, horizontal wells, and multi-lateral wells. Tight oil is a highlight in global unconventional oil, for which industrial breakthroughs have been achieved in North America. Tight oil reservoirs, as typical "man-made" oil reservoirs, are explored and developed by vertical well network fracturing and horizontal well volume fracturing in order to form "man-made permeability" and achieve substantial productivity. In this article, through case studies based on the tight oil practices in North America and China, major geological features of tight oil are identified and major parameters for the evaluation for "sweet area" are proposed. These can provide important reference for continuously promoting the exploration for this important

✉ Cai-Neng Zou
zcn@petrochina.com.cn

✉ Zhi Yang
yangzhi2009@petrochina.com.cn

1 PetroChina Research Institute of Petroleum Exploration and Development, Beijing 100083, China

Edited by Jie Hao

unconventional oil. "Sweet area" refers to the target area rich in unconventional tight oil which should be developed in priority under current economic and technical conditions.

2 Global tight oil exploration and development

Tight oil has become a new focus, after shale gas, in exploration and development of unconventional oil and gas around the world. The U.S. Energy Information Administration (EIA) predicted that the technically recoverable tight oil (shale oil) resources of 42 countries had reached 473×10^8 t in 2013, revealing great resource potential. At present, the exploration and development of tight oil are concentrated in North America and China. North America has achieved intensive exploration and development, while China is in the early stage of industrial exploration.

The US has repeated the shale gas success in the exploration and development of tight oil. Nearly 20 tight oil basins have so far been discovered, including Williston, Gulf Coast, and Fort Worth, with multiple producing zones like Bakken, Eagle Ford, Barnett, Woodford, and Marcellus-Utica (Schmoker 2002; Jarvie et al. 2007, 2010; Cander 2012; Corbett 2010; Camp 2011; EIA 2012, 2013; Lu et al. 2015). Since 2008, these discoveries have reversed the previous oil production decline in the US. On the whole, the marine tight oil of North America is predominantly distributed in three sets of shale formations, i.e., Devonian, Carboniferous, and Cretaceous, and occasionally in Cambrian, Ordovician, Permian, Jurassic, and Miocene. The US tight oil production amounted to 2.09×10^8 t by 2014, accounting for 36.2 % of total US oil production. EIA predicted in 2013 that the technically recoverable tight oil resources of the US would be 79.3×10^8 t, revealing good prospects for tight oil exploration and development. In addition to the US, tight oil has also been discovered in Canada, Argentina, Ecuador, the UK, Russia, etc.

In China, tight oil is extensively distributed in the continental strata of major oil and gas basins, as widely distributed tight sandstone oil or tight carbonate oil associated or in contact with lacustrine source rocks (Sun et al. 2011; Li et al. 2011; Li and Zhang 2011; Jia et al. 2012a, b; Kang 2012; Ma et al. 2012; Zhang 2012; Zou et al. 2012, 2013a, b, c; Guo et al. 2013). In recent years, strategic breakthroughs have been successively achieved in the Ordos and Songliao Basins (Yang et al. 2013; Yao et al. 2013; Huang et al. 2013; Zou et al. 2012, 2013a, b, c, 2015). By the end of 2013, total proven technically recoverable reserves of tight oil were 3.7×10^8 t in the Ordos, Songliao, Junggar, Bohai Bay, and Sichuan and Qaidam Basins (Liang et al. 2011; Kuang et al. 2012; Zhang et al. 2012; Liang et al. 2012; Song et al. 2013; Yang et al. 2013; Huang et al. 2013; Yao et al. 2013; Fu et al. 2013). Currently, the Chang7 oil layer in the Ordos Basin and the Fuyang oil layer in the Songliao Basin have been developed on a large scale. PetroChina Changqing Oilfield Company has developed the first 100-million-ton tight oil field—Xin'anbian Oilfield, making Changqing Oilfield form a capacity of 1 million tons per annum. With this discovery, which is a milestone in China's oil history, in 2014 tight oil was integrated officially into the category for reserves evaluation. At present, China is undertaking an in-depth study of technologies to evaluate tight oil "sweet area" and building test regions. Once breakthroughs are made in key technologies and more efforts are made in this aspect, the development and utilization of tight oil will be further accelerated.

3 Geological features of tight oil

The tight oil fields of North America and China have two essential features. Firstly, oil is extensively distributed, without clear trap boundaries. Secondly, there is no natural industrial oil production with unobvious Darcy flow (Zeng et al. 2010; Nelson 2009, 2011; Zou et al. 2012, 2013a, b, c; Chen et al. 2013; Gaswirth and Marra 2015; Peters et al. 2015; Li et al. 2015; Pang et al. 2015; Sun et al. 2014; Yuan et al. 2015; Wu et al. 2015; Li et al. 2015). Taking the continental tight oil of China as example, it has four typical geological features (Zou et al. 2012, 2013a, b, c) (Table 1; Fig. 1).

1. It is generally well preserved and distributed in depressions and slopes, where the structure is stable. In China, tight oil is predominantly distributed in depression and slope belts. For example, in the Ordos Basin, tight oil is distributed in the sandy debris flow and delta front sandbodies in the basin center; in the Junggar Basin, tight oil is distributed in the carbonate and hybrid sedimentary reservoirs in the depression center; in the Songliao Basin, tight oil is distributed in the delta front sandbodies in basin slope–depression locations.
2. Large-scale mature high-quality source rocks are extensively developed. In China, continental source rocks are frequently present in Mesozoic and Cenozoic strata of rift, depression, and foreland basins. Their TOC is in a wide range, generally 2 %–15 %, and the thermal evolution degree R_o is low, generally 0.6 %–1.0 %. Organic-rich shale is extensive in depression centers and slopes.
3. Tight oil reservoir space consists of micro- and nanopore throat systems, with poor physical properties. In China, continental tight oil reservoirs are

Table 1 Parameters of typical marine and continental tight oil reservoirs

Item	Evaluation parameters	Bakken	Eagle Ford	Lucaogou Formation, Junggar Basin	Chang7 Member, Ordos Basin
Source–reservoir assemblage	Contact area, km^2	>30,000	>30,000	2000	20,000
Source condition	TOC, %	11–>20	3–7	3–16	4–20
	Type of parent material	Type II	Type II	Type II	Type I, II_1
	R_o, %	0.7–1.0	0.5–1.5	0.6–0.9	0.7–1.1
	Thickness of source rocks, m	6.7	20–30	50–70	20–40
	Distribution area, km^2	58,000	52,000	1500	20,000
Reservoir condition	Petrology	Fine sandstone, dolomitic sandstone, dolomite	Marl	Dolomitic sandstone, dolomite	Silty-fine sandstone
	Reservoir space type	Fracture-pore type	Fracture-pore type	Fracture-pore type	Pore type
	Reservoir connection	Well connected	Well connected	Well connected	Well connected
	Porosity, %	5–12, 8 on average	5–12, 9 on average	6–16, 10 on average	7–14, 10 on average
	Mineral composition	Quartz + feldspar 40 %–60 %, carbonates 20 %–25 %, clay minerals 15 %–20 %	Quartz + feldspar 20 %–30 %, carbonates 50 %–70 %, clay minerals <10 %	Quartz + feldspar 20 %–50 %, carbonates 40 %–70 %, clay minerals <10 %	Quartz + feldspar 65 %–85 %, debris 15 %–35 %
	Buried depth, m	1370–2300	1300–3500	3000–4000	1900–2300
	Reservoir thickness, m	>30	30–90	25–35	20–30
	Distribution area, km^2	>40,000	>40,000	>5000	>30,000
Tectonic setting	Basin type	Craton basin	Craton basin	Depression basin	Craton basin
	Depositional environment	Marine facies	Marine facies	Continental facies	Continental facies
	Tectonic movement	Stable	Stable	Stable	Stable
	Formation pressure gradient, MPa/100 m	1.3–1.5	1.3–1.8	1.1–1.3	0.7–0.85
Transformative property	Natural fracture	Developed	Developed	Developed	Developed
	Poisson's ratio	0.22–0.29	0.24–0.26	0.23	0.21
Fluid condition	API	41–48	40–70	24–30	37–40
	Gas/oil ratio	500–1800	800–2000	~50	118
	Number of producing wells	Over 5000 wells	Over 5000 wells	20 wells	200 wells
	Individual well production, t/d	70	20–100	2–30	3–30
	Individual well cumulative production, 10^4 t	3–10	3–10	2–4 (predicted)	2–4 (predicted)

Table 1 continued

Item	Evaluation parameters	Bakken	Eagle Ford	Lucaogou Formation, Junggar Basin	Chang7 Member, Ordos Basin
	Current annual production, t	Over 60 million	66 million	3	70

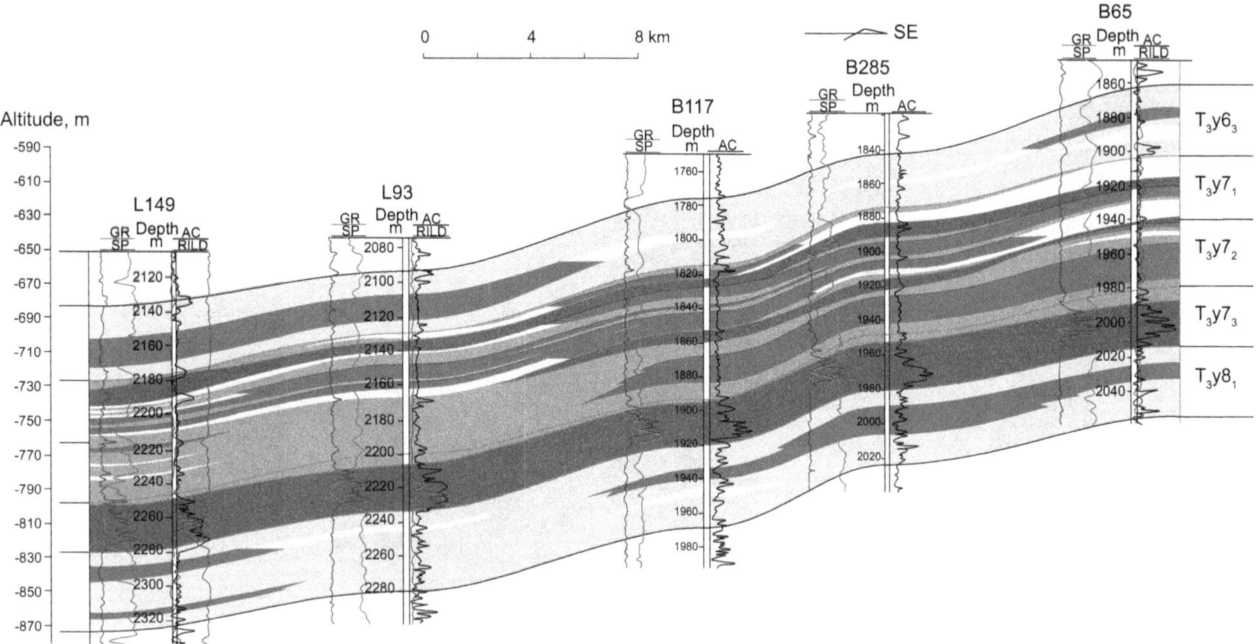

Fig. 1 Tight oil distribution of the Yanchang Formation, Ordos Basin

characterized by strong heterogeneity, rapid lateral variation, low porosity (5 %–12 %), and low matrix permeability, predominantly with micro- and nanopore throats.

4. Source rocks and reservoirs are in close contact. Tight oil with extensive distribution is accumulated near source rocks in a non-buoyant manner, with "sweet spots" present locally. In all major oil basins of China, tight oil is distributed close to source rocks in a large area and with large resource potential. For example, in the Ordos Basin, the contact area between the producing layer of the middle-upper Chang7 Member and the shale of the lower Chang7 Member is more than 5×10^4 km^2; in the Junggar Basin, the contact area between the tight oil of the Lucaogou and Fengcheng Formations and high-quality source rocks is large in depression center; in the Songliao Basin, the contact area between the Fuyang tight oil layer and underlying source rocks of the Qingshankou Formation is more than 3×10^4 km^2.

Compared with marine tight oil of the United States, continental tight oil of China is complex and special (Fig. 2) with six prominent features. First, its source rocks have a low degree of thermal evolution R_o (0.6 %–1.0 %). Second, reservoir porosity changes slightly (5 %–12 %). Third, oil saturation changes significantly (50 %–90 %). Fourth, reservoir fluid pressure changes significantly (both overpressure and negative pressure). Fifth, it is heavy oil with low gas/oil ratio. Sixth, the cumulative production of individual wells is low, generally 2×10–5×10^4 t. Its development and testing time are limited since it is still under pilot test phase at present. The individual well stable production of general horizontal segment after fracturing is 10–30 t/d. Industrial-scale development of continental tight oil in China faces great theoretical and technological challenges.

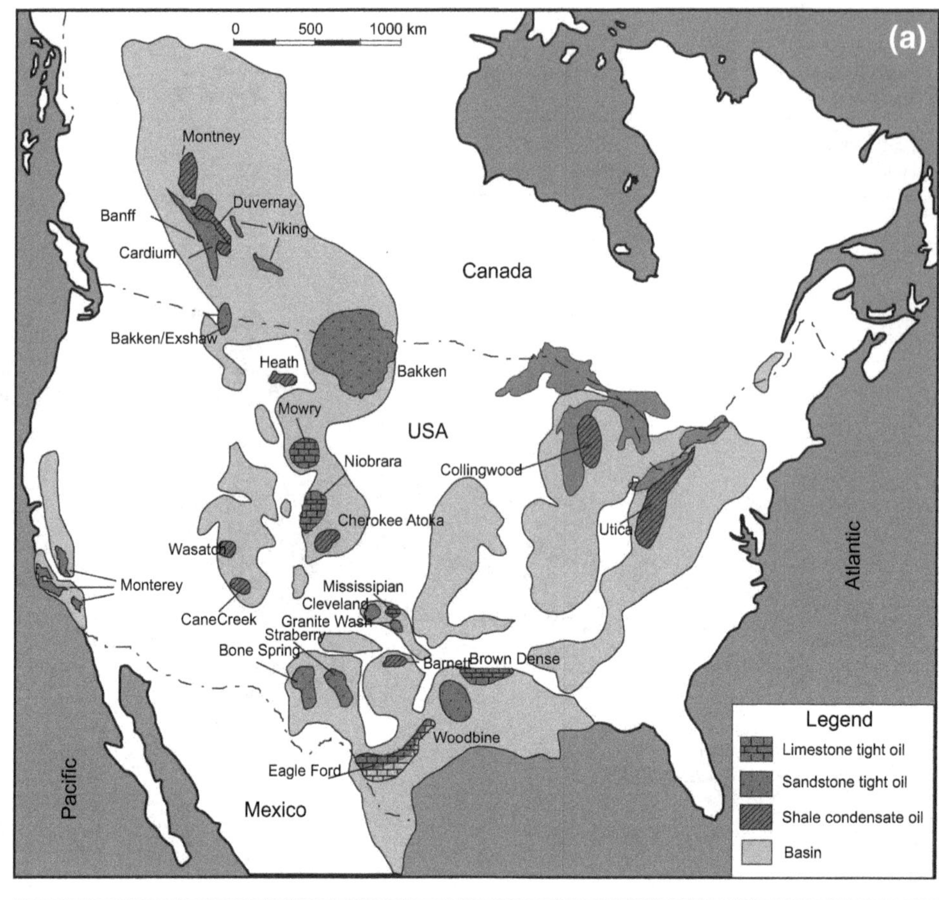

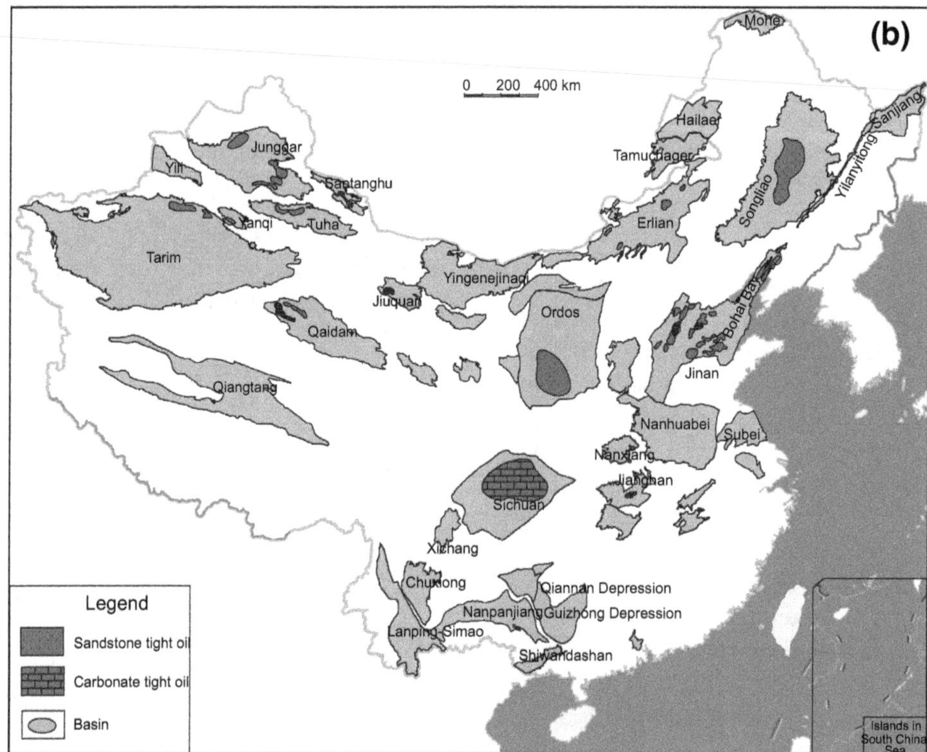

Fig. 2 Tight oil distribution in major oil/gas-bearing basins in North America and China

4 Tight oil "sweet area" evaluation

4.1 Parameters and criteria

A tight oil "sweet area" refers to an area rich in unconventional oil and gas, where test production and initial production of individual wells will both be high, thus its exploration and development can be conducted as a priority under current economic and technical conditions. Tight oil "sweet area" generally occur in the area where source rocks and reservoirs are associated and natural fractures and localized structures are present. "Sweet area" are characterized by wide distribution, large thickness, high-quality source rocks, relatively good reservoir physical properties, high oil and gas saturation, light oil, high formation energy (high gas/oil ratio, high formation pressure), and high brittleness index (Fig. 3). It should be noted that under current economic and technological conditions both domestic and overseas, a definite structural setting (which is favorable for long-term oil and gas accumulation and favorable for the development of natural fractures) and good fluidity are the prerequisite for the formation of tight oil "sweet area." For example, the Cretaceous Eagle Ford tight oil formation in Southwest Texas, US, has high-yield "sweet area" with good oil quality, high gas/oil ratio, and high formation pressure. These are concentrated in the inherited paleohigh crest and southwestern flanks, where natural fractures are frequently present.

Evaluating and selecting "sweet area" is the focus for unconventional tight oil research, which is being conducted throughout the entire exploration and development process. Unconventional tight oil sweet spots include geological, engineering, and economic sweet spots (Zou et al. 2012, 2013a, b, c). The evaluation should focus on relationships between "six features," namely source properties, lithology, physical properties, brittleness, hydrocarbon potential, and stress anisotropy, to evaluate source rock quality, reservoir quality, and engineering quality and determine the distribution scope of tight oil "sweet area." Eight evaluation indexes for tight oil "sweet area" are proposed, among which high TOC value, high porosity, and the development of micro-fractures are major controlling factors. Comprehensive evaluation should focus on source rock, reservoir, overpressure, and fracture for geological "sweet area," on buried depth, rock compressibility, and stress anisotropy for engineering "sweet area," and on resources scale, buried depth and surface conditions for economic "sweet area." At present, priority should be given to favorable source rock, reservoir, overpressure, fracture, and local structure of geological "sweet area," and pressure coefficient, brittleness, crustal stress, and buried depth of engineering "sweet area" (Table 2).

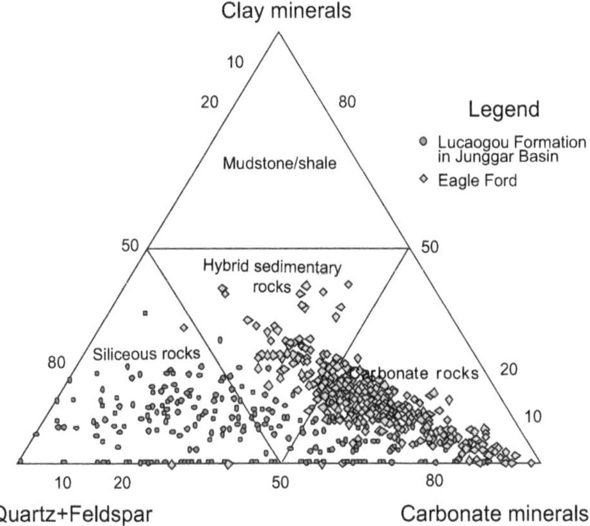

Fig. 3 Mineral composition of oil-bearing tight reservoirs of the Eagle Ford and the Lucaogou Formation in the Junggar Basin

According to the evaluation criteria of tight oil source rocks and reservoirs (Table 3), tight oil "sweet area." in China were evaluated. Areas with grade I and II source rocks and reservoirs are tight oil "sweet area." Finally, 20 favorable "sweet area" with an area of 5100 km^2 and resources of 45×10^8 t were identified in the Songliao, Ordos, and Junggar Basins. Based on the evaluation results, the tight oil "sweet area" of the Fuyu oil layer in the Songliao Basin have an area of 1.5×10^4 km^2 and resources of 25×10^8 t (Fig. 4); the tight oil "sweet area" of the Chang7_1 Member in the Ordos Basin have an area of 3.5×10^4 km^2 and resources of 14×10^8 t (Fig. 4); the tight oil "upper sweet section" of the Lucaogou Formation in the Jimsar Sag of the Junggar Basin have an area of 260 km^2 and resources of 2.5×10^8 t.

In China, tight oil is pervasive and diversified. It is predominantly tight sandstone oil and tight carbonate oil in contact with or associated with lacustrine source rocks. As estimated, major onshore tight oil basins in China have an area of 50×10^4 km^2, geological resources of about 200×10^8 t, and technically recoverable resources of (20–25) $\times 10^8$ t. The preliminarily proven technically recoverable reserves of tight oil are nearly 3.7×10^8 t in the tight sandstone of Cretaceous Qingshankou–Quantou Formation of the Songliao Basin, the tight sandstone of Triassic Chang7 Member of the Ordos Basin, the argillaceous dolomite of Permian Lucaogou Formation of the Junggar Basin, the Middle-Lower Jurassic tight limestone of the Sichuan Basin, the marl and tight sandstone of the Shahejie Formation of Bohai Bay Basin, the Cenozoic marl and tight sandstone of the Qaidam Basin, and the Cretaceous marl of the Jiuquan Basin.

Table 2 Evaluation criteria for tight oil "sweet spots"

Oil and gas types	Geological "sweet area"				Engineering "sweet area"				
	Source rock	Reservoir	Fracture	Local structure	Formation pressure gradient, MPa/100 m	Oil saturation, %	Brittleness index, %	Horizontal principal stress difference, MPa	Buried depth, m
Tight oil of North America									
Evaluation criteria	TOC is >4 %	Porosity>7 %	Micro-fractures are developed	Relatively high position	>1.3	>60	>50	<10	<3500
Eagle Ford liquid hydrocarbon of west of the Gulf of Mexico	Thickness is 20–60 m, TOC value is 4 %–8 %, R_o value is 0.9 %–1.5 %	Thickness is 10–30 m, porosity is 7 %–14 %, oil saturation is greater than 70 %	Micro-fractures are developed	Paleohigh crest and slope	>1.4	>60	>50	<10	2000–3500
Bakken tight oil of the Williston Basin	Thickness is >5 m, TOC value is >8 %, R_o value is >0.7 %	Thickness is 5–20 m, porosity is 7 %–12 %, oil saturation is greater than 70 %	Micro-fractures are developed	Relatively high position	>1.35	>60	>50	<10	1500–3500
Tight oil of China									
Evaluation criteria	TOC >2 %	Porosity >8 %	Micro-fractures are developed	Relatively high position	>1	>60	>40	<6	<4500
Tight oil of the Chang7 member, in the Ordos Basin	Thickness is 10–30 m, TOC value is 3 %–25 %, R_o value is 0.8 %–1.2 %, S_1 value is 1–8 mg/g	Thickness is 20–30 m, porosity is 10 %–15 %, oil saturation is greater than 60 %	Micro-fractures are developed	Relatively high position in slope	0.7–0.8	>70	>50	<10	1000–3000
Tight oil of the Lucaogou Formation in Jimsar Sag, Junggar Basin	Thickness is 100–130 m, TOC value is 5 %–6 %, R_o value is 0.8 %–1.0 %	Thickness is 25–50 m, porosity is 12 %–20 %, oil saturation is greater than 70 %	Micro-fractures are developed	Relatively high position in slope	1.2–1.5	>90	>50	<6	1000–4500

Table 3 Comprehensive evaluation of continental tight oil in China by source rocks and reservoirs

Item		Grade I	Grade II	Grade III
Source rocks	TOC, %	>2	1–2	<1
	$S_1 + S_2$, mg/g	>6	2–6	<2
	Organic matter type	I–II_1	II_1–II_2	II_2–III
	Maturity R_o, %	0.9–1.3	0.7–0.9	0.4–0.7
Reservoirs	Porosity, %	>8	5–8	<5
	Permeability, mD	>0.2	0.04–0.2	<0.04
	Oil saturation, %	>60	40–60	20–40
	Throat radius, mm	>0.15	0.05–0.15	0.01–0.05
	Thickness, m	>10	5–10	<5
	Brittleness index, %	>50	40–50	30–40

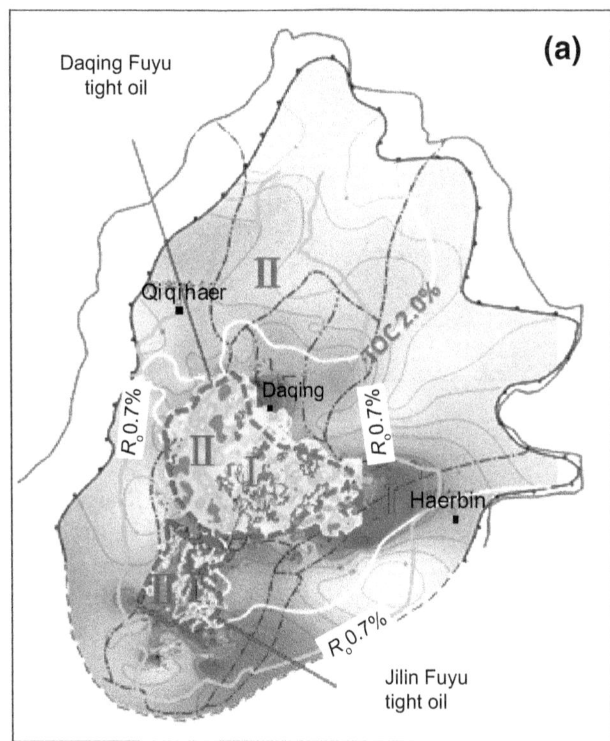

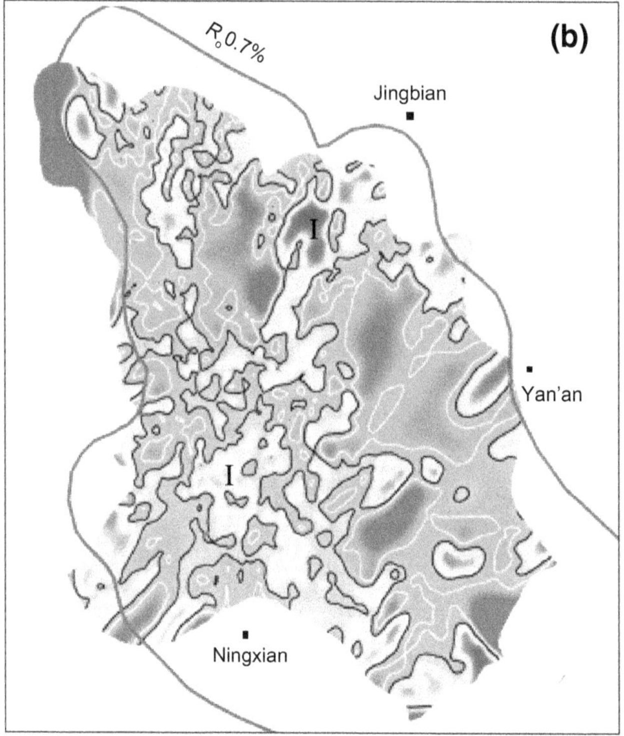

Fig. 4 Tight oil "sweet area" in China's major oil and gas basins. **a** Tight oil "sweet area" distribution of the Fuyu oil layer, Songliao Basin; **b** tight oil "sweet area" distribution of Chang 7_1 Member in the Ordos Basin

4.2 Case study

The Bakken tight oil region is located in the Williston Basin, crossing the United States and Canadian border, with an oil-bearing area of 7×10^4 km^2 (Fig. 5) (Sarg, 2012). The Upper Devonian–Lower Carboniferous strata can be divided into nine lithological units, with individual layer thicknesses of 5–15 m, cumulative thickness of 55 m, and buried depth of 2590–3200 m. They were deposited under offshore shelf–lower shoreface environments and are composed of dolomitic siltstone, bioclastic sandstone, and calcareous siltstone, with porosities of 2 %–9 % and an average permeability of 0.05 mD. Two sets of shale are present in the Bakken Formation, predominantly distributed in the northern-central basin, with thicknesses of 5–12 m, TOC of 10 %–14 %, and R_o of 0.6 %–0.9 %, and their recoverable resources are 68×10^8 t predicted by HIS. As of 2010, a total of 2362 tight oil wells were in production in the Bakken region of the US, with an average daily oil production of 12 t, with a maximum of 680 t. The crude oil density is 0.78–0.85 g/cm^3, which is light, the pressure coefficient is 1.15–1.84, gas/oil ratio is 53–160,

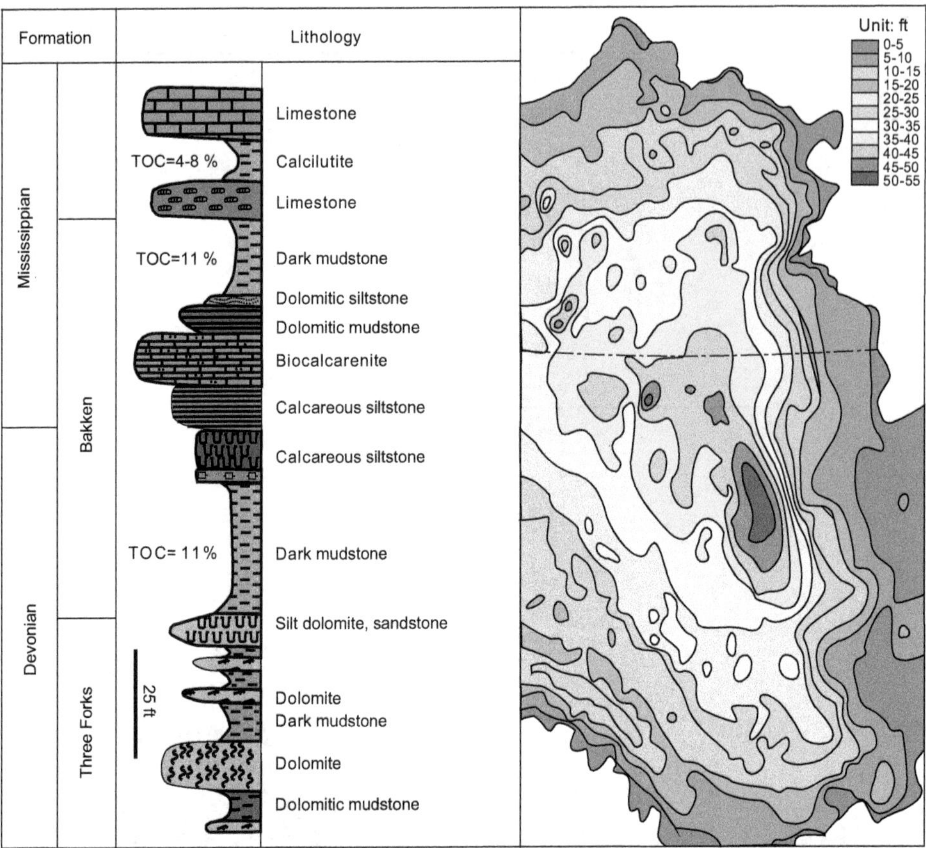

Fig. 5 Lithology and oil layer distribution of Bakken tight oil in the Williston Basin

and the annual oil production is greater than 5000×10^4 t. The tight reservoirs of the Bakken Member are bounded above and below by the source rocks of the Bakken, forming a good source–reservoir assemblage. The "sweet spots" are mainly controlled by the regional and local fracture systems arising from the tectonic background, and the superimposed areas of both organic-rich shales and thick dolomitic pore-type reservoirs.

The Eagle Ford tight oil region trending SW–NE is located in the Gulf Coast Basin, southern Texas. It is 440 km long and 80 km wide, with an area of 3×10^4 km^2. The tectonic setting of Eagle Ford is an NW–SE dipping slope, including three types of hydrocarbon maturity windows, i.e., crude oil, condensate oil–wet gas, and dry gas. Its formation thickness ranges from a few meters to more than one hundred meters (Li et al. 2011; Kevin 2010). Eagle Ford can be divided into upper and lower intervals, among which the lower interval, located between upper Austin limestone and lower Buda limestone, is the major target zone for oil and gas at present. So far, the exploration of Eagle Ford is focused on the liquid hydrocarbon-rich zone with high economic value. The maturity of source rocks ranges between 0.9 % and 1.5 %. The target zone is in the lower shale interval which is rich in organic matter. As of 2014, the daily tight oil production has exceeded 1×10^6 bbl/d. The distribution of the "sweet area" in the lower Eagle Ford major producing interval is predominantly under the control of maturity, formation thickness, API, gas/oil ratio, pressure coefficient, natural fracture, TOC, porosity, oil saturation, and other parameters. The developed area of "sweet area" generally has a high formation thickness (greater than 20 m) and TOC (greater than 4 %) value, high brittle mineral content (greater than 90 %), light oil (API of greater than 35°), high fluid pressure (pressure coefficient between 1.3 and 1.8), and gas/oil ratio (greater than 5000 scf/bbl), where natural fractures are extensive.

The "six features" evaluation parameters of the tight oil "sweet area" of the Lucaogou Formation in the Jimsar Sag are shown in Fig. 6. Source rocks are of high quality, with an average TOC of 5 %–6 %, R_o of 0.8 %–1.1 %, and type II kerogen. The reservoirs, composed of dolomitic sandstone, are also of high quality and have good physical properties, with a porosity of 6 %–20 % and a permeability

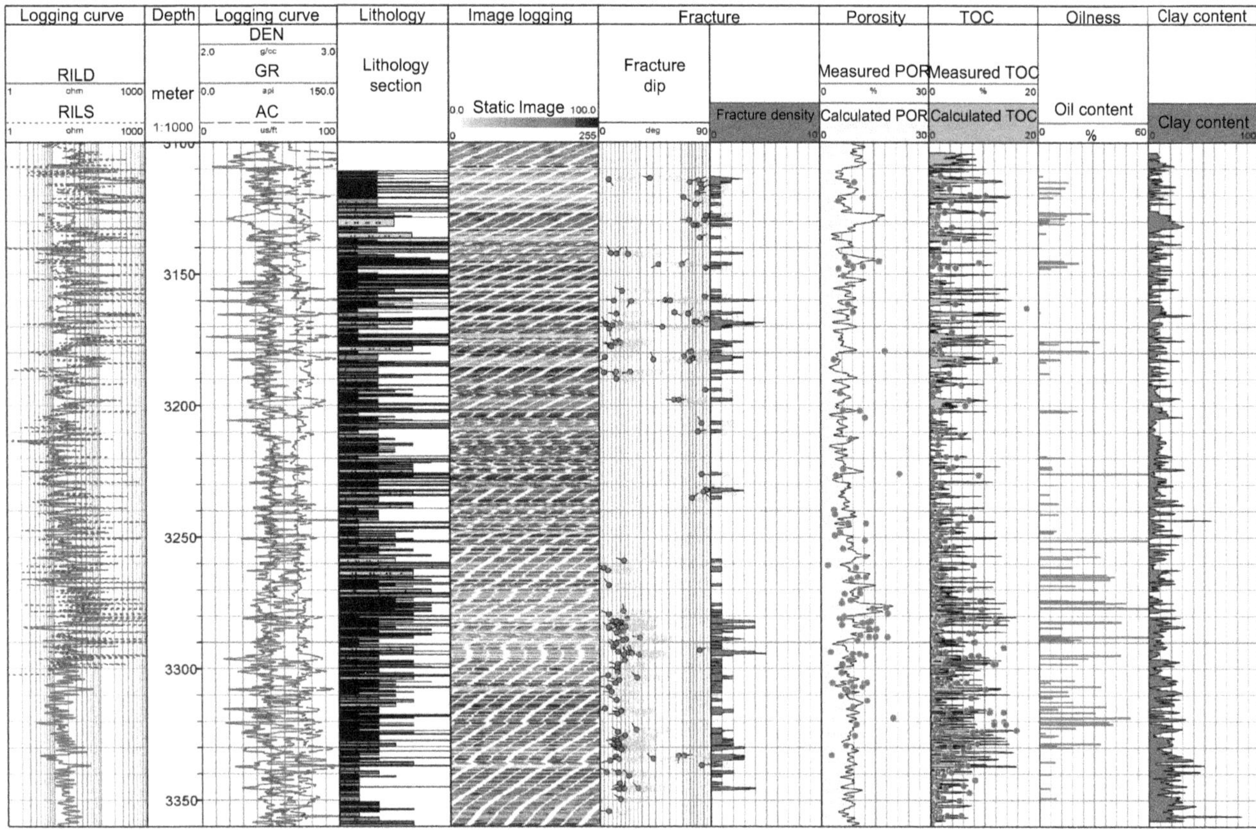

Fig. 6 Typical individual well comprehensive evaluation histogram of the Jimsar Sag, Junggar Basin

of generally lower than 1×10^{-3} μm^2, where matrix pores are frequently present, with well-connected micro-fine pores dominating. Oil potential is good, with oil saturation of generally greater than 70 % and crude density of 0.88–0.92 g/cm^3 without water cut. Reservoirs have a high brittle mineral content, with a brittleness index >60 %, elastic modulus $>1.0 \times 10^4$ MPa, and Poisson's ratio <0.35. The horizontal stress difference is small, generally less than 6 MPa, which is beneficial for volume fracturing. The oil-bearing properties of the tight reservoirs which are in the upper and lower members of the Lucaogou Formation in the Jimsar Sag are more closely related to the maturity of adjacent source rocks than reservoir porosity. The maturity of the shale in the lower member is higher than that of the shale in the upper member, and thus the tight oil/shale oil potential of the lower member of the Lucaogou Formation is significantly better than that of the upper member, since the tight oil/shale oil saturation of the lower member reaches over 90 %, basically without water cut. The major constraints of tight oil are low maturity, heavy oil quality, low gas content, low formation pressure, and poor fluid mobility as well as difficulty in exploiting and producing.

The Mesozoic Chang7 tight oil in the Ordos Basin is distributed in 11 enrichment regions, with an area of 3×10^4 km^2 and reserves of greater than 20×10^8 t. The Chang7 marl has a high abundance of organic matter, TOC value of 5 %–8 %, type I–II$_1$ kerogen, R_o of 0.7 %–1.2 %, and pyrolysis T_{max} of 435–455 °C (Fig. 7). Its tight reservoirs are predominantly composed of lithic feldspathic sandstone with primary and secondary pores. Reservoir physical properties are good, with a porosity of 7 %–13 %. The oil-bearing properties are good, with oil saturation of 60 %–80 %; brittle minerals are common, with a brittleness index of 35 %–45 %, and the horizontal stress difference is 5–7 MPa. The Chang7 tight oil is advantaged by light oil quality, high gas/oil ratio, good reservoir compressibility, extensive micro-fractures, and low water cut; however, its major constraint is low formation pressure.

5 Conclusions

Tight oil is an important type of unconventional oil resource. It is extensive in global oil and gas basins, especially in North America and China. In North America,

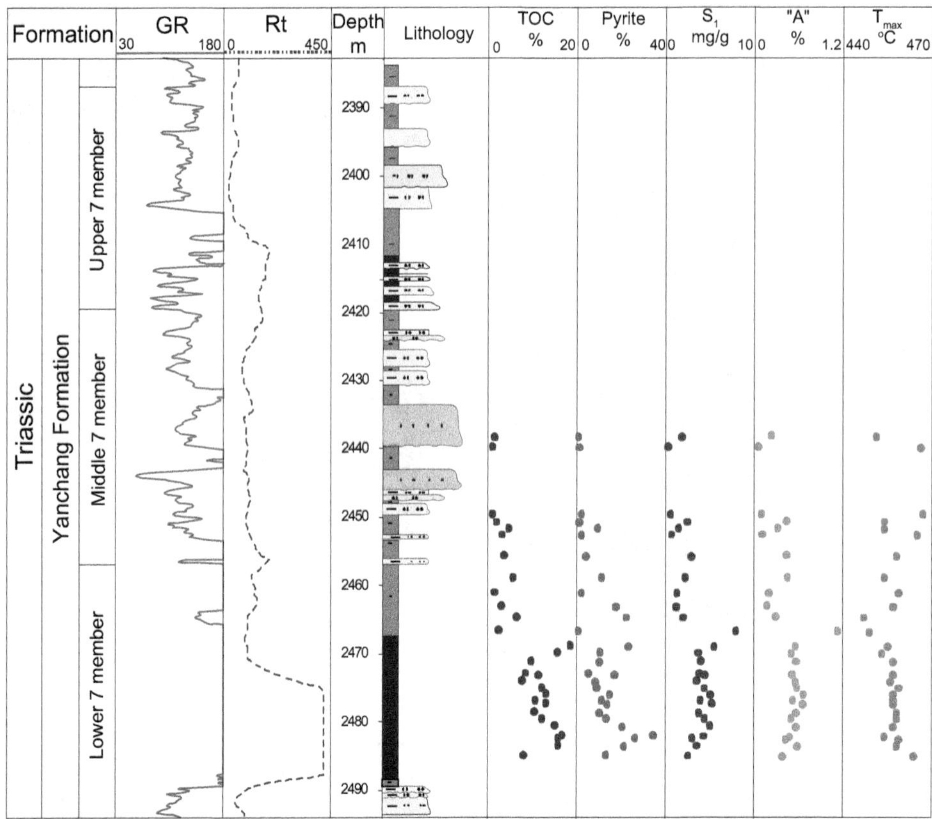

Fig. 7 Typical individual well-integrated histogram of the Triassic Chang7 Member, Ordos Basin

marine tight oil is intensively explored. In China, commercial tests are being steadily conducted for continental tight oil. Geologically, tight oil is characterized by distribution in depressions and slopes of basins, extensive, mature, and high-quality source rocks, large-scale reservoir space with micro- and nanopore throat systems, closely contacted source rocks and reservoirs with continuous distribution, and local "sweet area." The evaluation of the distribution of tight oil "sweet area" should focus on relationships between "six features," including source properties, lithology, physical properties, brittleness, hydrocarbon potential, and stress anisotropy. Continental tight oil is extensive in China, with preliminarily estimated technically recoverable oil resources of about (20–25) × 10^8 t.

Acknowledgments This work was supported by the National Key Basic Research and Development Program (973 Program), China (Grant 2014CB239000), and China National Science and Technology Major Project (Grant 2011ZX05001). This work could not have been achieved without the cooperation and support from PetroChina Research Institute of Exploration and Development. The authors appreciate both journal editors and anonymous reviewers for their precious time and useful suggestions.

References

Camp WK. Pore-throat sizes in sandstones, tight sandstones, and shales: discussion. AAPG Bull. 2011;95(8):1443–7.

Cander H. Sweet spots in shale gas and liquids plays: prediction of fluid composition and reservoir pressure. AAPG Search Discov. 2012; #40936.

Chen ZH, Osadetz KG. An assessment of tight oil resource potential in Upper Cretaceous Cardium Formation, Western Canada Sedimentary Basin. Pet Explor Dev. 2013;40(3):344–53.

Corbett K. Eagleford shale exploration models: depositional controls on reservoir properties. AAPG Search Discov. 2010; #10242.

EIA. Annual Energy Outlook 2012 with Projections to 2035. 2012. http://www.eia.gov/forecasts/aeo.

EIA. Status and outlook for shale gas and tight oil development in the U.S. 2013.

Fu ST, Zhang DW, Xue JQ, et al. Exploration potential and geological conditions of tight oil in the Qaidam Basin. Acta Sedimentol Sin. 2013;31(4):672–82 **(in Chinese)**.

Gaswirth SB, Marra KR. U.S. Geological Survey 2013 assessment of undiscovered resources in the Bakken and Three Forks Formations of the U.S. Williston Basin Province. AAPG Bull. 2015;99(4):639–60.

Guo QL, Chen NS, Wu XZ, et al. Method for assessment of tight oil resources. China Pet Explor. 2013;18(2):67–76 **(in Chinese)**.

Hao ZG, Fei HC, Hao QQ, et al. Major breakthroughs in geological theory, key techniques and exploration of tight oil in China. Acta Geol Sin. 2014;88(1):362–3.

Huang W, Liang JP, Zhao B, et al. Main controlling factors of tight oil accumulations in the Fuyu Layer of Cretaceous Quantou Formation in northern Songliao Basin. J Palaeogeogr. 2013;15(5):635–44 **(in Chinese)**.

Jarvie DM. Unconventional oil petroleum systems: shales and shale hybrids. AAPG Conference and Exhibition. Calgary, Alberta, Canada, September 12–15, 2010.

Jarvie DM, Hill RJ, Ruble TE, et al. Unconventional shale-gas systems: the Mississippian Barnett Shale of north-central Texas as one model for thermogenic shale-gas assessment. AAPG Bull. 2007;91(4):475–99.

Jia CZ, Zheng M, Zhang YF, et al. Unconventional hydrocarbon resources in China and the prospect of exploration and development. Pet Explor Dev. 2012a;39(2):139–46.

Jia CZ, Zou CN, Li JZ, et al. Assessment criteria, main types, basic features and resource prospects of the tight oil in China. Acta Pet Sin. 2012b;33(2):343–50 **(in Chinese)**.

Kang YZ. Characteristics and prospects of unconventional shale oil & gas reservoirs in China. Nat Gas Ind. 2012;32(4):1–5 **(in Chinese)**.

Kuang LC, Tang Y, Lei DW, et al. Formation conditions and exploration potential of tight oil in the Permian saline lacustrine dolomitic rock, Junggar Basin, NW China. Pet Explor Dev. 2012;39(6):657–67.

Li F, Martin R B, Thompson J W, et al. An integrated approach for understanding oil and gas reservoir potential in Eagle Ford shale Formation. Canadian Unconventional Resources Conference, 15–17 November, Calgary, Alberta, Canada. 2011; SPE 148751: 1–15.

Li HB, Guo HK, Yang ZM, et al. Tight oil occurrence space of Triassic Chang 7 Member in Northern Shaanxi Area, Ordos Basin, NW China. Pet Explor Dev. 2015;42(3):434–8.

Li YX, Zhang JC. Types of unconventional oil and gas resources in China and their development potential. Int Pet Econ. 2011;19(3):61–7 **(in Chinese)**.

Liang DG, Ran LH, Dai DS, et al. A re-recognition of the prospecting potential of Jurassic large-area and non-conventional oils in the central-northern Sichuan Basin. Acta Pet Sin. 2011;32(1):8–17 **(in Chinese)**.

Liang SJ, Huang ZL, Liu B, et al. Formation mechanism and enrichment conditions of Lucaogou Formation shale oil from Malang Sag, Santanghu Basin. Acta Pet Sin. 2012;33(4):588–94 **(in Chinese)**.

Lu JM, Ruppel SC, Rowe HD. Organic matter pores and oil generation in the Tuscaloosa marine shale. AAPG Bull. 2015;99(2):333–57.

Nelson PH. Pore-throat sizes in sandstones, tight sandstones, and shales: reply. AAPG Bull. 2011;95(8):1448–53.

Ma YS, Feng JH, Mou ZH, et al. Unconventional petroleum resource potential and exploration progress of SINOPEC. Eng Sci. 2012;14(6):22–30 **(in Chinese)**.

Nelson PH. Pore-throat sizes in sandstones, tight sandstones, and shales. AAPG Bull. 2009;93(3):329–40.

Pang XQ, Jia CZ, Wang WY. Petroleum geology features and research developments of hydrocarbon accumulation in deep petroliferous basins. Pet Sci. 2015;12(1):1–53.

Peters KE, Burnham AK, Walters CC. Petroleum generation kinetics: Single versus multiple heating-ramp open-system pyrolysis. AAPG Bull. 2015;99(4):591–616.

Sarg JF. The Bakken—an unconventional petroleum and reservoir system. Final Scientific/Technical Report. Office of Fossil Energy, Colorado School of Mines; 2012. p. 1–65.

Schmoker JW. Resource-assessment perspectives for unconventional gas systems. AAPG Bull. 2002;86(11):1993–9.

Song Y, Zhao MJ, Fang SH. Differential hydrocarbon accumulation controlled by structural styles along the southern and northern Tianshan Thrust Belt. Acta Geol Sin. 2013; 87(4):1109–19.

Sun ZD, Jia CZ, Li XF, et al. Unconventional petroleum exploration and development. Beijing: Petroleum Industry Press; 2011 **(in Chinese)**.

Sun L, Zou CN, Liu XL, et al. A static resistance model and the discontinuous pattern of hydrocarbon accumulation in tight oil reservoirs. Pet Sci. 2014;11(4):469–80.

Wu ST, Zhu RK, Cui JG, et al. Characteristics of lacustrine shale porosity evolution, Triassic Chang 7 Member, Ordos Basin, NW China. Pet Explor Dev. 2015;42(2):185–95.

Yang H, Li SX, Liu XY, et al. Characteristics and resource prospects of tight oil and shale oil in Ordos Basin. Acta Pet Sin. 2013;34(1):1–11 **(in Chinese)**.

Yao JL, Deng XQ, Zhao YD, et al. Characteristics of tight oil in Triassic Yanchang Formation, Ordos Basin. Pet Explor Dev. 2013;40(2):161–9 **(in Chinese)**.

Yuan XJ, Lin SH, Liu Q, et al. Lacustrine fine-grained sedimentary features and organic-rich shale distribution pattern: a case study of Chang 7 Member of Triassic Yanchang Formation in Ordos Basin. NW China. Pet Explor Dev. 2015;42(1):37–47.

Zeng JH, Cheng SW, Kong X, et al. Non-Darcy flow in oil accumulation (oil displacing water) and relative permeability and oil saturation characteristics of low-permeability sandstones. Pet Sci. 2010;7(1):20–30.

Zhang K. From tight oil & gas to shale oil & gas—approach for the development of unconventional oil & gas in China. Chin Geol Educ. 2012;21(2):9–15 **(in Chinese)**.

Zhang SW, Wang YS, Zhang LY, et al. Formation conditions of shale oil and gas in Bonan sub-sag, Jiyang Depression. Eng Sci. 2012;14(6):49–55 **(in Chinese)**.

Zou CN, Yang Z, Tao SZ, et al. Nano-hydrocarbon and the accumulation in coexisting source and reservoir. Pet Explor Dev. 2012;39(1):15–32.

Zou CN, Yang Z, Cui JW, et al. Formation mechanism, geological characteristics, and development strategy of nonmarine shale oil in China. Pet Explor Dev. 2013a;40(1):15–27.

Zou CN, Yang Z, Tao SZ, et al. Continuous hydrocarbon accumulation over a large area as a distinguishing characteristic of unconventional petroleum: The Ordos Basin, North-Central China. Earth Sci Rev. 2013b;126:358–69.

Zou CN, Zhang GS, Yang Z, et al. Geological concepts, characteristics, resource potential and key techniques of unconventional hydrocarbon: on unconventional petroleum geology. Pet Explor Dev. 2013c;40(4):413–28 **(in Chinese)**.

Zou CN, Yang Z, Dai JX, et al. The characteristics and significance of conventional and unconventional Sinian-Silurian gas systems in the Sichuan Basin, central China. Mar Pet Geol. 2015;64:386–402.

18

A new method for evaluating the injection effect of chemical flooding

Jian Hou[1,2] · Yan-Hui Zhang[3] · Nu Lu[2] · Chuan-Jin Yao[2] · Guang-Lun Lei[2]

Abstract Hall plot analysis, as a widespread injection evaluation method, however, often fails to achieve the desired result because of the inconspicuous change of the curve shape. Based on the cumulative injection volume, injection rate, and the injection pressure, this paper establishes a new method using the ratio of the pressure to the injection rate (RPI) and the rate of change of the RPI to evaluate the injection efficiency of chemical flooding. The relationship between the RPI and the apparent resistance factor (apparent residual resistance factor) is obtained, similarly to the relationship between the rate of change of the RPI and the resistance factor. In order to estimate a thief zone in a reservoir, the influence of chemical crossflow on the rate of change of the RPI is analyzed. The new method has been applied successfully in the western part of the Gudong 7th reservoir. Compared with the Hall plot analysis, it is more accurate in real-time injection data interpretation and crossflow estimation. Specially, the rate of change of the RPI could be particularly suitably applied for new wells or converted wells lacking early water flooding history.

✉ Jian Hou
houjian@upc.edu.cn

✉ Chuan-Jin Yao
ycj860714@163.com

1 State Key Laboratory of Heavy Oil Processing, China University of Petroleum, Qingdao 266580, Shandong, China

2 School of Petroleum Engineering, China University of Petroleum, Qingdao 266580, Shandong, China

3 CNOOC LTD., Tianjin Bohai Oilfield Institute, Tianjin 300452, China

Edited by Yan-Hua Sun

Keywords Ratio of the pressure to the injection rate · Rate of change of the RPI · Injection efficiency · Chemical crossflow

1 Introduction

Chemical flooding, a rapidly developed tertiary oil recovery technique, is applied successfully and widely both in the Daqing Oilfield and the Shengli Oilfield (Zhang et al. 2010; Hou et al. 2011, 2013; Shaker Shiran and Skauge 2013; Dag and Ingun 2014). Polymer flooding and surfactant-polymer flooding (SP flooding) are considered two of the most mature chemical methods (Chang et al. 2006; Vargo et al. 2000; Li et al. 2012; Delamaide et al. 2014; Sheng et al. 2015). Polymer flooding uses high-molecular weight polymers to increase the viscosity of the injection fluid, and to decrease the oil–water mobility ratio, while the SP flooding further improves the oil recovery by adding a surfactant to the injection fluid to reduce the interfacial tension and then increase the oil displacement efficiency (Shen et al. 2009; Urbissinova et al. 2010). However, due to the high prices of chemicals, in most cases, the implementation of chemical flooding is usually costly. Thus, a timely and accurate evaluation of the displacement efficiency of chemical floods, which can verify the validity of the chemical injection in advance, is urgently needed (Kaminsky et al. 2007; Dong et al. 2009; Seright et al. 2009; AlSofi and Blunt 2014; Ma et al. 2007).

Usually, the injection evaluation of chemical flooding is conducted by the variation of dynamic injection data such as the rise of the injection pressure and the drop of the injection rate. The increase in the flow resistance to the injection fluid indicates the increment of the chemical efficiency. After the chemical solution is injected into a reservoir, due to the high-molecular weight polymer dissolved in the injection fluid, the flow resistance increases, causing an increase in the injection pressure and a decline in the injection rate (Cheng et al. 2002; Li 2004). As a result, the variation of the injection pressure and the injection rate can be used to evaluate whether the chemical injection is effective.

The Hall plot describes the relationship between the time integral of the injection pressure difference and the cumulative injection volume. The Hall plot analysis was firstly used to evaluate the performance of waterflood wells (Hall 1963) and gradually was used as a simple, effective method for diagnosing the injection efficiency (DeMarco 1969). Since then, researchers have improved the Hall plot analysis. Based on the Hall plot analysis, the slope analysis method which produces an estimation of the pressure at the average water-bank radius in Hall's formula was put forward by Silin et al. (2005). The slope analysis is proved to be more accurate for all data needed is available from oilfields. After that, Izgec and Kabir (2009) presented a complete reformulation of the Hall plot analysis by updating the pressure at the average influence radius after each computing time step and studied the difference between the Hall plot and the derivative of the Hall slope, under the condition of both transient and pseudo-steady states. Compared with the Hall plot, the derivative of the Hall slope could overcome the smooth effect which is caused by integral involved in the Hall plot analysis.

With the development of chemical flooding, Hall plot analysis was used to evaluate the flow behavior of non-Newtonian fluids in the field of polymer flooding (Moffitt and Menzie 1978). Later, the analytical expressions of the Hall slope analysis for polymer floods, resistance factor, and the residual resistance factor were derived by Buell et al. (1990). Considering that the reservoir permeability is an important factor affecting the resistance factor, Kim and Lee (2014) used the effective permeability changing with the water saturation instead of regarding permeability as a constant, improving the accuracy of the Hall plot analysis. Li et al. (2011) investigated the application of the Hall plot analysis in gel flooding. Gradually, the resistance factor and the residual resistance factor have become quantitative indexes to evaluate the injection effect of chemical flooding (Honarpour and Tomutsa 1990; Sugai and Nishikiori 2006; Ghosh et al. 2012). The resistance factor is defined as the ratio of the Hall plot slope for the chemical flood to that for the water flood. As the change of the slope in the Hall plot is usually not significant in the early period of chemical flooding, it is hard to use the Hall plot analysis to obtain the resistance factor. In fact, the Hall plot analysis only represents the average injection effect over a period, rather than to reflect the real-time characteristics.

In view of this, a new method, derived from injection performance data, is proposed for the real-time characterization of the chemical flood. The assumption of the new method is the same as the Hall slope analysis [a two-phase, radial flow of Newtonian liquids (Buell et al. 1990)]. Compared to the Hall plot analysis, it is simple and accurate in parameter calculation and high permeability zone estimation. The first pilot test of SP flooding in China was implemented in the western part of the Gudong 7th reservoir and has achieved great success. In this paper, the dynamic data of this pilot test are used to verify the new method.

2 New approach for injection effect evaluation

2.1 Ratio of pressure to injection rate (RPI)

2.1.1 Theoretical basis

During the period of water flooding, the water injectivity index represents the real-time characteristics of injection wells. The water injectivity index is defined as the ratio of the injection rate to the injection pressure difference. In field applications, it is difficult to measure the injection pressure difference, so the wellhead pressure is often used to replace the injection pressure difference to determine the injectivity index. Similarly for chemical floods, after the chemical solutions are injected into the well, both the injection pressure and the injection rate will change due to an increase in flow resistance in reservoirs, and thus, the dynamic injection data in the chemical flooding can be characterized by the variation of the wellhead pressure and the injection rate. The higher the injection pressure is, the greater the flow resistance will be, and the better the displacement efficiency the chemical flood will achieve. Based on this, the RPI refers to the ratio of the wellhead pressure to the injection rate, describing as follows:

$$\beta = \frac{P_{\text{wh}}}{q}, \tag{1}$$

where β is the ratio of the wellhead pressure to the injection rate (RPI) in MPa d/m^3; P_{wh} is the wellhead pressure in MPa; and q is the injection rate in m^3/d.

Figure 1 shows the relationship between the RPI and the cumulative volume of fluids injected into two injection wells (well I34-3166 and well I30-146) in the western part of the Gudong 7th reservoir. As is shown, injection well I34-3166 has experienced three stages of displacement, including water flooding, chemical flooding, and subsequent water flooding; while injection well I30-146 is a converted well (an injection well converted from a production well) and only experienced chemical flooding stage

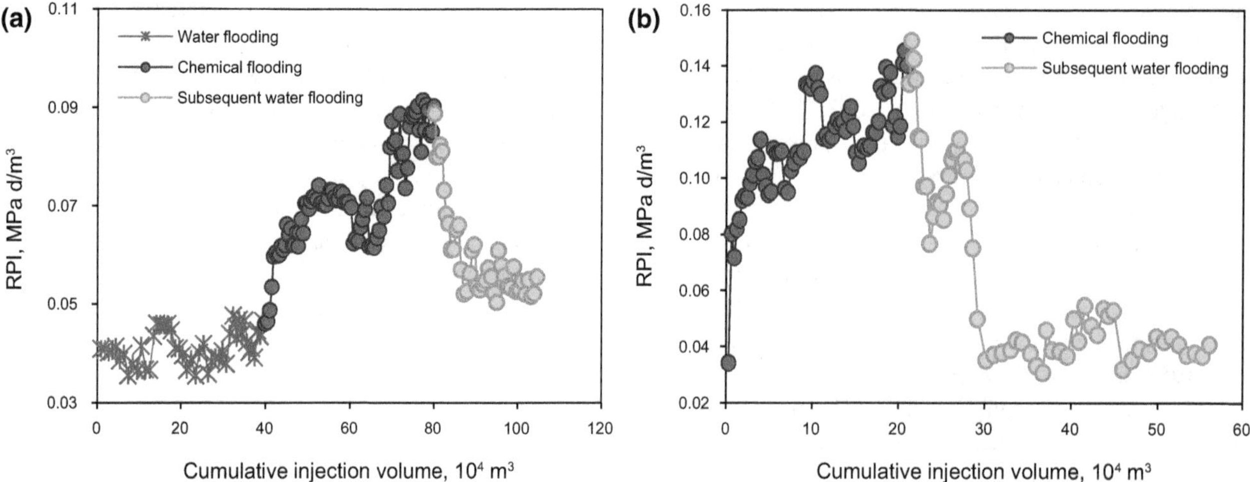

Fig. 1 Examples of the RPI under two different conditions. **a** Injection well I34-3166 with water flooding history. **b** Converted well I30-146 without water flooding history

and subsequent water flooding stage. It can be observed that at the stage of water flooding, the RPI fluctuates around a constant value; then, the curve rises rapidly and tends to vary gently with the increasing injection volume of the chemical solution; at the last stage of subsequent water flooding, the curve begins to fall fast, and eventually fluctuates steadily around a certain constant.

The variation of the RPI can be interpreted by the variation of resistance to flow in the reservoir. Due to the high viscosity of the polymer solution, the flow resistance will increase after the injection of the chemical solution. While at the stage of the subsequent water flooding, the flow resistance will decline gradually due to a decrease in the viscosity of the injection fluid. As a result, the RPI may be a reflection of the flow resistance in the reservoir at different displacement processes.

2.1.2 *Relationship with apparent (residual) resistance factor*

The RPI is defined as the ratio of the wellhead pressure to the injection rate in chemical flooding as well as in water flooding. It aims to calculate the apparent resistance factor and apparent residual resistance factor by the variation of the RPI, which is called the RPI method. The apparent resistance factor and the apparent residual resistance factor are used to characterize the difference of injectivity between water flooding and chemical flooding.

When the injection rate equals or nearly equals the production rate and the static reservoir pressure is constant or changes only slightly during the evaluation time, the bottom pressure can be approximated to the difference of the static water pressure and the friction pressure loss. The RPI can be presented as follows:

$$\beta \approx \frac{(P_{\mathrm{wh}} + \rho g H - \Delta P_{\mathrm{f}}) - P_{\mathrm{e}}}{q} = \frac{P_{\mathrm{wf}} - P_{\mathrm{e}}}{q}, \tag{2}$$

where ρ is the average fluid density in kg/m^3; g is the gravity acceleration in m/s^2, $g = 9.81$ m/s^2; H is the height of the wellhole liquid column in meters; ΔP_{f} is the friction pressure loss in MPa; P_{wf} is the bottom-hole pressure in MPa; and P_{e} is the static reservoir pressure in MPa.

Darcy's law for single-phase, steady-state, Newtonian flow was used to analyze the performance of the water injection wells by the Hall method. So, the RPI can be represented as follows:

$$\beta_{\mathrm{w}} \approx \frac{P_{\mathrm{wf}} - P_{\mathrm{e}}}{q} = \frac{1.867 B_{\mathrm{w}} \mu_{\mathrm{w}}}{K_{\mathrm{e}} h} \left(\ln \frac{R_{\mathrm{e}}}{R_{\mathrm{w}}} + S \right), \tag{3}$$

where β_{W} is the RPI of water flooding in MPa d/m^3; K_{e} is the effective permeability in 10^{-3} μm^2; R_{e} is the distance between the injector and the producer in meters; R_{w} is the wellhole radius in meters; S is the skin factor; B_{w} is the volume factor of water; μ_{w} is the water viscosity in mPa s; and h is the effective thickness in meters.

After a chemical flood is injected into a well, there will be several fluid banks between the injector and the producer. A simplified method to solve such a case is applying Darcy's law in a serial form and the resistance factor is defined to quantitatively characterize the variations of displacing fluid viscosity and reservoir permeability (Buell et al. 1990). Because of extensive water flooding in this reservoir, an oil bank does not form, and two banks, a water bank and polymer bank, can be taken into consideration. Thus, the RPI after the injection of the chemical fluid can be represented as

$$\beta_{\mathrm{p}} \approx \frac{1.867 B_{\mathrm{w}} \mu_{\mathrm{w}}}{K_{\mathrm{e}} h} \left\{ R_{\mathrm{f}} \left(\ln \frac{R_{\mathrm{b1}}}{R_{\mathrm{w}}} + S \right) + \ln \frac{R_{\mathrm{e}}}{R_{\mathrm{b1}}} \right\}, \tag{4}$$

where β_p is the RPI of chemical flooding in MPa d/m^3; R_f is the resistance factor; and R_{b1} is the radius of the chemical displacement front in meters.

Based on Eqs. (3) and (4), the ratio of the RPI of the chemical flooding to the RPI of the water flooding is described by Eq. (5):

$$\frac{\beta_p}{\beta_w} = \frac{R_f\left(\ln\frac{R_{b1}}{R_w} + S\right) + \ln\frac{R_e}{R_{b1}}}{\ln\frac{R_e}{R_w} + S}. \tag{5}$$

As shown in Eq. (5), under the condition that the radius of the chemical slug R_{b1} is equal to the distance between the injector and the producer R_e, the resistance factor equals the ratio of the RPI at the stage of the chemical flooding to the RPI at the stage of the water flooding. However, the size of the chemical slug is often less than the injector–producer distance, so the resistance factor does not equal the ratio of the RPI between the chemical flooding stage and the water flooding stage. Therefore, the apparent resistance factor is defined as

$$R_f' = \frac{\beta_p}{\beta_w}, \tag{6}$$

where R_f' is the apparent resistance factor. When the size of the chemical slug equals the injector–producer distance, the apparent resistance factor does equal the resistance factor.

The flow resistance in the subsequent water flooding process can also be calculated in series form. The enhanced oil recovery at this stage is due to the reservoir permeability reduction induced by the residual polymer, and therefore, the residual resistance factor is proposed to quantitatively characterize this mechanism. The residual resistance factor refers to the ratio of the reservoir permeability before chemical flooding to the permeability after the chemical injection. Thus, the RPI at the stage of subsequent water flooding can be represented as

$$\beta_{ww} \approx \frac{1.867 B_w \mu_w}{K_e h}\left\{R_{rf}\left(\ln\frac{R_{b2}}{R_w} + S\right) + R_f \ln\frac{R_{b1}}{R_{b2}} + \ln\frac{R_e}{R_{b1}}\right\}, \tag{7}$$

where β_{ww} is the RPI of subsequent water flooding in MPa d/m^3; R_{rf} is the residual resistance factor which is defined as the ratio of the absolute permeability before SP flooding to the absolute permeability after SP flooding (Buell et al. 1990); and R_{b2} is the radius of the displacement front of subsequent water flooding in meters.

The ratio of the RPI at the stage of subsequent water flooding to that at the stage of initial water flooding can be defined as

$$\frac{\beta_{ww}}{\beta_w} = \frac{R_{rf}\left(\ln\frac{R_{b2}}{R_w} + S\right) + R_f \ln\frac{R_{b1}}{R_{b2}} + \ln\frac{R_e}{R_{b1}}}{\ln\frac{R_e}{R_w} + S}. \tag{8}$$

As shown in Eq. (8), if the displacement radius R_{b2} of subsequent water flooding is equal to the injector–producer distance R_e, the residual resistance factor equals the ratio of the RPI at the stage of subsequent water flooding to that in the process of initial water flooding. The apparent residual resistance factor can be defined as

$$R_{rf}' = \frac{\beta_{ww}}{\beta_w}, \tag{9}$$

where R_{rf}' is the apparent residual resistance factor.

2.1.3 Advantages and disadvantages

The Hall plot analysis can only describe the average injection effect over a period of time, while the RPI is a reflection of the instantaneous value at every displacement moment. Thus, the RPI is more sensitive to the chemical injection and has more advantages over the Hall plot analysis, especially when the variation of Hall plot slope is not obvious, such as at the early stage of chemical flooding.

The disadvantage of the RPI is that it does not have a smoothing effect on the data when the injection pressure and rate have big and frequent fluctuations. An effective way to solve this problem is to apply some mathematical curve smoothing methods such as the data average method and the linear iterative method.

2.2 The rate of change of RPI

2.2.1 Theoretical basis

The RPI describes the variation range of the ratio of the injection pressure to the injection rate. However, the injection efficiency is related not only to the variation range of injection pressure and injection rate but also to the variation rate of injection pressure and injection rate. Under the condition of the same chemical injection rate, the faster the pressure increases, the more effective the chemical injection will be.

The rate of change of RPI refers to the variation of the RPI per unit injection volume. It can also be rearranged to the derivative value of the RPI to the cumulative injection volume:

$$\gamma = \frac{d\beta}{dW}, \tag{10}$$

where γ is the rate of change of the RPI in MPa d/m^6; and W is the cumulative injection volume in m^3.

If the injection rate between 2 months keeps constant or changes only slightly, the rate of change of the RPI can be obtained as follows:

$$\gamma = \frac{\beta_2 - \beta_1}{W_2 - W_1} = \frac{p_{\mathrm{wh2}} - p_{\mathrm{wh1}}}{q^2(t_2 - t_1)}, \tag{11}$$

where t_1 and t_1 are the injection time in days.

Based on Eq. (11), the rate of change of the RPI refers to the ratio of the wellhead pressure increment per unit time to the square of the injection rate.

The relationship between the RPI and the cumulative injection volume is illustrated in Fig. 2. It is observed that at the polymer pre-protection slug injection stage, the RPI increases along with an increase in the cumulative injection, while the rate of change of RPI behaves in a reverse manner. After curve fitting, there is a logarithmic relationship between the RPI and the cumulative injection volume, while the rate of change of the RPI has a linear relation with the reciprocal of the cumulative injection volume.

2.2.2 *Relationship with resistance factor*

Based on the definition Eq. (10) of the rate of change of the RPI and the expression Eq. (4) of the RPI during chemical flooding, the derivative of the RPI can be expressed as

$$\gamma = \frac{\mathrm{d}\beta}{\mathrm{d}W} = \frac{\mathrm{d}\left(\frac{1.867\mu_{\mathrm{w}}B_{\mathrm{w}}}{K_{\mathrm{e}}h}\left(R_{\mathrm{f}}(\ln(r/R_{\mathrm{w}}) + S) + \ln(R_{\mathrm{e}}/r)\right)\right)}{\mathrm{d}W}, \tag{12}$$

where r is the radius of the chemical displacement front in meters.

A small circular unit in a circular formation is selected as shown in Fig. 3. The circular unit formation can be

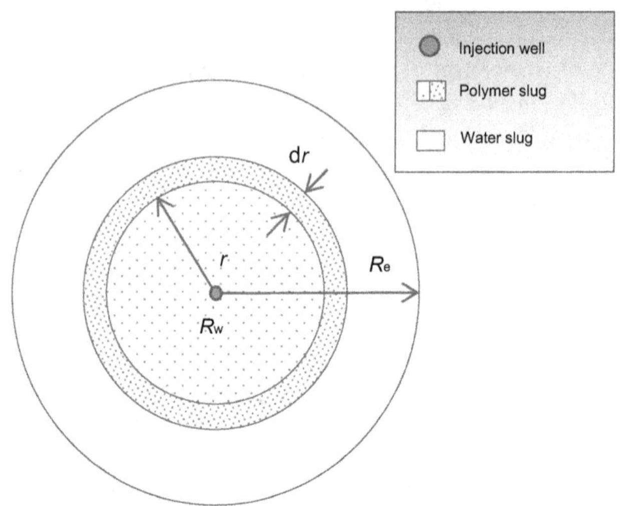

Fig. 3 Schematic diagram of the polymer injection model

regarded as a homogeneous formation with a constant thickness and constant effective permeability. The differential formula of the cumulative injection volume can be expressed as

$$\mathrm{d}W = 2\pi r\mathrm{d}r \cdot h \cdot \phi \cdot \phi_{\mathrm{D}}, \tag{13}$$

where ϕ is the formation porosity; and ϕ_{D} is the fraction of the accessible pore volume of the polymer solution.

Substituting Eq. (13) into Eq. (12) gives the simplified derivation:

$$\gamma = \frac{0.9335\mu_{\mathrm{w}}B_{\mathrm{w}}}{\pi K_{\mathrm{e}}h^2\phi \cdot \phi_{\mathrm{D}}} \cdot \frac{(R_{\mathrm{f}} - 1)}{r^2}. \tag{14}$$

The accumulative volume of the chemical solution injected into the reservoir is

$$W = \pi \cdot r^2 \cdot h \cdot \phi \cdot \phi_{\mathrm{D}}. \tag{15}$$

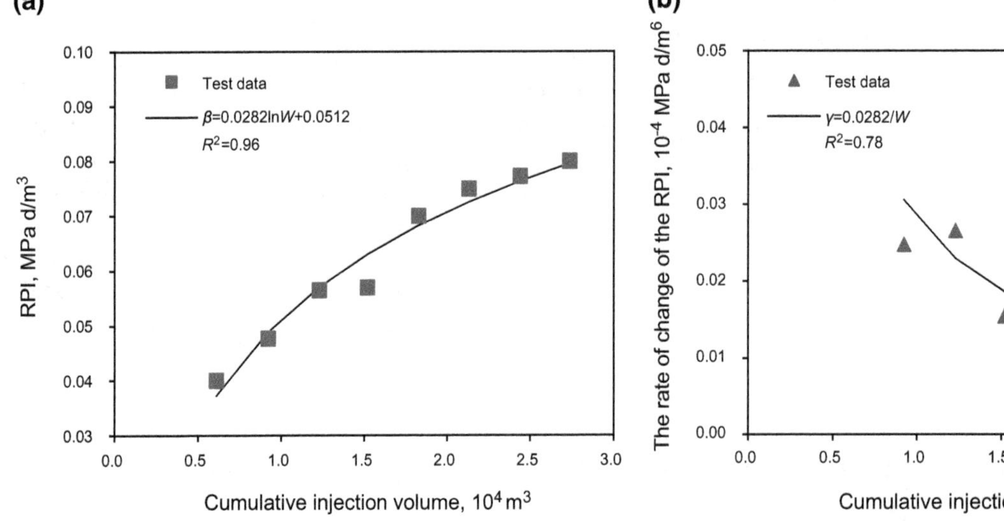

Fig. 2 Statistical relationship of new evaluation indexes and the cumulative injection volume. **a** RPI. **b** The rate of change of the RPI

The relationship between the rate of change of the RPI and the cumulative injection volume can be obtained by substituting Eq. (15) into Eq. (14):

$$\gamma = \frac{0.9335\mu_w B_w (R_f - 1)}{K_e h} \cdot \frac{1}{W}. \quad (16)$$

Equation (16) indicates that the rate of change of the RPI varies inversely with the cumulative injection volume, which also proves the validity of the regression expression in Fig. 2a. The expression of the resistance factor can be derived from Eq. (16):

$$R_f = 1 + \frac{K_e h}{0.9335\mu_w B_w} \cdot W \cdot \gamma. \quad (17)$$

Under the condition of the same cumulative injection volume, the greater the rate of change of the RPI is, the better the efficiency of the chemical injection.

2.2.3 Advantages

Dynamic data such as the injection pressure and the injection rate are indispensable to evaluate the resistance factor by both the Hall plot method and the RPI method during water flooding. However, for new wells or converted wells without water-flooding history, the rate of change of the RPI is the only effective way for injection effect evaluation.

3 Field applications

3.1 Field trial of SP flooding

The first pilot application of SP flooding in China is located in the western part of Gudong 7th reservoir, covering an area of 0.94 km^2, with nearly 277 × 10^4 tons oil in place. The target zone consists of three layers ($Ng5^4$, $Ng5^5$, and $Ng6^1$), with a depth of 1261–1294 m. The average porosity is 34 %, the permeability is 1320 × 10^{-3} μm^2, and the permeability variation coefficient is 0.58. Prior to production, the viscosity of the crude oil is 45 mPa s, and the initial oil saturation is 0.72. The initial reservoir pressure is 12.4 MPa, and the reservoir temperature is 68 °C, and the salinity of the formation water is 3152 mg/L.

A pilot water flood was initiated in this target zone in July 1986. Until August 2003, there were 21 production wells (20 of them were open) and 9 injection wells (8 of them were open), with an average daily fluid production rate of 123.2 t/d and an average daily injection rate of 205 m^3/d. The average daily oil production rate was 2.95 t/d and the average water cut was 97.6 %, while the average daily injection pressure was 11.6 MPa and the injection to production ratio was 0.88 with a cumulative injection to production ratio of 1.04. The oil recovery of reservoir was up to 34.5 %.

A pilot SP flood was started in September 2003 and ended in January 2010, including 27 wells, 17 production wells and 10 injection wells. As is shown in Fig. 4, a line drive pattern was adopted during oil production, while the row space is 300 m and the well space is 150 m.

The injection of the SP flood included four slugs: polymer slug (pre-slug), main SP slug I, main SP slug II, and polymer slug (post-slug). The pre-slug started in September 2003, while the main SP slug I started in June 2004, main SP slug II in June 2007, and post-slug in April 2009. After the SP flooding, the subsequent water flooding stage went into operation in January 2010. The total cumulative injection of chemicals was up to 0.635 PV, including polymer 5496 t, surfactant 8727 t, and auxiliary 3024 t. The detailed information is shown in Table 1.

3.2 Calculation of the apparent resistance factor by the RPI

The apparent resistance factor can be defined as the ratio of the RPI at the chemical flooding stage to that at the water flooding stage. Therefore, it is necessary to calculate the RPI at the water flooding stage. Since the RPI at the water flooding stage will not keep constant, as shown in Fig. 1, the average value over a period of time (in general, 1 month is enough) is used in calculation. Using the above method, the apparent resistance factor and the apparent residual resistance factor of injection well I34-3166 are shown in Fig. 5 and Table 2.

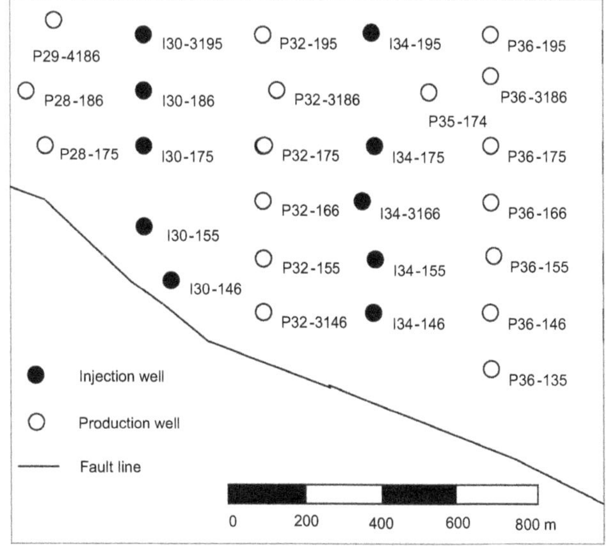

Fig. 4 Location of studied wells

Table 1 Injection data of SP flooding

Injection slug	Starting time	End time	Injection concentration			Slug size, PV
			Polymer, mg/L	Surfactant, mg/L	Auxiliary, mg/L	
Polymer slug (pre-slug)	Sep. 2003	May 2004	1934	0	0	0.078
Main SP slug I	Jun. 2004	May 2007	1856	4618	1610	0.302
Main SP slug II	Jun. 2007	Apr. 2009	1713	3056	1043	0.188
Polymer slug (post-slug)	Apr. 2009	Jan. 2010	1500	0	0	0.067

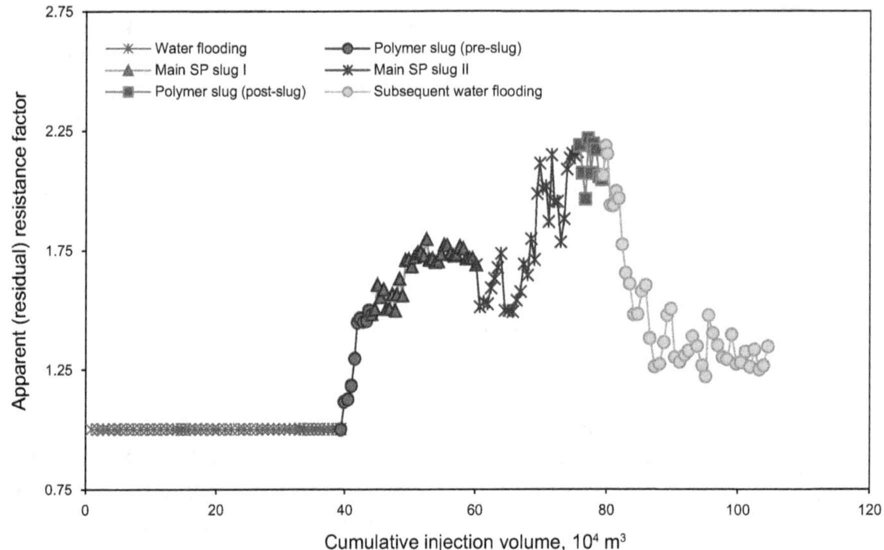

Fig. 5 Apparent (residual) resistance factor of injection well I34-3166 derived by the RPI method

Table 2 Apparent (residual) resistance factor of injection well I34-3166

Injection stage	Apparent (residual) resistance factor			
	RPI method			Hall plot analysis method
	Minimum	Maximum	Mean	
SP flooding				
Polymer slug (pre-slug)	1.1	1.5	1.35	1.65
Main SP slug I	1.5	1.8	1.67	
Main SP slug II	1.5	2.2	1.82	
Polymer slug (post-slug)	2.0	2.2	2.09	
Subsequent water flooding	1.2	2.2	1.48	1.40

Figure 5 indicates that the apparent resistance factor increases from 1.0 to 1.5 rapidly during the injection of the polymer slug (pre-slug), with an average value of 1.35. During the injection of the main SP slug I, the apparent resistance factor rises gradually from 1.5 to 1.8, and the average value is 1.67. However, when the main SP slug II is injected into the reservoir, the apparent resistance factor decreases a little with the cumulative injection volume first and then increases gradually to the maximum value of 2.2, with an average value of 1.82. During the injection of the polymer slug (post-slug), the apparent resistance factor changes relatively little, changing from 2.0 to 2.2, with an average value of 2.09. At last, during the subsequent water flooding, the residual apparent resistance factor decreases slowly until reaching a steady level near 1.3.

Meanwhile, the apparent resistance factors for the chemical flooding and the subsequent water flooding were also calculated from the Hall plot analysis, as shown in Fig. 6 and Table 2. There is a small difference in the apparent resistance factors calculated from these two

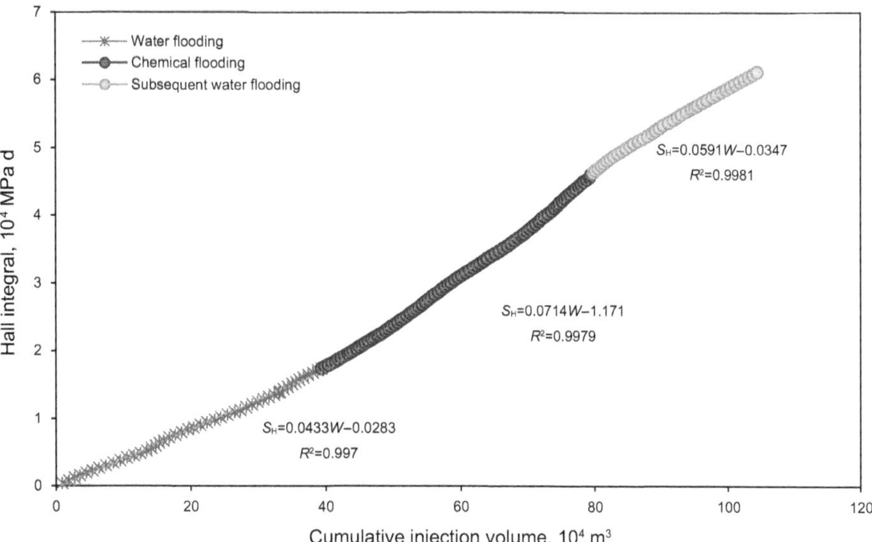

Fig. 6 Hall plot of injection well I34-3166

methods. In fact, both methods are based on the assumption of steady flow, but the RPI can reflect the instantaneous value at every displacement moment and be more sensitive than the Hall plot method. Thus, this method is more helpful for dynamic analysis, especially when the slope difference of the Hall plot is not obvious, i.e., at the early stage of chemical flooding.

3.3 Prediction of crossflow of the chemical floods by the RPI

If the concentration of chemicals fluctuates slightly in the process of the chemical injection, the RPI would increase gradually or remain unchanged with an increase in the injection volume. However, a sharp drop in the RPI indicates a decrease in the resistance to flow in the reservoir, and the possibility of chemical crossflow.

SP flooding pilot tests were conducted in the Shengli Oilfield, and Fig. 7 shows the RPI of injection well I34-175 and the concentration of polymer in the produced fluids from the surrounding production wells. When the cumulative volume of the SP solution injected into the reservoir reached 8×10^4 m^3, there was a sharp drop in the RPI, and 2 months after the drop, polymer was detected in the produced fluids from production well P32-3186 with a polymer concentration of 181 mg/L. A second drop in the RPI occurred when the cumulative injection volume came to 16×10^4 m^3, and 2 months later, polymer of high concentration was produced from wells P32-175 and P32-166. In addition, when the cumulative injection volume amounted to 30×10^4 m^3, a third drop in the RPI appeared and a significantly increase appeared in the concentration of polymers in the produced fluids from wells P32-175, P32-166, P35-174, P36-175, and P36-166.

Based on the analysis above, it is observed that after the RPI drops for some time, polymer will appear in production fluids. For a production well, both the early production of chemical fluids and the rapid growth of the polymer concentration might be caused by polymer crossflow. The existence of thief zones is the most common case causing chemical crossflow. To verify the prediction method by the RPI theoretically, a heterogeneous reservoir model with a production well and an injection well is shown in Fig. 8. The model is composed of two stratified homogeneous layers. The first layer has a low permeability k_1 and a thickness of h_1, while the second layer has a high permeability k_2 and a thickness of h_2. Figure 8 describes the flow without or with crossflow.

According to Darcy's law, the flow resistance can be defined as the ratio of the injection pressure difference to the injection rate. To compare the flow resistance in two cases, the reservoir is divided into four parts, and the length of each part is L_1, L_2, L_3, and L_4. Under the condition that no chemical crossflow exists, the flow resistance of these four part are R_1, R_2, R_3, and R_4, while with chemical crossflow, the flow resistance of these four part are R_1', R_2', R_3', and R_4'. Since under both conditions, a chemical solution is only in the first part with a length of L_1, R_1 is equal to R_1'. Similarly, for the fourth part with a length of L_4 in which only water flows, so R_4 is also equal to R_4'. When the cumulative injection volume of polymer is equal in two cases, the relationship between parameters can be described by Eq. (18):

$$L_2(h_1 + h_2)B\phi = (L_2 + L_3)h_2B\phi, \tag{18}$$

where B is the formation width in meters.

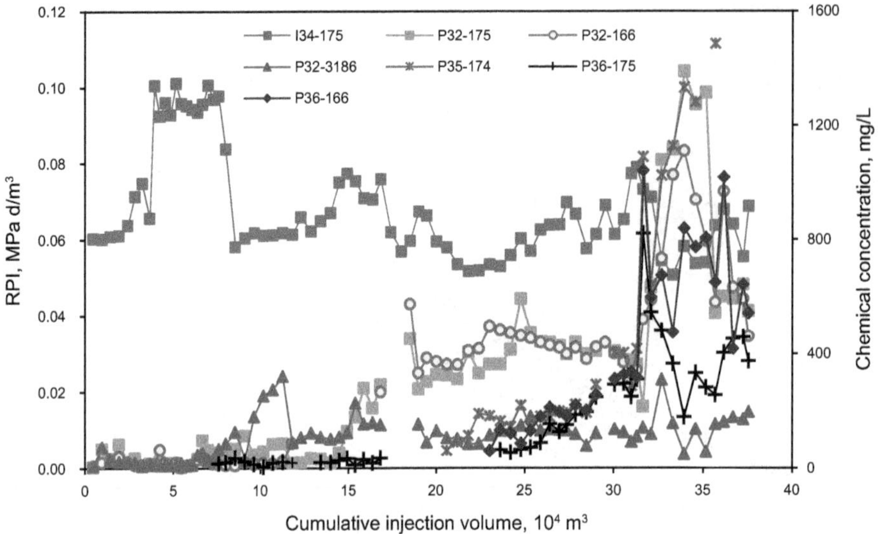

Fig. 7 RPI of injector I34-175 and chemical concentrations of surrounding producers

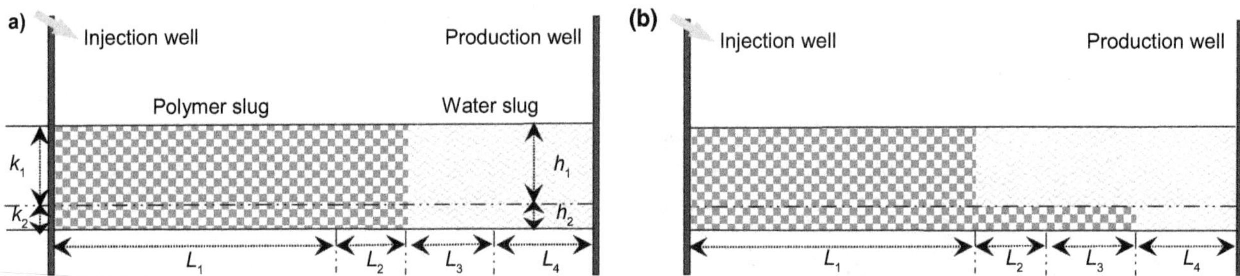

Fig. 8 Schematic diagram of the chemical flood in a reservoir. **a** No chemical crossflow. **b** Chemical crossflow

The difference of the flow resistance in two cases is expressed as

$$\begin{aligned} & R - R' \\ & = (R_1 + R_2 + R_3 + R_4) - \left(R_1' + R_2' + R_3' + R_4'\right) \\ & = R_2 + R_3 - R_2' - R_3' \\ & = \frac{1}{\frac{k_1 B_w h_1}{\mu_p L_2} + \frac{k_2 B_w h_2}{\mu_p L_2}} + \frac{1}{\frac{k_1 B_w h_1}{\mu_w L_3} + \frac{k_2 B_w h_2}{\mu_w L_3}} \\ & \quad - \frac{1}{\frac{k_1 B_w h_1}{\mu_w L_2} + \frac{k_2 B_w h_2}{\mu_p L_2}} - \frac{1}{\frac{k_1 B_w h_1}{\mu_w L_3} + \frac{k_2 B_w h_2}{\mu_p L_3}} \\ & = \frac{L_2 h_1 \left(\mu_p - \mu_w\right)\left(k_1 \mu_p - k_2 \mu_w\right)}{B_w (k_1 h_1 + k_2 h_2)\left(k_1 h_1 \mu_p + k_2 h_2 \mu_w\right)} \end{aligned} \quad (19)$$

where R is the flow resistance without chemical crossflow; R' is the flow resistances with chemical crossflow; and μ_p is the viscosity of the chemical fluid in mPa s.

In actual cases, because the viscosity ratio of the chemical fluid to water is larger than the permeability ratio of the two layers, R is greater than the R'. It can also be set out that when the chemical crossflow exists, the RPI decreases as a result of the drop in the flow resistance. Based on this theory, the chemical crossflow can be estimated by the variation of the RPI.

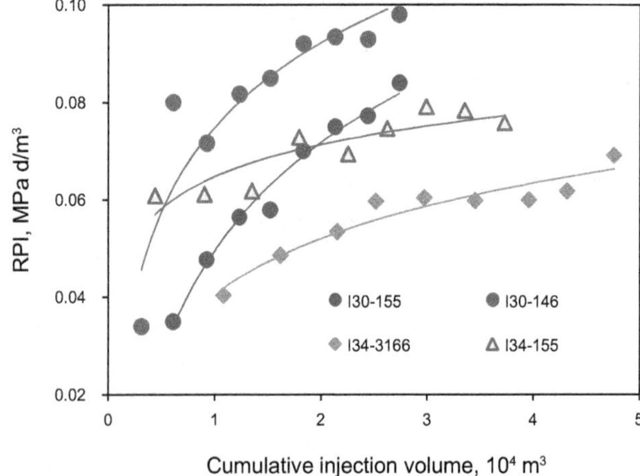

Fig. 9 The relationship between the RPI and the cumulative injection volume

Table 3 Input parameters and the (apparent) resistance factor

Injection well	Input parameters				(Apparent) resistance factor	
	Thickness, m	Water viscosity, mPa s	Permeability, μm^2	Regression coefficient	RPI change rate method	RPI method
I30-146	11.2	0.45	0.99	0.025	1.68	Conversion well (no water flooding history)
I30-155	11.4	0.45	1.40	0.033	2.23	
I34-155	18.0	0.45	0.78	0.009	1.30	1.41
I34-3166	17.5	0.45	0.51	0.016	1.34	1.35

3.4 Calculation of the resistance factor by the rate of change of RPI

The relationship between the resistance factor and the rate of change of the RPI is given by Eq. (17). By calculating the resistance factor of the polymer slug (pre-slug), the curve in Fig. 9 shows the RPI versus the cumulative injection volume of wells I30-155, I30-146, I34-3166, and I34-155. It is observed that the RPI of the four injection wells have a good logarithmic relationship with the cumulative injection volume, which further verifies Eq. (16).

The calculation of the apparent (residual) resistance factor is as follows:

Firstly, the regression expression of the RPI versus the cumulative injection volume is calculated:

Injection well I30-155

$$\beta = 0.033 \ln W + 0.050, \quad R^2 = 0.94 \tag{20}$$

Injection well I30-146

$$\beta = 0.025 \ln W + 0.075, \quad R^2 = 0.84 \tag{21}$$

Injection well I34-155

$$\beta = 0.009 \ln W + 0.064, \quad R^2 = 0.81 \tag{22}$$

Injection well I34-3166

$$\beta = 0.016 \ln W + 0.040, \quad R^2 = 0.91, \tag{23}$$

where R^2 is the coefficient of determination.

Secondly, the rate of change of the RPI versus the cumulative injection volume is calculated. For injection well I30-155, the equation is

$$\gamma = \frac{d\beta}{dW} = \frac{A}{W} = \frac{0.033}{W} \tag{24}$$

where A is the slope of the regression curve.

Lastly, the resistance factor is obtained based on the relational expression of the rate of change of the RPI and the cumulative injection volume (Table 3). For injection well I30-155, the expression of the resistance factor is

$$R_f = 1 + \frac{K_e h}{0.9335\mu_w B_w} \cdot A = 1 + 0.033\frac{K_e h}{0.9335\mu_w B_w} = 2.23. \tag{25}$$

Similarly, the resistance factor of the remaining three injection wells is shown in Table 3. Meanwhile, the RPI is also used to calculate the apparent resistance factor. Since both injection wells I30-146 and I30-155 in Table 3 are converted wells, the traditional Hall plot analysis is not applicable for the resistance factor calculation. The resistance factor of the remaining two wells shows no obvious difference between the RPI method and RPI change rate method. However, for new wells without water flooding history or converted wells, the RPI change rate method is an effective way for injectivity evaluation.

4 Conclusion

(1) A new method (RPI and RPI change rate) is developed for evaluating the efficiency of chemical flooding. It is proved to be a more sensitive and rapid method for calculating parameters than the Hall plot method. Also, it is proved to be an effective way in estimating the existence of chemical crossflow.

(2) The RPI characterizes the real-time injection performance of chemical flooding in a simple form. It is verified that the apparent resistance factor and the apparent residual resistance factor can be obtained by the RPI at different displacement stages, and this method is most attractive at the early stage of chemical injection or when no obvious effect is observed.

(3) The RPI will fall sharply when the resistance to flow declines (caused by chemical crossflow). According to this theory, the RPI can be used to indicate the existence of chemical crossflow.

(4) The rate of change of the RPI can quantitatively describe the response rate of chemical flooding. The greater the rate of change of the RPI is, the faster the

chemical flooding takes effect. Since only injection data of the chemical flooding stage is needed for evaluation, the rate of change of the RPI is most suitable for wells without water flooding history.

Acknowledgments The authors greatly appreciate the financial support from the National Natural Science Foundation of China (Grant No. 51574269), the Important National Science and Technology Specific Projects of China (Grant No. 2016ZX05011-003), the Fundamental Research Funds for the Central Universities (Grant No. 15CX08004A, 13CX05007A), and the Program for Changjiang Scholars and Innovative Research Team in University (Grant No. IRT1294).

References

AlSofi AM, Blunt MJ. Polymer flooding design and optimization under economic uncertainty. J Pet Sci Eng. 2014;124:46–59. doi:10.1016/j.petrol.2014.10.014.

Buell RS, Kazemi H, Poettmann FH. Analyzing injectivity of polymer solutions with the Hall plot. SPE Reserv Eng. 1990;5(1):41–6. doi:10.2118/16963-PA.

Chang HL, Zhang ZQ, Wang QM, et al. Advances in polymer flooding and alkaline/surfactant/polymer processes as developed and applied in the People's Republic of China. J Pet Technol. 2006;58(2):84–9. doi:10.2118/89175-JPT.

Cheng J, Wang D, Li Q. Field test performance of alkaline surfactant polymer flooding in the Daqing oil field. Acta Pet Sin. 2002;23(6):37–40 (in Chinese).

Dag CS, Ingun S. Literature review of implemented polymer field projects. J Pet Sci Eng. 2014;122:761–75. doi:10.1016/j.petrol.2014.08.024.

Delamaide E, Zaitoun A, Renard G, et al. Pelican Lake field: first successful application of polymer flooding in a heavy-oil reservoir. SPE Reserv Eval Eng. 2014;17(03):340–54. doi:10.2118/165234-PA.

DeMarco M. Simplified method pinpoints injection well problems. World Oil. 1969;168(5):97.

Dong M, Ma S, Liu Q. Enhanced heavy oil recovery through interfacial instability: a study of chemical flooding for Brintnell heavy oil. Fuel. 2009;88(6):1049–56. doi:10.1016/j.fuel.2008.11.014.

Ghosh B, Bemani AS, Wahaibi YM, et al. Development of a novel chemical water shut-off method for fractured reservoirs: laboratory development and verification through core flow experiments. J Pet Sci Eng. 2012;96:176–84. doi:10.1016/j.petrol.2011.08.020.

Hall HN. How to analyze waterflood injection well performance. World Oil. 1963;157(5):128–33.

Honarpour MM, Tomutsa L. Injection/production monitoring: an effective method for reservoir characterization. In: SPE/DOE enhanced oil recovery symposium, 22–25 April, Tulsa, Oklahoma; 1990. doi:10.2118/20262-MS.

Hou J, Du Q, Lu T, et al. The effect of interbeds on distribution of incremental oil displaced by a polymer flood. Pet Sci. 2011;8(2):200–6. doi:10.1007/s12182-011-0135-z.

Hou J, Pan G, Lu X, et al. The distribution characteristics of additional extracted oil displaced by surfactant–polymer flooding and its genetic mechanisms. J Pet Sci Eng. 2013;112:322–34. doi:10.1016/j.petrol.2013.11.021.

Izgec B, Kabir CS. Real-time performance analysis of water-injection wells. SPE Reserv Eval Eng. 2009;12(1):116–23. doi:10.2118/109876-PA.

Kaminsky RD, Wattenbarger RC, Szafranski RC, et al. Guidelines for polymer flooding evaluation and development. In: International petroleum technology conference, 4–6 Dec, Dubai, UAE; 2007. doi:10.2523/11200-MS.

Kim Y, Lee J. Evaluation of real-time injection performance for polymer flooding in heterogeneous reservoir. In: SPE the twenty-fourth international ocean and polar engineering conference, 15–20 June, Busan, Korea; 2014.

Li Z. Industrial test of polymer flooding in super high water cut stage of central No. 1 Block Gudao Oilfield. Pet Exp Dev. 2004;31(2):119–21 (in Chinese).

Li Y, Su Y, Ma K, et al. Study of gel flooding pilot test and evaluation method for conventional heavy oil reservoir in Bohai bay. In: SPE annual technical conference and exhibition, 30 Oct–2 Nov, Denver, Colorado, USA; 2011. doi:10.2118/146617-MS.

Li Z, Zhang A, Cui X, et al. A successful pilot of dilute surfactant-polymer flooding in Shengli oilfield. In: SPE improved oil recovery symposium, 14–18 April, Tulsa, Oklahoma, USA; 2012. doi:10.2118/154034-MS.

Ma S, Dong M, Li Z, et al. Evaluation of the effectiveness of chemical flooding using heterogeneous sandpack flood test. J Pet Sci Eng. 2007;55:294–300. doi:10.1016/j.petrol.2006.05.002.

Moffitt PD, Menzie DE. Well injection tests of non-newtonian fluids. In: SPE rocky mountain regional meeting, 17–19 May, Cody, Wyoming, USA; 1978.

Seright RS, Seheult JM, Talashek T. Injectivity characteristics of EOR polymers. In: SPE annual technical conference and exhibition, 21–24 Sept, Denver, Colorado, USA; 2009. doi:10.2118/115142-MS.

Shaker Shiran B, Skauge A. Enhanced oil recovery (EOR) by combined low salinity water/polymer flooding. Energy Fuels. 2013;27(3):1223–35. doi:10.1021/ef301538e.

Shen P, Wang J, Yuan S, et al. Study of enhanced-oil-recovery mechanism of alkali/surfactant/polymer flooding in porous media from experiments. SPE J. 2009;14(02):237–44. doi:10.2118/126128-PA.

Sheng JJ, Leonhardt B, Azri N. Status of polymer-flooding technology. J Can Pet Technol. 2015;54(2):116–26. doi:10.2118/174541-PA.

Silin DB, Holtzman R, Patzek TW. Monitoring waterflood operations: Hall method revisited. In: SPE western regional meeting, 30 Mar–1 April, Irvine, California, USA; 2005. doi:10.2118/93879-MS.

Sugai K, Nishikiori N. An integrated approach to reservoir performance monitoring and analysis. In: SPE Asia Pacific oil & gas conference and exhibition, 11–13 Sept, Adelaide, Australia; 2006. doi:10.2118/100995-MS.

Urbissinova TS, Trivedi JJ, Kuru E. Effect of elasticity during viscoelastic polymer flooding: a possible mechanism of increasing the sweep efficiency. J Can Pet Technol. 2010;49(12):49. doi:10.2118/133471-PA.

Vargo J, Turner J, Bob V, et al. Alkaline-surfactant-polymer flooding of the Cambridge Minnelusa field. SPE Reserv Eval Eng. 2000;3(06):552–8. doi:10.2118/68285-PA.

Zhang H, Dong M, Zhao S. Which one is more important in chemical flooding for enhanced court heavy oil recovery, lowering interfacial tension or reducing water mobility? Energy Fuels. 2010;24(3):1829–36. doi:10.1021/ef901310v.

Permissions

All chapters in this book were first published in PS, by Springer; hereby published with permission under the Creative Commons Attribution License or equivalent. Every chapter published in this book has been scrutinized by our experts. Their significance has been extensively debated. The topics covered herein carry significant findings which will fuel the growth of the discipline. They may even be implemented as practical applications or may be referred to as a beginning point for another development.

The contributors of this book come from diverse backgrounds, making this book a truly international effort. This book will bring forth new frontiers with its revolutionizing research information and detailed analysis of the nascent developments around the world.

We would like to thank all the contributing authors for lending their expertise to make the book truly unique. They have played a crucial role in the development of this book. Without their invaluable contributions this book wouldn't have been possible. They have made vital efforts to compile up to date information on the varied aspects of this subject to make this book a valuable addition to the collection of many professionals and students.

This book was conceptualized with the vision of imparting up-to-date information and advanced data in this field. To ensure the same, a matchless editorial board was set up. Every individual on the board went through rigorous rounds of assessment to prove their worth. After which they invested a large part of their time researching and compiling the most relevant data for our readers.

The editorial board has been involved in producing this book since its inception. They have spent rigorous hours researching and exploring the diverse topics which have resulted in the successful publishing of this book. They have passed on their knowledge of decades through this book. To expedite this challenging task, the publisher supported the team at every step. A small team of assistant editors was also appointed to further simplify the editing procedure and attain best results for the readers.

Apart from the editorial board, the designing team has also invested a significant amount of their time in understanding the subject and creating the most relevant covers. They scrutinized every image to scout for the most suitable representation of the subject and create an appropriate cover for the book.

The publishing team has been an ardent support to the editorial, designing and production team. Their endless efforts to recruit the best for this project, has resulted in the accomplishment of this book. They are a veteran in the field of academics and their pool of knowledge is as vast as their experience in printing. Their expertise and guidance has proved useful at every step. Their uncompromising quality standards have made this book an exceptional effort. Their encouragement from time to time has been an inspiration for everyone.

The publisher and the editorial board hope that this book will prove to be a valuable piece of knowledge for researchers, students, practitioners and scholars across the globe.

List of Contributors

P. Mansouri-Daneshvar, R. Moussavi-Harami, A. Mahboubi and M. H. M. Gharaie
Department of Geology, Faculty of Sciences, Ferdowsi University of Mashhad, Mashhad, Iran

A. Feizie
Exploration Directorate, National Iranian Oil Company, Tehran, Iran

Hai-Yu Jiang, Zu-Bin Chen, Xiao-Xian Zeng, Hao Lv and Xin Liu
Key Laboratory of Geo-Exploration and Instrumentation of Ministry of Education, College of Instrumentation and Electrical Engineering, Jilin University, Changchun 130026, Jilin, China

M. B. Clennell, D. N. Dewhurst, M. Pervukhina, Tongcheng Han and Junfang Zhang
CSIRO Energy Flagship, 26 Dick Perry Ave, Kensington, WA 6151, Australia

Keyu Liu
CSIRO Energy Flagship, 26 Dick Perry Ave, Kensington, WA 6151, Australia
Research Institute of Petroleum Exploration and Development, Petro China, Beijing 100083, China

Zhejun Pan
CSIRO Energy Flagship Ian Wark Laboratory, Bayview Avenue, Clayton, VIC 3169, Australia

Yan Song and Zhuo Li
State Key Laboratory of Petroleum Resource and Prospecting, China University of Petroleum, Beijing 102249, China
Unconventional Natural Gas Institute, China University of Petroleum, Beijing 102249, China

Lin Jiang and Feng Hong
Research Institute of Petroleum Exploration and Development, Petro China, Beijing 100083, China

Xi Yang, Qi Zhang and Si-Yuan Chen
Academy of Chinese Energy Strategy, China University of Petroleum (Beijing), Changping, Beijing 102249, China

Hong Wan and Jing-Cheng Zhou
Policy Research Office, China National Petroleum Corporation, Beijing 100007, China

Xue-Chun Wang, Jian-Hua Fang, Bo-Shui Chen, Jiu Wang, Jiang Wu and Di Xia
Department of Military Oil Application & Administration Engineering, Logistical Engineering University, Chongqing 401311, China

You-Long Zou and Ran-Hong Xie
State Key Laboratory of Petroleum Resources and Prospecting, China University of Petroleum, Beijing 102249, China

Alon Arad
Shell International Exploration and Production Inc., Houston 77079, TX, USA

Zhen-Xue Jiang, Zhuo Li, Feng Li and Wei Yang
State Key Laboratory of Petroleum Resources and Prospecting, China University of Petroleum, Beijing 102249, China
Unconventional Natural Gas Institute, China University of Petroleum, Beijing 102249, China

Xiong-Qi Pang, Luo-Fu Liu and Fu-Jie Jiang
State Key Laboratory of Petroleum Resources and Prospecting, China University of Petroleum, Beijing 102249, China
College of Geosciences, China University of Petroleum, Beijing 102249, China

Shereen Ezzat
Department of Mathematics, Faculty of Science and Letters in Al Bukayriyyah, Al-Qassim University, Al-Qassim, Saudi Arabia

Magdy Ezzat
Department of Mathematics, Faculty of Science and Letters in Al Bukayriyyah, Al-Qassim University, Al-Qassim, Saudi Arabia
Faculty of Education, Alexandria University, Alexandria, Egypt

Da-Zhen Tang, Chun-Miao Deng, Yan-Jun Meng, Zhi-Ping Li, Hao Xu, Shu Tao and Song Li
School of Energy Resources, China University of Geosciences, Beijing 100083, China
Coal Reservoir Laboratory of National Engineering Research Center of Coalbed Methane Development & Utilization, Beijing 100083, China

Cong-Sheng Bian, Wen-Zhi Zhao, Hong-Jun Wang, Zhi-Yong Chen, Ze-Cheng Wang, Chang-Yi Zhao, Zhao-Hui Xu, Yong-Xin Li and Lin Jiang
Research Institute of Petroleum Exploration & Development, CNPC, Beijing 100083, China

Guang-Di Liu
College of Geosciences, China University of Petroleum, Beijing 102249, China

Yun-Peng Wang
Guangzhou Institute of Geochemistry, Chinese Academy of Sciences, Guangzhou 510640, Guangdong, China

Lei-Lei Hu, Xiang-Feng Zhang, Run-Mei Yang, Hai-Ming Fan and Jie Geng
School of Petroleum Engineering, China University of Petroleum, Qingdao 266580, Shandong, China

Wan-Li Kang
School of Petroleum Engineering, China University of Petroleum, Qingdao 266580, Shandong, China
EOR Research Institute, China University of Petroleum (Beijing), Beijing 100249, China

Xian-Ming Xiao, Hai-Feng Gai and Hui Tian
State Key Laboratory of Organic Geochemistry, Guangzhou Institute of Geochemistry, Chinese Academy of Sciences, Guangzhou, Guangdong 510640, China

Teng-Fei Li, Mao-Lin Wang, Lei Pan, Ji Chen and Qiang Wei
State Key Laboratory of Organic Geochemistry, Guangzhou Institute of Geochemistry, Chinese Academy of Sciences, Guangzhou, Guangdong 510640, China
University of Chinese Academy of Sciences, Beijing 100049, China

Asma Larbi and Mortada Daaou
Laboratoire de Synthèse organique, Physico-chimie, Biomolcules et Environnement (LSPBE), Department of Industrial Organic Chemistry, Faculty of Chemistry, University of Sciences and Technology of Oran, Po. Box 1505, 31000 El-M'naouer, Algeria

Abassia Faraoun
Laboratoire de Synthèse organique, Physico-chimie, Biomolcules et Environnement (LSPBE), Department of Industrial Organic Chemistry, Faculty of Chemistry, University of Sciences and Technology of Oran, Po. Box 1505, 31000 El-M'naouer, Algeria
Laboratoire de Chimie Physique Macromolèculaire (LCPM), Department of Chemistry, Faculty of Sciences, University of Oran, Po. Box 1524, 31000 El-M'naouer, Algeria

De-Xiang Li, Liang Zhang, Yan-Min Liu, Wan-Li Kang and Shao-Ran Ren
School of Petroleum Engineering, China University of Petroleum, Qingdao 266580, Shandong, China

Ji-Dong Yang, Jian-Ping Huang, Xin Wang and Zhen-Chun Li
School of Geoscience, China University of Petroleum, Qingdao 266580, Shandong, China

Cai-Neng Zou, Zhi Yang, Lian-Hua Hou, Ru-Kai Zhu, Jing-Wei Cui, Song-Tao Wu, Sen-Hu Lin, Qiu-Lin Guo, She-Jiao Wang and Deng-Hua Li
Petro China Research Institute of Petroleum Exploration and Development, Beijing 100083, China

Jian Hou
State Key Laboratory of Heavy Oil Processing, China University of Petroleum, Qingdao 266580, Shandong, China
School of Petroleum Engineering, China University of Petroleum, Qingdao 266580, Shandong, China

Nu Lu, Chuan-Jin Yao and Guang-Lun Lei
School of Petroleum Engineering, China University of Petroleum, Qingdao 266580, Shandong, China

Yan-Hui Zhang
CNOOC LTD., Tianjin Bohai Oilfield Institute, Tianjin 300452, China

Index

O

P

Q

R

S

T

U

V

W